[Bolyai Janos Matematikai Tarsulat]

COLLOQUIA MATHEMATICA
SOCIETATIS JÁNOS BOLYAI, 39.

SEMIGROUPS

STRUCTURE AND UNIVERSAL ALGEBRAIC PROBLEMS

Edited by:

G. POLLÁK
Št. SCHWARZ
and
O. STEINFELD

NORTH-HOLLAND
AMSTERDAM – OXFORD – NEW YORK

Budapest, Hungary, 1985

ISBN North-Holland: 0 444 87553 0
ISBN Bolyai: 963 8021 69 1
ISSN Bolyai: 0139 3383

Joint edition published by

JÁNOS BOLYAI MATHEMATICAL SOCIETY
and
ELSEVIER SCIENCE PUBLISHERS B. V.
P. O. Box 1991
1000 BZ Amsterdam, The Netherlands

In the U. S. A. and Canada:

ELSEVIER SCIENCE PUBLISHING COMPANY INC.
52 Vanderbilt Avenue
New York, N. Y. 10017
U. S. A.

Printed in Hungary
Szegedi Nyomda, Szeged

PREFACE

The 11th conference in Szeged of the János Bolyai Mathematical Society was held from August 24 to 28, 1981, under the title "Semigroups. Structure and universal algebraic problems." There was a number of invited lectures and, besides, several 20 minute talks, most of which were connected with the topic given above. As in our earlier volumes, we publish also papers by authors who were intended to present them at the conference but could not attend. Also, there was a problem session; the problems posed there can be found at the end of the volume. We ask everybody who makes some progress in solving them to let know about it one (or more) of the editors, in order that we can mention the result in our next semigroup volume.

The Editors

CONTENTS

SCIENTIFIC PROGRAM

August 24, Monday
Morning

9.00– 9.20	Opening of the conference
9.30–10.20	N. Nivat: Infinitary languages and relations
10.40–11.30	L.N. Ševrin: Quasi-periodic semigroups

August 24, Monday
Afternoon

14.30–14.50	V.A. Baranskiĭ: Independence of the automorphism group and the ideal lattice of semigroups
14.50–15.10	F. Pastijn: Exchange of independent sets in semigroups
15.10–15.30	H.-J. Hoehnke: On varieties of *S*-systems
15.35–15.55	Ju.M. Važenin: Formal theories of finitely presented semigroups
15.55–16.15	A.V. Mihaljev: On wreath products of monoids with zero
16.45–17.05	B. Bosbach: On complete complementary semigroups
17.05–17.25	R. Strecker: Construction of *M*-radicals
17.25–17.45	S. Crvenković: Basis classes for some classes of semigroups
17.50–18.10	L.G. Mustafaev: Semigroups of linear continuous selfmaps
18.10–18.30	M.R. Reis: Dilworth-elements in groupoid lattices

August 25, Tuesday

Morning

9.00– 9.50	N.R. Reilly: Some recent developments in inverse semigroups
10.00–10.50	J. Meakin: The Rees construction in regular semigroups
11.00–11.50	J.M. Howie: Epimorphisms and amalgamation: a survey of recent progress

August 25, Tuesday

Afternoon

14.30–14.50	K.E. Osondu: Universal groups on reversible semigroups
14.50–15.10	I. Szabó: Matrix representations of primitive regular semigroups
15.10–15.30	R. Yoshida: Subdirect product representation of inverse semigroups
15.35–15.55	J. Zeleznikow: Regular semirings
15.55–16.15	C. Bonzini: The structure of some permutable semigroups
16.45–17.05	J.M. Howie: Embedding semigroups in semibands: some arithmetical results
17.05–17.25	A. Cherubini: Semigroups whose proper subsemigroups are quasicommutative
17.25–17.45	O. Grošek: A-ideals in semigroups, I
17.50–18.10	L. Satko: A-ideals in semigroups, II
18.10–18.30	L.A. Skornjakov: Representation of monoids from a universal point of view

August 26, Wendesday
Morning

9.00– 9.50	L.M. Martynov: Attainable classes of semigroups
10.00–10.50	P. Goralčík: Translations in semigroups
11.00–11.50	E.V. Suhanov: Recent results on semigroup varieties

August 26, Wendesday
Afternoon

Problem session

August 27, Thursday
Morning

9.00– 9.20	A.Ja. Aĭzenštat: Varieties whose lattices of subvarieties are Boolean algebras
9.20– 9.40	A.V. Tiščenko: On the join of semigroup varieties
9.40–10.00	G. Pollák: On some types of hereditarily finitely based varieties
10.05–10.25	A.M. Pachkoria: On the Schreier extensions of semimodules
10.25–10.45	J. Galanová: Codomain of the tensor product of semigroups
11.15–11.35	J.-C. Spehner: Intersection of two submonoids of a free monoid
11.35–11.55	I.S. Ponizovskiĭ: On the semisimplicity of semigroup rings
11.55–12.15	B. Pondělíček: A note on characterizations of semigroups
12.20–12.40	N. Celakoski: A representation of associatives into semigroups
12.40–13.00	K. Todorov: On the maximal subsemigroups of the ideals of the symmetric semigroup

August 28, Friday
Morning

9.00– 9.50 E.S. Ljapin: Collective identities

10.00–10.50 K.S.S. Nambooripad: Topics in the structure theory of regular semigroups

11.00–11.50 M.M. Lesohin: Approximation of semigroups by bilinear mappings

August 28, Friday
Afternoon

14.30–14.50 A.K. Slipenko: Operatives of mappings

14.50–15.10 H. Jürgensen: Disjunctive elements in Bruck–Reilly extensions

15.10–15.30 B. Trpenovski: Semigroups with *n*-property

15.40–16.00 V.B. Steinbuk: Semigroups and automata

16.00–16.20 M. Ito: On small categories whose objects form lattices

LIST OF PARTICIPANTS

AMIRÁS E., Department of Mathematics, Budapest Technical University, H-1111 Budapest, Műegyetem rkp. 9., H.V. 5, Hungary.

AĬZENŠTAT, A.JA., Institut Gercena, Kafedra Vysšeĭ Algebry, 191 186 Leningrad, Moĭka 48, USSR.

BABCSÁNYI I., H-9700 Szombathely, Révai u. 4. I. 16., Hungary.

BALCZA L., H-1088, Budapest, Rákóczi út 47. I. 9., Hungary

BARANSKIĬ, V.A., Uralskiĭ Gosudarstvennyĭ Universitet, Matfak, 620083 Sverdlovsk, ul. Lenina 51, USSR.

BECSKI G., H-1502 Budapest, Pf. 117, Hungary.

BONZINI, C., Istituto Matematico "Federigo Enriques", 20 133 Milano, Via C. Saldini 50, Italy.

BOSBACH, B., Gesamthochschule Kassel, Fachbereich Mathematik, 3500 Kassel, BRD.

CELAKOSKI, N., Matematicki fakultet, 9100 Skopje, P.O. Box 504, Yugoslavia.

CHERUBINI SPOLETINI, A., Istituto di Matematica, 20 133 Milano, Piazza Leonardo da Vinci 32, Italy.

CRVENKOVIĆ, S., Department of Mathematics, University of Novi Sad, Novi Sad, Yugoslavia.

CSÁKÁNY B., JATE Bolyai Intézet, H-6720 Szeged, Aradi vértanúk tere 1., Hungary.

FABRICI, I., Laliová 2, 812 00 Bratislava, Czechoslovakia.

GALANOVÁ, J., Katedra matematiky EF SVŠT, 880 49 Bratislava, Gottwaldovo nám. 19, Czechoslovakia.

GAVALCOVÁ, T., Katedra matematiky SF VŠT, 04187 Košice, Švermova 9, Czechoslovakia.

GORALČÍK, P., MFF UK, 18600 Praha 8-Karlín, Sokolovská 83, Czechoslovakia.

GROŠEK, O., Katedra matematiky EF SVŠT, 880 49 Bratislava, Gottwaldovo nám. 19, Czechoslovakia.

HOEHNKE, H.-J., Institut f. Mathematik, Akademie d. Wissenschaften d. DDR, 1080 Berlin, Mohrenstr. 39, GDR.

HOWIE, J.M., Mathematical Institute, North Haugh, St. Andrews KY16 9SS, UK.

HUHN A., JATE Bolyai Intézet, H-6720 Szeged, Aradi vértanúk tere 1., Hungary.

ITO, M., Faculty of Science, Kyoto Sangyo University, 603 Kyoto, Japan.

JÜRGENSEN, H., Institut f. Theoretische Informatik, Fb. Informatik, 6100 Darmstadt, Magdalenenstr. 11, BRD.

KISS E., Mathematical Institute, Hungarian Academy of Sciences, H-1053 Budapest, Reáltanoda u. 13–15., Hungary.

KLUKOVITS L., JATE Bolyai Intézet, H-6720 Szeged, Aradi vértanúk tere 1., Hungary.

KOZLOV, K.P., Institut Gercena, Kafedra Vysšeĭ Algebry, 191 186 Leningrad, Moĭka 48, USSR.

LAJOS S., Department of Mathematics, K. Marx University of Economics, H-1093 Budapest, Dimitrov tér 8., Hungary.

LENKEHEGYI A., JATE Bolyai Intézet, H-6720 Szeged, Aradi vértanúk tere 1., Hungary.

LESOHIN, M.M., Institut Gercena, Kafedra Vysšeĭ Algebry, 191 186 Leningrad, Moĭka 48, USSR.

LETIZIA, A., 73 100 Lecce, Via Luigi Corvaglia 32, Italy.

LJAPIN, E. S., 196 066 Leningrad, Moskovskiĭ pr. 208, kv. 16, USSR.

MAKARIDINA, V., Institut Gercena, Kafedra Vysšeĭ Algebry, 191 186 Leningrad, Moĭka 48, USSR.

MÁRKI L., Mathematical Institute, Hungarian Academy of Sciences, H-1053 Budapest, Reáltanoda u. 13–15., Hungary.

MARTYNOV, L.M., Omskiĭ Pedagogičeskiĭ Institut, Kafedra Matematiki, Omsk, USSR.

MEAKIN, J., Dept. of Mathematics and Statistics, University of Nebraska-Lincoln, Lincoln, Nebraska 68588, USA.

MEGYESI L., JATE Bolyai Intézet, H-6720 Szeged, Aradi vértanúk tere 1., Hungary.

MIHALEV, A. V., MGU, Kafedra Vysšeĭ Algebry, 117 234 Moskva, Leninskie Gory, USSR.

MILLER, D.D., Lucina 5913, Shell Point Village, Fort Myers, Fl. 33908 USA.

MUSTAFAEV, L.G., Institut Matematiki i Mehaniki AN AzSSR, Otdel algebry i topologii, Baku, USSR.

NAMBOORIPAD, K.S.S., Department of Mathematics, University of Kerala, Kariavattom 695 581, India.

NEMES A., H-1502 Budapest, Pf. 117, Hungary.

NIVAṬ, M., 75017 Paris, 167 Bld. Malesherbes, France.

OSONDU, K.E., Department of Mathematics, College of Technology, P.M.B. 1036, Owerri, Nigeria.

PAČKORIA, A.M., Institut Matematiki AN Gruzinskoĭ SSR, 380093 Tbilisi, ul. Z. Rukhadze 1, USSR.

PASTIJN, F., Dienst Hogere Meetkunde, Rijksuniversiteit te Gent, B-9000 Gent, Krijgslaan 271, Belgium.

POLLÁK GY., H-6720 Szeged, Somogyi u. 7., Hungary.

PONDĚLÍČEK, B., FEL ČVUT, 166 27 Praha 6, Suchbátarova 2, Czechoslovakia.

PONIZOVSKIĬ, I.S., 194100 Leningrad, Lesnoĭ pr. 61, kv. 120, USSR.

REILLY, N.R., Department of Mathematics, Simon Fraser University, Burnaby 2, B.C. V5A 1S6, Canada.

REIS, M.R., rua Tenente Raúl Cascais n°1-1°D, 1200 Lisboa, Portugal.

SALIĬ, V.N., 410 056 Saratov, ul. Pugačevskaja 80, kv. 81, USSR.

SATKO, L., Katedra matematiky EF SVŠT, 880 49 Bratislava, Gottwaldovo nám. 19, Czechoslovakia.

SCHMIDT T., Mathematical Institute, Hungarian Academy of Sciences, H-1053 Budapest, Reáltanoda u. 13–15., Hungary.

SCHWARZ, Š., Matematický ústav SAV, 886 25 Bratislava, Obrancov mieru 49, Czechoslovakia.

ŠEVRIN, L.N., Uralskiĭ Gosudarstvennyĭ Universitet, Matfak, 620 083 Sverdlovsk, ul. Lenina 51, USSR.

SKORNJAKOV, L.A., MGU, Kafedra Vysšeĭ Algebry, 117 234 Moskva, Leninskie Gory, USSR.

SLIPENKO, A.K., 340 017 Doneck, bulvar Ševčenko 44, kv. 30, USSR.

SPEHNER, J.-C., I.S.E.A., 68093 Mulhouse Cédex, 4 rue des Frères Lumière, France.

ŠTEĬNBUK, V.B., 226 069 Riga 69, ul. Rudzutaka 16, kv. 82, USSR

STRECKER, R., Pädagogische Hochschule "Liselotte Herrmann", DDR-26 Güstrow, Goldberger Str. 12, GDR.

STURM, T., FEL ČVUT, Katedra matematiky, 166 27 Praha 6, Suchbátarova 2, Czechoslovakia.

SUHANOV, E.V., Uralskiĭ Gosudarstvennyĭ Universitet, Matfak, 620 083 Sverdlovsk, ul. Lenina 51, USSR.

ŠULKA, R., Katedra matematiky EF SVŠT, 880 49 Bratislava, Gottwaldovo nám. 19, Czechoslovakia.

SULLIVAN, R.P., Mathematics Department, University of Western Australia, Nedlands, W.A. 6009, Australia.

SZABÓ I., H-1125 Budapest, Nógrádi u. 10., Hungary.

SZABÓ L., JATE Bolyai Intézet, H-6720 Szeged, Aradi vértanúk tere 1., Hungary.

SZENDREI J., Juhász Gyula Tanárképző Főiskola, H-6701 Szeged, Pf. 396, Hungary.

SZENDREI M.B., JATE Bolyai Intézet, H-6720 Szeged, Aradi vértanúk tere 1., Hungary.

TIŠČENKO, A.V., Vsesojuznyĭ Zaočnyĭ Institut Piščevoĭ Promyšlennosti, Kafedra Vysšeĭ Matematiki, Moskva, USSR.

TODOROV, K., Mathematical Faculty, 1126 Sofia, Bul. A. Ivanov 5, Bulgaria.

TRAHTMAN, A.N., 620 049 Sverdlovsk, ul. Kovalevskoĭ 1, kv. 99, USSR.

TRPENOVSKI, B., Matematicki fakultet, 9100 Skopje, P.O. Box 504, Yugoslavia.

VAŽENIN, J.M., Uralskiĭ Gosudarstvennyĭ Universitet, Matfak, 620 083 Sverdlovsk, ul. Lenina 51, USSR.

WIEGANDT R., Mathematical Institute, Hungarian Academy of Sciences, H-1053 Budapest, Reáltanoda u. 13–15., Hungary.

YOSHIDA, R., Dept. of Mathematics, Osaka College of Pharmacy, Matsubara City, Osaka Pref. 580, Japan.

ZELEZNIKOW, J., Dept. of Mathematics, Michigan State University, East Lansing, Michigan 48 824, USA.

COLLOQUIA MATHEMATICA SOCIETATIS JÁNOS BOLYAI
39. SEMIGROUPS, SZEGED (HUNGARY), 1981.

SOME VARIETIES WITH A DISTRIBUTIVE LATTICE OF SUBVARIETIES

A. JA. AĬZENŠTAT

Of late years, attention has been paid to finding varieties with a lattice of subvarieties having such-and-such properties. E.g. in [3] chain varieties of semigroup (i.e. varieties such that the lattice of their subvarieties is a chain) have been described, up to varieties of groups. Minimal non-chain varieties are described in the same announcement. In [4] Ševrin initiated the investigation of semigroup varieties mentioned in the title. In the present note we give some of these with finite lattice of subvarieties, among others all varieties which have Boolean algebras for their lattices of subvarieties: these are the varieties generated by the union of some finite set of atoms (i.e. the finite joins of atoms). We have as a corollary that all finite joins of atoms form an ideal in the lattice of all varieties of semigroups. This ideal is then countable and isomorphic to the lattice of all finite subsets of a countable set.

We shall consider semigroup identities over a countable alphabet $\{x_1, x_2, \ldots\}$. The letters of this alphabet will be denoted also by $x, y, z, \ldots$. If Φ is a system of identities, we shall denote by $V(\Phi)$ the variety defined by Φ. For a word u, $l(u)$ denotes the length and $l_x(u)$

the number of occurrences of the letter x in u. The identity $u = v$ is said to be *normal,* if u and v contain the same variables, and *anomalous* else. For the atoms of the lattice of semigroup varieties we adopt the notation $V_0 = V(x_1x_2 = x_3x_4)$, $V_l = V(xy = x)$, $V_r = V(xy = y)$, $V_s = (x^2 = x, xy = yx)$, $V_p = V(x^p y = y, xy = yx)$, where p is an arbitrary prime. It is known that the identities $u = v$ that hold in V_0 are those satisfying $l(u) > 1$, $l(v) > 1$; in $V_l[V_r]$ those, in which the first [the last] letters of u and v coincide; in V_s the normal identities; and in V_p those such that $p \mid l_x(u) - l_x(v)$ for all x ([1]).

Let $W_k = V(xyzt = xzyt, xy^{k+1} = x^{k+1}y = xy)$, where k is a natural number. It is easy to see that the identity $u = v$ holds in W_k iff it is normal, $l(u) \geqslant 2$, $l(v) \geqslant 2$, the first letters of u and v as well as their last letters coincide, and $k \mid l_x(u) - l_x(v)$ for all x.

Theorem 1. *All subvarieties of* W_k *are the following ones:*

$$V(\sigma, x^{n+1}y = xy^{n+1} = xy); \quad V(\sigma, x^{n+1} = x);$$

$$V(\sigma, xy^n z = xz); \quad V(\sigma, x^{n+1} = x, xy^n z = xz),$$

where σ *is one of the identities* $xyzt = xzyt$, $xyz = yxz$, $xyz = xzy$, $xy = yx$, *and* $n \mid k$.

Let us prove first a number of auxiliary propositions.

Proposition 1. *If* V *is a subvariety of* W_k, *but does not contain* V_l *[or* V_r*], then the identity* $xyz = yxz$ $[xyz = xzy$, *resp.*$]$ *is fulfilled in* V.

Proof. As V does not contain V_l, some identity

(1) $\qquad xf_1(x, y) = yf_2(x, y)$

must hold there. Multiplying (1) with x on the right and using $xyzt = xzyt$, we get

(2) $\qquad x^a y^b x = y^c x^d.$

Note that if $k > 1$ then the identity

(3) $\qquad x^{2k} = x^k$

holds in W_k and, consequently, also in V (because $x^{2k} = x^{k-1}x^{k+1} = x^{k-1}x = x^k$). Hence, substituting x by x^k and y by y^k in (2) and applying (3), we obtain

$$x^k y^k x^k = y^k x^k$$

(for $k = 1$ this identity obviously holds, too). Multiplying this latter identity by xyz on the right, we find

$$x^k y^k xyz = y^k xyz.$$

This and the medial identity $xyzt = xzyt$ yield $xyz = yxz$.

Corollary 1. *If* $V \subseteq W_k$, *and* $V \not\supseteq V_l, V_r$, *then* V *is commutative.*

Proof. The identities $xyz = xzy$, $xyz = yxz$ imply $xyz = yzx$. Replacing z by y^k, we get $xy^{k+1} = y^{k+1}x$ which is equivalent to commutativity.

Proposition 2. *If* $V \subseteq W_k$ *does not contain* V_s *then* $xy^k z = xz$ *holds in* V.

Proof. An anomalous identity $x^a = f(x, y)$ holds in V. Multiply it by x on the left and by z on the right and use mediality. This gives $x^a z = x^b y^c z$. Substituting x and y by x^k and y^k, respectively, and multiplying once more with x on the left, we obtain $xy^k z = xz$.

Proposition 3. *If* $V \subseteq W_k$ *does not contain* V_0 *then* $x^{k+1} = x$ *holds in* V.

Proof. Some identity $x = x^{a+1}$ holds in V. Besides, $x^{k+2} = x^2$ holds in V, too. Hence $x = x^{d+1}$ (d = g.c.d. (a, k)) which yields the assertion.

Proposition 4. *If some identity* $u = v$ *with* $k \nmid l_x(u) - l_x(v)$ *for some* x *holds in* $V \subseteq W_k$, *then also* $x^{d+1}y = xy^{d+1} = xy$ *holds there for some proper divisor* d *of* k.

Proof. Let d be the g.c.d. of all differences $l_x(w_1) - l_x(w_2)$, where $V \models w_1 = w_2$ and x is an arbitrary variable. Then $V \models x^{m+d} = x^m$ for some m. Also, $V \models x^{2+k} = x^2$ whence $V \models x^{2+d} = x^2$. Besides $d \mid \text{g.c.d.}\,(k, l_x(u) - l_x(v)) \neq k$, whence d is a proper divisor of k. Furthermore, $xy = x^{2k+1}y = x^{2k+1+d}y = x^{d+1}y$. Analogously one obtains $xy = xy^{d+1}$.

PROOF OF THEOREM 1

Let V be an arbitrary subvariety of W_k, and let n be the minimal positive integer such that $V \models x^{n+1}y = xy^{n+1} = xy$. Then for every identity $u = v$ satisfied in V and for arbitrary x we have $n \mid l_x(u) - l_x(v)$. We consider the following possibilities.

1. If V contains both V_0 and V_s then

$$V = V(\sigma, xy^{n+1} = x^{n+1}y = xy).$$

Indeed, every $u = v$ satisfied in V is normal, and $l(u) \geqslant 2$, $l(v) \geqslant 2$. If V contains also V_l and V_r, then the first letters of u and v, as well as the last ones, coincide and, since $n \mid l_x(u) - l_x(v)$ for all x, we have $W_n = V$. If V contains V_l but not V_r then, in virtue of Proposition 1, $V \models xyz = xzy$, and u and v begin with the same letter. An argument analogous to the above one yields $V = V(xyz = xzy, x^{n+1}y = xy^{n+1} = xy)$. In the dual case $V = V(xyz = yxz, x^{n+1}y = xy^{n+1} = xy)$. Finally, if neither of V_l, V_r lies in V, Corollary 1 yields $V = V(xy = yx, x^{n+1}y = xy^{n+1} = xy)$.

2. If $V_s \subseteq V$, $V_0 \nsubseteq V$ then

$$V = V(\sigma, x^{n+1} = x).$$

Indeed, all identities of V are normal and $V \subseteq W_n$. By Proposition 3, $V \models x^{n+1} = x$. If $V_l, V_r \subseteq V$ then u and v start (and end) with the same letter for every identity which holds in V. Furthermore, $n \mid l_x(u) - l_x(v)$ for every x. Hence $V = V(xyzt = xzyt, x^{n+1} = x)$.

Analogously, we obtain $V = V(xyz = yxz, x^{n+1} = x)$ if $V_r \subseteq V$, $V_l \nsubseteq V$, and the dual statement. Finally, if V does not contain either V_l or V_r, we have $V = V(xy = yx, x^{n+1} = x)$.

3. V contains V_0, but it does not contain V_s. In this case

$$V = V(\sigma, xy^n z = xz).$$

Indeed, for every identity $u = v$ which is fulfilled in V, we have $l(u) \geqslant 2$, $l(v) \geqslant 2$. If $V_l, V_r \subseteq V$ then the first letters of u and v, as well as the last ones, coincide. As $V_s \not\subseteq V \subseteq W_n$, $V \models xy^n z = xz$ according to Proposition 2. As, furthermore, $n \mid l_x(u) - l_x(v)$ for every identity $u = v$ which holds in V and for every x, we have $V = V(xyzt = xzyt, xy^n z = xz)$. The remaining cases can be dealt with analogously.

4. If $V_0 \not\subseteq V$, $V_s \not\subseteq V$ then $V \models x^{n+1} = x$, $V \models xy^n z = xz$. The rest is analogous to the above considerations. In this case we obtain

$$V = V(\sigma, x^{n+1} = x, xy^n z = xz).$$

Corollary 2. *The lattice of subvarieties of* W_k *is isomorphic to the dirᵉct product of the lattice of subsets of a four-element set and the lattice of divisors of* k, *and so it is distributive.*

This can be immediately concluded from the fact that every subvariety of W_k is completely characterized by some divisor of k and by the sct of those of the atoms V_0, V_s, V_l, V_r which are contained in it.

Theorem 2. *Let* V *be a variety of semigroups. The lattice of subvarieties of* V *is a Boolean algebra if and only if* V *is a finite join of atoms.*

Proof. It is well known that every variety contains some atom. If V contains an infinite set of them then it contains the variety of commutative semigroups and, consequently, the lattice $\mathscr{L}(V)$ of its subvarieties is not distributive ([2]). Suppose V contains only the atoms $V_{i(1)}, \ldots, V_{i(t)}$. If V is not their join, then this join has no complement in $\mathscr{L}(V)$. Thus, $\mathscr{L}(V)$ is not a Boolean algebra. On the other hand, if $V = \bigvee_{j=1}^{t} V_{i(j)}$ then, obviously, V is contained in some W_k. As $\mathscr{L}(W_k)$ is distributive, $\mathscr{L}(V)$ is isomorphic to the Boolean algebra of subsets of $\{i(1), \ldots, i(t)\}$.

Corollary 3. *If the lattice of subvarieties of the variety* V *is a Boolean algebra, then it is finite.*

REFERENCES

[1] J. Kalicki – D. Scott, Equational completeness of abstract algebras, *Nederl. Akad. Wetensch. Proc. Ser. A*, 58 (1955), 650–659.

[2] R. Schwabauer, A note one commutative semigroups, *Proc. Amer. Math. Soc.*, 20 (1969), 503–504.

[3] E.V. Suhanov, On varieties of periodic semigroups, *10. All-Union Conference on Algebra*, Krasnojarsk, 1979 (in Russian).

[4] *Sverdlovskaja Tetrad'*, Sverdlovsk, 1979 (in Russian).

A.Ja. Aĭzenštat

Gercen Pedagogical Institute, 191 186 Leningrad, Moĭka 48, USSR.

COLLOQUIA MATHEMATICA SOCIETATIS JÁNOS BOLYAI
39. SEMIGROUPS, SZEGED (HUNGARY), 1981.

MEDIAL PERMUTABLE SEMIGROUPS

C. BONZINI – A. CHERUBINI

SUMMARY

A permutable semigroup is a semigroup whose congruences are pairwise permutable. In this paper we give a characterization of medial permutable semigroups.

INTRODUCTION

A permutable semigroup is a semigroup whose congruences are pairwise permutable. A subclass of such semigroups is given by the semigroups whose congruences form a chain with respect to the inclusion (Δ-semigroups) which was introduced by Schein and Tamura and studied by several authors.

Hamilton [5] characterized commutative permutable semigroups. In [1] we gave a construction of $E-2$ permutable semigroups without zero.

In this paper we give a characterization of medial permutable semigroups. First we deduce from [1] the classification of archimedean medial permutable semigroups. Then we examine medial permutable semigroups which are ideal extension of a nilsemigroup N by a semilattice of order 2; we prove that such semigroups are Δ-semigroups and we give their structure which we shall utilize in the sequel, since every non-archimedean medial permutable semigroup has one of the above semigroups as a homomorphic image. Finally we prove that a medial permutable non-archimedean semigroup S is an ideal extension of a nilsemigroup S_0 by $S_1 \cup 0$ where S_1 is a suitable rectangular group, and S satisfies one of the following conditions: $a \in S_1 a S_1$ for every $a \in S$, or $a \in S_1 a$ and $aS_1 = 0$ for every $a \in S_0$, or $a \in aS_1$ and $S_1 a = 0$ for every $a \in S_0$. This enables us to prove Theorems 3.4, 3.5, 3.6, 3.7 which provide a characterization of medial permutable non-archimedean semigroups.

1.

Definition 1.1. A semigroup S is called *medial* if, for every $a, b, x, y \in S$, it satisfies $axyb = ayxb$.

Definition 1.2. A semigroup S which satisfies the identity $(ab)^2 = a^2 b^2$ is called an $E-2$ *semigroup.*

Definition 1.3. A Δ-*semigroup* is a semigroup whose lattice of congruences is a chain with respect to the inclusion.

Definition 1.4. A *permutable semigroup* is a semigroup S in which any pair of congruences permute (i.e. for every congruences ρ, σ of S, $\rho \cdot \sigma = \sigma \cdot \rho$).

Let us now recall the following well-known propositions (see [4]):

(a) A Δ-semigroup is permutable.

(b) The ideals of a permutable semigroup form a chain with respect to the inclusion.

(c) Every homomorphic image of a permutable semigroup is permutable.*

*Some other properties of permutable semigroups can be found in [5] and in [1].

Furthermore we remark that a medial semigroup is an $E-2$ semigroup and we can therefore utilize the results of paper [1] which concerns $E-2$ permutable semigroups.

Lemma 1.1. *A permutable medial nilsemigroup S is commutative.*

In fact, since the ideals of S form a chain by (b), S is a Δ-semigroup (see [10], Lemma 1.4), so that S is commutative (see [4], Theorem 3.33).

From the above lemma and from Theorem 5.1 of [1] we can easily deduce the following

Corollary 1.2. *S is an archimedean medial permutable semigroup if and only if*

(i) *S is a commutative nilsemigroup whose principal ideals form a chain with respect to the inclusion, or*

(ii) *$S = I \times G \times \Lambda$ where I is a left-zero semigroup with $|I| \leqslant 2$, G is an abelian group, Λ is a right-zero semigroup with $|\Lambda| \leqslant 2$.*

Definition 1.5. Let S_0 be a nilsemigroup, G a group, I a left-zero semigroup, Λ a right-zero semigroup, S_1 the rectangular group $I \times G \times \Lambda$. A semigroup is called *of type* a if it is an ideal extension of S_0 by $S_1 \cup 0$ where $|I|, |\Lambda| \leqslant 2$. *

Remark 1.1. Let S be a semigroup of type a. S is not archimedean * and it can be immediately verified that *the zero of* S_0 *is also the zero of* S. Moreover, *the relation* $a = xay$ *with* $a \in S$, $x, y \in S^1$ ** *implies* $a = 0$ *if either* x *or* y *belongs to* S_0. In fact from $a = xay$ we deduce $a = x^h a y^h$ for any positive integer h and, S_0 being a nilsemigroup, it implies $a = 0$ if either x or y belongs to S_0.

Remark 1.2. Let S be a permutable semigroup of type a and J an ideal of S. By (b), we have either $S_0 \subseteq J$ or $J \subseteq S_0$. If $S_0 \subseteq J$ then

*It is easy to verify that a semigroup of type a is a semilattice of two archimedean semigroups S_0 and S_1. Moreover we remark that the decomposition of a semigroup in a semilattice of archimedean semigroups is unique, if it exists.

**$S^1 = S$ if S is a semigroup with identity, otherwise $S^1 = S \cup 1$.

$J = S_0$ or $J = S$ (in fact, the Rees congruence $\bmod J$ must be permutable with the congruence whose classes are S_0 and S_1). Then S_0 *is the unique maximal ideal of* S *whence, for any* $x \in S_1$, *we have* $SxS = S$.

From Lemma 1.2 and 2.2 of [1], remarking that if $|S_0| \neq 1$, the $E-2$ semigroups described in Theorems 5.2, 8.2 and in Remark 6.2 of [1] are not medial, we obtain immediately the following

Corollary 1.3. *A permutable medial non-archimedean semigroup is a semigroup of type* a. *

2.

Definition 2.1. By a semigroup of *type* α we mean a semigroup of type a with $|S_1| = 1$, and we shall denote the element of S_1 by e.

First we shall give a characterization of permutable semigroups of type α (Theorem 2.3); then we will determine the structure of medial ones (Theorem 2.4). These results will be utilized later in the paper.

Lemma 2.1. *Let* S *be a permutable semigroup of type* α, a *an element of* S_0. *If there is a congruence* ρ *of* S *which has* $\{a, ea\}$ *as a class, then* $ea \in \{0, a\}$.

Proof. We suppose $ea \neq a$, whence $a \neq 0$ (otherwise the lemma is obvious). Putting $J(ea) = S^1 eaS^1$, we have $a \notin J(ea)$, $J(ea) \subseteq a\rho = \{a, ea\}$ (it is easy to verify that $a \in J(ea)$ implies the contradiction $ea = a$; moreover, if $J(ea) \nsubseteq a\rho$, then there exists an element b of S with $b \in J(ea)$, $b \notin a\rho$, and, denoting the Rees congruence $\bmod J(ea)$ by σ, we have $(a, b) \in \rho \cdot \sigma$, $(a, b) \notin \sigma \cdot \rho$, another contradiction). From these relations $J(ea) = ea$ follows, whence $ea = 0$.

Repeating the proof used by Trotter in [10] for proving Lemma 3.3, we can deduce from Lemma 2.1 the following

Lemma 2.2. *If* S *is a permutable semigroup of type* α, *then* S_0 *is a* Δ*-semigroup (whence* S_0 *itself is permutable).*

*It is necessary to remark that, if S is a medial semigroup of type a, then G is an abelian group.

Theorem 2.3. *A semigroup of type* α *is permutable if and only if it is a* Δ*-semigroup.*

Proof. Let S be a permutable semigroup of type α and ρ a congruence of S such that $(e,a) \in \rho$ for some $a \in S_0$. Then $\rho = S \times S$. In fact, denoting by σ the Rees congruence mod S_0, we have $(e,b) \in \rho \cdot \sigma$ for every $b \in S_0$, and S being permutable, this implies $(e,b) \in \sigma \cdot \rho$, from which it follows $(e,b) \in \rho$, and so $\rho = S \times S$. Hence, recalling that S_0 is a Δ-semigroup (Lemma 2.2), we can easily deduce that S itself is a Δ-semigroup. The converse is well known.

Now we suppose that S is a medial permutable semigroup of type α. The nilsemigroup S_0 is permutable (Lemma 2.2) and therefore commutative (Lemma 1.1). Consequently, since the principal ideals of S_0 form a chain, S_0 is a totally ordered semigroup with respect to the relation $a \leqslant b$ if and only if $bS_0^1 \subseteq aS_0^1$ $(a, b \in S_0)$* (see [9]). We remark that we have $a \leqslant b$ if and only if $b = ax$ for some $x \in S_0^1$ and moreover, if $a, b, x \in S_0$, then the relations $xa = xb \neq 0$ imply $a = b$ (see [8], Theorem 3.39).

Now we prove the following

Theorem 2.4. S *is a medial permutable semigroup of type* α *if and only if*

(i) $S = S_0^1$ *where* S_0 *is a commutative nilsemigroup whose principal ideals form a chain with respect to the inclusion, or*

(ii) S *is one of the following semigroups*

	e	a	0
e	e	a	0
a	0	0	0
0	0	0	0

	e	a	0
e	e	0	0
a	a	0	0
0	0	0	0

Proof. It is easily verified that the semigroups described in (i) and (ii) are medial permutable semigroups of type α.**

*It is easy to verify that the relation $aS_0^1 = bS_0^1$ implies $a = b$ $(a, b \in S_0)$.

**If S is a semigroup of type (i), then S is a Δ-semigroup (see [8]), whence S is permutable.

Conversely, let S be a medial permutable semigroup of type α. From the relation $S = SeS$ (Remark 1.2) we can easily deduce that, if $a \in S_0$, the possible cases are

$$a = ea = ae \quad \text{or} \quad a = ea \neq ae \quad \text{or} \quad a = ae \neq ea.$$

First, we remark that

(1) $$a \in S_0,\ a = ea \neq ae \ \ [a = ae \neq ea] \quad \text{imply} \quad a = 0,\ ae = 0 \ \ [ea = 0].$$

In fact, S being an exponential Δ-semigroup, we have $ae \in S_0^1 a S_0^1$ (see [8], Theorem 3.4), whence there exists some $x \in S_0$ such that $ae = ax$; hence $ae = axe = x(ae)$ and $ae = 0$ (Remark 1.1).

Now we prove that

(2) $$\text{if } \exists a \in S_0,\ a \neq 0 \text{ such that } a = ea = ae \text{ then we have } n\,O\,en = ne,\ \forall n \in S_0.$$

Let n be an element of S_0. If $n \leqslant a$ we have $a = nx$, $x \in S_0^1$, from which it follows that $a = ea = (en)x = x(en)$, $a = ae = (nx)e = x(ne)$. Then, since $a \neq 0$, we have $x(en) = x(ne) \neq 0$, whence $en = ne$. If $a < n$ then $n = ax$, $x \in S_0^1$, from which it follows that $en = eax = ax = n$, $ne = axe = xae = xa = ax = n$, whence again $en = ne$.

Thus we verify that

(3) $$\text{if } \exists a \in S_0 \text{ such that } a = ea \neq ae \ \ [a = ae \neq ea] \text{ then we have } S_0 = \{0, a\}.$$

Let $n \in S_0$, $n \neq a$. If $n < a$ then $a = nx$, $x \in S_0$, $ea = enx = (ex)n = (ne)x = x(ne)$, $ae = nxe = x(ne)$; then the contradiction $ae = ea$ follows. Therefore we have $a < n$ and this implies $n = ax$, $x \in S_0$, $ne = a(xe) = (xe)a = xa = ax = n$, $en = eax = ax = n$, $n = en = ne$. Consequently, if $n \neq 0$, the contradiction $a = ea = ae$ follows by (2), so that we obtain $n = 0$.

Then (1), (2), (3) immediately imply the statement of the theorem.

3.

Our aim is now to characterize medial non-archimedean permutable semigroups.

Definition 3.1. A semigroup S is called *right* [*left*]-*commutative* if, for every $a, x, y \in S$, we have $axy = ayx$ $[xya = yxa]$.

A right- or left-commutative semigroup is obviously medial. Moreover we remark that in (ii) of Theorem 2.4 the first semigroup is left-commutative while the second one is right-commutative.

If S is a non-archimedean right [left]-commutative permutable semigroup, then S is a semigroup of type a (Corollary 1.3) where $S_1 = I \times G$ $[S_1 = G \times \Lambda]$, G is an abelian group.

We prove the following

Lemma 3.1. *Let S be a non-archimedean right [left]-commutative permutable semigroup. Then*

$$S_1 a = 0 \text{ for every } a \in S_0 \text{ or } a \in S_1 a \text{ for every } a \in S.$$

$$[aS_1 = 0 \text{ for every } a \in S_0 \text{ or } a \in aS_1 \text{ for every } a \in S].$$

Proof. We suppose that S is right-commutative and we distinguish two cases:

1. Let $|G| = 1$. If also $|I| = 1$, then S is a semigroup of type α and the lemma follows immediately from Theorem 2.4. Then we shall suppose $|I| = 2$. Putting $S_1 = I = \{i, j\}$, the relation $S = SiS$ (Remark 1.2) implies

(4) $\quad xi = xj = x$ for every $x \in S$.

Now let a be an element of S_0. It can be immediately verified that $0 \in S_1 a$ implies $S_1 a = 0$. Then we suppose $0 \notin S_1 a$. Hence we obtain $ia \neq 0$, $a \neq 0$. If $ia = a$, then we have $a \in S_1 a$. If $ia \neq a$ then, using the relation (4), we prove that $A = \{a, ia\}$ is a normal complex of S.* Then,

*Following Ljapin, a normal complex A of a semigroup S is a non-empty subset of S such that $xAy \cap A \neq \phi$ implies $xAy \subseteq A$ $(x, y \in S^1)$. We recall that if A is a normal complex of a semigroup S then there exists a congruence of S which has A as a class (see [6], pp. 284–285).

if σ is the Rees congruence modulo $S^1 iaS^1$ and ρ is a congruence which has A as a class, we have $(a,0) \in \rho \cdot \sigma$; moreover, S being permutable, it implies $(a,0) \in \sigma \cdot \rho$ whence $a \in S^1 iaS^1$ (otherwise we have $a\sigma = a$ and, since $(a,0) \notin \rho$, it follows that $(a,0) \notin \sigma \cdot \rho$). Then, as $a \neq 0$, $a \neq ia$, the equality $a = ja$ easily follows from the relation $a \in S^1 iaS^1$, that is, $a \in S_1 a$. Thus, we have proved that

(5) $\quad \forall a \in S_0$ we have $S_1 a = 0$ or $a \in S_1 a$.

Now we put $J = \{s \in S_0 : S_1 s = 0\}$ and we prove that

(6) $\quad S_0 \setminus J \neq \phi$ implies $J = 0$.

If there exists $a \in S_0 \setminus J$ (whence $S_1 a \neq 0$, $a \in S_1 a$), then, since J is an ideal of S, by (b) we deduce $J \subseteq S^1 a S^1 = S_0 a \cup aS_0 \cup S_1 a$. Let $s \in J$. If $s \in S_1 a$ we have $is \in iS_1 a = ia$, $js \in jS_1 a = ja$ whence the contradiction $ia = ja = 0$ is implied because $is = js = 0$. If $s \in aS_0$ then we have $s = ax$ $(x \in S_0)$ and, since $a \in S_1 a$, we obtain $s = ax \in S_1 ax = S_1 s = 0$. Hence we must conclude that $J \subseteq S_0 a$ and hence $s = xa$ with $x \in S_0$. If $x \notin J$ then we have (from (5)) $x \in S_1 x$ whence it follows $s = xa \in S_1 xa = S_1 s = 0$; if $x \in J \subseteq S_0 a$, then $x = ya$, $s = xa = ya^2$ $(y \in S_0)$ whence it follows $s = 0$ if $y \notin J$, or $s = za^3$ with $z \in S_0$ if $y \in J$. Since S is a nilsemigroup, repeating the same procedure, we again conclude $s = 0$. Thus we proved (6).

From (5) and (6), remarking that for any $a \in S_1$, we have $a \in S_1 a$, the statement of the theorem follows immediately.

2. Now suppose $|G| \geqslant 1$. Let u be the identity of G and i an element of I. From the relation $S = S(i,u)S$ (Remark 1.2) it immediately follows that $x = x(i,u)$ for any $x \in S$. Hence it is easy to verify that

(a) the relation $\tau = \{(x,y) \in S \times S : x \in y(\{i\} \times G)\}$ is a congruence such that $x\tau = x(\{i\} \times G)$, $\forall x \in S$;

(b) the semigroup S/τ satisfies the conditions of the case 1. Moreover, if φ is the canonical homomorphism of S onto S/τ then we have

$$(S_1\varphi)(a\tau) = 0\tau, \ \forall a \in S_0 \quad \text{or} \quad a\tau \in (S_1\varphi)(a\tau), \ \forall a \in S.$$

In the first case, for any $x \in S_1$ we have $0\tau = (x\tau)(a\tau) = (xa)\tau$, that is, $xa \in 0\tau$ and, since $0\tau = 0$, it implies $xa = 0$ whence $S_1 a = 0$. In the second case there exists $x \in S_1$ such that $a\tau = (x\tau)(a\tau) = (xa)\tau$; hence $a \in (xa)\tau = (xa)(\{i\} \times G) = x(\{i\} \times G)a \subseteq S_1 a$. Thus the statement is proved.

Lemma 3.2. *Let S be a medial semigroup of type a and we suppose that for any $a \in S_0$ we have $a \in S_1 a$, $aS_1 = 0$ $[a \in aS_1, S_1 a = 0]$. The relation*

$$\rho = \{(x,y) \in S \times S\colon x,y \in S_1 \text{ or } S_1 x = S_1 y\}$$

$$[\rho = \{(x,y) \in S \times S\colon x,y \in S_1 \text{ or } xS_1 = yS_1\}]$$

is the least congruence of S which has S_1 as a class and, moreover we have for any $a \in S_0$, $s \in S_1$

$$\text{(7)} \quad \begin{array}{l} a\rho = S_1 a, \quad (a\rho)(s\rho) = 0\rho \quad (s\rho)(a\rho) = a\rho \\ [a\rho = aS_1, \quad (s\rho)(a\rho) = 0\rho \quad (a\rho)(s\rho) = a\rho]. \end{array}$$

Proof. Let us suppose that S is a semigroup of type a such that $a \in S_1 a$, $aS_1 = 0$ for any $a \in S_0$. First we remark that, if $x \in S_1$ and $y \in S_0$ then we have $S_1 x \subseteq S_1$, $S_1 y \subseteq S_0$ whence $(x,y) \notin \rho$. Then it is easy to verify that ρ is an equivalence relation which has S_1 as a class.

Now let $(x,y) \in \rho$ and $z \in S$. If $x,y,z \in S_1$ we have $xz, yz, zx, zy \in S_1$. If $x,y \in S_1$ and $z \in S_0$ then $zx = zy = 0$; moreover, since S_1 is simple, we have $x = syt$ $(s,t \in S_1)$ whence we have $S_1 xz = S_1 sytz = S_1 styz \subseteq S_1 yz$; analogously we obtain $S_1 yz \subseteq S_1 xz$ and therefore $S_1 xz = S_1 yz$. If $x,y \in S_0$, we have $S_1 x = S_1 y$, whence we obtain $S_1 xz = S_1 yz$; moreover, since $x \in S_1 x$, we have $x = sy$ $(s \in S_1)$ and $S_1 zx = S_1 zsy = S_1 szy \subseteq S_1 zy$; in an analogous way it is verified that $S_1 zy \subseteq S_1 zx$ whence $S_1 zx = S_1 zy$. Thus ρ is a congruence.

Let σ be a congruence which has S_1 as a class. From the relation $(x,y) \in \rho$ with $x,y \in S_0$, recalling that $x \in S_1 x$, $y \in S_1 y$, we deduce $x = sy$, $y = ty$ $(s,t \in S_1)$ whence $(s,t) \in \sigma$, from which it follows $(sy, ty) \in \sigma$. Hence $\rho \subseteq \sigma$.

Finally, we verify the first relation of (7) (the other two relations are obvious). Let $x \in S_1 a$, $a \in S_0$. We can write $x = (i, g, \lambda)a$ $(i \in I, g \in G, \lambda \in \Lambda)$. Moreover, since $a \in S_1 a$, we have $a = (j, h, \mu)a$ $(j \in I, h \in G, \mu \in \Lambda)$ whence

$$(j, g^{-1}, \lambda)x = (j, g^{-1}, \lambda)(i, g, \lambda)(j, h, \mu)a =$$

$$= (j, h, \mu)a = a.$$

Therefore the relations $x \in S_1 a$, $a \in S_0$ imply $a \in S_1 x$, from which $a\rho = S_1 a$ easily follows for any $a \in S_0$.

The following proposition whose proof is analogous to that of Lemma 3.2 will be useful.

Lemma 3.3. *Let* S *be a medial semigroup of type* a *and suppose that for any* $a \in S$ *we have* $a \in S_1 a S_1$. *The relation*

$$\rho = \{(x, y) \in S \times S \colon S_1 x S_1 = S_1 y S_1\}$$

is the least congruence of S *which has* S_1 *as a class, and moreover for every* $a \in S$, $s \in S_1$ *we have*

$$a\rho = S_1 a S_1, \quad (a\rho)(s\rho) = (s\rho)(a\rho) = a\rho.$$

Now it is convenient to introduce the following

Definition 3.2. A semigroup S is called *of first kind* if

(i) S is of type a,

(ii) $a \in S_1 a S_1$ for every $a \in S$.

S is called *of second kind* [*third kind*] if S satisfies (i) and if moreover

(iii) $a \in S_1 a$ and $a S_1 = 0$ [$a \in a S_1$ and $S_1 a = 0$] for every $a \in S_0$.

Then we are able to prove the following

Theorem 3.4. *If* S *is a non-archimedean medial permutable semigroup, then* S *is of first, second or third kind.*

Proof. If S is a non-archimedean medial permutable semigroup, it is of type a by Corollary 1.3. Now let us prove that S satisfies one of conditions (ii) and (iii) of Definition 3.2. Let $\Gamma(S) = \{\varphi_a \colon a \in S\}$ the semigroup of inner left translations of S.* Since S is a medial permutable semigroup and $\Gamma(S)$ is a homomorphic image of S, $\Gamma(S)$ is a right-commutative permutable semigroup. Moreover it can be immediately verified that $\Gamma(S)$ is non-archimedean, whence, from Lemma 3.1 it is easy to deduce that we have

$$(8) \quad \begin{aligned} &\varphi_{sa} = \varphi_0, \ \forall s \in S_1 \ \text{ and } \ \forall a \in S_0, \ \text{ or} \\ &\forall a \in S \ \exists s \in S_1 \ \text{ such that } \ \varphi_a = \varphi_{sa}. \end{aligned}$$

Analogously, if $\Delta(S) = \{\psi_a \colon a \in S\}$ is the left-commutative semigroup of inner right translations of S, then we have

$$(9) \quad \begin{aligned} &\psi_{as} = \psi_0, \ \forall s \in S_1 \ \text{ and } \ \forall a \in S_0, \ \text{ or} \\ &\forall a \in S \ \exists s \in S_1 \ \text{ such that } \ \psi_a = \psi_{as}. \end{aligned}$$

We verify that

$$(10) \quad \begin{aligned} &\forall a \in S \ \exists s \in S_1 \ \text{ such that } \ \varphi_a = \varphi_{sa} \\ &\text{implies } \ a \in S_1 a, \ \forall a \in S. \end{aligned}$$

Let $a \in S$ and $s \in S_1$. Since $S = SsS$ (Remark 1.2), we can write $a = ysz$ $(y, z \in S)$; moreover, by assumption, there exists $s_1 \in S_1$ such that $\varphi_y = \varphi_{s_1 y}$, whence for any $x \in S$ the equality $yx = s_1 yx$ follows. We therefore have $a = y(sz) = s_1 y(sz) = s_1 a \in S_1 a$.

Now we prove that

$$(11) \quad \begin{aligned} &\varphi_{sa} = \varphi_0 \ \ \forall s \in S_1 \ \text{ and } \ \forall a \in S_0 \ \text{ implies } \ S_1 a = 0 \ \ \forall a \in S_0 \\ &\text{or } \ aS = 0, \ \forall a \in S_0. \end{aligned}$$

Let $a \in S_0$ and $s \in S_1$. Recalling Remark 1.2, we may suppose $a = ysz$ $(y, z \in S)$ where either y or z belongs to S_0. If $y \in S_0$, we have $\varphi_{sy} = \varphi_0$ whence $syx = 0$ $\forall x \in S$. In particular, $syz = 0$ and hence

*We recall that, for any $a \in S$, the map defined by $x\varphi_a = ax$ $[x\psi_a = xa]$ for any $x \in S$ is called the inner left [right] translation of S induced by a. The set $\Gamma(S)$ $[\Delta(S)]$ of all inner left [right] translations of S is a semigroup with respect to multiplication $\varphi_a \varphi_b = \varphi_{ab}$ $[\psi_a \psi_b = \psi_{ab}]$, $a, b \in S$ (see [7], p. 61).

$sa = sysz = s(syz) = 0$. Then, let s_1 be another element of S_1; S_1 being simple, we can write $s_1 = vxw$ $(v, w \in S_1)$ and we have $s_1 a = vswa = vw(sa) = 0$, so that $S_1 a = 0$. If $z \in S_0$ we have $\varphi_{sz} = \varphi_0$ whence $szx = 0$, $\forall x \in S$. Therefore it follows that $ax = y(szx) = 0$ so that $aS = 0$. We have therefore proved that $\varphi_{sa} = \varphi_0$, $\forall s \in S_1$ and $\forall a \in S_0$, implies $S_1 a = 0$ or $aS = 0$ for any $a \in S_0$. That being stated, we examine the set $J = \{x \in S_0 \colon S_1 x = 0\}$. Making use of the procedure analogous to the proof of Lemma 3.1 we obtain that $S_0 \setminus J = \phi$ implies $J = 0$ *, whence the statement (11) follows.

In the same way we prove that

(12) $\forall a \in S$ $\exists s \in S_1$ such that $\psi_a = \psi_{as}$ implies $a \in aS_1$, $\forall a \in S$;

(13) $\psi_{as} = \psi_0$, $\forall s \in S_1$ and $\forall a \in S_0$ implies $aS_1 = 0$, $\forall a \in S_0$, or $Sa = 0$, $\forall a \in S_0$.

Then, we remark that

(14) $aS = 0$, $\forall a \in S_0$ implies $a \in S_1 a$, $\forall a \in S_0$.

Let $a \in S_0$, $s \in S_1$. By Remark 1.2, we have $a = xsy$ $(x, y \in S)$. If $x \in S_0$ we have $xs = 0$ whence $0 = a \in S_1 a$. If $x \in S_1$, putting $xs = (i, g, \lambda)$ $(i \in I,\ g \in G,\ \lambda \in \Lambda)$ and denoting by u the identity of G, we have $(i, u, \lambda)a = (i, u, \lambda)xsy = (i, u, \lambda)(i, g, \lambda)y = (i, g, \lambda)y = xsy = a$ whence $a \in S_1 a$.

In the same way we can verify that

(15) $Sa = 0$, $\forall a \in S_0$ implies $a \in aS_1$, $\forall a \in S_0$.

The statement of the theorem easily follows from (8)-(15).

Theorem 3.5. *A medial semigroup of first kind is permutable if and only if S/ρ is a Δ-semigroup where ρ is the least congruence of S which has S_1 as a class.*

*Let $a \in S_0 \setminus J$ (therefore $S_1 a \neq 0$, $aS = 0$). Since J is an ideal of S, we have $J \subseteq S^1 a S^1 = \{a\} \cup S_0 a \cup S_1 a$. If $x \in J$ then $x \neq a$; if $x \in S_1 a$ then $x = (i, g, \lambda)a$ $(i \in I,\ g \in G,\ \lambda \in \Lambda)$ and, denoting by u the identity of G, we have $0 = (i, u, \lambda)x = (i, g, \lambda)a = x$. Thus $J \subseteq S_0 a$ whence the statement follows.

Proof. The necessity of the condition immediately follows from (c) of Section 1 and Theorem 2.3. Then let us prove the sufficiency. Let S be a medial semigroup of first kind, such that S/ρ is a Δ-semigroup. Then S/ρ is a medial semigroup of type α with identity (Lemma 3.3); moreover S/ρ (being by assumption a Δ-semigroup) is permutable (Theorem 2.3), whence we have $S/\rho = \Sigma_0^1$ where $\boldsymbol{\Sigma}_0$ is a commutative nilsemigroup whose principal ideals form a chain with respect to the inclusion (Theorem 2.4). Therefore S/ρ has only Rees congruences (see [10], proof of Lemma 1.4(iii)). Now let σ be a congruence of S. In S/ρ we examine the relation

$$\sigma^* = \{(a\rho, b\rho) \in S/\rho \times S/\rho \colon \exists x \in a\rho,\ y \in b\rho \text{ such that } x\,\sigma\,y\}.$$

It is easy to show that σ^* is a congruence. We shall here verify the transitivity of σ^* only. We suppose that $(a\rho)\,\sigma^*\,(b\rho)$, $(b\rho)\,\sigma^*\,(c\rho)$ $(a, b, c \in S)$. There exist $x \in a\rho$, $y, v \in b\rho$, $w \in c\rho$ such that $x\,\sigma\,y$, $v\,\sigma\,w$. Since $y, v \in b\rho = S_1 b S_1$ (Lemma 3.3) and S_1 is simple, we have $y = s_1 b s_2$, $v = t_1 b t_2$ $(s_1, s_2, t_1, t_2 \in S_1)$, $t_1 = m_1 s_1 n_1$, $t_2 = m_2 s_2 n_2$ $(m_1, n_1, m_2, n_2 \in S_1)$ and we obtain $m_1 n_1 y m_2 n_2 = m_1 n_1 s_1 b s_2 m_2 n_2 = m_1 s_1 n_1 b m_2 s_2 n_2 = t_1 b t_2 = v$. We therefore have $(m_1 n_1 x m_2 n_2)\,\sigma\,v\,\sigma\,w$ and since $m_1 n_1 x m_2 n_2 \in S_1 a S_1 = a\rho$ we obtain $(a\rho)\,\sigma^*\,(c\rho)$.

Now we verify that

(16) $\qquad x\rho \subseteq 0\sigma \Leftrightarrow x \in 0\sigma \Leftrightarrow x\rho \in (0\rho)\sigma^*.$

It is obvious that $x\rho \subseteq 0\sigma$ implies $x \in 0\sigma$ and $x\rho \in (0\rho)\sigma^*$. We prove that $x\rho \in (0\rho)\sigma^*$ implies $x\rho \subseteq 0\sigma$. We suppose $(x\rho)\,\sigma^*\,(0\rho)$ and $a \in x\rho$. There exists $y \in x\rho$ such that $y\,\sigma\,0$ (we recall that $0\rho = S_1 0 S_1 = 0$). Since $a, y \in x\rho = S_1 x S_1$ and S_1 is simple, we can write $a = s_1 x s_2$, $y = t_1 x t_2$, $s_1 = m_1 t_1 n_1$, $s_2 = m_2 t_2 n_2$ $(s_1, s_2, t_1, t_2, m_1, n_1, m_2, n_2 \in S_1)$ whence $m_1 n_1 y m_2 n_2 = a$ follows. Therefore $y\,\sigma\,0$ implies $(m_1 n_1 y m_2 n_2)\,\sigma\,0$ whence it follows that $a\,\sigma\,0$, that is, $a \in 0\sigma$.

Now let $x \in S$ and suppose $x \notin 0\sigma$. First we remark that $x\sigma \subseteq S_1 x S_1$. In fact $x\rho \notin (0\rho)\sigma^*$ (see (16)) and, since σ^* is a Rees congruence mod $(0\rho)\,\sigma^*$, we have $(x\rho)\sigma^* = x\rho$; therefore if $y \in x\sigma$ we can write $(y\rho)\,\sigma^*\,(x\rho)$ whence we obtain $y\rho = x\rho$ from which it follows that $y \in x\rho = S_1 x S_1$. Moreover, since $x \in S_1 x S_1$, there exist $i_x \in I$, $\lambda_x \in \Lambda$ such that

$$(i_x, u, \lambda_x)x = x(i_x, u, \lambda_x) = x$$

where u is the identity of G. Put

$$M_x = \{g \in G\colon (i_x, g, \lambda_x)x \,\sigma\, x\},$$

$$A_1 = (\{i_x\} \times M_x \times \{\lambda_x\})x,$$

$$A_2 = (I \times M_x \times \{\lambda_x\})x,$$

$$A_3 = (\{i_x\} \times M_x \times \{\lambda_x\})x(\{i_x\} \times \{u\} \times \Lambda),$$

$$A_4 = (I \times M_x \times \{\lambda_x\})x(\{i_x\} \times \{u\} \times \Lambda).$$

We prove that $x\sigma$ is one of the classes A_h $(h = 1, 2, 3, 4)$. First we note that, for any $z, v \in A_1$, we have $z \,\sigma\, v$ and $x \in A_1$. Thus $A_1 \subseteq x\sigma$. Let us suppose $A_1 \subset x\sigma$ and $a \in x\sigma \setminus A_1$. Since $x\sigma \subseteq S_1 x S_1$ it follows that $a = (i_a, g_a, \lambda_x)x(i_x, u, \lambda_a)$ where $i_a \in I$, $\lambda_a \in \Lambda$ and $g_a \in M_x$. There are several possible cases. First we suppose $i_a \neq i_x$, $\lambda_a = \lambda_x$ (whence $I = \{i_a, i_x\}$). Then $a \in A_2$ and the elements of A_2 are pairwise equivalent by means of σ.* Therefore $A_2 \subseteq x\sigma$. In the same way we can show that $i_a = i_x$, $\lambda_a \neq \lambda_x$ imply $A_3 \subseteq x\sigma$ and that $i_a \neq i_x$, $\lambda_a \neq \lambda_x$ imply $A_4 = x\sigma$.** Then, if we have $A_2 \subset x\sigma$ $(A_3 \subset x\sigma)$, by means of the above procedure we obtain $A_4 = x\sigma$.

Similarly, if τ is another congruence of S and $x \notin 0\tau$ $(x \in S)$, $x\tau$ is one of the following classes

$$B_1 = (\{i_x\} \times N_x \times \{\lambda_x\})x,$$

$$B_2 = (I \times N_x \times \{\lambda_x\})x,$$

$$B_3 = (\{i_x\} \times N_x \times \{\lambda_x\})x(\{i_x\} \times \{u\} \times \Lambda),$$

$$B_4 = (I \times N_x \times \{\lambda_x\})x(\{i_x\} \times \{u\} \times \Lambda)$$

where $N_x = \{g \in G\colon (i_x, g, \lambda_x)x \,\tau\, x\}$.

*We have $a = (i_a, g_a, \lambda_x)x$, $a \,\sigma\, x$, $a \,\sigma\, (i_a, u, \lambda_x)x$. Let $y \in A_2$. We can put $y = (i, g, \lambda_x)x$ $(i \in I,\ g \in M_x)$ and we have $(i_x, g, \lambda_x)x \,\sigma\, x$. Therefore, if $i = i_x$ then $a \,\sigma\, y$; if, on the contrary, $i = i_a$ then we have $(i_a, u, \lambda_x)(i_x, g, \lambda_x)x \,\sigma\, (i_a, u, \lambda_x)x$ whence we also have $a \,\sigma\, y$.

**Repeating the proof described in the previous foot-note we obtain $A_4 \subseteq x\sigma$. Moreover, if $b \in x\sigma \subseteq S_1 x S_1$ then we have $b = (i_b, g_b, \lambda_x)x(i_x, u, \lambda_b)$ $(i_b \in I,\ g_b \in G,\ \lambda_b \in \Lambda)$ and from $b \,\sigma\, x$ it follows that $(i_x, g_b, \lambda_x)x \,\sigma\, x$, that is, $g_b \in M_x$; therefore $b \in A_4$ whence $x\sigma \subseteq A_4$.

From (16) it easily follows that we have $0\sigma \subseteq 0\tau$ if and only if $(0\rho)\sigma^* \subseteq (0\rho)\tau^*$. Therefore, since the ideals of S/ρ form a chain, the set $\{0\sigma \mid \sigma \text{ is a congruence of } S\}$ is totally ordered by inclusion.

Now we are able to prove that S is permutable. Let σ and τ be two congruences of S and $0\tau \subseteq 0\sigma$ (in the analogous way we could study the case $0\sigma \subseteq 0\tau$). If $a, b \in S$ with $(a, b) \in \sigma \cdot \tau$ then there exists $x \in S$ such that $a\, \sigma\, x$, $x\, \tau\, b$. We shall consider various cases:

1. If $x \in 0\tau$ then we have $a, b \in 0\sigma$ whence $(a, b) \in \tau \cdot \sigma$.

2. If $x \in 0\sigma \setminus 0\tau$ then we have $b\rho = x\rho \in (0\rho)\sigma^*$ whence also $a, b \in 0\sigma$.

3. If $x \notin 0\sigma$, then $x\sigma$ is one of the classes A_h, $x\tau$ is one of the classes B_k $(h, k = 1, 2, 3, 4)$ and therefore it is easy to verify that $(a, b) \in \tau \cdot \sigma$. *

Therefore we have proved that $(a, b) \in \sigma \cdot \tau$ always implies $(a, b) \in$ $\in \tau \cdot \sigma$, hence the statement is proved.

*Let us suppose for example that $x\sigma = A_4$, $x\tau = B_4$. Then $a = (i_a, g_a, \lambda_x)x(i_x, u, \lambda_a)$, $b = (i_b, g_b, \lambda_x)x(i_x, u, \lambda_b)$ $(i_a, i_b \in I;\ \lambda_a, \lambda_b \in \Lambda;\ g_a \in M_x;\ g_b \in N_x)$. The relations $a\,\sigma\, x$, $x\,\tau\, b$ imply

$$(\alpha)\quad a\,\tau\,(i_a, g_a g_b, \lambda_x)x(i_x, u, \lambda_a);\ (i_b, g_a g_b, \lambda_x)x(i_x, u, \lambda_b)\,\sigma\, b.$$

Since all the elements $(i_a, g_a, \lambda_x)x$, $(i_b, g_b, \lambda_x)x$, $x(i_x, u, \lambda_a)$, $x(i_x, u, \lambda_b)$ belong to A_4, we have

$$(\beta)\quad (i_a, g_a, \lambda_x)x\,\sigma\,(i_b, g_a, \lambda_x)x,$$

$$(\gamma)\quad x(i_x, u, \lambda_a)\,\sigma\, x(i_x, u, \lambda_b).$$

From (β) it follows that

$$(i_a, g_a, \lambda_x)x(i_a, g_b, \lambda_x)(i_x, u, \lambda_b)\,\sigma\,(i_b, g_a, \lambda_x)x(i_a, g_b, \lambda_x)(i_x, u, \lambda_b)$$

and from this, S being medial, we deduce

$$(i_a, g_a g_b, \lambda_x)x(i_x, u, \lambda_b)\,\sigma\,(i_b, g_a g_b, \lambda_x)x(i_x, u, \lambda_b).$$

From (γ) we have

$$(i_a, g_a g_b, \lambda_x)x(i_x, u, \lambda_a)\,\sigma\,(i_a, g_a g_b, \lambda_x)x(i_x, u, \lambda_b)$$

whence

$$(i_a, g_a g_b, \lambda_x)x(i_x, u, \lambda_a)\,\sigma\,(i_b, g_a g_b, \lambda_x)x(i_x, u, \lambda_b)$$

so that, from (α) it follows immediately that $(a, b) \in \tau \cdot \sigma$.

Now let us consider a medial semigroup of second kind and let x be an element of S_0. Since $x \in S_1 x$, there exist $i_x \in I$ and $\lambda_x \in \Lambda$ such that $(i_x, u, \lambda_x)x = x$, where u is the identity of G. Then, it is immediate that the set $H_x = \{g \in G\colon (i_x, g, \lambda_x)x = x\}$ is a subgroup of G.

Now we are able to prove the following

Theorem 3.6. *A medial semigroup S of second kind is permutable if and only if*

(i) *S/ρ is a Δ-semigroup where ρ is the least congruence of S which has S_1 as a class,*

(ii) *if $|I| = 2$, every subgroup K of G, which satisfies the condition $H_x \subset K$ for some $x \in S_0$, contains two elements k_1 and k_2 such that $(i_1, k_1, \lambda_x)x = (i_2, k_2, \lambda_x)x$, where $I = \{i_1, i_2\}$.*

Proof. The conditions are necessary. Let S be a medial permutable semigroup of second kind. Condition (i) immediately follows from (c) of Section 1 and Theorem 2.3. Let us prove that (ii) holds. Then, let $|I| = 2$ and K be a subgroup of G such that $H_x \subset K$ for some $x \in S_0$. First let us prove that the relation $(\{i\} \times g_1 K \times \{\lambda_x\})x \cap (I \times g_2 K \times \{\lambda_x\})x \neq$ $\neq \phi$ $(i \in I,\ g_1, g_2 \in G)$ implies $g_1 K = g_2 K$. In fact, if $(i, g_1 k_1, \lambda_x)x =$ $= (j, g_2 k_2, \lambda_x)x$ $(j \in I,\ k_1, k_2 \in K)$ we have $x = (i_x, g_1^{-1} g_2 k_1^{-1} k_2, \lambda_x)x$, hence $g_1^{-1} g_2 k_1^{-1} k_2 \in H_x$; thus, since $H_x \subset K$, we have $g_1^{-1} g_2 \in K$. Let

$$(17) \qquad (\{i_1\} \times K \times \{\lambda_x\})x \cap (\{i_2\} \times K \times \{\lambda_x\})x = \phi.$$

The partition of S whose classes are $(I \times \{u\} \times \{\lambda_x\})x$ and the single elements of S is a congruence σ of S. Moreover, it is easily seen that the subsets of S $(\{i_1\} \times K \times \{\lambda_x\})x$, $(\{i_2\} \times K \times \{\lambda_x\})x$, $(I \times gK \times \{\lambda_x\})x$ with $g \in G \setminus K$, and the remaining single elements of S form a partition of S which turns out to be a congruence τ, which is not permutable with σ. Then the relation (17) cannot hold and therefore there exist k_1 and $k_2 \in K$ such that $(i_1, k_1, \lambda_x)x = (i_2, k_2, \lambda_x)x$.

Now we shall prove the sufficiency of the conditions. Let S be a medial semigroup of second kind which satisfies the conditions (i) and (ii). Let us suppose that S_0 contains at least one element $x \neq 0$ (if $S_0 = 0$, we have a particular case of that treated in the preceding theorem and therefore S is permutable.*). Since $x \in S_1 x$, we have $S_1 x \neq 0$. Hence the semigroup S/ρ is a medial permutable semigroup of type α (Theorem 2.3). Therefore it is the first semigroup of (ii) of Theorem 2.4. The elements of S/ρ are exactly $S_1 0, S_1 x, S_1$ (Lemma 3.2), therefore we have $S_0 = \{0\} \cup S_1 x$ $(x \in S_0 \setminus 0)$. Let σ be a congruence of S. We may easily prove the following two propositions:

The relations $a \in S_0$, $s \in S_1$, $(a, s) \in \sigma$ imply $\sigma = S \times S$. **

If $\sigma \neq S \times S$, we have $0\sigma = 0$ or $0\sigma = S_0$. ***

Let us suppose $0\sigma = 0$ and $x \in S_0 \setminus 0$. We have $x\sigma \subseteq S_1 x$. Moreover, since $x \in S_1 x$, there are $i_x \in I$, $\lambda_x \in \Lambda$ such that $(i_x, u, \lambda_x)x = x$, where u is the identity of G. Let us put

$$M_x = \{g \in G\colon (i_x, g, \lambda_x)x \;\sigma\; x\}.$$

It can be immediately verified that M_x is a subgroup of G, and it is obvious that $H_x \subseteq M_x$. We may verify, analogously to the preceding theorem, that $x\sigma$ is one of the following classes:

$$(\{i_x\} \times M_x \times \{\lambda_x\})x, \qquad (I \times M_x \times \{\lambda_x\})x.$$

Hence,

$$(18) \qquad \begin{aligned} &\text{if } H_x = M_x \text{ then } x\sigma = \{x\} \text{ or } x\sigma = (I \times M_x \times \{\lambda_x\})x, \\ &\text{if } H_x \subset M_x \text{ then } x\sigma = (I \times M_x \times \{\lambda_x\})x. \text{****} \end{aligned}$$

*If $S_0 = 0$, we have (using the notation of the preceding case) $(0\rho)\sigma^* = 0\rho$ or $(0\rho)\sigma^* = S/\rho$; whence it follows that $0\sigma = 0$ or $0\sigma = S$.

**In fact, since there exists $e \in S_1$ such that $se = s$ and $ae = 0$, the relation $a \;\sigma\; s$ implies $0 \;\sigma\; s$. Furthermore, S_1 being simple, for every $t \in S_1$ there exist $m, n \in S_1$ such that $t = msn$; thus from $0 \;\sigma\; s$ it follows that $0 \;\sigma\; t$ whence $0 \;\sigma\; tx$ and $\sigma = S \times S$.

***In fact, if 0σ contains an element $a \neq 0$ then $a = sx$ $(s \in S_1)$ and, if $b = tx$ $(t \in S_1)$ is another element of S_0 with $b \neq 0$, we have $t = msn$ $(m, n \in S_1)$, $mna = mnsx = msnx = tx = b$, whence $b \in 0\sigma$.

****If $|I| = 1$ then we have $x\sigma = (I \times M_x \times \{\lambda_x\})x$. If $|I| = 2$ and $H_x \subset M_x$ then, by assumption there are $m_1, m_2 \in M_x$ such that $(i_x, m_1, \lambda_x)x = (i, m_2, \lambda_x)x$ with $i \in I$, $i \neq i_x$, thus we have $x \;\sigma\; (i, m_2, \lambda_x)x$, whence it follows that $y \;\sigma\; x$ for any $y \in (I \times M_x \times \{\lambda_x\})x$.

Now can verify that S is permutable. Let σ, τ be two congruences of S different from $S \times S$, a, b two elements of S such that $(a, b) \in \in \sigma \cdot \tau$. Then there exists an element $x \in S$ such that $a \, \sigma \, x$, $x \, \tau \, b$. From the above we infer either $a, b, x \in S_0$ or $a, b, x \in S_1$. In the latter case, since S_1 is permutable (Corollary 1.2), we obtain $(a, b) \in \tau \cdot \sigma$. If, on the contrary, $a, b, x \in S_0$, and either $0\sigma = S_0$ or $0\tau = S_0$ then we obviously have $(a, b) \in \tau \cdot \sigma$; similarly, if $0\sigma = 0\tau = 0$ and either $x\sigma = \{x\}$ or $x\tau = \{x\}$ then we again have $(a, b) \in \tau \cdot \sigma$. Then, let us examine the case $0\sigma = 0\tau = 0$ with $x\sigma, x\tau \neq \{x\}$. By (18), we may write $a = (i_a, m, \lambda_x)x$ $(i_a \in I,\ m \in M_x)$, $b = (i_b, n, \lambda_x)x$ $(i_b \in I,\ n \in N_x = \{g \in G\colon (i_x, g, \lambda_x)x \, \tau \, x\})$, whence $a \, \tau \, (i_a, mn, \lambda_x)x$, $(i_b, mn, \lambda_x)x \, \sigma \, b$.

If $i_a = i_b$ (as it is implied for instance when $|I| = 1$) we have $(a, b) \in \tau \cdot \sigma$. Let us suppose $i_a \neq i_b$ and consider the class $a\tau$. If $a\tau = \{a\}$ then $(i_a, mn, \lambda_x)x = a = (i_a, m, \lambda_x)x$, whence $(i_x, n, \lambda_x)x = x$; thus $n \in H_x \subseteq M_x$, $b = (i_b, n, \lambda_x)x \in (I \times M_x \times \{\lambda_x\})x = x\sigma$, whence $(a, b) \in \sigma$ and consequently $(a, b) \in \tau \cdot \sigma$. If $a\tau = (I \times N_a \times \{\lambda_x\})a$ where $N_a = \{g \in G\colon (i_a, g, \lambda_a)a \, \tau \, a\}$ then from $a\tau(i_a, mn, \lambda_x)x$ we may deduce $(i_b, u, \lambda_x)(i_a, mn, \lambda_x)x \, \tau \, (i_b, u, \lambda_x)a \, \tau \, a$, that is, $a\tau(i_b, mn, \lambda_x)x \, \sigma \, b$. Thus $(a, b) \in \tau \cdot \sigma$ and the statement is proved.

Finally, let us consider a medial semigroup of third kind. For every $x \in S_0$ we have $x(i_x, u, \lambda_x) = x$, where u is the identity of G and $i_x \in I$, $\lambda_x \in \Lambda$. Let ${}_xH = \{g \in G\colon x(i_x, g, \lambda_x) = x\}$. Then we can prove the following theorem similarly to the preceding case.

Theorem 3.7. *A medial semigroup S of third kind is permutable if and only if*

(i) *S/ρ is a Δ-semigroup where ρ is the least congruence of S which has S_1 as a class.*

(ii) *if $|\Lambda| = 2$, every subgroup K of G satisfying the condition ${}_xH \subset K$ for some $x \in S_0$, contains two elements k_1 and k_2 such that $x(i_x, k_1, \lambda_1) = x(i_x, k_2, \lambda_2)$, where $\Lambda = \{\lambda_1, \lambda_2\}$.*

REFERENCES

[1] C. Bonzini – A. Cherubini Spoletini, Sugli $E-2$ semigruppi permutabili, *Istituto Lombardo (Rend. Sc.)* A, 115 (1981), to appear.

[2] J.L. Chrislock, On medial semigroups, *J. Algebra,* 12 (1969), 1–9.

[3] A.H. Clifford – G.B. Preston, *The algebraic theory of semigroups,* vol. I, Math. Surveys No. 7, Amer. Math. Soc., Providence, R. I., 1961.

[4] A. Etterbeeck, *Semigroups whose lattice of congruences form a chain,* Dissertation of University of California, Davis, 1970.

[5] H. Hamilton, Permutability of congruences on commutative semigroups, *Semigroup Forum,* 10 (1975), 55–66.

[6] E.S. Ljapin, *Semigroups,* Translation of Mathematical Monographs, Amer. Math. Soc., Providence, R. I. 1974.

[7] M. Petrich, *Introduction to semigroups,* Merrill Research and Lectures series, 1973.

[8] M. Satyanarayana, *Positively ordered semigroups,* Lecture Notes in Pure and Applied Mathematics, 42, M. Dekker, Inc., New York – Basel, 1979.

[9] T. Tamura, Commutative semigroups whose lattice of congruences is a chain, *Bull. Soc. Math. France,* 97 (1969), 369–380.

[10] P.G. Trotter, Exponential Δ-semigroups, *Semigroup Forum,* 12 (1976), 312–331.

Celestina Bonzini

Istituto Matematico "F. Enriques", Via Saldini 50, 20133 Milano, Italy.

Alessandra Cherubini

Istituto di matematica Politecnico, Piazza L. da Vinci 32, 20133 Milano, Italy.

COLLOQUIA MATHEMATICA SOCIETATIS JÁNOS BOLYAI
39. SEMIGROUPS, SZEGED (HUNGARY), 1981.

ON TRANSLATIONS, EXTENSIONS AND RESHAPING OF SEMIGROUPS

M. DEMLOVÁ – P. GORALČÍK – V. KOUBEK

The semigroup theorists of all times only described semigroups; we want to change them. To change a thing means to destroy part of its structure and then either to leave it as it is or to add some new structure, according to the goal we pursue. Some destruction, rejection of the old in favour of the new, is necessary in any process of restructuring. However, the extent of the destruction necessary to achieve the goal, may it be radical or moderate, should be adequate. For, if ever a goal has been achieved by inadequate destruction, the price usually was exorbitant.

Also reshaping of a semigroup, resulting in a new semigroup, will go through a destructive and reconstructive stage. To partly destroy a semigroup $S = (X, \cdot)$ will mean here to pass to a partial subsemigroup, or a "fragment", $P = (X, \circ)$, with the same underlying set X (which thus remains unaffected by the destruction) but whose operation "$\circ$" is an arbitrary restriction of the (total) operation "$\cdot$" of S. If S is given by a multiplication table then we obtain the partial table, or table-fragment, of P by wiping off some of the inner entries from the table of S (the outer entries by which the rows and columns of the table are labelled remain unaffected). The case when only one entry in the table is erased has been dealt with in [4].

Any partial semigroup originated in the above way from a total semigroup will be called authentic. A partial semigroup (i.e. a partial groupoid satisfying $x(yz) = (xy)z$ whenever both sides are defined) need not be authentic and its testing for authenticity may prove hard. We shall refer to this as the authenticity problem.

The authenticity problem relates to the reconstructive stage in which we want a given fragment $P = (X, \circ)$ to be accommodated in a total semigroup $T = (X, *)$ with specified properties. How to know that a filling of a particular entry in a particular way into an authentic table fragment would not spoil it and might lead to an authentic table fragment again? To avoid false steps in a successive reconstruction we need advice: a plan, strategy, theorems, which could help to improve the art of producing semigroups with interesting properties.

Let $S = (X, \cdot)$ be a group with a subgroup $H = (Y, \square)$. If H is not a normal subgroup of S we may wish to reshape S into a group $T = (X, *)$ in which H is preserved as a subgroup, but the rest of the multiplication is redefined in such a way that H becomes a normal subgroup of T. We may also prescribe the quotient T/H to be isomorphic to some group Q (of order equal to the index of H in S). To reshape S in the required way, we certainly can first destroy all multiplication outside H and wind up with the fragment $(X, \square)$ in which only products for couples of elements of Y are defined, but it will be more adequate to go only as far as the fragment $P = (X, \circ)$ in which $a \circ b$ is defined iff $a \in Y$ and $b \in X$. It is even advantageous (as always with a wise and cautious destruction) that we have in P preserved the left action of H on S, for, as Schreier did in the middle of the twenties, we can utilize this action for a description of all possible reconstructions of P in the shape of groups.

There has been a number of attempts to directly generalize the classical Schreier group extensions to semigroups (mostly with an identity). Through the reconstruction of partial semigroups (in particular of the inner translations, i.e. authentic partial semigroups in which only multiplication on the left by a single element is specified, cf. [1], [3]) we came to see extensions as a special type of two-step reconstruction. The first step consists in selecting a suitable congruence θ on the given partial semigroup $P = (X, \circ)$ ("suitable" is connected with the possibility of coordinatization of θ-blocks in terms of P, i.e. of localizing individual elements in the blocks). The second step consists in selecting a suitable "block-semigroup" $Q = (X/\theta, */\theta)$. Once congruence θ and quotient Q fixed, we have set up an extension setting by which we split off, from the general reconstruction problem, a particular extension problem, of finding a total semigroup $T = (X, *)$ with P a partial subsemigroup, θ a congruence, and $Q = T/\theta$ a quotient.

Another method of reconstruction is based on combining several authentic fragments, possibly borrowed from different semigroups. Let $P = (X, \circ)$ and $R = (X, \square)$ be two authentic partial semigroups. We say that P and R are coherent if there exists a total semigroup $T = (X, *)$ accommodating both P and R as its partial subsemigroups. Clearly, coherent fragments P, R must be compatible in the sense that $x \circ y = x \square y$ whenever both sides are defined. When reconstructing a table-fragment, each new entry we are going to fill in must be coherent with the fragment. Let us illustrate the idea of combining fragments for the purpose of reshaping a semigroup by the following

Statement. *Let $S = (X, \cdot)$ be a semigroup possessing two subsemigroups $E, F \subseteq X$ and a retraction (i.e. an idempotent endomorphism) φ of S onto its ideal I such that*

(a) $\varphi(E) \subseteq E$, $\varphi(F) \subseteq F$,

(b) $E \cdot (X - E) \subseteq X - E$, $E \cdot (X - F) \subseteq X - F$,

(c) $(X - E) \cdot F \subseteq X - E$, $(X - F) \cdot F \subseteq X - F$.

Then the fragment $P = (X, \circ)$ in which $x \circ y$ is defined by $x \circ y = x \cdot y$ iff $x \in E$ or $y \in F$, is coherent with any compatible fragment $R = (X, \square)$ such that $(I, \square)$ is a total semigroup and $x \square y$ is defined in R iff both $x \in I$ and $y \in I$. Moreover, P and R can then both be accommodated in a total semigroup $T = (X, *)$ with I an ideal and φ a retraction onto I.

Proof. Denote $\varphi(x) = \bar{x}$ for $x \in X$ and define $T = (X, *)$ by

$$(1) \qquad x * y = \begin{cases} x \cdot y & \text{if } x \in E \text{ or } y \in F, \\ \bar{x} \mathbin{\square} \bar{y} & \text{otherwise.} \end{cases}$$

We first show that it holds in T that

$$(2) \qquad \overline{x * y} = \bar{x} * \bar{y} = \bar{x} * y = x * \bar{y}$$

for any $x, y \in X$. Indeed, it is easy to see that $X * I \subseteq I$ and $I * X \subseteq I$, that is to say, I is an ideal of T. If $x \in E$ or $y \in F$ then by (a) also $\bar{x} \in E$ or $\bar{y} \in F$, hence by (1), $\overline{\overline{x * y}} = \overline{\bar{x} * \bar{y}} = \overline{\bar{x} * y} = \overline{x * \bar{y}}$, and, since $\overline{x * y}$, $\bar{x} * \bar{y}$, $\bar{x} * y$, $x * \bar{y} \in I$, we get (2) by idempotency of φ. If $x \notin E$ then $\bar{x} \notin E$; for assuming $x \notin E$ and $\bar{x} \in E$, we have $\bar{x} \cdot x \notin E$ by (b), however, $\bar{x} \cdot x \in I$, thus $\bar{x} \cdot x = \overline{\bar{x} \cdot x} = \bar{x} \cdot \bar{x} \in E$, a contradiction. Likewise, if $y \notin F$ then $\bar{y} \notin F$. So if $x \notin E$ and $y \notin F$ then also $\bar{x} \notin E$ and $\bar{y} \notin F$ and we get (2) straight by (1).

Turning to the proof of associativity of T, note that by compatibility of P and R we have

$$(3) \qquad (\bar{x} * \bar{y}) * \bar{z} = \bar{x} * (\bar{y} * \bar{z})$$

for any $x, y, z \in X$. We have to consider eight cases:

I. If $x \in E$, $y \in E$, and z is arbitrary then

$$x * y \overset{(1)}{=} x \cdot y \in E$$

and

$$(x * y) * z \overset{(1)}{=} (x \cdot y) \cdot z = x \cdot (y \cdot z) \overset{(1)}{=} x * (y * z).$$

II. If $x \in E$, $y \notin E$, $z \in F$ then

$$(x * y) * z \overset{(1)}{=} (x \cdot y) \cdot z = x \cdot (y \cdot z) \overset{(1)}{=} x * (y * z).$$

III. If $x \in E$, $y \notin E$, $z \notin F$ then by (b) we have $x * y \notin E$, hence

$$(x * y) * z \overset{(1)}{=} \overline{(x * y)} * \bar{z} \overset{(2)}{=} (\bar{x} * \bar{y}) * \bar{z} \overset{(3)}{=} \bar{x} * (\bar{y} * \bar{z}) \overset{(2)}{=}$$

$$= \bar{x} * \overline{(y * z)} \overset{(2)}{=} x * \overline{(y * z)} \overset{(2)}{=} x * (\bar{y} * \bar{z}) \overset{(1)}{=} x * (y * z).$$

IV. If $x \notin E$, $y \in E - F$, and z is arbitrary then by (b) and (c) we have $y * z \notin F$, hence

$$(x * y) * z \overset{(1)}{=} (\bar{x} * \bar{y}) * z \overset{(2)}{=} (\overline{x * y}) * z \overset{(2)}{=} (\overline{x * y}) * \bar{z} \overset{(2)}{=}$$

$$= (\bar{x} * \bar{y}) * \bar{z} \overset{(3)}{=} \bar{x} * (\bar{y} * \bar{z}) \overset{(2)}{=} \bar{x} * (\overline{y * z}) \overset{(1)}{=} x * (y * z).$$

V. If $x \notin E$, $y \in F$, $z \notin F$ then by (c) we have $x * y \notin E$, hence

$$(x * y) * z \overset{(1)}{=} (\overline{x * y}) * \bar{z} \overset{(2)}{=} (\bar{x} * \bar{y}) * \bar{z} \overset{(3)}{=} \bar{x} * (\bar{y} * \bar{z}) \overset{(2)}{=} \bar{x} * (\overline{y * z});$$

now, if $y \in E$ then by (b) we have $y * z \notin F$, thus

$$\bar{x} * (\overline{y * z}) = x * (y * z),$$

while for $y \notin E$ we have

$$\bar{x} * (\overline{y * z}) \overset{(2)}{=} x * (\overline{y * z}) \overset{(2)}{=} x * (\bar{y} * \bar{z}) \overset{(1)}{=} x * (y * z).$$

VI. If $x \notin E$, $y \in F$, $z \in F$ then $y * z \overset{(1)}{=} y \cdot z \in F$, hence

$$(x * y) * z \overset{(1)}{=} (x \cdot y) \cdot z = x \cdot (y \cdot z) \overset{(1)}{=} x * (y * z).$$

VII. If $x \notin E$, $y \notin E \cup F$, $z \notin F$ then

$$(x * y) * z \overset{(1)}{=} (\bar{x} * \bar{y}) * z \overset{(2)}{=} (\overline{x * y}) * z \overset{(2)}{=} (\overline{x * y}) * \bar{z} \overset{(2)}{=}$$

$$= (\bar{x} * \bar{y}) * \bar{z} \overset{(3)}{=} \bar{x} * (\bar{y} * \bar{z}) \overset{(2)}{=} \bar{x} * (\overline{y * z}) \overset{(2)}{=}$$

$$= x * (\overline{y * z}) \overset{(2)}{=} x * (\bar{y} * \bar{z}) \overset{(1)}{=} x * (y * z).$$

VIII. If $x \notin E$, $y \notin E \cup F$, $z \in F$ then by (c) we have $y * z \notin F$, hence

$$(x * y) * z \overset{(1)}{=} (\bar{x} * \bar{y}) * z \overset{(2)}{=} (\overline{x * y}) * z \overset{(2)}{=} (\overline{x * y}) * \bar{z} \overset{(2)}{=}$$

$$= (\bar{x} * \bar{y}) * \bar{z} \overset{(3)}{=} \bar{x} * (\bar{y} * \bar{z}) \overset{(2)}{=} \bar{x} * (\overline{y * z}) \overset{(1)}{=} x * (y * z).$$

The reconstruction used in the proof amounts to a sort of "cautious inflation" of the ideal I; indeed, the missing products are defined so as to make φ an inflation endomorphism with respect to them.

The statement remains true if E or F or both are empty. In the latter case fragment P is empty, so what is asserted is only the authenticity of R with $(I, \square)$ completely arbitrary. The authentication is done by an inflation of I determined by φ as the inflation endomorphism.

If $E \cup F \neq \phi$ then $(I, \square)$ can no more be quite arbitrary if R is to be compatible with P. Put otherwise, $(I, \square)$ must be a reconstruction of the subfragment $(I, \circ)$ of P. The partial table of $(I, \circ)$ contains the whole rows (i.e. left inner translations) corresponding to the elements of $\varphi(E)$ and the whole columns (i.e. right inner translations) corresponding to the elements of $\varphi(F)$. The statement does not say how to reconstruct $(I, \circ)$ into $(I, \square)$, this is a problem in itself; some clues as to how to go about it are given in [1]. The statement rather stresses the possibility to turn $(I, \circ)$ into $(I, \square)$ in a completely arbitrary way and still to get $R = (X, \square)$ coherent with P.

Let us now indicate some situations to which the statement applies.

Every element c of a semigroup $S = (X, \cdot)$ creates by its action on S some very natural equivalences. Let us consider, more specifically, the left translation by c and let us call the left connectedness by c the equivalence $\sim_c$ defined by

$$x \sim_c y \Leftrightarrow \exists m, n \quad (c^m \cdot x = c^n \cdot y),$$

the left stratification by c the equivalence $\approx_c$ defined by

$$x \approx_c y \Leftrightarrow \exists m \qquad (c^m \cdot x = c^m \cdot y)$$

(m, n denote positive integers).

It is easily observed that $\approx_c$ is a refinement of $\sim_c$ and both are right congruences. In some cases $\sim_c$ or $\approx_c$ may even be congruences, e.g. if S is commutative. In the last section of [3] we have proved that if S has an identity and $\sim_c$ is a congruence than so is $\approx_c$.

If the element c is periodic (i.e. the semigroup $\langle c \rangle$ generated by c is finite) then the left translation by c restricts to the subset $C_c = \bigcap_{n=1}^{\infty} c^n \cdot X$ of S as bijection and each $\approx_c$-class meets C_c in exactly one point. More generally, define the left core of an arbitrary element c of S as the set $C_c = \bigcap_{n=1}^{\infty} c^n \cdot X$. Equivalently, C_c is the biggest subset of S on which the left translation by c is surjective. Call c left quasiperiodic if each class of $\approx_c$ meets C_c in exactly one point (this inevitably means that C_c meets every $\sim_c$-class and the left translation by c is (injective and thus) bijective on C_c).

The left core C_c of any element c of S is a right ideal of S; in some cases (e.g. in the commutative case) C_c is a two-sided ideal. If this is the case and if, moreover, $\approx_c$ is a congruence then the assignment $x \mapsto \bar{x}$, where $\bar{x}$ is uniquely determined as the element in the intersection of C_c with the $\approx_c$-class of x, defines a retraction φ of S onto C_c.

The $\sim_c$-class of c, denoted by E_c, is easily shown, for arbitrary c, to be a subsemigroup of S (if $x \sim_c c \sim_c y$ then $x \cdot y \sim_c c \cdot y \sim_c$ $\sim_c y \sim_c c$). Moreover, if $x \in E_c$ and $y \notin E_c$ then $x \cdot y \notin E_c$ (if $x \sim_c c$ then $x \cdot y \sim_c c \cdot y \sim_c y$), hence $E_c \cdot (X - E_c) \subseteq X - E_c$.

The above considerations lead straight to the following

Corollary. *If* c *is a quasiperiodic element of a semigroup* $S =$ $= (X, \cdot)$, C_c *is an ideal of* S, *and* $\approx_c$ *is a congruence on* S, *then the fragment* $P = (X, \circ)$, *with* $x \circ y$ *defined iff* $x \in E_c$, *is coherent with any compatible fragment* $R = (X, \square)$, *with* $x \square y$ *defined iff* $x, y \in C_c$.

In the situation of this corollary we may find it easier to obtain suitable fragments R, since any reconstruction $(C_c, \square)$ of the partial semigroup $(C_c, \blacksquare)$, with $x \blacksquare y$ defined iff $x = \bar{c}$, will be compatible with $(C_c, \circ)$. The table-fragment describing $(C_c, \blacksquare)$ contains a single row – the left translation by $\bar{c}$. If we wish to reconstruct $(C_c, \blacksquare)$ to $(C_c, \square)$ in such a way that the left connectedness by $\bar{c}$ becomes a congruence on $(C_c, \square)$ then the results of the forthcoming paper [3] enable us to do it in all possible ways.

The corollary applies most smoothly to quasiperiodic commutative semigroup (i.e. semigroups with every element quasiperiodic). Any element c of such a semigroup S meets the assumptions and thus may be used to the purpose of reshaping S, unless $S = E_c$ since then there would be no choice for R.

REFERENCES

[1] M. Demlová – P. Goralčík, Translation constructions for monoids, *Algebraic Theory of Semigroup* (Proc. Conf. Szeged, 1976), Coll. Math. Soc. J. Bolyai, vol. 20, North-Holland, Amsterdam, 1979, 61–72.

[2] P. Goralčík – V. Koubek, Translational extensions of semigroups, *Algebraic Theory of Semigroups* (Proc. Conf. Szeged, 1976), Coll. Math. Soc. J. Bolyai, vol. 20, North-Holland, Amsterdam, 1979, 173–218.

[3] M. Demlová – P. Goralčík – V. Koubek, Inner injective transextensions of semigroups, *Acta Sci. Math.*, 44 (1982), 215–237.

[4] K.P. Kozlov, On the rigid cells of a subsemigroup of a semigroup, *Algebraic Theory of Semigroups* (Proc. Conf. Szeged, 1976), Coll. Math. Soc. J. Bolyai, vol. 20, North-Holland, Amsterdam, 1979, 299–307.

M. Demlová

Elektrotechnická fakulta ČVUT, Katedra matematiky, Suchbatarová 2, 16627 Praha 6, Czechoslovakia.

P. Goralčík

MFF UK, Sokolovská 83, 18600 Praha 8-Karlín, Czechoslovakia.

V. Koubek

VC UK, Malostranské nám. 25, 11800 Praha 1, Czechoslovakia.

COLLOQUIA MATHEMATICA SOCIETATIS JÁNOS BOLYAI
39. SEMIGROUPS, SZEGED (HUNGARY), 1981

ON VARIETIES OF S-SYSTEMS

H.-J. HOEHNKE

A. Fröhlich [2], [3] considered the correspondence between varieties of groups over a d. g. near-ring and of associative algebras over a commutative ring and certain variety functors. We establish the analogous correspondence for the subvarieties of the variety S_r of all unital S-(right) systems in the sense of the author [4] over a monoid S. S_r at the same time denotes the category of all unital S-systems as objects together with their homomorphism $\varphi\colon M \to N$ $(M, N \in S_r)$.

For $(a, b) \in M \times M$ set $\varphi(a, b) := (\varphi(a), \varphi(b)) =: (\varphi a, \varphi b)$; i.e. we write $\varphi(a, b)$ instead of $(\varphi \times \varphi)(a, b)$. Let $\operatorname{Con} M$ be the lattice of all congruences of $M \in S_r$ with the least element $\mathbf{0}$. Let

$$(\varphi \times \varphi)^{-1}\mathbf{0} := \{(m_1, m_2) \in M \times M,\ \varphi m_1 = \varphi m_2\} \in \operatorname{Con} M.$$

We write $\varphi^{-1}\mathbf{0}$ instead of $(\varphi \times \varphi)^{-1}\mathbf{0}$. For $M \in S_r$ set

$$VM = \bigcap_{\substack{\varphi : M \to N \in S_r, \\ N \in \mathbf{V}}} \varphi^{-1}\mathbf{0} \quad \text{and} \quad UM = M / VM$$

where $\mathbf{V}$ denotes a fixed subvariety of S_r.

1. For $M \in S_r$, VM is a congruence on M. If VM at the same time is considered as an S-subsystem of the S-system $M \times M$ then V and U constitute functors $S_r \to S_r$, the subfunctor and the quotient functor associated with the variety $\mathbf{V}$. We have the commutative diagrams (φ: $M \to N \in S_r$, can canonical projection)

$$\begin{array}{ccc} VM & \xrightarrow{V\varphi} & VN \\ \cap\!\mid \downarrow & & \downarrow \mid\!\cap \\ M \times M & \xrightarrow[\varphi \times \varphi]{} & N \times N \end{array} \qquad \begin{array}{ccc} M & \xrightarrow{\varphi} & N \\ \text{can} \downarrow & & \downarrow \text{can} \\ UM & \xrightarrow[U\varphi]{} & UN \end{array}$$

2.

(i) *For all* $M \in S_r$ *we have* $VM = \mathbf{0} \Leftrightarrow UM = M \Leftrightarrow M \in \mathbf{V}$.

(ii) *For all* $M \in S_r$ *we have* $UM \in \mathbf{V}$.

(iii) *The subfunctor* V *associated with the variety* $\mathbf{V}$ *preserves surjective homomorphisms* $\in S_r$.

Proof. (i) and (ii) follow immediately from the definitions.

(iii): Let χ: $M \to L \in S_r$. Then $V\chi$: $VM \to VL$ is induced by the restriction to $VM \subseteq M \times M$ of the homomorphism $\chi \times \chi$: $M \times M \to$ $\to L \times L$. The functor property of V (which is easily checked) implies $\operatorname{Im} V\chi := \chi(VM) \subseteq VL$. Since $\operatorname{Im} V\chi \subseteq \langle \operatorname{Im} V\chi \rangle$ where $\langle \ldots \rangle$ denotes the congruence generated by $\ldots$, χ induces a homomorphism $M/VM \to$ $\to L/\langle \operatorname{Im} V\chi \rangle$ which is surjective if χ is. In this case since $M/VM \in \mathbf{V}$ we have $L/\langle \operatorname{Im} V\chi \rangle \in \mathbf{V}$ and hence (since $\langle \operatorname{Im} V\chi \rangle \subseteq VL$) $\langle \operatorname{Im} V\chi \rangle = VL$; this is meant by the expression "preserves" surjective homomorphisms.

Definition. A functor F: $S_r \to S_r$ is called a variety functor if

(i) for all $M \in S_r$, $FM \in \operatorname{Con} M$,

(ii) F is a subfunctor of the diagonal functor

$$D: S_r \to S_r: M \mapsto M \times M,$$

(iii) F preserves surjective homomorphisms $\in S_r$.

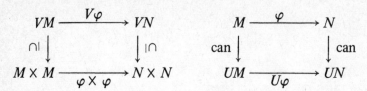

3. *If* $F: S_r \to S_r$ *is a variety functor then*

$$\mathbf{V}_F = \{M \in S_r,\ FM = \mathbf{0}\}$$

is a subvariety of S_r *and the subfunctor* V *associated with the variety* $\mathbf{V}_F$ *is equal to* F.

Proof. Let N be an S-subsystem of M. Since F is a subfunctor of D we have the commutative diagram

$$\begin{array}{ccc} FN & \longrightarrow & FM \\ \cap\!\!\mid\ \downarrow & & \downarrow\ \mid\!\!\cap \\ N \times N & \underset{\subseteq}{\longrightarrow} & M \times M \end{array}\ .$$

Hence $FN \subseteq FM$ and if $M \in \mathbf{V}_F$ then $FM = \mathbf{0}$. Thus $FN = \mathbf{0}$ and $N \in \mathbf{V}_F$. Let $\chi: M \to L$ be a surjective homomorphism. Then $\langle \operatorname{Im} F\chi \rangle = FL$. If $M \in \mathbf{V}_F$ then $FM = \mathbf{0}$ and $\operatorname{Im} F\chi = \chi(FM) = \chi\mathbf{0} = \mathbf{0}$. Hence $\mathbf{0} = \langle \mathbf{0} \rangle = FL$ and $L \in \mathbf{V}_F$. Let M be the cartesian product of a family $\{M_\lambda\}$ of S-systems $M_\lambda \in \mathbf{V}_F$. Let the π_λ be the defining projections of this product. By assumption,

$$F\pi_\lambda: FM \to FM_\lambda = FM \to \mathbf{0}.$$

But $F\pi_\lambda$ is the restriction of $\pi_\lambda \times \pi_\lambda$ to FM. Hence $FM \subseteq \bigcap_\lambda \pi_\lambda^{-1}\mathbf{0} = \mathbf{0}$ and $M \in \mathbf{V}_F$.

Consider the surjective homomorphisms

$$\pi: M \to M/FM, \qquad \psi: M \to M/VM.$$

Since F is a subfunctor of D and preserves surjective homomorphisms the following diagram is commutative.

$$\begin{array}{ccc} FM & \overset{F\pi}{\longrightarrow} & F(M/FM) = \langle F\pi(FM) \rangle = \langle (\pi \times \pi)(FM) \rangle = \mathbf{0} \\ \cap\!\!\mid\ \downarrow & & \downarrow\ \mid\!\!\cap \\ M \times M & \underset{\pi \times \pi}{\longrightarrow} & (M/FM) \times (M/FM) \end{array}$$

Hence $M/FM \in \mathbf{V}_F$ and therefore $VM \subseteq FM$. On the other hand, by 2(ii), $M/VM \in \mathbf{V}_F$ and thus $F(M/VM) = \mathbf{0}$, i.e. $F\psi$ is zero, i.e. $FM \subseteq VM$, according to the following commutative diagram

$$\begin{array}{ccc} FM & \xrightarrow{F\psi} & F(M/VM) = \mathbf{0} \\ \cap\big\downarrow & & \big\downarrow\cap \\ M \times M & \xrightarrow[\psi \times \psi]{} & (M/VM) \times (M/VM) \end{array}$$

Therefore the equality $FM = VM$ holds for all $M \in S_r$.

4. For any subvariety $\mathbf{V}$ of S_r, the $\mathbf{V}$-free S-systems are precisely the S-systems of the form M/VM where M is a free unital S-system and V is the variety functor associated with $\mathbf{V}$.

Definition. A congruence $C \in \operatorname{Con} M$ $(M \in S_r)$ is called fully invariant if it is mapped into itself by $\operatorname{End} M$ (the monoid of all endomorphisms of M).

5.

(i) *If V is a variety functor and $M \in S_r$ then VM is a fully invariant congruence on M.*

(ii) *Any fully invariant congruence C on a projective S-system M determines a variety $\mathbf{V} \subseteq S_r$ by the law:*

$$\mathbf{V} = \{N \in S_r,\ (\forall \varphi\colon M \to N \in S_r)\ C \subseteq \varphi^{-1}\mathbf{0}\}.$$

(iii) *Let $V = V_{\mathbf{V}}$ be the subfunctor associated with the variety* $\mathbf{V}$. *Then $C = VM$.*

Proof.

(i) follows from the subfunctor property of a variety functor.

(ii): It is easy to check that $\mathbf{V}$ is closed with respect to S-subsystems. Further for $N \in \mathbf{V}$ let $\psi\colon N \to L$ be a surjective homomorphism. Since M is projective any $\varphi\colon M \to L$ factorizes through ψ according to $\varphi = \chi\psi$ with $\chi\colon M \to N$. Hence $\varphi C = \psi(\chi C) \subseteq \psi\mathbf{0} = \mathbf{0}$ and $L \in \mathbf{V}$. Let $N = \bigtimes_{\lambda} N_\lambda$ with $N_\lambda \in \mathbf{V}$ and with the defining projections $\pi_\lambda\colon N \to N_\lambda$. Let $\varphi\colon M \to N \in S_r$. Then, for each λ, $\varphi\pi_\lambda\colon M \to N_\lambda$

satisfies $\varphi\pi_\lambda(C) \subseteq \mathbf{0}$ whence $\varphi C \subseteq \bigcap_\lambda \pi_\lambda^{-1}\mathbf{0} = \mathbf{0}$ and $N \in \mathbf{V}$. Therefore $\mathbf{V}$ is a variety.

(iii): Since $M/VM \in \mathbf{V}$ and $VM = \pi^{-1}\mathbf{0}$ where $\pi: M \to M/VM$ is the canonical projection we have $C \subseteq VM$. Consider the canonical projection $\psi: M \to M/C$. Since M is projective every $\varphi: M \to M/C$ factorizes through ψ according to $\varphi = \chi\psi$ with $\chi: M \to M$. Thus we get $\varphi C = \psi(\chi C) \subseteq \psi C \subseteq \mathbf{0}$ and $M/C \in \mathbf{V}$, i.e. $VM \subseteq C$. Therefore the equality $C = VM$ holds.

6. Every possible identity for a unital S-system M is of one of the forms $(s, t \in S)$:

$$(\forall x \in M) \;\; xs = xt; \qquad (\forall x, y \in M) \;\; xs = yt.$$

Thus let T_0, W_0 be any two subsets of $S \times S$. Then all subvarieties of S_r are of one of the forms

$$\mathbf{V}_{T_0} = \{M \in S_r,\ MT_0 \subseteq \mathbf{0}\},$$

$$\mathbf{V}_{W_0} = \{M \in S_r,\ (M \times M)W_0 \subseteq \mathbf{0}\},$$

$$\mathbf{V}_{T_0, W_0} = V_{T_0} \cap V_{W_0}.$$

The associated subfunctors V_{T_0}, V_{W_0}, V_{T_0,W_0} are given respectively by

$$V_{T_0}M = \langle MT_0 \rangle, \quad V_{W_0}M = \langle (M \times M)W_0 \rangle,$$

$$V_{T_0, W_0} = \langle MT_0 \rangle \vee \langle (M \times M)W_0 \rangle.$$

7. Let $M \in S_r$. Then

$$M^{-1}\mathbf{0} = \{(s, t) \in S \times S,\ M(s, t) \subseteq \mathbf{0}\}$$

is a congruence on the monoid S,

$$I_M = \{s \in S,\ \operatorname{card} Ms = 1\}$$

is a two-sided ideal of S and

$$(M \times M)^{-1}\mathbf{0} = \{(s, t) \in S \times S,\ (M \times M)(s, t) \subseteq \mathbf{0}\}$$

is a congruence on the semigroup I_M. Consequently, for a subvariety $\mathbf{V}$ of S_r,

$$T_{\mathbf{V}} = \bigcap_{M \in \mathbf{V}} M^{-1}\mathbf{0}$$

is a congruence on the monoid S,

$$I_{\mathbf{V}} = \bigcap_{M \in \mathbf{V}} I_M$$

is a two-sided ideal of S and

$$W_{\mathbf{V}} = \bigcap_{M \in \mathbf{V}} (M \times M)^{-1}\mathbf{0}$$

is a congruence on $I_{\mathbf{V}}$ ($\mathbf{0} \cup W_{\mathbf{V}}$ a congruence on S).

Moreover, $T_{\mathbf{V}}$ and $W_{\mathbf{V}}$ have certain obvious maximality properties with respect to $\mathbf{V}$. Further if $\mathbf{V} = \mathbf{V}_{T_0}$, $\mathbf{V} = \mathbf{V}_{W_0}$, $\mathbf{V} = \mathbf{V}_{T_0, W_0}$ then we have

$$\mathbf{V}_{T_{\mathbf{V}}} = \mathbf{V}, \quad \mathbf{V}_{W_{\mathbf{V}}} = \mathbf{V}, \quad \mathbf{V}_{T_{\mathbf{V}}, W_{\mathbf{V}}} = \mathbf{V},$$

respectively. Hence

$$V_{T_{\mathbf{V}}} = V_{T_0}, \quad V_{W_{\mathbf{V}}} = V_{W_0}, \quad V_{T_{\mathbf{V}}, W_{\mathbf{V}}} = V_{T_0, W_0}$$

and, in particular,

$$T_{\mathbf{V}} = \langle ST_{\mathbf{V}} \rangle_r = \langle ST_0 \rangle_r = \langle T_0 \rangle_c,$$

$$\langle W_{\mathbf{V}} \rangle_r = \langle (S \times S) W_{\mathbf{V}} \rangle_r = \langle (S \times S) W_0 \rangle_r = \langle (S \times S) W_0 \rangle_c$$

where $\langle \ldots \rangle_r$ and $\langle \ldots \rangle_c$, respectively, denote the right congruence and the congruence on the monoid S generated by . . . .

8. As E. Hotzel observed (to whom the author is indebted for useful discussions) I_M (for $M \in S_r$) is saturated with respect to the congruence $M^{-1}\mathbf{0}$ on S, i.e. I_M consists of full $M^{-1}\mathbf{0}$-classes of S, and moreover $(M \times M)^{-1}\mathbf{0}$ can be expressed by I_M and $M^{-1}\mathbf{0}$ according to

(1) $$(M \times M)^{-1}\mathbf{0} = (I_M \times I_M) \cap M^{-1}\mathbf{0}.$$

The $M^{-1}\mathbf{0}$-classes into which I_M decomposes are obviously left ideals of S and I_M is empty or necessarily contains all those $M^{-1}\mathbf{0}$-classes of S which are left ideals of S. Equality (1) implies

$$(2) \qquad W_{\mathbf{V}} = (I_{\mathbf{V}} \times I_{\mathbf{V}}) \cap T_{\mathbf{V}}$$

and again $I_{\mathbf{V}}$ is empty or decomposes into all of those $T_{\mathbf{V}}$-classes of S which are left ideals of S.

After this article was finished the author was pointed out by V. Steinbuk to another approach [6]. Theorem 1 of that paper seems to be essentially identical with the subsequent Theorem:

9. *Because of the validity of* $\mathbf{V}_{T_{\mathbf{V}}, W_{\mathbf{V}}} = \mathbf{V}$ (cf. 7) *every subvariety* $\mathbf{V}$ *of* S_r *is uniquely determined by the pair* $(T_{\mathbf{V}}, W_{\mathbf{V}})$ *(or equivalently by* $(T_{\mathbf{V}}, I_{\mathbf{V}})$*). Conversely let* T *be any congruence on the monoid* S *and let* I *be the two-sided ideal of* S *which is the union of all those* T*-classes of* S *which are left ideals of* S *or, if this union is empty, the empty set. Put* $W = (I \times I) \cap T$. *Then the subvariety*

$$\mathbf{V} = \mathbf{V}_{T,W} = \{M \in S_r,\ MT \cup (M \times M)W \subseteq \mathbf{0}\}$$

of S_r *has the property that* $T = T_{\mathbf{V}}$, $I = I_{\mathbf{V}}$, *and hence* $W = W_{\mathbf{V}}$.

Proof. By definition $T \subseteq T_{\mathbf{V}}$, $I \subseteq I_{\mathbf{V}}$ and $W \subseteq W_{\mathbf{V}}$ hold. We prove the converse inclusions. Let F_X be the free unital S-system freely generated by a set X, i.e. $F_X = \dot{\bigcup}_{x \in X} xS = X \times S$ with the operation $(x, s)t = (x, st)$ $(x \in X;\ s, t \in S)$. We consider the fully invariant congruence C_X on F_X which is generated by $F_X T \cup (F_X \times F_X)W$ and claim that $C_X = A_X \cup B_X$, where A_X and B_X are the fully invariant congruences on F_X given by

$$A_X = \{\langle (x, t_1), (x, t_2) \rangle,\ x \in X;\ (t_1, t_2) \in T\},$$

$$B_X = \{\langle (x, t_1), (y, t_2) \rangle,\ x, y \in X;\ (t_1, t_2) \in W\}.$$

The full invariance of A_X and B_X is ensured by the implications

$$\langle (x, t_1), (x, t_2) \rangle \in A_X,\quad s \in S \Rightarrow (st_1, st_2) \in T,$$

$$\langle (x, st_1), (x, st_2) \rangle \in A_X$$

and

$$\langle (x, t_1), (y, t_2) \rangle \in B_X, \quad r, s \in S \Rightarrow (t_1, t_2) \in W, \quad t_{1T} = t_{2T}$$

(a T-class which is a left ideal of S),

$$rt_1 \in t_{1T}, \quad st_2 \in t_{2T}, \quad (rt_1)_T = t_{1T} = t_{2T} = (st_2)_T,$$

$$(rt_1, st_2) \in W, \quad \langle (x, rt_1), (y, st_2) \rangle \in B_X,$$

respectively. One can easily see that the set-theoretical union $A_X \cup B_X$ is already a congruence on F_X which is fully invariant. Hence $C_X = A_X \cup B_X$. As it is well known every free unital S-system of $\mathbf{V}$ has to be of the form F_X/C_X and every element $M \in \mathbf{V}$ is the image $(F_X/C_X)\varphi$ of F_X/C_X by a suitable homomorphism of unital S-systems $\varphi\colon F_X/C_X \to M$. Let $(a, b), (u, v) \in S \times S$ such that

(3) $\quad (F_X/C_X)(a, b) \subseteq \mathbf{0}$,

(4) $\quad (F_X/C_X \times F_X/C_X)(u, v) \subseteq \mathbf{0}$.

Then for all $M\,(= (F_X/C_X)\varphi)$ of $\mathbf{V}$ one has $M(a, b) \subseteq \mathbf{0}$ and $(M \times M)(u, v) \subseteq \mathbf{0}$. Hence, in order to obtain $T_{\mathbf{V}}$ and $W_{\mathbf{V}}$, one has to consider the conditions (3), (4) or, equivalently,

(3′) $\quad (\forall x \in X,\ s \in S)\ \langle (x, sa), (x, sb) \rangle \in C_X$,

(4′) $\quad (\forall x, y \in X;\ s, t \in S)\ \langle (x, su), (y, tv) \rangle \in C_X$.

(3′) implies $(sa, sb) \in T$ for all $s \in S$, i.e. $(a, b) \in T$ and $T_{\mathbf{V}} \subseteq T$. (4′) implies that $(su, tv) \in T$ for all $s, t \in S$. Hence the T-class u_T is a left ideal of S. The question arises whether u and v are elements of I? We assume that card $X \geqslant 2$. Then the answer is 'yes' for the following reason. Let x be a fixed element of X. Then there is an element $y \in X$ different from x. Condition (4′) implies that $(su, tv) \in W$. Thus $(u, v) \in W$ and $W_{\mathbf{V}} \subseteq W$.

10. (Analogue of (6.5) in [2] and of (4.2) in [3].) *Let T be a congruence on the monoid S. Then, for every $M \in S_r$, we have*

$$M/\langle MT \rangle \cong M \otimes S/T.$$

Proof. By definition (cf. M. Delorme [1], M. Kil'p [5], and B. Stenström [6]) the tensor product $M \otimes S/T$ of the unital S-system $M \in S_r$ and S/T, considered as a unital S-left system, is expressible as $M \otimes S/T = (M \times S/T)/(\equiv)$ where the equivalence $(\equiv)$ is generated by the pairs $\langle (au, s_T), (a, us_T) \rangle$ ($a \in M$; $s, t \in S$; s_T the T-class of S containing s). We consider $M \times S/T$ as a (unital) S-system according to the composition $(a, s_T)t = (a, (st)_T)$. Since T is a congruence on S this definition is correct. Then the equivalence $(\equiv)$ turns out to be a congruence on the S-system $M \times S/T$. Indeed since

$$(au, s_T)t = (au, st_T) \equiv (a, (ust)_T) = (a, us_T)t$$

the generating relation of $(\equiv)$ is compatible with the composition and hence so is $(\equiv)$. Thus the tensor product $M \otimes S/T$ itself is a unital S-system.

Consider the map $\varphi\colon M \times S/T \to M/\langle MT \rangle$ given by

$$(a, s_T) \mapsto (as)_{\langle MT \rangle}.$$

This map is well defined. For if $s_T = t_T$, i.e. $(s, t) \in T$, then $a(s, t) \in \in MT$ and $(as)_{\langle MT \rangle} = (at)_{\langle MT \rangle}$. Further φ is a homomorphism of S-systems because of

$$((a, s_T)t)\varphi = (a, (st)_T)\varphi = (ast)_{\langle MT \rangle} = (as)_{\langle MT \rangle}t = ((a, s_T)\varphi)t$$

for all $t \in S$. Moreover $(\equiv)$ is contained in the kernel $\varphi^{-1}\mathbf{0}$ of the map φ because of

$$(au, s_T)\varphi = (aus)_{\langle MT \rangle} = (a, us_T)\varphi.$$

Therefore φ induces a homomorphism of S-systems

$$\varphi\colon M \otimes S/T \to M/\langle MT \rangle.$$

We shall see that φ is invertible. Consider the map

$$\psi\colon M \to M \otimes S/T\colon a \mapsto (a, 1_T)_{(\equiv)}.$$

This map is a homomorphism of S-systems since

$$(as)\psi = (as, 1_T)_{(\equiv)} = (a, s_T)_{(\equiv)} = (a, 1_T)_{(\equiv)}s = (a\psi)s$$

$(a \in M,\ \ s \in S)$. The kernel $\psi^{-1}\mathbf{0}$ of ψ contains $\langle MT \rangle$ since $a \in M$, $(s, t) \in T$ implies

$$(as)\psi = (as, 1_T)_{(\equiv)} = (a, s_T)_{(\equiv)} = (a, t_T)_{(\equiv)} =$$
$$= (at, 1_T)_{(\equiv)} = (at)\psi.$$

Hence ψ induces a homomorphism of S-systems

$$\psi\colon M/\langle MT \rangle \to M \otimes S/T.$$

We have

$$(a, s_T)_{(\equiv)}\varphi\psi = (as)_{\langle MT \rangle}\psi = (as, 1_T)_{(\equiv)} = (a, s_T)_{(\equiv)}$$

and

$$a_{\langle MT \rangle}\psi\varphi = (a, 1_T)_{(\equiv)}\varphi = a_{\langle MT \rangle}.$$

Thus $\varphi\psi = 1$ and $\psi\varphi = 1$. Therefore φ is an isomorphism with the inverse ψ.

The question seems to be open how Theorem 10 can be extended to the case of $\langle MT \cup (M \times M)W \rangle$ instead of $\langle MT \rangle$. In the special case where the ideal I has an identity e we can derive such an extension (cf. Theorem 12 below).

11. *Let T be a congruence on the monoid S (identity 1) and let I be a two-sided ideal of S with an identity e such that I is the union of some (not necessarily all) of those T-classes of S which are left ideals of S. Set $W = (I \times I) \cap T$, $W^0 = W \cup \mathbf{0}$ $(\mathbf{0} = \{(s, s),\ s \in S\})$.*

Then, for every $M \in S_r$, we have

$$M/\langle (M \times M)W \rangle \cong M/\langle (M \times M)e \rangle \otimes S/W^0.$$

Proof. It was already indicated under 7 that W^0 is a congruence on S. Indeed, let $(u, v) \in W^0$, i.e.

(i) $u = v$ or

(ii) $u, v \in I$ and $(u, v) \in W$.

Let $t \in S$. Then in case (i) $ut = vt$ and in case (ii) $ut, vt \in It \subseteq I$.

Since ut, vt belong to the same T-class of S, one has $(ut, vt) \in W \subseteq W^0$. Hence W^0 is a right congruence on S. Let $s \in S$. Then in case (i) $su = sv$ and in case (ii) $su, sv \in I$. Since su, sv belong to the same T-class of S one has $(su, sv) \in W \subseteq W^0$. Hence S/W^0 can be considered as a unital S-left system by the composition $us_{W^0} = (us)_{W^0}$ which is well defined. By definition of the tensor product,

$$M/\langle (M \times M)e \rangle \otimes S/W^0 = (M/\langle (M \times M)e \rangle \times S/W^0)/(\equiv)$$

where $(\equiv)$ denotes the equivalence on $M/\langle (M \times M)e \rangle \times S/W^0$ which is generated by the pairs $\langle (\overline{au}, s_{W^0})\ (\bar{a}, us_{W^0}) \rangle$ $(\bar{a} = a_{\langle (M \times M)e \rangle},\ a \in M;$ $u, s \in S)$. Again we consider $M/\langle (M \times M)e \rangle \times S/W^0$ as a (unital) S-system with respect to the operation $(\bar{a}, s_{W^0})t = (\bar{a}, (st)_{W^0})$ $(a \in M;\ s, t \in S)$ which is well defined. As before, with respect to this operation, the equivalence $(\equiv)$ is a congruence on S-systems and the tensor product $M/\langle (M \times M)e \rangle \otimes S/W^0$ is a unital S-system. Consider the map

$$\sigma\colon M/\langle (M \times M)e \rangle \times S/W^0 \to M/C$$

where $C = \langle (M \times M)W \rangle$, given by

$$(\bar{a}, s_{W^0}) \mapsto (as)_C.$$

This map is well defined. For if $\bar{a} = \bar{b}$ and $s_{W^0} = t_{W^0}$ then $a = b$ or $aj = bj$ for some $j \in I$ and further

(i) $s = t,\ (as, bt) \in C$ or

(ii) $s, t \in I,\ (s, t) \in W,\ (as, bt) \in C.$

Moreover σ is a homomorphism of S-systems because of

$$((\bar{a}, s_{W^0})t)\sigma = (\bar{a}, (st)_{W^0})\sigma = (ast)_C = (as)_C t = (\bar{a}, s_{W^0})\sigma t$$

for all $t \in S$. The congruence $(\equiv)$ is contained in the kernel $\sigma^{-1}0$ of σ because of

$$(\overline{au}, s_{W^0})\sigma = (aus)_C = (\bar{a}, us_{W^0})\sigma.$$

Hence σ induces a homomorphism of S-systems

$$\sigma: M/\langle (M \times M)e \rangle \otimes S/W^0 \to M/C.$$

We prove that σ is invertible. Consider the map

$$\tau: M \to M/\langle (M \times M)e \rangle \otimes S/W^0 : a \mapsto (\bar{a}, 1_{W^0})_{(\equiv)}.$$

This map is a homomorphism of S-systems since

$$(as)\tau = (\overline{as}, 1_{W^0})_{(\equiv)} = (\bar{a}, s_{W^0})_{(\equiv)} = (\bar{a}, 1_{W^0})_{(\equiv)} s = (a\tau)s.$$

C is contained in the kernel $\tau^{-1}0$ of τ since $a, b \in M$, $(s, t) \in W$ implies $s = es$, $t = et$, $\overline{ae} = \overline{be}$ and

$$(as)\tau = (\overline{as}, 1_{W^0})_{(\equiv)} = (\overline{ae}, s_{W^0})_{(\equiv)} = (\overline{be}, t_{W^0})_{(\equiv)} =$$

$$= (\overline{bt}, 1_{W^0})_{(\equiv)} = (bt)\tau.$$

Thus τ induces a homomorphism of S-systems

$$\tau: M/C \to M/\langle (M \times M)e \rangle \otimes S/W^0.$$

We have

$$(\bar{a}, s_{W^0})_{(\equiv)} \sigma\tau = (as)_C \tau = (\bar{a}, s_{W^0})_{(\equiv)}$$

and

$$a_C \tau\sigma = (\bar{a}, 1_{W^0})_{(\equiv)} \sigma = a_C.$$

Thus $\sigma\tau = 1$ and $\tau\sigma = 1$, i.e. σ is an isomorphism with the inverse τ.

As a corollary we get:

12. *Suppose that* I *has an identity element* e. *Then for every* $M \in S_r$,

$$M/\langle MT \cup (M \times M)W \rangle = M/\langle (M \times M)e \rangle \otimes S/(I \cup T) \otimes S/W^0$$

holds where, for brevity, $(I \times I) \cup (((S \setminus I) \times (S \setminus I)) \cap T)$ *is denoted by* $I \cup T$.

Proof. Set $\bar{M} = M/\langle MT \rangle$. By 11 we have

$$\bar{M}/\langle(\bar{M} \times \bar{M})W\rangle \cong \bar{M}/\langle(\bar{M} \times \bar{M})e\rangle \otimes S/W^0.$$

On the other hand we have

$$\langle(\bar{M} \times \bar{M})W\rangle = \langle MT \cup (M \times M)W\rangle/\langle MT\rangle$$

whence

$$\bar{M}/\langle(\bar{M} \times \bar{M})W\rangle \cong M/\langle MT \cup (M \times M)W\rangle.$$

Further we have $(S/T)e \cong I/W$ (we may identify both sides) and, by 10, $\bar{M} \cong M \otimes S/T$. Therefore

$$\bar{M}/\langle(\bar{M} \times \bar{M})e\rangle \cong$$

$$\cong (M \times S/T)_{(\equiv)}/\langle((M \times S/T)_{(\equiv)} \times (M \times S/T)_{(\equiv)})e\rangle =$$

$$= (M \otimes S/T)/\langle(M \otimes I/W) \times (M \otimes I/W)\rangle =$$

$$= (M \otimes S/T)/(M \otimes I/W) \cong M/\langle(M \times M)e\rangle \otimes S/(I \cup T).$$

The latter isomorphy can be proved by applying the pair of maps defined by

$$(a, s_T)_{(\equiv)} \mapsto (\bar{a}, s_{I \cup T})_{(\doteq)}, \quad (\bar{a}, s_{I \cup T}) \mapsto (a, s_T)_{(\equiv)(M \otimes I/W)}$$

where $(\doteq)$ is a certain congruence relation which appears in the definition of the latter tensor product.

REFERENCES

[1] M. Delorme, Sur la platitude des demi-groupes de fractions, *C. R. Acad. Sci. Paris, Sér. A*, 269 (1969), 630–632.

[2] A. Fröhlich, On groups over a d.g. near-ring (II): Categories and functors, *Quart. J. Math. Oxford Ser.* (2), 2 (1960), 211–228.

[3] A. Fröhlich, Baer invariants of algebras, *Trans. Amer. Math. Soc.*, 109 (1963), 221–244.

[4] H.-J. Hoehnke, Structures of semigroups, *Canad. J. Math.*, 18 (1966), 449–491.

[5] M. Kil'p, On homological classification of monoids, *Sibirsk. Mat. Ž.*, 13 (1972), 578–586 (in Russian).

[6] B.I. Plotkin – C.E. Dididze – E.M. Kublanova, On varieties of automata, *Dokl. Akad. Nauk SSSR,* 221 (1975), 1284–1287 (in Russian); English transl.: Soviet Math. Dokl. 16 (1975), 537–541.

[7] B. Stenström, Flatness and localization over monoids, *Math. Nachr.*, 48 (1970), 315–334.

H.-J. Hoehnke

Institut für Mathematik, Akademie der Wissenschaften der DDR, 1086 Berlin, Mohrenstr. 39, DDR.

COLLOQUIA MATHEMATICA SOCIETATIS JÁNOS BOLYAI
39. SEMIGROUPS, SZEGED (HUNGARY), 1981.

EPIMORPHISMS AND AMALGAMATIONS: A SURVEY OF RECENT PROGRESS

JOHN M. HOWIE

INTRODUCTION

It was only after I had given my title to the conference organisers that I realised what a huge subject I had volunteered to tackle. I soon decided that a comprehensive historical survey would prove somewhat indigestible; what I shall attempt instead is to explain what seem to have been the most fruitful concepts in the subject, to describe a few of the results obtained, and to propose a new method of proof for one of the deepest theorems.

Both Clifford and Preston [1] and Howie [7] have sections on amalgamations, and there is also a most useful survey article [4] by T.E. Hall. I shall, however, begin at the beginning.

1. THE MAIN IDEAS

First let me introduce the two technical terms in the title. Much of what I have to say at this stage can be said in fairly general categorical terms, but I shall confine myself to the category **Sg** of semigroups and to various subcategories of **Sg**.

Let $\mathbf{K}$ be a category of semigroups, let A, B belong to $\mathbf{K}$ and let $\alpha: A \to B$ be a homomorphism in $\mathbf{K}$. We say that α is an *epimorphism* in $\mathbf{K}$ if for all C in $\mathbf{K}$ and all homomorphisms $\beta: B \to C$, $\gamma: B \to C$ in $\mathbf{K}$,

$$\alpha\beta = \alpha\gamma \Rightarrow \beta = \gamma.$$

(Notice that I am writing mapping symbols on the right, so that $\alpha\beta$ means α followed by β.) If $\alpha: A \to B$ is *onto* then α is an epimorphism; for

$$\alpha\beta = \alpha\gamma \Rightarrow (\forall x \in A)x\alpha\beta = x\alpha\gamma \Rightarrow (\forall y \in B)y\beta = y\gamma \Rightarrow \beta = \gamma.$$

The converse, however, is not in general true.

Turning now to the other technical term, let us consider semigroups U, S, T and monomorphisms $\alpha: U \to S$, $\beta: U \to T$. Such an assembly is called a (semigroup) *amalgam* and will be written as $[S, T; U; \alpha, \beta]$ or just as $[S, T; U]$. The semigroup U is called the *core* of the amalgam. (More informally, an amalgam may be thought of as consisting of two semigroups S and T with a common subsemigroup U.) We say that the amalgam $\mathfrak{A} = [S, T; U; \alpha, \beta]$ is *(strongly) embeddable* (or is *(strongly) embedded in* Q) if there exists a semigroup Q and monomorphisms $\theta: S \to Q$, $\varphi: T \to Q$ such that

(i) the diagram

$$\begin{array}{ccc} U & \xrightarrow{\alpha} & S \\ \beta \downarrow & & \downarrow \theta \\ T & \xrightarrow[\varphi]{} & Q \end{array}$$

commutes; and

(ii) $S\theta \cap T\varphi = U\alpha\theta \ (= U\beta\varphi)$.

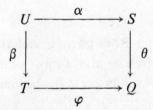

If (i) holds but not (ii) we say that $\mathfrak{A}$ is *weakly* embedded in Q. Informally, if $\mathfrak{A}$ is strongly embedded in Q then Q contains isomorphic copies of S and T intersecting precisely in an isomorphic copy of U.

In considering whether an amalgam $[S, T; U; \alpha, \beta]$ is embeddable the most natural first step is to consider the *pushout* P of the diagram

$$\begin{array}{ccc} U & \xrightarrow{\ \alpha\ } & S \\ {\scriptstyle\beta}\downarrow & & \\ T & & \end{array}$$

This is the semigroup characterized up to isomorphism by the properties:

(i) there exist homomorphisms $\lambda: S \to P$, $\mu: T \to P$ such that the diagram

$$\begin{array}{ccc} U & \xrightarrow{\ \alpha\ } & S \\ {\scriptstyle\beta}\downarrow & & \downarrow{\scriptstyle\lambda} \\ T & \xrightarrow[\ \mu\ ]{} & P \end{array} \tag{1.1}$$

commutes;

(ii) for every commutative diagram

$$\begin{array}{ccc} U & \xrightarrow{\ \alpha\ } & S \\ {\scriptstyle\beta}\downarrow & & \downarrow{\scriptstyle\theta} \\ T & \xrightarrow[\ \varphi\ ]{} & Q \end{array}$$

there exists a unique $\gamma: P \to Q$ such that the diagram

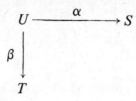

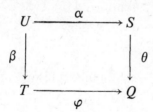

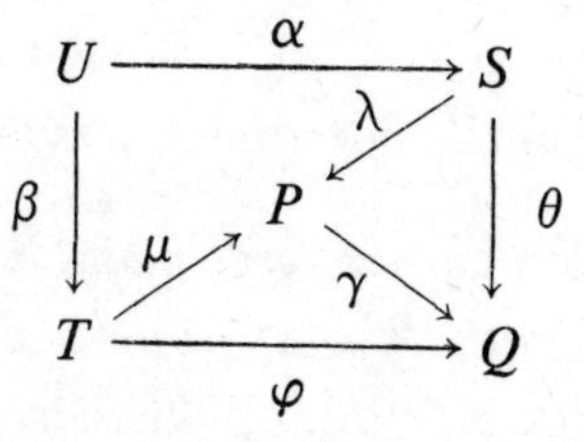

commutes.

In some categories there can be problems about the existence of a pushout P, but if for the moment we confine ourselves to the category **Sg** of all semigroups the pushout P can be taken as $(S * T)/\rho$, where $S * T$ is the free product of S and T and ρ is the congruence on $S * T$ generated by $\{(u\alpha, u\beta)\colon u \in U\}$.

The maps λ, μ of diagram (1.1) are given by $s\lambda = s\rho$, $t\mu = t\rho$ $(s \in S,\ t \in T)$.

If in the pushout diagram (1.1) λ and μ are monomorphisms and $S\lambda \cap T\mu = U\alpha\lambda\ (= U\beta\mu)$ we say that the amalgam $[S, T; U; \alpha, \beta]$ is *naturally embedded in* P. Using the second half of the pushout property it is not hard to show that $[S, T; U; \alpha, \beta]$ *is embeddable if and only if it is naturally embedded in* P.

The connection between epimorphisms and amalgamations is not immediately obvious. The linking idea, that of *dominion,* was introduced by Isbell [9] in 1965. Let **K** be a category of semigroups and let U be a subsemigroup of S, where $U, S \in \mathbf{K}$. We say that an element d of S is *dominated by* U if, for all semigroups T in **K** and for all homomorphisms $\beta, \gamma\colon S \to T$ in **K**,

$$[(\forall u \in U) u\beta = u\gamma] \Rightarrow d\beta = d\gamma.$$

The set of elements of S dominated by U is called the (**K**-) *dominion* of U in S and is denoted by $\mathrm{Dom}_S^{\mathbf{K}} U$. If $\mathbf{K} = \mathbf{Sg}$ we talk simply of the *dominion* and write $\mathrm{Dom}_S U$.

The connection between epimorphisms and dominions is easily stated (and easily proved):

Theorem 1.2. *If* $\varphi\colon S \to T$ *is an epimorphism then*

$$\mathrm{Dom}_T(\mathrm{Im}\,\varphi) = T.$$

The connection with amalgamations is less elementary:

Theorem 1.3. *Let* S *be a semigroup and let* U *be a subsemigroup of* S. *Let* S' *be a semigroup disjoint from* S *and let* $\alpha\colon S \to S'$ *be an isomorphism. Let* P *be the pushout of the diagram*

$$\begin{array}{ccc} U & \xrightarrow{\quad\iota\quad} & S \\ {\scriptstyle \alpha|U}\big\downarrow & & \\ S' & & \end{array}$$

where ι *is the inclusion map and let* $\mu\colon S \to P$, $\mu'\colon S' \to P$ *be the associated homomorphisms. Then* μ, μ' *are monomorphisms and*

$$(S\mu \cap S'\mu')\mu^{-1} = \mathrm{Dom}_S U.$$

2. SOME RESULTS ON AMALGAMATION, AND A NEW PROOF

So far we have considered only amalgams of two semigroups. It is of course possible to generalise the notions of strong and weak embeddability to an amalgam

$$\mathfrak{A} = [\{S_i\colon i \in I\}; U; \{\alpha_i\colon i \in I\}].$$

If a class $\mathbf{K}$ of semigroups has the property that every amalgam $\mathfrak{A}$ for which $U, S_i \in \mathbf{K}$ is strongly embeddable [weakly embeddable] in a semigroup from $\mathbf{K}$ we say that $\mathbf{K}$ has the *strong* [*weak*] *amalgamation property*. If $U \in \mathbf{K}$ is such that every amalgam in $\mathbf{K}$ with U as core is strongly [weakly] embeddable in a semigroup from $\mathbf{K}$ we say that U is a *strong* [*weak*] *amalgamation base* in $\mathbf{K}$.

Then, for example, we have

Theorem 2.1 (Schreier [11]). *The class* $\mathbf{G}$ *of groups has the strong amalgamation property.*

Theorem 2.2 (Howie [5]). *Every group is a strong amalgamation base in the class of semigroups.*

Both of these results have been generalized:

Theorem 2.3 (Hall [2]). *The class* $\mathbf{I}$ *of inverse semigroups has the strong amalgamation property.*

Theorem 2.4 (Howie [6], Hall [3]). *Every inverse semigroup* *is a strong amalgamation base in the class of semigroups.*

These results both appear to be quite deep. Hall's proof of Theorem 2.4 is much shorter than mine, but is by no means easy and involves some fairly complex operations with representations. Realising how the concept of module has simplified and clarified the theory of group representations I recently tried to recast Hall's proof in terms of S-systems. I believe that this does help and so I would like to spend a little time explaining. For simplicity I shall consider an amalgam $[S, T; U; \alpha, \beta]$ and I shall suppose that S, T, U are all monoids in which 1 is 'isolated':

$$xy = 1 \Rightarrow x = y = 1;$$

also I shall suppose that

$$1_U\alpha = 1_S,\ 1_U\beta = 1_T.$$

It is in any event easy to adjoin extra identity elements so as to obtain these properties.

If S is a monoid, we say that a set X is a *(unitary right) S-system* if there is a map $(x, s) \mapsto xs$ from $X \times S$ into X such that

$$(xs)t = x(st) \qquad (x \in X,\ s, t \in S),$$

$$x1 = x \qquad (x \in X).$$

If X, Y are S-systems then $\varphi\colon X \to Y$ is called an *S-homomorphism,* or just an S-map, if (for all x in X and s in S)

$$(xs)\varphi = (x\varphi)s.$$

We shall say that U is a *perfect* submonoid of S if whenever we have an S-system X, a U-system Y and a U-monomorphism $\lambda\colon X \to Y$ there exists an S-system Z, a U-monomorphism $\theta\colon Y \to Z$ and an S-monomorphism $\varphi\colon X \to Z$ such that the diagram

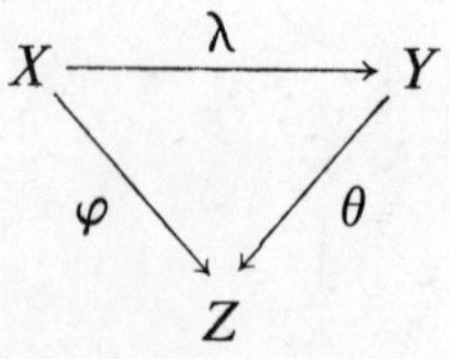

of U-monomorphisms commutes. (Informally, we have that Y is bigger, X is better, while Z is both bigger and better.)

We then have

Lemma 2.5. *Let S be a monoid and let U be an inverse submonoid of S. Then U is perfect.*

The proof of this is not easy. It is included as an appendix to this article.

Theorem 2.4 will follow immediately if we can show:

Lemma 2.6. *Let $\mathfrak{A} = [S, T; U; \alpha, \beta]$ be a monoid amalgam. If $U\alpha$ is a perfect submonoid of S and $U\beta$ is a perfect submonoid of T then $\mathfrak{A}$ is strongly embeddable.*

Proof. Both S and T may be regarded as U-systems if we define (for s in S and t in T)

(2.7) $\quad su = s(u\alpha), \qquad tu = t(u\beta).$

In the category of U-systems we have a pushout P for the diagram

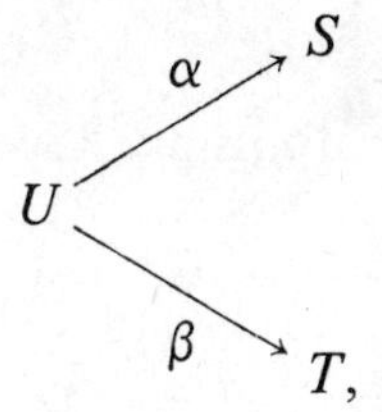

giving a commutative diagram

(2.8)

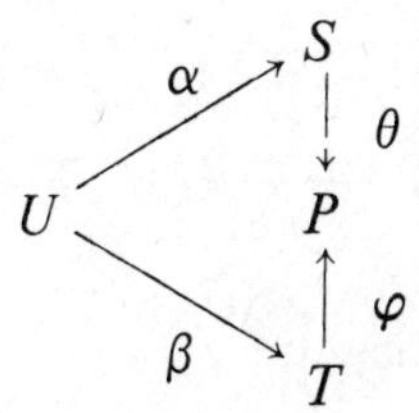

in which all the maps are U-monomorphisms. (We have in fact a 'concrete' realisation of P as $U \cup (S \setminus U\alpha) \cup (T \setminus U\beta)$ with the obvious U-system structure and the obvious maps θ, φ.)

We now have a U-system P, an S-system S and a U-monomorphism $\theta: S \to P$. Hence by the 'perfect' property there exists an S-system Q_1, an S-monomorphism $\lambda_1: S \to Q_1$ and a U-monomorphism $\alpha_1: P \to Q_1$ such that $\lambda_1 = \theta\alpha_1$. By the same token there exists a T-system R_1, a T-monomorphism $\mu_1: T \to R_1$ and a U-monomorphism $\beta_1: P \to R_1$ such that $\mu_1 = \varphi\beta_1$. We thus have a commutative diagram

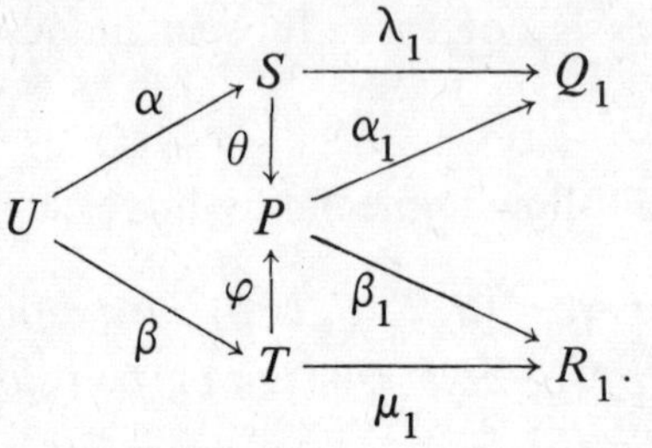

Now take P_1 as the pushout of the diagram

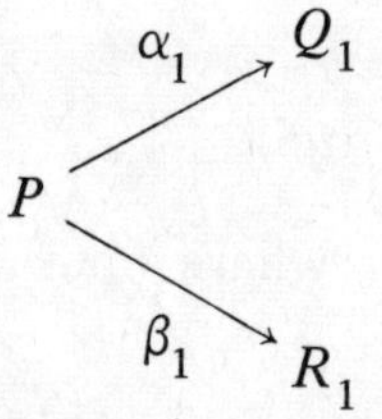

of U-systems and U-monomorphisms. Repeating the argument we obtain a commutative diagram

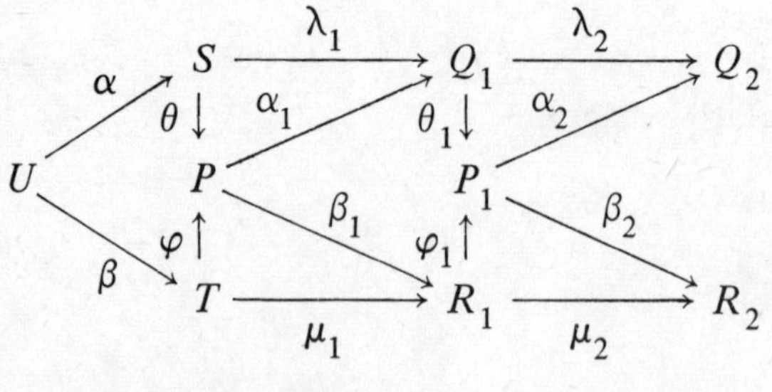

Indeed we have an infinite diagram

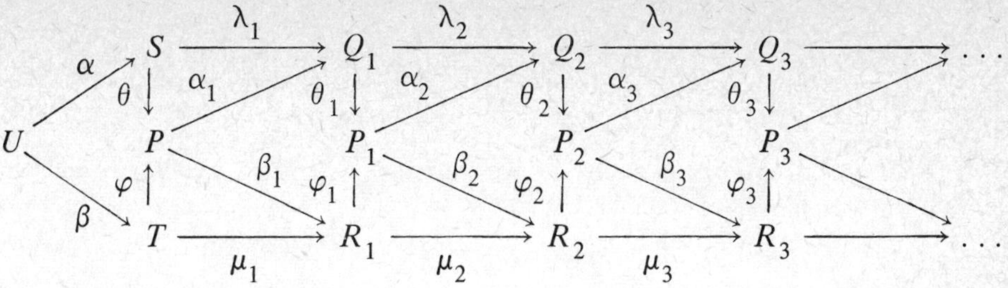

in which $\lambda_1, \lambda_2, \lambda_3, \ldots$ are S-monomorphisms, $\mu_1, \mu_2, \mu_3, \ldots$ are T-monomorphisms and all other maps are U-monomorphisms. Writing $\alpha\theta = \gamma$ and $\alpha_i\theta_i = \gamma_i$ $(i = 1, 2, \ldots)$, we thus have three sequences

$$U \xrightarrow{\gamma} P \xrightarrow{\gamma_1} P_1 \xrightarrow{\gamma_2} P_2 \xrightarrow{\gamma_3} \ldots,$$

$$S \xrightarrow{\lambda_1} Q_1 \xrightarrow{\lambda_2} Q_2 \xrightarrow{\lambda_3} Q_3 \xrightarrow{\lambda_4} \ldots,$$

$$T \xrightarrow{\mu_1} R_1 \xrightarrow{\mu_2} R_2 \xrightarrow{\mu_3} R_3 \xrightarrow{\mu_4} \ldots.$$

Let us denote the direct limits of these sequences by X, Y, Z respectively; thus X is a U-system, Y is an S-system and Z is a T-system.

It is fairly easily deduced from the existence of the 'cross' maps $\theta_i, \varphi_i, \alpha_i, \beta_i$ that as U-systems X, Y and Z are isomorphic. Indeed the properties of direct limits ensure that we have a commutative diagram

(2.9)

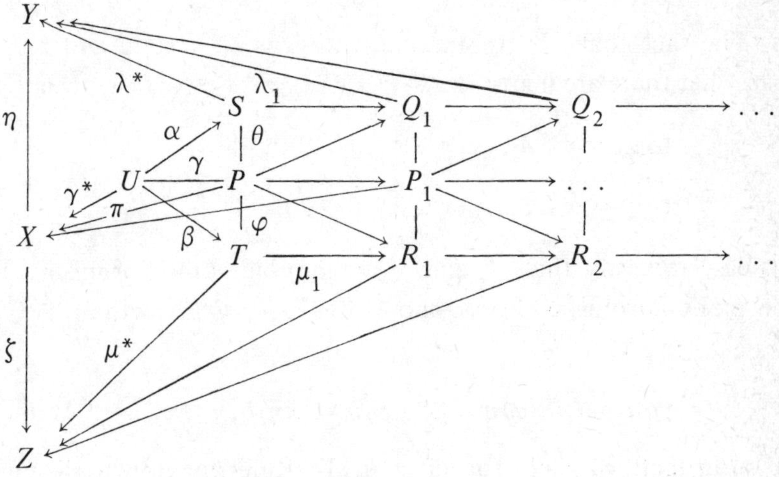

Here η and ζ are U-isomorphisms, $\gamma^*: U \to X$ and $\pi: P \to X$ are U-monomorphisms, λ^* is an S-monomorphism and μ^* is a T-monomorphism.

If for x in X and s in S we define

$$x \circ s = ((x\eta)s)\eta^{-1}$$

then we make X into an S-system. Similarly, we make X into a T-system by defining

$$x \circ t = ((x\zeta)t)\zeta^{-1}.$$

Notice that the possibility of regarding the S-system Y as a U-system depends in effect upon identifying $u\alpha$ with u for each u in U (see (2.7)); thus the U-map property of $\eta: X \to Y$ is to be interpreted as saying that for all x in X and all u in U,

$$(xu)\eta = (x\eta)(u\alpha);$$

equally the U-map property of $\zeta: X \to Z$ means

$$(xu)\zeta = (x\zeta)(u\beta).$$

Hence

$$\begin{aligned} x \circ (u\alpha) &= ((x\eta)(u\alpha))\eta^{-1} = (xu)\eta\eta^{-1} = xu = \\ &= (xu)\zeta\zeta^{-1} = ((x\zeta)(u\beta))\zeta^{-1} = x \circ (u\beta). \end{aligned} \tag{2.10}$$

The fact that X is simultaneously an S-system and a T-system means that there are maps $\psi: S \to \mathcal{T}(X)$, $\chi: T \to \mathcal{T}(X)$ given by

$$x(a\psi) = x \circ a \qquad (x \in X,\ a \in S),$$

$$x(b\chi) = x \circ b \qquad (x \in X,\ b \in T).$$

It is easy to verify that ψ and χ are monoid homomorphisms. They are even monomorphisms. For suppose that $a\psi = b\psi$, with $a, b \in S$. Then for all x in X

$$(x\eta)a = (x \circ a)\eta = (x \circ b)\eta = (x\eta)b,$$

and so in fact $ya = yb$ for all y in Y. Referring now to diagram (2.9), we see that

$$a\lambda^* = (1a)\lambda^* = (1\lambda^*)a = (1\lambda^*)b = \ldots = b\lambda^*$$

and hence that $a = b$. A similar argument applies to χ.

We now have that $[S, T; U; \alpha, \beta]$ is weakly embedded in $\mathscr{T}(X)$. To see that the embedding is strong notice first that for all u in U and all x in X

$$x((u\alpha)\psi) = x \circ (u\alpha) = x \circ (u\beta) = \quad \text{(by (2.10))}$$
$$= x((u\beta)\chi).$$

Thus $u\alpha\psi = u\beta\chi$ and so we have a commutative diagram

$$(2.11) \qquad \begin{array}{ccc} U & \xrightarrow{\ \alpha\ } & S \\ \beta \downarrow & & \downarrow \psi \\ T & \xrightarrow[\ \chi\ ]{} & \mathscr{T}(X) \end{array}$$

of monoid monomorphisms. Finally, if $x \in S$ and $t \in T$ are such that $s\psi = t\chi$ then for all x in X we have

$$((x\eta)s)\eta^{-1} = ((x\zeta)t)\zeta^{-1}.$$

Looking again at diagram (2.9) we deduce that

$$s\theta\pi = s\lambda^*\eta^{-1} = ((1_S\lambda^*)s)\eta^{-1} = ((1_U\alpha\lambda^*)s)\eta^{-1} =$$
$$= ((1_U\gamma^*\eta)s)\eta^{-1} = ((1_U\gamma^*\zeta)t)\zeta^{-1} = ((1_U\beta\mu^*)t)\zeta^{-1} =$$
$$= ((1_T\mu^*)t)\zeta^{-1} = t\mu^*\zeta^{-1} = t\varphi\pi;$$

hence $s\theta = t\varphi$ since π is a monomorphism. This takes us back to the simple U-system pushout diagram (2.8). Taking the 'concrete' form of P as the disjoint union

$$U \cup (S \setminus U\alpha) \cup (T \setminus U\beta),$$

we have that

$$s\theta = \begin{cases} s & \text{if } s \in S \setminus U\alpha \\ s\alpha^{-1} & \text{if } s \in U\alpha, \end{cases}$$

$$t\varphi = \begin{cases} t & \text{if } t \in T \setminus U\beta \\ t\beta^{-1} & \text{if } t \in U\beta, \end{cases}$$

and we readily see that $s\theta = t\varphi$ only if there exists u in U such that $s = u\alpha$, $t = u\beta$.

We conclude that in the diagram (2.11) $S\psi \cap T\chi = U\alpha\psi$ $(= U\beta\chi)$ and so finally obtain that the amalgam $[S, T; U; \alpha, \beta]$ is strongly embedded in $\mathcal{T}(X)$.

3. MISCELLANEOUS RESULTS

We have four properties which may be possessed by a class **K** of semigroups:

(A) the strong amalgamation property;

(B) the weak amalgamation property;

(C) the special amalgamation property;

(D) epimorphisms are onto.

The only one of these not yet defined is (C): we say that **K** has the *special amalgamation property* if the strong embedding property holds for every amalgam $[S, T; U; \alpha, \beta]$ in **K** such that there is an isomorphism $\varphi\colon S \to T$ making

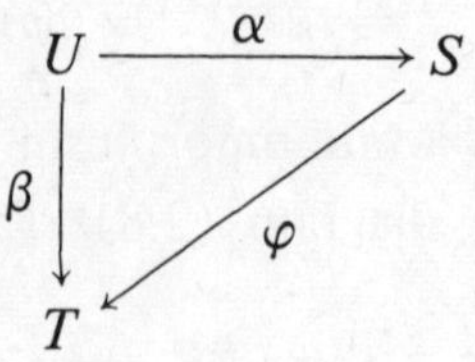

commute.

Looking at the interrelationships of these four properties, we notice first that (obviously): (A) ⇒ (B), (A) ⇒ (C). Also it is not too hard to show that (B) and (C) ⇒ (A); this holds in fact in any reasonable class of algebras. Hall [3] has shown that any weak amalgamation base in the class **Sg** of all semigroups is also a strong amalgamation base in **Sg**. The same implication holds in the class of all commutative semigroups. It appears to be still an open question whether (B) ⇒ (A) for varieties of semigroups.

By Theorems 1.2 and 1.3 it follows that (C) ⇒ (D).

Finally, the implications (A) ⇒ (C), (C) ⇒ (D) are not reversible: Imaoka [8] shows that the class of *left regular bands* $(xy = xyx)$ satisfies (C) but not (A); while Scheiblich [10] shows that the class of *finite bands* satisfies (D) but not (C).

APPENDIX: PROOF OF LEMMA 2.5

Following Hall's approach we establish the result in two main stages. First, we have

Lemma A. *Let S be a monoid with isolated identity, let U be an inverse submonoid of S and let Y be a right U-system. Then the U-map $\tau_Y: y \to y \otimes 1$ from Y into $Y \otimes_U S$ is a monomorphism.*

Proof. Let $y, z \in Y$ be such that $y \otimes 1 = z \otimes 1$. Then by the definition of tensor products in U-systems there exist $u_1, \ldots, u_{2n}$ in U, $y_1, \ldots, y_n$ in Y and $s_1, \ldots, s_{n-1}$ in S such that

$$
\begin{aligned}
&y = y_1 u_1, && u_1 = u_2 s_1,\\
&y_1 u_2 = y_2 u_3, && u_3 s_1 = u_4 s_2,\\
&\ldots\ldots\\
&y_{i-1} u_{2i-2} = y_i u_{2i-1}, && u_{2i-1} s_{i-1} = u_{2i} s_i \quad (i = 2, \ldots, n-1),\\
&\ldots\ldots\\
&y_{n-1} u_{2n-2} = y_n u_{2n-1}, && u_{2n-1} s_{n-1} = u_{2n},\\
&y_n u_{2n} = z.
\end{aligned}
\tag{1}
$$

It helps at this stage to introduce some more notation: let

(2)
$$w_1 = u_1^{-1}, \quad w_2 = u_1^{-1}u_2u_3^{-1}, \ldots,$$
$$w_{i+1} = w_iu_{2i}u_{2i+1}^{-1} \qquad (i = 1, \ldots, n-1);$$

let

(3)
$$z_0 = 1, \quad z_1 = u_{2n-2}u_{2n-1}^{-1},$$
$$z_i = u_{2n-2i}u_{2n-2i+1}^{-1}z_{i-1} \qquad (i = 1, \ldots, n-1).$$

Notice that

(4) $$w_iz_{n-i} = w_n \qquad (i = 1, \ldots, n).$$

Now, by induction on i one can prove that

$$y = y_i(w_i^{-1}w_iu_{2i-1}s_{i-1});$$

then putting $i = n$ gives

(5) $$y = y_n(w_n^{-1}w_n)u_{2n}.$$

Again by induction on i one has

$$z = y_{n-i}(z_iz_i^{-1}u_{2n-2i}s_{n-i});$$

putting $i = n - 1$ and making use of (4) gives

(6) $$z = yw_nw_n^{-1}.$$

A similar inductive argument gives

$$w_nu_{2n} = w_nz_i^{-1}u_{2n-2i}s_{n-i};$$

putting $i = n - 1$ gives

(7) $$w_nu_{2n} = w_nw_n^{-1}.$$

Finally, we show inductively that

$$yw_n = y_iw_i^{-1}w_n,$$

which when $i = n$ gives

(8) $$yw_n = y_nw_n^{-1}w_n.$$

It now follows that

$$y = y_n(w_n^{-1}w_n)u_{2n} = \qquad \text{by (5)},$$

$$= yw_nu_{2n} = \qquad \text{by (8)},$$

$$= yw_nw_n^{-1} = \qquad \text{by (7)}$$

$$= z \qquad \text{by (6)},$$

and so τ_Y is one-one as required.

The property described in Lemma A is in effect what Hall [3] calls the *representation extension property*. The next main stage is to show that an inverse submonoid has Hall's *free representation extension property*. In our language this is:

Lemma B. *Let* S *be a monoid with isolated identity element* 1 *and let* U *be an inverse submonoid of* S. *Then* S *is a 'flat' left* U*-system: i.e., for all right* U*-systems* X *and* Y *and all right* U*-monomorphisms* $\lambda: X \to Y$ *the induced right* S*-map* $\lambda^*: X \otimes_U S \to Y \otimes_U S$ *given by* $(x \otimes s)\lambda^* = (x\lambda) \otimes s$ *is a monomorphism.*

Proof. Let $x, x' \in X$, $x, t \in S$ and let $(x\lambda) \otimes s = (x'\lambda) \otimes t$ in $Y \otimes_U S$. Then there exist $u_1, \ldots, u_{2n}$ in U, $y_1, \ldots, y_n$ in Y, $s_1, \ldots, s_{n-1}$ in S such that

$$x\lambda = y_1u_1, \qquad u_1s = u_2s_1,$$

$$y_1u_2 = y_2u_3, \qquad u_3s_1 = u_4s_2,$$

$$\cdots\cdots$$

(9) $$y_{i-1}u_{2i-2} = y_iu_{2i-1}, \qquad u_{2i-1}s_{i-1} = u_{2i}s_i \qquad (i = 2, \ldots, n-1)$$

$$\cdots\cdots$$

$$y_{n-1}u_{2n-2} = y_nu_{2n-1}, \quad u_{2n-1}s_{n-1} = u_{2n}t,$$

$$y_nu_{2n} = x'\lambda.$$

Notice the strong formal similarity to equations (1). We aim to show that

(10) $$x \otimes s = x' \otimes t$$

in $X \otimes_U S$, and this will follow if we can show that the equations down the left hand side of (9) can be replaced by

$$x = x_1 u_1, \quad x_{i-1} u_{2i-2} = x_i u_{2i-1} \qquad (i = 2, \ldots, n),$$
$$x_n u_{2n} = x'$$

for some $x_1, \ldots, x_n$ in X.

With the same definition of w_i as before (see formula (2)) define

(11) $\qquad p_i = y_i w_i^{-1} w_i \qquad (i = 1, \ldots, n).$

Then it is not hard to show that

(12) $\qquad p_{i-1} u_{2i-2} = p_i u_{2i-1} \qquad (i = 2, \ldots, n).$

Also, $p_1 = y_1 u_1 u_1^{-1} = (x\lambda) u_1^{-1} = (x u_1^{-1})\lambda \in X\lambda$ and (for $i = 2, \ldots, n$)

$$p_i = y_i w_i^{-1} w_i = y_i u_{2i-1} u_{2i-2}^{-1} w_{i-1}^{-1} w_{i-1} u_{2i-2} u_{2i-1}^{-1} =$$
$$= y_{i-1} u_{2i-2} u_{2i-2}^{-1} w_{i-1}^{-1} w_{i-1} u_{2i-2} u_{2i-1}^{-1} =$$
$$= y_{i-1} w_{i-1}^{-1} w_{i-1} u_{2i-2} u_{2i-1}^{-1} = p_{i-1} u_{2i-2} u_{2i-1}^{-1};$$

hence by induction $p_i \in X\lambda$ for $i = 1, \ldots, n$. Define $x_i = p_i \lambda^{-1}$ (uniquely, since λ is one-one); then (12) gives

(13) $\qquad x_{i-1} u_{2i-2} = x_i u_{2i-1} \qquad (i = 2, \ldots, n).$

These equations, together with those on the right hand side of (9), show that in $X \otimes_U S$

$$x_1 u_1 \otimes s = x_n u_{2n} \otimes t.$$

Now

$$x_1 u_1 = (p_1 u_1)\lambda^{-1} = (y_1 u_1 u_1^{-1} u_1)\lambda^{-1} =$$
$$= (y_1 u_1)\lambda^{-1} = x\lambda\lambda^{-1} = x,$$

and

$$x_n u_{2n} = (p_n u_{2n})\lambda^{-1} = (y_n w_n^{-1} w_n u_{2n})\lambda^{-1} =$$
$$= (y_n u_{2n} u_{2n}^{-1} w_n^{-1} w_n u_{2n})\lambda^{-1} = x' u_{2n}^{-1} w_n^{-1} w_n u_{2n};$$

thus we have shown that in $X \otimes_U S$

(14) $$x \otimes s = x' u_{2n}^{-1} w_n^{-1} w_n u_{2n} \otimes t.$$

Now, equations (9) provide a link from x to x', but we may of course also regard them as providing a link from x' to x. If we start from the other end we obtain a result, analogous to (14), to the effect that in $X \otimes_U S$

(15) $$x w_n u_{2n} u_{2n}^{-1} w_n^{-1} \otimes s = x' \otimes t.$$

We are of course aiming to prove (10) and so neither (14) nor (15) is quite what we require. However, the same technique that gave formulae (7) and (8) as conseqences of equations (1) also yields

(16) $$w_n u_{2n} t = w_n w_n^{-1} s,$$

(17) $$(x\lambda) w_n = y_n w_n^{-1} w_n$$

as consequences of equations (9). Hence, premultiplying (16) by $u_{2n}^{-1} w_n^{-1}$, we obtain

(18) $$u_{2n}^{-1} w_n^{-1} w_n u_{2n} t = u_{2n}^{-1} w_n^{-1} s;$$

also, postmultiplying (17) by $u_{2n} u_{2n}^{-1} w_n^{-1}$ and mapping by λ^{-1}, we obtain

(19) $$x w_n u_{2n} u_{2n}^{-1} w_n^{-1} = x' u_{2n}^{-1} w_n^{-1}.$$

Hence, finally, in $X \otimes_U S$,

$$x \otimes s = x' u_{2n}^{-1} w_n^{-1} w_n u_{2n} \otimes t = \qquad \text{(by (14))}$$

$$= x' \otimes u_{2n}^{-1} w_n^{-1} w_n u_{2n} t =$$

$$= x' \otimes u_{2n}^{-1} w_n^{-1} s = \qquad \text{(by (18))}$$

$$= x' u_{2n}^{-1} w_n^{-1} \otimes s =$$

$$= x w_n u_{2n} u_{2n}^{-1} w_n^{-1} \otimes s = \qquad \text{(by (19))}$$

$$= x' \otimes t. \qquad \text{(by (15))}$$

This completes the proof of Lemma B.

Let us turn now to the proof of Lemma 2.5. Suppose, therefore, that we have an S-system X, a U-system Y and a U-monomorphism $\lambda: X \to Y$. By Lemma A the U-maps $\tau_Y: Y \to Y \otimes_U S$, $\tau_X: X \to X \otimes_U S$ are monomorphisms. By Lemma B the induced S-map $\lambda^*: X \otimes_U S \to$ $\to Y \otimes_U S$ is also a monomorphism. So if we could show that the one-one mapping $\tau_X \lambda^*: X \to Y \otimes_U S$ was an S-map then $Y \otimes_U S$ would be the Z we require. Unfortunately we see that

$$(xs)\tau_X \lambda^* = (xs)\lambda \otimes 1, \quad (x\tau_X \lambda^*)s = (x\lambda) \otimes s;$$

and these are in general distinct members of $Y \otimes_U S$.

For brevity let us write $Q = Y \otimes_U S$. Let

$$\begin{aligned} \sigma_X &= \{(x_1\lambda \otimes s_1,\ x_2\lambda \otimes s_1): \\ (20) \qquad & x_1, x_2 \in X,\ s_1, s_2 \in S,\ x_1 s_1 = x_2 s_2\}; \end{aligned}$$

then $\sigma = \sigma_X \cup 1_Q$ is an S-congruence on Q. Let $Z = Q/\sigma$; then Z is an S-system, with

$$((y \otimes s)\sigma)t = (y \otimes st)\sigma \qquad (y \in Y,\ s, t \in S).$$

Consider now the sequence

$$Y \xrightarrow{\ \tau_Y\ } Q \xrightarrow{\ \sigma^\natural\ } Z$$

of U-maps. If $y_1 \tau_Y \sigma^\natural = y_2 \tau_Y \sigma^\natural$ then $(y_1 \otimes 1, y_2 \otimes 1) \in \sigma$. Hence either $y_1 \otimes 1 = y_2 \otimes 1$, giving $y_1 = y_2$; or $y_1 = x_1\lambda$, $y_2 = x_2\lambda$ for some x_1, x_2 in X and

$$(x_1\lambda \otimes 1,\ x_2\lambda \otimes 1) \in \sigma_X.$$

By (20) this again gives $y_1 = y_2$. Thus the U-map $\tau_Y \sigma^\natural: Y \to Z$ is a monomorphism.

The same argument shows that the composition

$$X \xrightarrow{\ \tau_X\ } X \otimes_U S \xrightarrow{\ \lambda^*\ } Y \otimes_U S \xrightarrow{\ \sigma^\natural\ } Z$$

maps X in one-one fashion into Z. Moreover $\psi = \tau_X \lambda^* \sigma^\natural$ is an S-map, since for all x in X and s in S

$$(xs)\psi = ((xs)\lambda \otimes 1)\sigma^{\natural} =$$

$$= ((x\lambda) \otimes s)\sigma^{\natural} = \qquad \text{(by (20))}$$

$$= (((x\lambda) \otimes 1)s)\sigma^{\natural} =$$

$$= (x\psi)s.$$

This completes the proof of Lemma 2.5.

REFERENCES

[1] A.H. Clifford – G.B. Preston, *The algebraic theory of semigroups,* vol. 2, Math. Surveys of the American Math. Soc., Providence, R. I., 1967.

[2] T.E. Hall, Free products with amalgamation of inverse semigroups, *J. Algebra,* 34 (1975), 375–385.

[3] T.E. Hall, Representation extension and amalgamation for semigroups, *Quart. J. Math. Oxford,* (2) 29 (1978), 309–334.

[4] T.E. Hall, Inverse and regular semigroups and amalgamation: a brief survey, *Symposium on Regular Semigroups,* Northern Illinois University, 1979.

[5] J.M. Howie, Embedding theorems with amalgamation for semigroups, *Proc. London Math. Soc.,* (3) 12 (1962), 511–534.

[6] J.M. Howie, Semigroup amalgams whose cores are inverse semigroups, *Quart. J. Math. Oxford,* (2) 26 (1975), 23–45.

[7] J.M. Howie, *An introduction to semigroup theory,* Academic Press, London – New York – San Francisco, 1976.

[8] T. Imaoka, Free products with amalgamation of bands, *Mem. Fac. Sci. Shimane Univ.,* 10 (1976), 7–17.

[9] J.R. Isbell, Epimorphism and dominions, *Proc. Conference on Categorical Algebra,* (La Jolla 1965), Springer, 1966, 232–246.

[10] H.E. Scheiblich, On epics and dominions of bands, *Semigroup Forum,* 13 (1976), 103–114.

[11] O. Schreier, Die Untergruppen der freien Gruppen, *Hamburg Abh.,* 5 (1927), 161–183.

John M. Howie

Mathematical Institute, St. Andrews, Scotland.

COLLOQUIA MATHEMATICA SOCIETATIS JÁNOS BOLYAI
39. SEMIGROUPS, SZEGED (HUNGARY), 1981.

ON SMALL CATEGORIES WHOSE SETS OF OBJECTS FORM LATTICES

M. ITO

0. INTRODUCTION

Many classes of algebraic systems can be considered as partially ordered sets by homomorphism relation. And some of them form lattices. For instance, in [2]-[4], we have treated classes of automata, and proved that some of them, e.g., the class of quasiperfect automata, the class of perfect automata and the class of strongly cofinal automata, form lattices.

In the present paper, we discuss relationships between classes of an algebraic system which form lattices and their subclasses.

1. STRONG SUBCATEGORIES AND LATTICES

In the present section, union and intersection of subcategories of a given small category are dealt with. Let $\mathscr{A}$ be a small category, and Obj $\mathscr{A}$ and $\mathscr{A}(A, B)$ be, respectively, the set of all objects of $\mathscr{A}$ and the set of all morphisms from A to B.

Definition 1.1. Let A and B be two elements of Obj $\mathscr{A}$. Then, $A \leqslant B$ means $\mathscr{A}(B, A) \neq \phi$. Moreover, we denote $A = B$ if $A \leqslant B$ and

$B \leqslant A$. Identifying equal elements, the set Obj $\mathscr{A}$ can be considered as a partially ordered set in which the partial order is given by $\leqslant$.

In what follows, we assume that Obj $\mathscr{A}$ is countable and

$$\# \{B \in \text{Obj } \mathscr{A} \mid \mathscr{A}(A, B) \neq \phi\} < +\infty,$$

where $\# K$ denotes the cardinality of K.

Definition 1.2. A subcategory $\mathscr{A}'$ of $\mathscr{A}$ is said to be *strong* if the following conditions are satisfied:

(i) $\{B \in \text{Obj } \mathscr{A} \mid \mathscr{A}(A, B) \neq \phi, \ A \in \text{Obj } \mathscr{A}'\} \subseteq \text{Obj } \mathscr{A}'$,

(ii) $\mathscr{A}'(A, B) = \mathscr{A}(A, B)$ for all $A, B \in \text{Obj } \mathscr{A}'$.

Obj $\mathscr{A}'$ can be considered as a subpartially ordered set of Obj $\mathscr{A}$.

Definition 1.3. Let $\mathscr{A}'_\gamma$ $(\gamma \in \Gamma)$ be a family of subcategories of $\mathscr{A}$. Then, the subcategory $\bigcup_{\gamma \in \Gamma} \mathscr{A}'_\gamma$ of $\mathscr{A}$ is defined as follows:

(i) $\text{Obj}\left(\bigcup_{\gamma \in \Gamma} \mathscr{A}'_\gamma\right) = \bigcup_{\gamma \in \Gamma} \text{Obj } \mathscr{A}'_\gamma$,

(ii) $\left(\bigcup_{\gamma \in \Gamma} \mathscr{A}'_\gamma\right)(A, B) = \bigcup_{\gamma \in \Gamma} \mathscr{A}'_\gamma(A, B)$ for all $A, B \in \text{Obj}\left(\bigcup_{\gamma \in \Gamma} \mathscr{A}'_\gamma\right)$.

Moreover, if $\bigcap_{\gamma \in \Gamma} \text{Obj } \mathscr{A}'_\gamma \neq \phi$, then the subcategory $\bigcap_{\gamma \in \Gamma} \mathscr{A}'_\gamma$ of $\mathscr{A}$ is defined as follows:

(i) $\text{Obj}\left(\bigcap_{\gamma \in \Gamma} \mathscr{A}'_\gamma\right) = \bigcap_{\gamma \in \Gamma} \text{Obj } \mathscr{A}'_\gamma$,

(ii) $\left(\bigcap_{\gamma \in \Gamma} \mathscr{A}'_\gamma\right)(A, B) = \bigcap_{\gamma \in \Gamma} \mathscr{A}'_\gamma(A, B)$ for all $A, B \in \text{Obj}\left(\bigcap_{\gamma \in \Gamma} \mathscr{A}'_\gamma\right)$.

The following results can easily be proved.

Proposition 1.1. *Let* $\mathscr{A}'_\gamma$ $(\gamma \in \Gamma)$ *be a family of strong subcategories of* $\mathscr{A}$. *Then* $\bigcup_{\gamma \in \Gamma} \mathscr{A}'_\gamma$ *and* $\bigcap_{\gamma \in \Gamma} \mathscr{A}'_\gamma$ *are strong subcategories of* $\mathscr{A}$.

Propostion 1.2. *Let* $\mathscr{A}'$ *and* $\mathscr{A}'_\gamma$ $(\gamma \in \Gamma)$ *be strong subcategories of* $\mathscr{A}$. *If* $\text{Obj } \mathscr{A}' = \bigcup_{\gamma \in \Gamma} \text{Obj } \mathscr{A}'_\gamma$, *then* $\mathscr{A}' = \bigcup_{\gamma \in \Gamma} \mathscr{A}'_\gamma$. *Moreover, if* $\text{Obj } \mathscr{A}' = \bigcap_{\gamma \in \Gamma} \text{Obj } \mathscr{A}'_\gamma$, *then* $\mathscr{A}' = \bigcap_{\gamma \in \Gamma} \mathscr{A}'_\gamma$:

Now, consider the case where the set of all objects of a subcategory of $\mathscr{A}$ forms a lattice by the partial order $\leqslant$. Let $\mathscr{A}'$ be a subcategory of $\mathscr{A}$ such that Obj $\mathscr{A}'$ forms a lattice. Then, by $[A \circ B]_{\mathscr{A}'}$ and $[A * B]_{\mathscr{A}'}$, we denote, respectively, the least upper bound of $\{A, B\}$ in Obj $\mathscr{A}'$ and the greatest lower bound of $\{A, B\}$ in Obj $\mathscr{A}'$.

Theorem 1.1. *Let* $\mathscr{A}'_\gamma$ $(\gamma \in \Gamma)$ *be a family of strong subcategories of* $\mathscr{A}$ *such that* Obj $\mathscr{A}'_\gamma$ $(\gamma \in \Gamma)$ *forms a lattice. Moreover, let* $\mathscr{A}'$ *be a subcategory of* $\mathscr{A}$ *such that* Obj $\mathscr{A}'$ *forms a lattice and* $\bigcup_{\gamma \in \Gamma} \text{Obj } \mathscr{A}'_\gamma \subseteq \text{Obj } \mathscr{A}'$. *Then,* $\bigcap_{\gamma \in \Gamma} \mathscr{A}'_\gamma$ *can be defined and it is a strong subcategory of* $\mathscr{A}$ *such that* $\text{Obj}\left(\bigcap_{\gamma \in \Gamma} \mathscr{A}'_\gamma\right)$ *forms a sublattice of* Obj $\mathscr{A}'$.

Proof. First, we prove that $\bigcap_{\gamma \in \Gamma} \text{Obj } \mathscr{A}'_\gamma \neq \phi$. Let ξ be a fixed element of Γ and A_ξ be an element of Obj $\mathscr{A}'_\xi$. Suppose that there exists a $\mu \in \Gamma$ such that $A_\xi \notin \text{Obj } \mathscr{A}'_\mu$. Let $A_\mu \in \text{Obj } \mathscr{A}'_\mu$. Then, there exists the element $A_{\xi\mu} = [A_\xi * A_\mu]_{\mathscr{A}'}$. Thus, we have $A_{\xi\mu} \in$ $\in \text{Obj } \mathscr{A}'_\xi \cap \text{Obj } \mathscr{A}'_\mu$ because of $A_\xi \geqslant A_{\xi\mu}$ and $A_\mu \geqslant A_{\xi\mu}$. Now, replace A_ξ by $A_{\xi\mu}$ and continue this process for another $\mu' \in \Gamma$ such that $A_\xi =$ $= A_{\xi\mu} \notin \text{Obj } \mathscr{A}'_{\mu'}$. By the finiteness of

$$\#\{B \in \text{Obj } \mathscr{A}'_\xi \mid \mathscr{A}'_\xi(A, B) \neq \phi\},$$

it can easily be seen that there exists an element $A \in \text{Obj } \mathscr{A}'_\xi$ such that for all $\mu \in \Gamma$ we have $A \in \text{Obj } \mathscr{A}'_\mu$. That is, we have $A \in \bigcap_{\gamma \in \Gamma} \text{Obj } \mathscr{A}'_\gamma \neq$ $\neq \phi$. Thus, $\bigcap_{\gamma \in \Gamma} \mathscr{A}'_\gamma$ is defined and, by Proposition 1.1, it becomes a strong subcategory of $\mathscr{A}$.

Now, we prove that there exists the $[A \circ B]_{\bigcap_{\gamma \in \Gamma} \mathscr{A}'_\gamma}$ for all $A, B \in$ $\in \bigcap_{\gamma \in \Gamma} \text{Obj } \mathscr{A}'_\gamma$. Let $C = [A \circ B]_{\mathscr{A}'}$, where $A, B \in \bigcap_{\gamma \in \Gamma} \text{Obj } \mathscr{A}'_\gamma$. Note that, for all $\gamma \in \Gamma$, we have $A, B \in \text{Obj } \mathscr{A}'_\gamma$ and $[A \circ B]_{\mathscr{A}'_\gamma} \in \text{Obj } \mathscr{A}'_\gamma \subseteq$ $\subseteq \text{Obj } \mathscr{A}'$. By $A, B \leqslant [A \circ B]_{\mathscr{A}'_\gamma}$, we have $C = [A \circ B]_{\mathscr{A}'} \leqslant [A \circ B]_{\mathscr{A}'_\gamma}$. Since $\mathscr{A}'_\gamma$ is a strong subcategory of $\mathscr{A}$, we have $C \in \text{Obj } \mathscr{A}'_\gamma$. Consequently, $C \in \bigcap_{\gamma \in \Gamma} \text{Obj } \mathscr{A}'_\gamma$. On the other hand, by the definition of $[A \circ B]_{\mathscr{A}'}$, we can see that, for all $C' \in \bigcap_{\gamma \in \Gamma} \text{Obj } \mathscr{A}'_\gamma$ such that $A, B \leqslant C'$, we have $C \leqslant C'$. Therefore, there exists the $[A \circ B]_{\bigcap_{\gamma\in\Gamma} \mathscr{A}'_\gamma}$ and it equals $[A \circ B]_{\mathscr{A}'}$.

Finally, we prove that there exists the $[A * B]_{\bigcap_{\gamma\in\Gamma} \mathscr{A}'_\gamma}$ for all $A, B \in \bigcap_{\gamma\in\Gamma} \text{Obj}\, \mathscr{A}'_\gamma$. Let $C = [A * B]_{\mathscr{A}'}$, where $A, B \in \bigcap_{\gamma\in\Gamma} \text{Obj}\, \mathscr{A}'_\gamma$. Since $C \leqslant A, B$, we have $C \in \bigcap_{\gamma\in\Gamma} \text{Obj}\, \mathscr{A}'_\gamma$. Suppose $C' \leqslant A, B$, where $C' \in \bigcap_{\gamma\in\Gamma} \text{Obj}\, \mathscr{A}'_\gamma$. Then, by the definition of $[A * B]_{\mathscr{A}'}$, $C' \leqslant C$. This means that $[A * B]_{\bigcap_{\gamma\in\Gamma} \mathscr{A}'_\gamma}$ exists and it equals $[A * B]_{\mathscr{A}'}$. Q.E.D.

Corollary 1.1. *Let* $\mathscr{A}'$ *and* $\mathscr{B}'$ *be strong subcategories of* $\mathscr{A}$ *whose sets of all objects form lattices. If* $\text{Obj}\, \mathscr{B}' \subseteq \text{Obj}\, \mathscr{A}'$, *then* $\text{Obj}\, \mathscr{B}'$ *is a sublattice of* $\text{Obj}\, \mathscr{A}'$.

Theorem 1.2. *Let* $\mathscr{A}'_\gamma$ $(\gamma \in \Gamma)$ *be a family of strong subcategories of* $\mathscr{A}$ *such that* $\text{Obj}\, \mathscr{A}'_\gamma$ $(\gamma \in \Gamma)$ *forms a lattice. Then, the following two conditions are equivalent:*

(i) $\bigcup_{\gamma\in\Gamma} \text{Obj}\, \mathscr{A}'_\gamma$ *is a lattice,*

(ii) *For all* $\delta, \mu \in \Gamma$, $\mathscr{A}'_\delta \cap \mathscr{A}'_\mu$ *can be defined and* $\text{Obj}\, (\mathscr{A}'_\delta \cap \mathscr{A}'_\mu)$ *forms a lattice.*

Moreover, for all $A, B \in \bigcup_{\gamma\in\Gamma} \text{Obj}\, \mathscr{A}'_\gamma$, *there exists some* $\xi \in \Gamma$ *such that* $A, B \in \text{Obj}\, \mathscr{A}'_\xi$.

Proof. First, assume that (i) holds. Let $\delta, \mu \in \Gamma$. Since $\text{Obj}\, \mathscr{A}'_\delta \cup \text{Obj}\, \mathscr{A}'_\mu \subseteq \bigcup_{\gamma\in\Gamma} \text{Obj}\, \mathscr{A}'_\gamma$ and $\bigcup_{\gamma\in\Gamma} \text{Obj}\, \mathscr{A}'_\gamma$ forms a lattice, by Theorem 1.1, we can see that $\mathscr{A}'_\delta \cap \mathscr{A}'_\mu$ is defined and $\text{Obj}\, (\mathscr{A}'_\delta \cap \mathscr{A}'_\mu)$ forms a lattice. Next, let $A, B \in \bigcup_{\gamma\in\Gamma} \text{Obj}\, \mathscr{A}'_\gamma$. Since $\bigcup_{\gamma\in\Gamma} \text{Obj}\, \mathscr{A}'_\gamma$ forms a lattice, we have $A, B \leqslant [A \circ B]_{\bigcup_{\gamma\in\Gamma} \mathscr{A}'_\gamma} \in \text{Obj}\left(\bigcup_{\gamma\in\Gamma} \mathscr{A}'_\gamma\right)$. Therefore, there exists some $\xi \in \Gamma$ such that $[A \circ B]_{\bigcup_{\gamma\in\Gamma} \mathscr{A}'_\gamma} \in \text{Obj}\, \mathscr{A}'_\xi$. As $\mathscr{A}'_\xi$ is a strong subcategory of $\mathscr{A}$, we have $A, B \in \text{Obj}\, \mathscr{A}'_\xi$.

Now, assume that (ii) holds. Let $A, B \in \bigcup_{\gamma\in\Gamma} \text{Obj}\, \mathscr{A}'_\gamma$. Then, there exists some $\xi \in \Gamma$ such that $A, B \in \text{Obj}\, \mathscr{A}'_\xi$. Put $C = [A \circ B]_{\mathscr{A}'_\xi}$. Let

$C' \in \bigcup_{\gamma\in\Gamma} \text{Obj } \mathscr{A}'_\gamma$ such that $C' \geqslant A, B$. Then, there exists some $\eta \in \Gamma$ such that $C' \in \text{Obj } \mathscr{A}'_\eta$. It is obvious that $C' \geqslant [A \circ B]_{\mathscr{A}'_\eta}$. Since $C, C' \in \bigcup_{\gamma\in\Gamma} \text{Obj } \mathscr{A}'_\gamma$, there exists some $\theta \in \Gamma$ such that $C, C' \in \text{Obj } \mathscr{A}'_\theta$. Then, we have $A, B \leqslant [A \circ B]_{\mathscr{A}'_\xi} \in \text{Obj } \mathscr{A}'_\theta$. Thus, we have $A, B \leqslant [A \circ B]_{\mathscr{A}'_\theta} \leqslant [A \circ B]_{\mathscr{A}'_\xi} = C$. However, $\mathscr{A}'_\xi$ is a strong subcategory of $\mathscr{A}$. Therefore, $[A \circ B]_{\mathscr{A}'_\theta} \in \text{Obj } \mathscr{A}'_\xi$. Thus, $[A \circ B]_{\mathscr{A}'_\xi} \leqslant [A \circ B]_{\mathscr{A}'_\theta}$. This means that $[A \circ B]_{\mathscr{A}'_\theta} = [A \circ B]_{\mathscr{A}'_\xi}$. On the other hand, since $A, B \leqslant C' \in \text{Obj } \mathscr{A}'_\theta$, we have $[A \circ B]_{\mathscr{A}'_\theta} \leqslant C'$. Consequently, $C = [A \circ B]_{\mathscr{A}'_\xi} = [A \circ B]_{\mathscr{A}'_\theta} \leqslant C'$. Thus, there exists the $[A \circ B]_{\bigcup_{\gamma\in\Gamma} \mathscr{A}'_\gamma}$ and it equals C. We prove now that there exists the $[A * B]_{\bigcup_{\gamma\in\Gamma} \mathscr{A}'_\gamma}$ for all $A, B \in \bigcup_{\gamma\in\Gamma} \text{Obj } \mathscr{A}'_\gamma$. Let $A, B \in \bigcup_{\gamma\in\Gamma} \text{Obj } \mathscr{A}'_\gamma$. Then, there exists some $\xi \in \Gamma$ such that $A, B \in \text{Obj } \mathscr{A}'_\xi$. Put $C = [A * B]_{\mathscr{A}'_\xi}$. Let $C' \in \bigcup_{\gamma\in\Gamma} \text{Obj } \mathscr{A}'_\gamma$ such that $C' \leqslant A, B$. It is obvious that $C' \in \text{Obj } \mathscr{A}'_\xi$. Therefore, $C' \leqslant [A * B]_{\mathscr{A}'_\xi} = C$. This means that there exists the $[A * B]_{\bigcup_{\gamma\in\Gamma} \mathscr{A}'_\gamma}$ and it equals C. Consequently, $\bigcup_{\gamma\in\Gamma} \mathscr{A}'_\gamma$ becomes a lattice. Q.E.D.

We consider now the case $\Gamma = N$, where N is the set of all natural numbers. Let $\{U_k \mid k \in N\}$ be a family of sets. Then, we define $\lim_{k\to\infty} U_k = \{x \mid \text{there exists some } k(x) \in N \text{ such that for all } k \geqslant k(x)\ (k \in N) \text{ we have } x \in U_k\}$.

Theorem 1.3. *Let* $\mathscr{A}'_i$ $(i \in N)$ *be a family of strong subcategories of* $\mathscr{A}$ *whose sets of all objects form lattices such that, for all* $i, j \in N$, $\mathscr{A}'_i \cap \mathscr{A}'_j$ *is defined and* $\text{Obj}(\mathscr{A}'_i \cap \mathscr{A}'_j)$ *forms a lattice. Then,* $\bigcup_{i\in N} \text{Obj } \mathscr{A}'_i$, *is a lattice if and only if there exists a sequence of natural numbers* $\{i_k\}_{k\in N}$ *such that* $\bigcup_{i\in N} \text{Obj } \mathscr{A}'_i = \lim_{k\to\infty} \text{Obj } \mathscr{A}'_{i_k}$.

Proof. Assume that $\bigcup_{i\in N} \text{Obj } \mathscr{A}'_i$ is a lattice. Since $\text{Obj } \mathscr{A}$ is a countable set, so is $\text{Obj } \mathscr{A}'_i$ for any $i \in N$. Thus, we can give a number for every element of $\bigcup_{i\in N} \text{Obj } \mathscr{A}'_i$. Put $\bigcup_{i\in N} \text{Obj } \mathscr{A}'_i = \{A_i \mid i \in N\}$. First, we choose an i_1 such that $A_1 \in \mathrm{Obj}\,\mathscr{A}'_{i_1}$. Next, we choose an i_2 such that $A_1, A_2 \in \mathrm{Obj}\,\mathscr{A}'_{i_2}$. There exists such an i_2. For instance, an i_2 satisfying the condition $[A_1 \circ A_2]_{\bigcup_{i \in N} \mathscr{A}'_i} \in \mathrm{Obj}\,\mathscr{A}'_{i_2}$ is such one. Now, we choose an i_3 such that $[[A_1 \circ A_2]_{\bigcup_{i \in N} \mathscr{A}'_i} \circ A_3]_{\bigcup_{i \in N} \mathscr{A}'_i} \in \mathrm{Obj}\,\mathscr{A}'_{i_3}$. Then, we have $A_1, A_2, A_3 \in \mathrm{Obj}\,\mathscr{A}'_{i_3}$. By the same way, we can choose an i_n such that $A_1, A_2, A_3, \ldots, A_n \in \mathrm{Obj}\,\mathscr{A}'_{i_n}$ for all $n \in N$. Thus, there exists a sequence of natural numbers $\{i_k\}_{k \in N}$ such that $\bigcup_{i \in N} \mathrm{Obj}\,\mathscr{A}'_i = \lim_{k \to \infty} \mathrm{Obj}\,\mathscr{A}'_{i_k}$.

Now, assume that there exists a sequence of natural numbers $\{i_k\}_{k \in N}$ such that $\lim_{k \to \infty} \mathrm{Obj}\,\mathscr{A}'_{i_k} = \bigcup_{i \in N} \mathrm{Obj}\,\mathscr{A}'_i$. Put $A, B \in \bigcup_{i \in N} \mathrm{Obj}\,\mathscr{A}'_i$. Then, there exists some $k' \in N$ such that for all $k > k'$ $(k \in N)$ we have $A \in \in \mathrm{Obj}\,\mathscr{A}'_{i_k}$. And there exists some $k'' \in N$ such that for all $k > k''$ $(k \in N)$ we have $B \in \mathrm{Obj}\,\mathscr{A}'_{i_k}$. Therefore, for all $k > \max\{k', k''\}$ we have $A, B \in \mathrm{Obj}\,\mathscr{A}'_{i_k}$. Using Theorem 1.2, we can see that $\bigcup_{i \in N} \mathscr{A}'_i$ is a lattice. Q.E.D.

Corollary 1.2. *Let* Δ *be a finite set and* $\mathscr{A}'_\delta$ $(\delta \in \Delta)$ *be a family of strong subcategories of* $\mathscr{A}$ *such that, for all* $\delta, \delta' \in \Delta$, $\mathrm{Obj}\,\mathscr{A}'_\delta \cap \cap \mathrm{Obj}\,\mathscr{A}'_{\delta'}$ *is a lattice. Then, there exists some* $\xi \in \Delta$ *such that* $\mathrm{Obj}\,\mathscr{A}'_\xi = = \bigcup_{\delta \in \Delta} \mathrm{Obj}\,\mathscr{A}'_\delta$ *if* $\bigcup_{\delta \in \Delta} \mathrm{Obj}\,\mathscr{A}'_\delta$ *is a lattice.*

Proof. The corollary is obvious from the finiteness of Δ and Theorem 1.3. Q.E.D.

From the above result, the possibility that we obtain a subcategory of $\mathscr{A}$ whose set of all objects forms a new lattice from a finite number of strong subcategories of $\mathscr{A}$ by union operation is denied.

2. REPRESENTATION OF SUBCATEGORIES

In this section, a problem that we represent a strong subcategory of $\mathscr{A}$ by union of some simpler subcategories is considered. Representation theorem (Theorem 2.1) is the main result of this section.

Definition 2.1. Let $\mathscr{A}'$ and $\mathscr{A}'_\gamma$ $(\gamma \in \Gamma)$ be strong subcategories of $\mathscr{A}$ whose sets of all objects form lattices. If $\mathscr{A}' = \bigcup_{\gamma \in \Gamma} \mathscr{A}'_\gamma$ holds, we say that $\mathscr{A}'$ is *represented* by $\{\mathscr{A}'_\gamma\}_{\gamma \in \Gamma}$. Furthermore, we denote $\mathscr{A}' = [\mathscr{A}'_\gamma;\ \gamma \in \Gamma]_{\mathscr{A}'}$ (or briefly $[\mathscr{A}'_\gamma;\ \gamma \in \Gamma]$) instead of $\mathscr{A}' = \bigcup_{\gamma \in \Gamma} \mathscr{A}'_\gamma$ and the right hand side is called a *representation of* $\mathscr{A}'$.

Definition 2.2. Let $[\mathscr{A}'_\gamma;\ \gamma \in \Gamma]$ and $[\mathscr{B}'_\delta;\ \delta \in \Delta]$ be two representations of $\mathscr{A}'$. Then, these two representations are said to be *equivalent* if for all $\gamma \in \Gamma$ there exists some $\delta \in \Delta$ such that Obj $\mathscr{A}'_\gamma \subseteq$ $\subseteq$ Obj $\mathscr{B}'_\delta$ and for all $\delta \in \Delta$ there exists some $\gamma \in \Gamma$ such that Obj $\mathscr{B}'_\delta \subseteq$ $\subseteq$ Obj $\mathscr{A}'_\gamma$.

Definition 2.3. Let $[\mathscr{A}'_i;\ i \in N]$ be a representation of $\mathscr{A}'$. Then, this representation is said to be *of finite type* if every Obj $\mathscr{A}'_i$ $(i \in N)$ is a finite set.

For a finite type representation, we have the following result.

Lemma 2.1. *Let* $[\mathscr{A}'_i;\ i \in N]_{\mathscr{A}'}$ *be a finite type representation. Then, for all* $l, m \in N$ *there exists some* $n \in N$ *such that* Obj $\mathscr{A}'_l \cup$ $\cup$ Obj $\mathscr{A}'_m \subseteq$ Obj $\mathscr{A}'_n$.

Proof. By $\#\,\text{Obj}\,(\mathscr{A}'_l \cup \mathscr{A}'_m) < +\infty$, we can assume that there exists some $s \in N$ such that $\text{Obj}\,(\mathscr{A}'_l \cup \mathscr{A}'_m) = \{A_1, A_2, \ldots, A_s\}$. By Theorems 1.2 and 1.3, there exists a sequence of natural numbers $\{i_k\}_{k \in N}$ such that $\text{Obj}\,\mathscr{A}' = \lim_{k \to \infty} \text{Obj}\,\mathscr{A}'_{i_k}$. Thus, for all t $(1 \leqslant t \leqslant s)$ there exists some $k_t \in N$ such that for all $k > k_t$ we have $A_t \in \text{Obj}\,\mathscr{A}'_{i_k}$. Here, we put $n = i_{k'}$, where $k' > \max\{k_1, k_2, \ldots, k_s\}$. Then, we have Obj $\mathscr{A}'_l \cup$ Obj $\mathscr{A}'_m \subseteq$ Obj $\mathscr{A}'_n$. Q.E.D.

Proposition 2.1. *Let* $[\mathscr{A}'_i;\ i \in N]$ *and* $[\mathscr{B}'_j;\ j \in N]$ *be two finite type representations of* $\mathscr{A}'$. *Then, these two representations are equivalent.*

Proof. By Corollary 1.2, the proof can easily be carried out for the case $\#\,\text{Obj}\,\mathscr{A}' < +\infty$. Now, let Obj $\mathscr{A}'$ be an infinite set. We shall prove that for all $n \in N$ there exists some $m \in N$ such that Obj $\mathscr{A}'_n \subseteq$ $\subseteq$ Obj $\mathscr{B}'_m$. Since $\mathscr{A}' = \mathscr{A}' \cup \mathscr{A}'_n = \left(\bigcup_{j \in N} \mathscr{B}'_j\right) \cup \mathscr{A}'_n$, $[\mathscr{A}'_n, \mathscr{B}'_j;\ j \in N]$

is also a finite type representation of $\mathscr{A}'$. Furthermore, by $\# \text{Obj } \mathscr{A}'_n < < +\infty$, there exists some $l \in N$ such that $\text{Obj } \mathscr{B}'_l - \text{Obj } \mathscr{A}'_n \neq \phi$. By Lemma 2.4, there exists some $m \in N$ such that $\text{Obj } \mathscr{B}'_l \cup \text{Obj } \mathscr{A}'_n \subseteq \subseteq \text{Obj } \mathscr{B}'_m$. Consequently, we have $\text{Obj } \mathscr{A}'_n \subseteq \text{Obj } \mathscr{B}'_m$. By the same way, for all $m \in N$ there exists some $n \in N$ such that $\text{Obj } \mathscr{B}'_m \subseteq \text{Obj } \mathscr{A}'_n$. Q.E.D.

Definition 2.4. For any $A \in \text{Obj } \mathscr{A}$, the subcategory $\mathscr{A}(A)$ of $\mathscr{A}$ is defined as follows:

(i) $\text{Obj } \mathscr{A}(A) = \{B \in \text{Obj } \mathscr{A} \mid \mathscr{A}(A, B) \neq \phi\}$,

(ii) $\mathscr{A}(A)(B, C) = \mathscr{A}(B, C)$ for all $B, C \in \text{Obj } \mathscr{A}(A)$.

Then, it is obvious that $\mathscr{A}(A)$ is a finite strong subcategory of $\mathscr{A}$.

The following result can easily be proved.

Lemma 2.2. *Let $\mathscr{A}'$ be a strong subcategory of $\mathscr{A}$ such that $\text{Obj } \mathscr{A}'$ forms a lattice, and A be an element of $\text{Obj } \mathscr{A}'$. Then, $\text{Obj } \mathscr{A}(A)$ is a finite sublattice of $\text{Obj } \mathscr{A}'$.*

Theorem 2.1 (Representation Theorem). *Let $\mathscr{A}'$ be a strong subcategory of $\mathscr{A}$ such that $\text{Obj } \mathscr{A}'$ forms a lattice. Then, $\mathscr{A}'$ has a unique (up to equivalence) finite type representation.*

Proof. By Proposition 2.1, for the proof, we need only to show that $\mathscr{A}'$ has a finite type representation. Since $\text{Obj } \mathscr{A}$ is countable, so is $\text{Obj } \mathscr{A}'$. Put $\text{Obj } \mathscr{A}' = \{A_i \mid i \in N\}$. Then, it can easily be seen that $\mathscr{A}' = \bigcup_{i \in N} \mathscr{A}(A_i)$ holds. Consequently, $\mathscr{A}'$ has a finite type representation $[\mathscr{A}(A_i);\ i \in N]$. Q.E.D.

3. REPRESENTATION SPACE

Let $\mathscr{A}'$ be a strong subcategory of $\mathscr{A}$ such that $\text{Obj } \mathscr{A}'$ forms a lattice. Then, $\mathscr{A}'$ has a unique (up to equivalence) finite type representation. There is a possibility that $\mathscr{A}'$ has other representations. We investigate in this section the set of all representations $\mathscr{R}(\mathscr{A}')$ of $\mathscr{A}'$.

Definition 3.1. Let $\mathscr{A}'$ be a strong subcategory of $\mathscr{A}$ such that Obj $\mathscr{A}'$ forms a lattice, and $[\mathscr{A}'_\gamma;\ \gamma\in\Gamma]_{\mathscr{A}'}$ and $[\mathscr{B}'_\delta;\ \delta\in\Delta]_{\mathscr{A}'}$ be two representations of $\mathscr{A}'$. If for all $\gamma\in\Gamma$ there exists some $\delta\in\Delta$ such that $\text{Obj}\,\mathscr{A}'_\gamma\subseteq\text{Obj}\,\mathscr{B}'_\delta$, then we denote $[\mathscr{A}'_\gamma;\ \gamma\in\Gamma]_{\mathscr{A}'}\preceq$ $\preceq[\mathscr{B}'_\delta;\ \delta\in\Delta]_{\mathscr{A}'}$. When $[\mathscr{A}'_\gamma;\ \gamma\in\Gamma]_{\mathscr{A}'}\preceq[\mathscr{B}'_\delta;\ \delta\in\Delta]_{\mathscr{A}'}$ and $[\mathscr{B}'_\delta;\ \delta\in\Delta]_{\mathscr{A}'}\preceq[\mathscr{A}'_\gamma;\ \gamma\in\Gamma]_{\mathscr{A}'}$ hold, these two representations are equivalent. If two equivalent representations are considered as the same element, then the set of all representations of $\mathscr{A}'$ forms a partially ordered set by relation $\preceq$. This set is called the *representation space* of $\mathscr{A}'$ and denoted by $\mathscr{R}(\mathscr{A}')$.

Proposition 3.1. *$\mathscr{R}(\mathscr{A}')$ has a minimum element v_0 and a maximum element v_I.*

Proof. It can easily be seen that v_0 is the finite type representation and v_I is the representation $[\mathscr{A}'_\gamma;\ \gamma\in\Gamma]_{\mathscr{A}'}$ ($\mathscr{A}'_\gamma=\mathscr{A}'$ for all $\gamma\in\Gamma$). Q.E.D.

We have the following results. The proofs are rather monotone, but not difficult. So, they are omitted.

Proposition 3.2. *$\mathscr{R}(\mathscr{A}')$ is a distributive lattice.*

Proposition 3.3. $\text{Atom}\,(\mathscr{R}(\mathscr{A}'))=\{[\mathscr{A}'_i,\mathscr{U}';\ i\in N]\,|\,[\mathscr{A}'_i;\ i\in N]=$ $=v_0$ *and $\mathscr{U}'$ is an infinite strong subcategory of $\mathscr{A}$ whose set of all objects forms a sublattice of* Obj $\mathscr{A}'$ *such that for any infinite strong subcategory $\mathscr{W}'$ of $\mathscr{A}$ whose set of all objects forms a sublattice of* Obj $\mathscr{A}'$ *if* $\text{Obj}\,\mathscr{W}'\subseteq\text{Obj}\,\mathscr{U}'$ *then* $\mathscr{W}'=\mathscr{U}'\}$. *Here,* Atom $(\mathscr{R}(\mathscr{A}'))$ *denotes the set of all atomic elements of $\mathscr{R}(\mathscr{A}')$.*

Proposition 3.4. *We have $\#\mathscr{R}(\mathscr{A}')=+\infty$ if and only if for all $n\in N$ there exists some $[\mathscr{A}'_\gamma;\ \gamma\in\Gamma]\in\mathscr{R}(\mathscr{A}')$ (if $\gamma\neq\mu$ then $\mathscr{A}'_\gamma\neq$ $\neq\mathscr{A}'_\mu$) such that $\#\widetilde{\Gamma}\geqslant n$, where $\widetilde{\Gamma}=\{\gamma\,|\,\gamma\in\Gamma,\ \#\,\text{Obj}\,\mathscr{A}'_\gamma=+\infty\}$.*

4. LATTICE OF SUBCATEGORIES

In the present section, we investigate the structure of all strong subcategories of a given strong subcategory of $\mathscr{A}$.

Definition 4.1. Let $\mathscr{A}'$ be a strong subcategory of $\mathscr{A}$ such that Obj $\mathscr{A}'$ forms a lattice. Next, we put $\mathscr{L}(\mathscr{A}') = \{\mathscr{U}' \mid \mathscr{U}'$ is a strong subcategory of $\mathscr{A}$ such that Obj $\mathscr{U}'$ forms a sublattice of Obj $\mathscr{A}'\}$. Moreover, we put $\widetilde{\mathscr{L}}(\mathscr{A}') = \{\widetilde{\mathscr{U}}' \mid \mathscr{U}' \in \mathscr{L}(\mathscr{A}')\}$. Here, by $\widetilde{\mathscr{U}}'$ ($\mathscr{U}' \in \mathscr{L}(\mathscr{A}')$), we denote $\mathscr{U}'$ when Obj $\mathscr{U}'$ is an infinite set, and 0 when Obj $\mathscr{U}'$ is a finite set. Now, let ψ be the mapping such that $\psi(0) = \phi$ and $\psi(\widetilde{\mathscr{U}}') =$ $=$ Obj $\mathscr{U}'$ for all $\widetilde{\mathscr{U}}' \in \widetilde{\mathscr{L}}(\mathscr{A}') - \{0\}$. Then, we can introduce a partial order as follows: $\widetilde{\mathscr{U}}'$ is smaller than or equal to $\widetilde{\mathscr{V}}'$ if and only if $\psi(\widetilde{\mathscr{U}}') \subseteq \psi(\widetilde{\mathscr{V}}')$.

Thus, we have a partially ordered set $\widetilde{\mathscr{L}}(\mathscr{A}')$.

Lemma 4.1. $\widetilde{\mathscr{L}}(\mathscr{A}')$ *forms a lattice.*

Proof. Let $\mathscr{U}', \mathscr{V}' \in \mathscr{L}(\mathscr{A}')$ such that $\#\,\text{Obj}\,\mathscr{U}' = \#\,\text{Obj}\,\mathscr{V}' = +\infty$. Put $\{\mathscr{W}'_\gamma \mid \gamma \in \Gamma\} = \{\mathscr{W}' \mid \mathscr{W}'$ is a strong subcategory of $\mathscr{A}$ such that Obj $\mathscr{W}'$ forms a sublattice of Obj $\mathscr{A}'$ and Obj $\mathscr{U}'$, Obj $\mathscr{V}' \subseteq$ Obj $\mathscr{W}'\}$. Then, $\{\mathscr{W}'_\gamma \mid \gamma \in \Gamma\} \neq \phi$ because of $\mathscr{A}' \in \{\mathscr{W}'_\gamma \mid \gamma \in \Gamma\}$. Since Obj $\mathscr{U}'$, Obj $\mathscr{V}' \subseteq \bigcap_{\gamma \in \Gamma}$ Obj $\mathscr{W}'_\gamma$ and $\bigcup_{\gamma \in \Gamma}$ Obj $\mathscr{W}'_\gamma \subseteq$ Obj $\mathscr{A}'$, by Theorem 1.1, $\bigcap_{\gamma \in \Gamma} \mathscr{W}'_\gamma$ can be defined and Obj $\left(\bigcap_{\gamma \in \Gamma} \mathscr{W}'_\gamma\right)$ forms a sublattice of Obj $\mathscr{A}'$. From this result, it is obvious that $[\widetilde{\mathscr{U}}' \circ \widetilde{\mathscr{V}}']_{\widetilde{\mathscr{L}}(\mathscr{A}')}$ exists and it equals $\bigcap_{\gamma \in \Gamma} \mathscr{W}'_\gamma$. For any $\widetilde{\mathscr{U}}' \in \widetilde{\mathscr{L}}(\mathscr{A}')$, it is obvious that $[0 \circ \widetilde{\mathscr{U}}']_{\widetilde{\mathscr{L}}(\mathscr{A}')} = [\widetilde{\mathscr{U}}' \circ 0]_{\widetilde{\mathscr{L}}(\mathscr{A}')} = \widetilde{\mathscr{U}}'$. Next, for all $\widetilde{\mathscr{U}}', \widetilde{\mathscr{V}}' \in \widetilde{\mathscr{L}}(\mathscr{A}')$ ($\mathscr{U}', \mathscr{V}' \in \mathscr{L}(\mathscr{A}')$), it can easily be proved that $[\widetilde{\mathscr{U}}' * \widetilde{\mathscr{V}}']_{\widetilde{\mathscr{L}}(\mathscr{A}')}$ exists and it equals $\widetilde{\mathscr{U}' \cap \mathscr{V}'}$. Q.E.D.

Now, we study the relationship between $\mathscr{R}(\mathscr{A}')$ and $\widetilde{\mathscr{L}}(\mathscr{A}')$. Note that, by Proposition 3.4, the conditions $\#\,\mathscr{R}(\mathscr{A}') < +\infty$ and $\#\,\widetilde{\mathscr{L}}(\mathscr{A}') <$ $< +\infty$ are equivalent.

Theorem 4.1. $\widetilde{\mathscr{L}}(\mathscr{A}')$ *is an image of* $\mathscr{R}(\mathscr{A}')$ *by an upper semilattice homomorphism.*

Proof. Let $[\mathscr{A}'_\gamma;\ \gamma \in \Gamma]$ be a representation of $\mathscr{A}'$. The mapping Ψ of $\mathscr{R}(\mathscr{A}')$ onto $\widetilde{\mathscr{L}}(\mathscr{A}')$ can be defined as follows: $\Psi([\mathscr{A}'_\gamma;\ \gamma \in \Gamma]) = \bigcap_{\mathscr{W}'} \left\{\widetilde{\mathscr{W}}' \mid \mathscr{W}' \in \mathscr{L}(\mathscr{A}'),\ \text{Obj}\,\mathscr{W}' \supseteq \bigcup_{\gamma \in \widetilde{\Gamma}} \text{Obj}\,\mathscr{A}'_\gamma\right\}$, where $\widetilde{\Gamma} = \{\gamma \in \Gamma \mid \# \text{Obj } \mathscr{A}'_\gamma = +\infty\}$. Then, it is not difficult to prove that Ψ is an upper semilattice homomorphism of $\mathscr{R}(\mathscr{A}')$ onto $\widetilde{\mathscr{L}}(\mathscr{A}')$, i.e., $\Psi([u \circ v]_{\mathscr{R}(\mathscr{A}')}) = [\Psi(u) \circ \Psi(v)]_{\widetilde{\mathscr{L}}(\mathscr{A}')}$ for all $u, v \in \mathscr{R}(\mathscr{A}')$. Q.E.D.

A lattice $(L, \leqslant)$ satisfying the condition $[A \circ [B * C]] = [C * [A \circ B]]$ for all $A, B, C \in L$ such that $A \leqslant C$ is called a *modular lattice*. There is a case where Obj $\mathscr{A}'$ becomes a modular lattice. For a modular lattice, we have the following result.

Theorem 4.2. *Let* $\mathscr{A}'$ *be a strong subcategory of* $\mathscr{A}$ *such that* Obj $\mathscr{A}'$ *forms a modular lattice. If* $\# \widetilde{\mathscr{L}}(\mathscr{A}') < +\infty$, *then we have* $\# \text{Atom}(\widetilde{\mathscr{L}}(\mathscr{A}')) = 1$ *and* $\widetilde{\mathscr{L}}(\mathscr{A}')$ *is also a modular lattice.*

Proof. First, we show that $\# \text{Atom}(\widetilde{\mathscr{L}}(\mathscr{A}')) = 1$. Suppose $\# \text{Atom}(\widetilde{\mathscr{L}}(\mathscr{A}')) > 1$. Let $\mathscr{U}'$ and $\mathscr{W}'$ be two distinct atomic elements of $\widetilde{\mathscr{L}}(\mathscr{A}')$. Thus, we have $\widetilde{\mathscr{U}}' = \mathscr{U}'$, $\widetilde{\mathscr{W}}' = \mathscr{W}'$ and $\# \text{Obj}(\mathscr{U}' \cap \mathscr{W}') < +\infty$. Then, there exist two sequences $\{A_i\}_{i \in N}$ and $\{B_j\}_{j \in N}$ satisfying the following conditions:

(i) $A_1 \leqslant A_2 \leqslant A_3 \leqslant \ldots (A_i \in \text{Obj } \mathscr{U}' - \text{Obj } \mathscr{W}',\ A_i \neq A_j$ if $i \neq j)$ and $B_1 \leqslant B_2 \leqslant B_3 \leqslant \ldots (B_j \in \text{Obj } \mathscr{W}' - \text{Obj } \mathscr{U}',\ B_i \neq B_j$ if $i \neq j)$.

(ii) $\mathscr{U}' = \bigcup_{i \in N} \mathscr{A}(A_i)$ and $\mathscr{W}' = \bigcup_{j \in N} \mathscr{A}(B_j)$.

By $[A_i * B_j]_{\mathscr{A}'} \in \text{Obj}(\mathscr{U}' \cap \mathscr{W}')$, we have $\#\{[A_i * B_j]_{\mathscr{A}'} \mid i, j \in N\} < +\infty$. Now, we prove that, for all $n \in N$, $\text{Obj}\left(\bigcup_{j \in N} \mathscr{A}([A_n \circ B_j]_{\mathscr{A}'})\right)$ forms a lattice. Put $D, D' \in \text{Obj}\left(\bigcup_{j \in N} \mathscr{A}([A_n \circ B_j]_{\mathscr{A}'})\right)$. Then, there exist some $j', j'' \in N$ such that $D \leqslant [A_n \circ B_{j'}]_{\mathscr{A}'}$ and $D' \leqslant [A_n \circ B_{j''}]_{\mathscr{A}'}$. Consequently, we have $D, D' \leqslant [A_n \circ B_{\max\{j', j''\}}]_{\mathscr{A}'}$. That is, we have $D, D' \in \text{Obj}(\mathscr{A}([A_n \circ B_{\max\{j', j''\}}]_{\mathscr{A}'})$. Therefore, by Theorems 1.1 and 1.2, $\bigcup_{j \in N} \mathscr{A}([A_n \circ B_j]_{\mathscr{A}'})$ becomes a lattice. It is obvious that this is a strong subcategory of $\mathscr{A}$. Moreover, by $\text{Obj}\left(\bigcup_{j \in N} \mathscr{A}([A_n \circ B_j]_{\mathscr{A}'})\right) \supseteq \bigcup_{j \in N} \text{Obj } \mathscr{A}(B_j) = \text{Obj } \mathscr{W}'$, it becomes an infinite subcategory. Next, we prove that for all $n \in N$ there exists some $k \in N$ $(k > n)$ such that $\text{Obj}\left(\bigcup_{j \in N} \mathscr{A}([A_n \circ B_j]_{\mathscr{A}'})\right) \subset \text{Obj}\left(\bigcup_{j \in N} \mathscr{A}([A_k \circ B_j]_{\mathscr{A}'})\right)$.

It is obvious that for all $k \in N$ $(k > n)$ we have $A_k \in \in \text{Obj}\left(\bigcup_{j \in N} \mathscr{A}([A_k \circ B_j]_{\mathscr{A}'})\right)$. On the other hand, if $A_k \in \in \text{Obj}\left(\bigcup_{j \in N} \mathscr{A}([A_n \circ B_j]_{\mathscr{A}'})\right)$ then there exists some $j \in N$ such that $A_k \leqslant [A_n \circ B_j]_{\mathscr{A}'}$. Consequently, we have $A_k = [A_k \circ A_k]_{\mathscr{A}'} \leqslant \leqslant [A_k * [A_n \circ B_j]_{\mathscr{A}'}]_{\mathscr{A}'} = [A_n \circ [A_k * B_j]_{\mathscr{A}'}]_{\mathscr{A}'}$ (modular condition). Since n is fixed and $\#\{[A_k * B_j]_{\mathscr{A}'} \mid k \in N\} < +\infty$ holds, the above inequality is not satisfied for a sufficiently large number k. Thus, there exists some $k \in N$ $(k > n)$ such that $A_k \notin \text{Obj}\left(\bigcup_{j \in N} \mathscr{A}([A_n \circ B_j]_{\mathscr{A}'})\right)$. That is, we have $\text{Obj}\left(\bigcup_{j \in N} \mathscr{A}([A_n \circ B_j]_{\mathscr{A}'})\right) \subset \text{Obj}\left(\bigcup_{j \in N} \mathscr{A}([A_k \circ B_j]_{\mathscr{A}'})\right)$. Therefore, we have $\#\left\{\bigcup_{j \in N} \mathscr{A}([A_n \circ B_j]_{\mathscr{A}'}) \mid n \in N\right\} = +\infty$. This contradicts $\# \widetilde{\mathscr{L}}(\mathscr{A}') < +\infty$. Consequently, we have $\# \text{Atom}(\widetilde{\mathscr{L}}(\mathscr{A}')) = 1$.

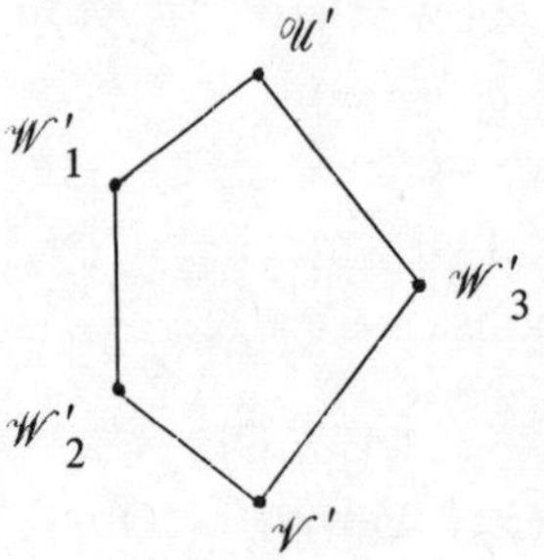

Figure 4.1

Now, we show that $\widetilde{\mathscr{L}}(\mathscr{A}')$ is a modular lattice. Suppose that $\widetilde{\mathscr{L}}(\mathscr{A}')$ is not a modular lattice. Then, there is a partial Hasse diagram of $\widetilde{\mathscr{L}}(\mathscr{A}')$ as Fig. 4.1. Here, by $\# \text{Atom}(\widetilde{\mathscr{L}}(\mathscr{A}')) = 1$, we have $\widetilde{\mathscr{U}}' = \mathscr{U}'$, $\widetilde{\mathscr{W}}'_i = \mathscr{W}'_i$ $(i = 1, 2, 3)$, $\widetilde{\mathscr{V}}' = \mathscr{V}'$ $(\# \text{Obj}\, \mathscr{V}' = +\infty)$ and $\mathscr{V}' = = \mathscr{W}'_1 \cap \mathscr{W}'_3 = \mathscr{W}'_2 \cap \mathscr{W}'_3$. Put $\mathscr{W}' = \bigcup_{A \in \text{Obj}\, \mathscr{W}'_2,\, B \in \text{Obj}\, \mathscr{W}'_3} \mathscr{A}([A \circ B]_{\mathscr{A}'})$.

First, we have $\text{Obj}\, \mathscr{U}' \supseteq \text{Obj}\, \mathscr{W}' \supseteq \text{Obj}\, \mathscr{W}'_2, \text{Obj}\, \mathscr{W}'_3$. Moreover, for all $C, C' \in \text{Obj}\, \mathscr{W}'$ there exist some $A, A' \in \text{Obj}\, \mathscr{W}'_2$ and $B, B' \in \in \text{Obj}\, \mathscr{W}'_3$ such that $C \leqslant [A \circ B]_{\mathscr{A}'}$ and $C' \leqslant [A' \circ B']_{\mathscr{A}'}$. Here, we put $A'' = [A \circ A']_{\mathscr{A}'}$ and $B'' = [B \circ B']_{\mathscr{A}'}$. Then, by Corollary 1.1, we have $A'' \in \text{Obj}\, \mathscr{W}'_2$ and $B'' \in \text{Obj}\, \mathscr{W}'_3$. Moreover, we have $C, C' \leqslant [A'' \circ B'']_{\mathscr{A}'}$. Consequently, we have $C, C' \in \in \text{Obj}\, \mathscr{A}([A'' \circ B'']_{\mathscr{A}'})$. By Theorem 1.2, this means that $\text{Obj}\, \mathscr{W}'$ is a lattice. And it is obvious that $\mathscr{W}'$ is a strong subcategory of $\mathscr{A}$.

Thus, $\mathscr{U}' = \bigcup_{A \in \text{Obj}\,\mathscr{W}'_2,\, B \in \text{Obj}\,\mathscr{W}'_3} \mathscr{A}([A \circ B]_{\mathscr{A}'})$ holds. On the other hand, since $\text{Obj}\,\mathscr{W}'_2 \subset \text{Obj}\,\mathscr{W}'_1$, there exist some $C' \in \text{Obj}\,\mathscr{W}'_1 - \text{Obj}\,\mathscr{W}'_2$, $A \in \text{Obj}\,\mathscr{W}'_2$ and $B \in \text{Obj}\,\mathscr{W}'_3$ such that $C' \in \text{Obj}\,\mathscr{A}([A \circ B]_{\mathscr{A}'})$. That is, we have $C' \leqslant [A \circ B]_{\mathscr{A}'}$. Next, we put $C = [C' \circ A]_{\mathscr{A}'}$. Then, we have $A \leqslant C$. Furthermore, since $A \in \text{Obj}\,\mathscr{W}'_2$ and $[C * B]_{\mathscr{A}'} \in$ $\in \text{Obj}\,(\mathscr{W}'_1 \cap \mathscr{W}'_3) = \text{Obj}\,(\mathscr{W}'_2 \cap \mathscr{W}'_3)$, we have $[A \circ [C * B]_{\mathscr{A}'}]_{\mathscr{A}'} \in$ $\in \text{Obj}\,\mathscr{W}'_2$. On the other hand, since $C \geqslant C'$ and $[A \circ B]_{\mathscr{A}'} \geqslant C'$, we have $[C * [A \circ B]_{\mathscr{A}'}]_{\mathscr{A}'} \geqslant C'$. Consequently, we have $[C * [A \circ B]_{\mathscr{A}'}]_{\mathscr{A}'} \notin \text{Obj}\,\mathscr{W}'_2$. For, if $[C * [A \circ B]_{\mathscr{A}'}]_{\mathscr{A}'} \in \text{Obj}\,\mathscr{W}'_2$ then $C' \in \text{Obj}\,\mathscr{W}'_2$ holds, and this is a contradiction. Thus, we have $A \leqslant C$ and $[C * [A \circ B]_{\mathscr{A}'}]_{\mathscr{A}'} \neq [A \circ [C * B]_{\mathscr{A}'}]_{\mathscr{A}'}$. This means that $\text{Obj}\,\mathscr{A}'$ is not a modular lattice. This contradicts the assumption of the theorem. Therefore, $\widetilde{\mathscr{L}}(\mathscr{A}')$ must be a modular lattice. Q.E.D.

5. CONCLUSION

To conclude this paper, we give an example from the algebraic automata theory. Let $V_\Sigma(\text{Au})$ be the set of all Σ-automata (i.e., their input sets are Σ). Then, $V_\Sigma(\text{Au})$ can be considered as the set of objects of a small category $\mathscr{A}$. Namely, $\text{Obj}\,\mathscr{A} = V_\Sigma(\text{Au})$ and $\mathscr{A}(A,B) =$ $= \text{Hom}\,(A,B)$ $(A, B \in V_\Sigma(\text{Au}))$, where $\text{Hom}\,(A,B)$ means the set of all homomorphisms of A onto B. In this case, it is obvious that $\text{Obj}\,\mathscr{A}$ is a countable set and $\#\{B \in \text{Obj}\,\mathscr{A} \mid \mathscr{A}(A,B) \neq \phi\} < +\infty$ for all $A \in \text{Obj}\,\mathscr{A}$. There are many strong subcategories of $\mathscr{A}$ whose sets of all objects form lattices, for instance, $\mathscr{A}'$ such that $\text{Obj}\,\mathscr{A}' = V_\Sigma(\text{Pe})$, i.e., the set of all perfect Σ-automata, and $\mathscr{A}''$ such that $\text{Obj}\,\mathscr{A}'' =$ $= V_\Sigma(\text{Scf})$, i.e., the set of all strongly cofinal Σ-automata. There are also some subcategories of $\mathscr{A}$ whose sets of all objects form modular lattices, for instance, $\mathscr{A}'$ and $\mathscr{A}'''$ such that $\text{Obj}\,\mathscr{A}''' = V_\Sigma(\text{Qp})$, i.e., the set of all quasiperfect Σ-automata.

Besides classes of automata, there may be many algebraic systems to which our theory can be applied.

REFERENCES

[1] G. Birkhoff, *Lattice Theory,* Amer. Math. Soc., Providence, R.I., 1967.

[2] M. Ito, Some classes of automata as partially ordered sets, *Mathematical Systems Theory,* 15 (1982), 357–370.

[3] M. Ito, Some classes of automata as partially ordered sets (Abstract), *Papers on Automata Theory III,* Karl Marx University of Economics, Budapest, 1981, DM 81-2, 41–52.

[4] M. Ito – G. Tanaka, Cartesian composition of classes of automata, *Papers on Automata Theory III,* Karl Marx University of Economics, Budapest, 1981, DM 81-2, 53–75.

M. Ito

Faculty of Science, Kyoto Sangyo University, 603 Kyoto, Japan.

COLLOQUIA MATHEMATICA SOCIETATIS JÁNOS BOLYAI
39. SEMIGROUPS, SZEGED (HUNGARY), 1981.

WEAKLY FREE SEMIGROUPS IN IDENTITY INCLUSIVE VARIETIES

E.S. LJAPIN

The well-known notion of free semigroups within a given class is extremely important in the theory of semigroup varieties. However, if a class is not a variety then it may contain no free semigroups at all. This is the case e.g. in identity inclusive varieties [4], which are a natural generalization of semigroup varieties. In the present paper we generalize the notion of free semigroups within a class in a fairly natural way. We call these generalized free semigroups weakly free and prove that they exist in a broad class of identity inclusive varieties. We shall also see that they have one of the basic properties which make free semigroups so important for the theory of varieties: namely, we shall show that every semigroup in the concerned identity inclusive varieties is a homomorphic image of a weakly free semigroup.

As was kindly pointed out by the referee, the main result of this paper (essentially together with its proof) can be extended to more general classes of semigroups than identity iclusive varieties. He also noticed that the basic notions we investigate here can be defined for arbitrary algebras, and many of our results can be obtained at that level of generality by argumentations more or less near to the ones in the present paper.

1. RELATIONS IN SEMIGROUPS

1.1. Let X be an arbitrary finite set. Any finite sequence of elements of X written in the form $w = x_1 x_2 \dots x_n$ $(x_i \in X)$ will be called a word of length n over X $(n \geqslant 1)$. One-letter words will not be distinguished from the corresponding elements of X. The set of all words over X will be denoted by $\mathfrak{W}(X)$. In $\mathfrak{W}(X)$ we can take the operation of concatenation, which assigns to the words $u = x_{i_1} \dots x_{i_p}$ and $v = x_{j_1} \dots x_{j_q}$ the word $x_{i_1} \dots x_{i_p} x_{j_1} \dots x_{j_q} \in \mathfrak{W}(X)$. This operation is associative, so it makes $\mathfrak{W}(X)$ a semigroup which contains X as a generating set.

If $X \subset S$ for a semigroup S then each word $w = x_1 x_2 \dots x_n \in \in \mathfrak{W}(X)$ has a value which is an element of S:

$$\mathrm{val}_S w = x_1 \cdot x_2 \cdot \ldots \cdot x_n \in S.$$

We extend every mapping $\varphi: X \to X'$ in the canonical way to a mapping of the subsets of X into the subsets of X' by putting $\varphi \{x_i\}_{i \in I} = \{\varphi x_i\}_{i \in I}$. We do the same with the words: for $w = x_1 x_2 \dots x_n \in \mathfrak{W}(X)$ we put $\varphi w = (\varphi x_1)(\varphi x_2) \dots (\varphi x_n) \in \mathfrak{W}(X')$.

This extension will be tacitly performed for mappings φ, and the extended mapping will be denoted by the same letter.

For a mapping $\varphi: X \to \mathfrak{W}(Y)$ into a set of words we shall also consider its extension $\tilde{\varphi}: \mathfrak{W}(X) \to \mathfrak{W}(Y)$ defined for $w = x_1 x_2 \dots x_n \in \mathfrak{W}(X)$ and $\varphi x_i = y_{i1} \dots y_{is_i}$ $(i = 1, 2, \dots, n)$ by $\tilde{\varphi} w = \tilde{\varphi}(x_1 x_2 \dots x_n) = y_{11} \dots y_{1s_1} y_{21} \dots y_{2s_2} \dots y_{n1} \dots y_{ns_n} \in \mathfrak{W}(Y)$.

1.2. By a relation over X we mean any pair of words $\alpha = (u, v) \in \in \mathfrak{W}(X) \times \mathfrak{W}(X)$. If $X \subset S$ for a semigroup S, then a relation $\alpha = (u, v)$ over X is said to be valid in S if $\mathrm{val}_S u = \mathrm{val}_S v$.

For a mapping $\varphi: X \to X'$ and a relation $\alpha = (u, v)$ over X we shall use the notation $\varphi\alpha = (\varphi u, \varphi v)$ for the "image relation" over X'.

1.3. Let α be a relation and Φ be a set of relations over X. The relation α is said to be a consequence of Φ if for every mapping $\eta: X \to S$ into any semigroup, if all the relations $\eta\beta$ $(\beta \in \Phi)$ are valid in

S then so is $\eta\alpha$. It is well known (see e.g. [3]) how α can be expressed in this case by relations belonging to Φ.

The family of all those relations over X which are consequences of Φ, will be denoted by $\bar{\Phi}$. It is easy to see that $\bar{\Phi}$ is an equivalence in $\mathfrak{W}(X)$. This equivalence is stable from both sides under concatenation, hence it is a congruence in $\mathfrak{W}(X)$ if the latter is considered as a semigroup with respect to concatenation.

The validity of the following assertions can be immediately checked:

(1) $\bar{\Phi} \supset \Phi$,

(2) $\bar{\bar{\Phi}} = \bar{\Phi}$,

(3) $\Phi_1 \subset \Phi_2$ implies $\overline{\Phi_1} \subset \overline{\Phi_2}$,

(4) if Φ is empty then $\bar{\Phi}$ is the equality relation (the diagonal) in $\mathfrak{W}(X)$.

The families of relations Φ_1 and Φ_2 over X are said to be equivalent if $\overline{\Phi_1} = \overline{\Phi_2}$.

1.4. As is well known (see e.g. [3]), for any X and $\Phi \subset \mathfrak{W}(X) \times \mathfrak{W}(X)$ we construct the semigroup $\mathfrak{W}(X)/\Phi$. The elements of this semigroup are the equivalence classes of words over X under $\bar{\Phi}$. Multiplication is concatenation of representing elements from the classes. Clearly, $\mathfrak{W}(X)/\Phi = \mathfrak{W}(X)/\bar{\Phi}$, and here $=$ means concrete equality of the two semigroups, not only isomorphism.

Obviously, the set of those classes which contain one-letter words (i.e. elements of X) form a generating set of the semigroup $\mathfrak{W}(X)/\Phi$.

However, notice that different elements of X may be equivalent under $\bar{\Phi}$. If this is not the case then it is convenient (and we shall do so) to identify the class containing $x \in X$ with x itself, and thus to consider X as a subset of the semigroup $\mathfrak{W}(X)/\Phi$. Of course, X is then a generating set in this semigroup. In this case we shall say that X is embedded into $\mathfrak{W}(X)/\Phi$.

1.5. Let $\Phi \subset \Phi'$, where $\Phi, \Phi' \subset \mathfrak{W}(X) \times \mathfrak{W}(X)$, and consider the mapping η: $\mathfrak{W}(X)/\Phi \to \mathfrak{W}(X)/\Phi'$ which assigns to a class of words K which is an element of $\mathfrak{W}(X)/\Phi$, the class ηK which is the element of $\mathfrak{W}(X)/\Phi'$ containing K. Clearly, η is a homomorphism.

If X is embedded into both of these semigroups then the homomorphism η is an extension of the identical mapping of X onto itself.

1.6. For a subset X in a semigroup S, denote by $\Omega^S(X)$ the collection of all those relations over X which are valid in S.

Let $\alpha \in \overline{\Omega^S(X)}$. Since all the relation from $\Omega^S(X)$ are valid in S, the same must hold for α, too, i.e. $\alpha \in \Omega^S(X)$. Hence $\overline{\Omega^S(X)} = \Omega^S(X)$. In what follows we shall tacitly use this equality.

1.7. Let $S = \mathfrak{W}(X)/\Phi$ and suppose that X is embedded into S. We are going to show that in this case $\Omega^S(X) = \bar{\Phi}$.

Let $(x_1 \ldots x_p, y_1 \ldots y_q) \in \Omega^S(X)$ $(x_i, x_j \in X)$, i.e. $\mathrm{val}_S x_1 x_2 \ldots \ldots x_p = \mathrm{val}_S y_1 y_2 \ldots y_q$. This means that $x_1 \cdot \ldots \cdot x_p = y_1 \cdot \ldots \cdot y_q$ holds in S, where $x_1 \cdot \ldots \cdot x_p$ stands for the class of words from $\mathfrak{W}(X)$ which are equivalent to the word $x_1 \ldots x_p$. This equality in S says that the two words are equivalent modulo $\bar{\Phi}$, whence $\Omega^S(X) \subset \bar{\Phi}$.

Conversely, if $(x_1 \ldots x_p, y_1 \ldots y_q) \in \bar{\Phi}$ then the class of words equivalent to $x_1 \ldots x_p$ under $\bar{\Phi}$ is the same as the class of words equivalent to $y_1 \ldots y_q$. Therefore $x_1 \cdot \ldots \cdot x_p = y_1 \cdot \ldots \cdot y_q$ in S, i.e. $(x_1 \ldots x_p, y_1 \ldots y_q) \in \Omega^S(X)$, whence $\bar{\Phi} \subset \Omega^S(X)$.

1.8. Let X be a generating set of the semigroup S. Assigning to every element $x \in X$ the class of words equivalent to x under $\Omega^S(X)$, we obtain a mapping φ: $S \to \mathfrak{W}(X)/\Omega^S(X)$ which is easily seen to be an isomorphism. This isomorphism extends the identical mapping of X (considered as a subset of S) onto itself (considered in turn as a subset of $\mathfrak{W}(X)/\Omega^S(X)$).

Let Φ be a family of relations equivalent to $\Omega^S(X)$. Then S is isomorphic with $\mathfrak{W}(X)/\Phi = \mathfrak{W}(X)/\Omega^S(X)$. In this case Φ is said to be a defining family of relations for S over X.

By 1.7, if X is embedded into a semigroup $\mathfrak{W}(X)/\Phi$ then Φ is a defining family for this semigroup. Therefore we shall often say that the semigroup $\mathfrak{W}(X)/\Phi$ is defined by a defining family Φ of relations over X.

1.9. Definition. Let Γ be a class of semigroup. A family Φ of relations over the set X is called a Γ-family if $\mathfrak{W}(X)/\Phi \in \Gamma$.

1.10. Let the class Γ be closed under isomorphisms and X be a generating set for a semigroup S. Since S is isomorphic to $\mathfrak{W}(X)/\Omega^S(X)$ (see 1.8), the family of relations $\Omega^S(X)$ is a Γ-family if and only if $S \in \Gamma$.

1.11. A class Γ of semigroups is said to be closed under inclusions if $S \in \Gamma$ implies $S' \in \Gamma$ for all subsemigroups S' of S.

Γ is homomorphically closed if $S \in \Gamma$ implies $S' \in \Gamma$ whenever there is a surjective homomorphism φ: $S \to S'$.

1.12. Let Γ be a homomorphically closed class of semigroups and Φ, Φ' be two families over X such that $\Phi \subset \Phi'$. By 1.5 and 1.10, if Φ is a Γ-family then Φ' is also.

2. WEAKLY FREE SEMIGROUPS

2.1. **Definition.** Let Γ be a class of semigroups which is closed under isomorphisms. A semigroup $F \in \Gamma$ is said to be weakly free of rank m in Γ, if F has a generating set B of cardinality m and a defining family Φ of relations over B (which is clearly a Γ-family for $\mathfrak{W}(B)/\Phi$ is isomorphic to F by 1.8, 1.9, 1.10) such that every Γ-family Φ' which is a subset of Φ is equivalent with Φ (i.e., by 1.3, $\bar{\Phi}' = \bar{\Phi}$).

2.2. Notice that the equality $\bar{\Phi}' = \bar{\Phi}$ occurring in the above definition can be replaced by $\mathfrak{W}(B)/\Phi' = \mathfrak{W}(B)/\Phi$ (where the latter sign of equality means that the two semigroups are in fact the same, not only isomorphic).

Taking into consideration that in a semigroup $\mathfrak{W}(C)/\Psi$ not only the relations from Ψ but all those from $\bar{\Psi}$ are valid, the requirements on Φ in Definition 2.1 can be formulated in such a way that the validity of the relations from Φ should involve a condition of minimality (in the obvious sense) for a semigroup to belong to the class Γ (of course, we mean relative but not universal minimality).

2.3. As shown by the following theorem, the notion of weak freeness generalizes the well-known concept of freeness. The latter can be approached both by means of homomorphisms and by relations over generating sets (see e.g. [3], Ch. IX, §4). The naturality of our generalization is best seen if we make use of relations.

2.4. Theorem. *Let* Γ *be a class of semigroups which is closed under isomorphisms and suppose that* H *is a free semigroup of rank* m *in* Γ *(i.e.* H *has a generating set* C *of cardinality* m *which is free with respect to* Γ*). Then* H *is a weakly free semigroup of rank* m *in* Γ.

Proof. By 1.8 there is an isomorphism of H onto $H_0 = \mathfrak{W}(C)/\Omega^H(C)$ which extends the identical mapping of C (for C is embedded into H_0). Since C is free in H with respect to Γ, it will be so in H_0, too.

By 1.6 and 1.7 we have $\Omega^{H_0}(C) = \overline{\Omega^H(C)} = \Omega^H(C)$. We shall show that the generating set C of H_0 and the family of relations $\Omega^H(C) = \Omega^{H_0}(C)$ satisfy the conditions in Definition 2.1. This will imply that H_0, and then also H, is a weakly free semigroup of rank m in Γ.

Let Φ' be a Γ-family such that $\Phi' \subset \Omega^H(C) = \Omega^{H_0}(C)$. Since C is embedded into H_0, it is then embedded into $H' = \mathfrak{W}(C)/\Phi'$, too. Since H_0 is a free semigroup in Γ, the identical mapping of C onto itself can be extended to a homomorphism of H_0 onto H'. Hence by 1.7, we have $\Omega^{H_0}(C) \subset \Omega^{H'}(C) = \overline{\Phi'} \subset \overline{\Omega^{H_0}(C)} = \Omega^{H_0}(C)$.

Thus $\overline{\Omega^{H_0}(C)} = \overline{\Phi'}$, hence $\Omega^{H_0}(C)$ and Φ' are equivalent, and this means that C and $\Omega^{H_0}(C)$ satisfy in H_0 the conditions in Definition 2.1.

2.5. As is well known, free semigroups of any rank exist in every semigroup variety. It is a very important property of them that every semigroup is a homomorphic image of some free one. However, in other classes free semigroups may not exist. Then, having the same property in mind, considering weakly free semigroups can prove very useful, especially if every semigroup of the given class turns out to be a homomorphic image of a weakly free semigroup. This is the case e.g. if our class is homomorphically closed.

2.6. Example. For a fixed natural number n, denote by Γ_n the class of those semigroups which have at most n elements. Every semigroup $F \in \Gamma_n$ with exactly n elements is weakly free in Γ_n. In fact, consider its generating set consisting of all elements. If $\Phi' \subset \Omega^F(F)$ and $\overline{\Phi'} \neq$ $\neq \overline{\Omega^F(F)} = \Omega^F(F)$ then the semigroup $F' = \mathfrak{W}(F)/\overline{\Phi'}$ has a homomorphism onto the semigroup $(F)/\Omega^F(F)$, which is isomorphic to F by 1.5. Since $\overline{\Phi'} \neq \overline{\Omega^F(F)}$, this homomorphism is not injective. Hence F' has more elements than F has, so $F' \notin \Gamma_n$ and Φ' is not a Γ_n-family.

It is easily seen, for instance, by using the construction of inflation ([1], 3.2) that every semigroup in Γ_n is a homomorphic image of a semigroup with n elements.

Our example shows also that non-isomorphic weakly free semigroups of the same rank can exist.

Notice also that, for $n > 1$, Γ_n has no free semigroups. In fact, if $F \in \Gamma_n$ were such with free generating set B, then by considering a homomorphism of F into the cyclic group of order n which takes an element $b \in B$ into a generator of the group, we would obtain that b is a group element of order n. But then $B = F$, and the cyclic group B has no homomorphism into a non-group monogenic semigroup B' with n elements which extends the mapping carrying b into the generator of B'.

3. EXISTENCE OF WEAKLY FREE SEMIGROUPS IN IDENTITY INCLUSIVE VARIETIES

3.1. Let X be a countable (infinite) alphabet. Following [4], a pair $\sigma = (u, V)$ where $u \in \mathfrak{W}(X)$ and $V \subset \mathfrak{W}(X)$, $V \neq \phi$, is called a semigroup identity inclusion. For such a pair we shall use the notation $u \dot{\in} V$. If V is a finite set of words then σ is said to be of finite length.

We shall say that σ holds in S if for every mapping $\varphi \colon X \to S$, the value of φu in S equals the value of one of the φv, $v \in V$.

In case V consists of one word, we obtain the usual notion of semigroup identity.

Let Φ be a family of semigroup identity inclusions. The class of those semigroups in which all $\sigma \in \Phi$ are valid, will be denoted by $\Pi\Phi$ and called the identity inclusive variety defined by the family Φ of identity inclusions. Varieties in the usual sense are, of course, special cases of this notion.

As is immediately seen, every identity inclusive variety is homomorphically closed and closed under inclusion. Some other properties of them are investigated in [4].

3.2. In order to shed light on the role of the notion of identity inclusive varieties, we present several examples of important classes of semigroups which are such.

$\Pi(xy \dot{\in} \{x, y\})$ is the class of semigroups in which every subset is a subsemigroup.

$\Pi(xyz \dot{\in} \{x, y, z\})$ is the class of semigroups in which every subset is ternary closed.

$\Pi(xyz \dot{\in} \{xy, yz, xz\})$ is the class of exclusive semigroups.

$\Pi(xy \dot{\in} \{y, y^2, y^3, \ldots\})$ is the class of semigroups in which every subsemigroup is a left ideal.

$\Pi(xy \dot{\in} \{x, y, y^2, \ldots\},\ \ xy \dot{\in} \{y, x, x^2, \ldots\})$ is the class of semigroups in which every subsemigroup is unit ideal.

$\Pi(xy \dot{\in} \{x, x^2, x^3, \ldots, y, y^2, y^3, \ldots\})$ is the class of semigroups in which the union of any two subsemigroups is a subsemigroup.

The structure and properties of the semigroups belonging to these classes have been studied in various papers (see e.g. [2], [5], [6], [7], [8], [9]).

3.3. Lemma. *In order that the identity inclusion* $u \dot{\in} V$ *be valid in a semigroup* S, *it is necessary and sufficient that for every mapping* $\psi\colon X \to \mathfrak{W}(X)$ *there be a word* $v \in V$ *such that* $\mathrm{val}_S \tilde{\psi} u = \mathrm{val}_S \tilde{\psi} v$, *where* $\mathrm{val}_S \tilde{\psi} w$ *denotes the value of* $\tilde{\psi} w$ *in* S *for any word* $w \in \mathfrak{W}(X)$.

Proof.

(1) Let $u \dot{\in} V$ be valid in S, and consider the mapping $\varphi\colon X \to S$ defined by $\varphi x = \mathrm{val}_S \psi x$ $(x \in X)$. Since $u \dot{\in} V$ is satisfied in S, there is a $v \in V$ such that $\mathrm{val}_S \varphi u = \mathrm{val}_S \varphi v$.

For $u = x_1 x_2 \ldots x_n$ $(x_i \in S)$ we have now:

$$\mathrm{val}_S \varphi u =$$

$$= \mathrm{val}_S[(\varphi x_1) \ldots (\varphi x_n)] = \mathrm{val}_S[(\mathrm{val}_S \psi x_1) \ldots (\mathrm{val}_S \psi x_n)] =$$

$$= (\mathrm{val}_S \psi x_1) \cdot \ldots \cdot (\mathrm{val}_S \psi x_n) = \mathrm{val}_S \tilde{\psi}(x_1 \ldots x_n) = \mathrm{val}_S \tilde{\psi} u.$$

Similarly we have $\mathrm{val}_S \varphi v = \mathrm{val}_S \tilde{\psi} v$, hence $\mathrm{val}_S \tilde{\psi} u = \mathrm{val}_S \tilde{\psi} v$.

(2) Suppose that the above condition is satisfied for every mapping $\psi\colon X \to \mathfrak{W}(S)$. Every $\varphi\colon X \to S$ is such a mapping, hence for φ there is a $v \in V$ such that $\mathrm{val}_S \tilde{\varphi} u = \mathrm{val}_S \tilde{\varphi} v$. But in our case $\tilde{\varphi} w = \varphi w$ for all $w \in \mathfrak{W}(S)$, and therefore $\mathrm{val}_S \varphi u = \mathrm{val}_S \varphi v$, i.e. $u \dot{\in} V$ holds in S.

3.4. Theorem. *Let* Γ *be an identity inclusive variety defined by semigroup identity inclusions of finite lengths. Then for every semigroup* $S \in \Gamma$ *which has a generating set of cardinality* m, *there is a weakly free semigroup* F *of rank* m *in* Γ *such that* S *is a homomorphic image of* F.

Proof. Let $\Gamma = \Pi\{\sigma_i\}_{i \in I}$, $\sigma_i = (u_i \dot{\in} \{v_{i1}, v_{i2}, \ldots, v_{im_i}\})$ $(i \in I)$, where the u_i and the v_{ij} are words over an alphabet X. Let $S \in \Gamma$ have a generating set A of cardinality m.

Consider the pairs (i, ψ) for $i \in I$ and $\psi: X \to \mathfrak{W}(A)$. Well-order their set by assigning to (i, ψ) the ordinal number $\tau(i, \psi)$, and do this so that there be a greatest one, say τ^*, among these ordinal numbers. We shall work with the $\widetilde{\psi}$ obtained from the ψ as described in 1.1.

For any ordinal number $\gamma = \tau(i, \psi) \leqslant \tau^*$, by γ-rows we mean those sequences of m_i elements in which the k-th element $(1 \leqslant k \leqslant m_i)$ is either the zero symbol 0 or the relation $(\widetilde{\psi}u_i, \widetilde{\psi}v_{ik})$ over A. For every $\gamma \leqslant \tau^*$ we define an order in the finite set of all γ-rows by putting $P \leqslant Q$ for two γ-rows P, Q if for each k, either the k-th element of Q is different from zero or the k-th elements of both P and Q are zero. If $P \leqslant Q$ then we say that P precedes Q. If, in addition, $P \neq Q$, then we say that P strictly precedes Q and write $P < Q$.

Since the set of all γ-rows (for fixed γ) is finite, each of its subfamilies has a minimal element with respect to the order defined above. This property will often be made use of in what follows.

By a scheme we mean a transfinite sequence of γ-rows (γ runs from 1 to τ^*). We define an order in the set of all schemes by putting $\Sigma \leqslant \Sigma'$ for two schemes Σ and Σ' if for every $\mu \leqslant \tau^*$, the μ-row in Σ precedes the μ-row in Σ'. We denote by Σ^* the scheme in which every element is different from 0; this scheme is the greatest in the given order in the set of all schemes. The scheme all of whose elements are zero, is the smallest scheme.

Given a non-empty family $\{\Sigma^{(s)}\}$ of schemes, it must have its infimum $\Sigma' = \operatorname{Inf}_s \Sigma^{(s)}$. Let $\mu = \tau(i, \psi)$. The k-th element in the μ-row of Σ' is the relation $(\widetilde{\psi}u_i, \widetilde{\psi}v_{ik})$ if the same relation is the k-th element in the μ-row of each $\Sigma^{(s)}$, and 0 otherwise.

We shall denote by $\Psi(\Sigma)$ the set of those relations over A which occur (as elements) in the scheme Σ.

In what follows it will often be important for us to know whether the following properties hold in the scheme Σ we consider.

(1) All rows of Σ are non-zero (i.e., every row has at least one element different from 0).

(2) All relations over A which occur in $\Psi(\Sigma)$, are valid in S.

(3) If a relation α is the k-th element in the γ-row of Σ^* and α is a consequence of $\Psi(\Sigma)$, then α is the k-th element of the γ-row of Σ.

Denote by Σ_0 the scheme in which the k-th element of the γ-row, where $\gamma = \tau(i, \psi)$, is the relation $(\tilde{\psi}u_i, \tilde{\psi}v_{ik})$ if this relation holds in S, and 0 otherwise. From this definition it follows that Σ_0 has property (2).

Since σ_i is valid in S, 3.3 implies that for every $\gamma = \tau(i, \psi) \leqslant \tau^*$, at least one of the equalities

$$\mathrm{val}_S\, \tilde{\psi}u_i = \mathrm{val}_S\, \tilde{\psi}v_{ir} \qquad (r = 1, 2, \ldots, m_i)$$

must hold in S. This means that the γ-row of Σ_0 has a non-zero element, hence Σ_0 possesses property (1).

Let $\alpha \in \overline{\Psi(\Sigma_0)}$ be the k-th element in the γ-row of Σ^*. Since all the relations from $\Psi(\Sigma_0)$ are valid in S and α is a consequence of them, also α must hold in S. But then α is the k-th element of the γ-row of Σ_0, whence Σ_0 has property (3).

Clearly, every scheme which has all of the properties (1), (2), (3) must precede Σ_0, so Σ_0 is their supremum.

We are going to construct, by transfinite induction, scheme Σ_γ $(1 \leqslant \gamma \leqslant \tau^*)$ which have (1), (2), (3), and also the following three properties.

(4.γ) If $\delta < \gamma$ then $\Sigma_\delta \geqslant \Sigma_\gamma$.

(5.γ) If $\delta < \gamma$ then the δ-row of Σ_γ coincides with the δ-row of Σ_δ.

(6.γ) If a scheme Σ' has the properties (1), (2), (3), (4.γ), (5.γ), the last two with Σ' in place of Σ_γ, and the γ-row P of Σ' precedes the γ-row Q of Σ_γ, then $P = Q$.

$\gamma = 1$. Consider the set of all schemes which have the properties (1), (2), (3) (such a scheme is e.g. Σ_0). As Σ_1 fix a scheme from this set in which the 1-row is minimal (we have seen above that there is such a scheme). It is clear that (4.1), (5.1), (6.1) hold in Σ_1.

Let γ be a non-limit ordinal such that $1 < \gamma \leq \tau^*$. Consider the set of schemes having properties (1), (2), (3), (4.γ), (5.γ). This set is not empty, in fact $\Sigma_{\gamma-1}$ is easily seen to belong to it. As Σ_γ we fix a scheme from this set in which the γ-row is minimal. By its choice, Σ_γ has the properties (1), (2), (3), (4.γ), (5.γ), (6.γ).

Let γ be a limit ordinal such that $1 < \gamma < \tau^*$. Consider the set of those schemes which have properties (1), (2), (3), (4.γ), (5.γ). Next we show that the scheme $\Sigma = \operatorname*{Inf}_{\lambda<\gamma} \Sigma_\lambda$ belongs to this set.

For any ordinal number $\mu \leq \tau^*$, consider the μ-rows of the schemes Σ_λ $(\lambda < \gamma)$. They are all non-zero and there are finitely many of them, hence there is a minimal one among them, say, the one belonging to the scheme Σ_{λ_0}. Since Σ_{λ_0} has property (4.γ), $\Sigma_{\lambda_0} \leq \Sigma_\lambda$ holds for all $\lambda \leq \lambda_0$. Therefore the μ-row of Σ_{λ_0} precedes the μ-rows of all Σ_λ $(\lambda \leq \lambda_0)$. For $\lambda \geq \lambda_0$ $(\lambda < \gamma)$ we have $\Sigma_\lambda \leq \Sigma_{\lambda_0}$, hence the μ-rows of these Σ_λ precede that of Σ_{λ_0}. In view of the minimality of the μ-row of Σ_{λ_0}, herefrom we infer that the μ-row of every Σ_λ $(\lambda \geq \lambda_0)$ coicnides with that of Σ_{λ_0}. By the definition of the scheme $\operatorname*{Inf}_{\lambda<\gamma} \Sigma_\lambda$, all this implies that the μ-row of Σ is the same as that of Σ_{λ_0}. Therefore it is non-zero, whence Σ has property (1).

Since all the relations which occur in $\Psi(\Sigma)$ belong to $\Psi(\Sigma_1)$, they are all valid in S, hence Σ has property (2).

Let the relation $\alpha \in \overline{\Psi(\Sigma)}$ be the k-th element of the μ-row of Σ^* $(\mu \leq \tau^*)$. Since $\Sigma_\lambda \geq \Sigma$ for $\lambda < \gamma$, the relation α is a consequence of $\Psi(\Sigma_\lambda)$, and as Σ_λ has property (3), α is the k-th element of the μ-row of Σ_λ. By the construction of Σ, the k-th element of its μ-row is then also α, which means that Σ has property (3).

Let $\delta < \gamma$. Since $\Sigma \leqslant \Sigma_\delta$, Σ has property $(4.\gamma)$. If $\delta \leqslant \lambda < \gamma$, then by the validity of $(5.\lambda)$ for Σ_λ we infer that the δ-row of Σ_λ coincides with the δ-row of Σ_δ. If $\lambda < \delta$ then the δ-row of Σ_δ precedes that of Σ_λ, for Σ_δ has property $(4.\delta)$. This implies that the δ-row of Σ must be the same as that of Σ_δ, whence Σ has property $(5.\gamma)$.

Thus we have shown that the set of schemes which have properties (1), (2), (3), $(4.\gamma)$, $(5.\gamma)$ is non-empty, in fact, it contains Σ. The γ-rows of the schemes belonging to this set are non-zero, and there is a minimal one among them. For Σ_γ we choose a scheme from this set with a minimal γ-row, and this scheme has all the desired properties.

As a result of the above construction we obtain the system $\Psi(\Sigma_{\tau*})$ of relations over A. Consider the semigroup $F = \mathfrak{W}(A)/\Psi(\Sigma_{\tau*})$. Its elements are equivalence classes of words over A modulo $\overline{\Psi(\Sigma_{\tau*})}$. Let $a_1 \neq a_2$ $(a_1, a_2 \in A)$. The relation (a_1, a_2) is not contained in $\Omega^S(A)$. Since $\Sigma_{\tau*}$ has property (2), we have $\Psi(\Sigma_{\tau*}) \subset \Omega^S(A)$. Then, by 1.3 and 1.6, $\overline{\Psi(\Sigma_{\tau*})} \subset \overline{\Omega^S(A)} = \Omega^S(A)$, whence $(a_1, a_2) \notin \overline{\Psi(\Sigma_{\tau*})}$, which means that the two elements of F which are the classes of words containing a_1 and a_2, respectively, are different. Therefore we can identify the element of F which is the class containing the one-letter word a, with the element $a \in A$ itself. So we shall assume $A \subset F$, and for every word w over A, its value in F is the class of words over A equivalent to w under $\overline{\Psi(\Sigma_{\tau*})}$. Thus we see, in particular, that the set A, which is of cardinality m, is a generating set for F.

Since $\Psi(\Sigma_{\tau*})$ has property (2), we have $\Psi(\Sigma_{\tau*}) \subset \Omega^S(A)$, which implies by 1.5 the existence of a homomorphism of the semigroup $F = \mathfrak{W}(A)/\Psi(\Sigma_{\tau*})$ onto $\mathfrak{W}(A)/\Omega^S(A)$ which extends the identical mapping of A (considered as a subset of F) onto itself (considered in turn as a subset of $\mathfrak{W}(A)/\Omega^S(A)$). There is also an isomorphism of $\mathfrak{W}(A)/\Omega^S(A)$ onto S which extends the identical mapping of A onto itself.

Thus we have a homomorphism of F onto S which extends the identical mapping of A onto itself.

Next we show that $F \in \Gamma$.

Let $i \in I$ and $\varphi \colon X \to F$. For every $x \in X$ we fix a word $\psi x \in$ $\in \mathfrak{W}(A)$ in the class which is the element φx of F. Thus we obtain a mapping $\psi \colon X \to \mathfrak{W}(A)$ and also a mapping $\widetilde{\psi} \colon \mathfrak{W}(X) \to \mathfrak{W}(A)$.

Further we define the mapping $\xi \colon \mathfrak{W}(A) \to F$ which assigns to each word from $\mathfrak{W}(A)$ the class of those words which are equivalent to it under $\overline{\Psi(\Sigma_{\tau^*})}$. If consider $\mathfrak{W}(A)$ and F as semigroups (see Section 1) then ξ becomes a homomorphism, and taking A as a subset of F we obtain for any $w = a_1 \dots a_n \in \mathfrak{W}(A)$:

$$\xi w = \xi(a_1 a_2 \dots a_n) = (\xi a_1) \cdot (\xi a_2) \cdot \dots \cdot (\xi a_n) =$$
$$= a_1 \cdot a_2 \cdot \dots \cdot a_n = \mathrm{val}_F w.$$

By the definition of ψ we have $\xi \psi x = \varphi x$ $(x \in X)$. Recalling the definition of $\widetilde{\psi}$ from 1.1 and using the homomorphic property of ξ, we have for every word $z = x_1 x_2 \dots x_m \in \mathfrak{W}(X)$:

$$\xi \widetilde{\psi} z = \xi \widetilde{\psi}(x_1 x_2 \dots x_m) = (\xi \psi x_1) \cdot (\xi \psi x_2) \cdot \dots \cdot (\xi \psi x_m) =$$
$$= (\varphi x_1) \cdot (\varphi x_2) \cdot \dots \cdot (\varphi x_m) =$$
$$= \mathrm{val}_F [(\varphi x_1)(\varphi x_2) \dots (\varphi x_m)] = \mathrm{val}_F \varphi z.$$

Consider the γ-row in the scheme Σ_{τ^*}, where $\gamma = \tau(i, \psi)$. Since Σ_{τ^*} has property (2), there is a non-zero element in this γ-row, say $(\widetilde{\psi} u_i, \widetilde{\psi} v_{it})$. By 1.7, $(\widetilde{\psi} u_i, \widetilde{\psi} v_{it}) \in \Psi(\Sigma_{\tau^*})$ means that $\xi \widetilde{\psi} u_i = \xi \widetilde{\psi} v_{it}$ holds, whence $\mathrm{val}_F \varphi u_i = \mathrm{val}_F \varphi v_{it}$.

The validity of this equality shows that the identity inclusion σ_i holds in F, hence $F \in \Gamma$ and $\Psi(\Sigma_{\tau^*})$ is a Γ-family.

The proof of the theorem will be complete if we show that the semigroup F is weakly free in Γ. Consider its generating set A and its defining family of relations $\Psi(\Sigma_{\tau^*})$, we are going to exhibit that they satisfy the condition in Definition 2.1.

Let $\Phi' \subset \Psi(\Sigma_{\tau^*})$ and $H = \mathfrak{W}(A)/\Phi' \in \Gamma$.

Since A is a subset of F, for any $a_1 \neq a_2$ $(a_1, a_2 \in A)$ the one-letter words a_1 and a_2 are not equivalent under $\overline{\Sigma_{\tau*}}$ hence also not under $\overline{\Phi'}$. Therefore A is embedded in H (i.e. it is a subset of H) and it is a generating set there.

Put $\Phi'' = \Psi(\Sigma^*) \cap \overline{\Phi'}$ and construct a new scheme Σ'' as follows. Let $i \in I$, $\psi\colon X \to \mathfrak{W}(A)$, $\gamma = \tau(i, \psi)$, $\alpha = (\widetilde{\psi} u_i, \widetilde{\psi} v_{ik}) \in \Psi(\Sigma^*)$. The k-th element of the γ-row of Σ'' be α if the latter is a consequence of Φ' (i.e. if $\alpha \in \Phi''$) and 0 otherwise. Clearly, $\Psi(\Sigma'') = \Phi''$.

For $\gamma = \tau(i, \psi)$, consider the γ-row of Σ''. Since Φ' is a Γ-family, σ_i holds in H. By 3.3 this implies $\mathrm{val}_H \widetilde{\psi} u_i = \mathrm{val}_H \widetilde{\psi} v_{it}$ for some $t \leqslant m_i$, i.e. $(\widetilde{\psi} u_i, \widetilde{\psi} v_{it}) \in \Omega^H(A) = \overline{\Phi'}$. In view of $(\widetilde{\psi} u_i, \widetilde{\psi} v_{it}) \in \Psi(\Sigma^*)$ this means that $(\widetilde{\psi} u_i, \widetilde{\psi} v_{it}) \in \Phi''$, hence $(\widetilde{\psi} u_i, \widetilde{\psi} v_{it})$ is the t-th element of the γ-row of Σ'', so Σ'' has property (1).

Since $\Phi'' \subset \overline{\Phi'} \subset \overline{\Psi(\Sigma_{\tau*})}$ and $\Psi(\Sigma_{\tau*}) \subset \Omega^H(A) = \overline{\Omega^H(A)}$ (for $\Psi(\Sigma_{\tau*})$ has property (2)), we obtain $\Phi'' \subset \Omega^H(A)$, whence Σ'' has property (2).

Let the relation α be the k-th element in the γ-row of Σ^* and suppose that α is a consequence of $\Psi(\Sigma'') = \Phi''$. Then $\alpha \in \overline{\overline{\Phi'}} = \overline{\Phi'}$, hence $\alpha \in \Phi''$. According to the construction of Σ'', the relation α is therefore the k-th element of the γ-row of Σ'', which means that Σ'' has property (3).

Suppose that the relation β is the k-th element of the γ-row of Σ''. Then $\beta \in \Phi''$ and so β is a consequence of Φ'. Since $\Phi' \subset \Psi(\Sigma_{\tau*})$, we have $\beta \in \overline{\Psi(\Sigma_{\tau*})}$. As $\Sigma_{\tau*}$ has property (3), the relation β must be the k-th element of the γ-row of $\Sigma_{\tau*}$. Thus $\Sigma'' \leqslant \Sigma_{\tau*}$.

Suppose now $\Sigma'' < \Sigma_{\tau*}$.

Denote by μ the smallest ordinal number for which the μ-row of Σ'' strictly precedes the μ-row of $\Sigma_{\tau*}$. Since $\Sigma_{\tau*}$ has property (5.τ*), the μ-row of $\Sigma_{\tau*}$ coincides with the μ-row of Σ_μ. So we see that the μ-row of Σ'' strictly precedes the μ-row of Σ_μ.

Let $\delta < \mu$. Since $\Sigma_\delta \geqslant \Sigma_{\tau^*} > \Sigma''$, the scheme Σ'' has property (4.μ).

As Σ_{τ^*} has property (5.τ^*), the δ-row of Σ_{τ^*} must coincide with the δ-row of Σ_δ and also with the δ-row of Σ_μ, for Σ_μ has property (5.μ). By the choice of μ, the δ-row of Σ'' is the same as the δ-row of Σ_{τ^*}, hence also the same as the δ-row of Σ_μ. This shows that Σ'' has property (5.μ).

Altogether, the scheme Σ'' has properties (1), (2), (3), (4.μ), (5.μ), but its μ-row strictly precedes the μ-row of Σ_μ, contrary to Σ_μ having property (6.μ).

This contradiction implies that $\Sigma'' = \Sigma_{\tau^*}$. Since $\overline{\Psi(\Sigma_{\tau^*})} = \overline{\Psi(\Sigma'')} = \overline{\Phi''} \subset \overline{\Phi'}$ and $\Phi' \subset \Psi(\Sigma_{\tau^*})$, it holds $\overline{\Phi'} = \Psi(\Sigma_{\tau^*})$. This means that A and $\Psi(\Sigma_{\tau^*})$ in F satisfy the requirements in Definition 2.1, whence the semigroup $F = \mathfrak{W}(A)/\Psi(\Sigma_{\tau^*})$ is weakly free in Γ.

REFERENCES

[1] A.H. Clifford – G.B. Preston, *The Algebraic Theory of Semigroups,* I, Amer. Math. Soc., Providence, R.I., 1961.

[2] N. Kimura – T. Tamura – R. Merkel, Semigroups in which all subsemigroups are left ideals, *Canad. J. Math.,* 17 (1965), 52–62.

[3] E.S. Ljapin, *Semigroups,* Translations of Mathematical Monographs, 3, Amer. Math. Soc., Providence, R.I., 1974.

[4] E.S. Ljapin, Atoms of the lattice of identity inclusive varieties of semigroups, *Sibirsk. Mat. Ž.,* 16 (1975), 1224–1230 (in Russian).

[5] E.S. Ljapin, Identity inclusions in semigroups in which every subset is a subsemigroup, *Modern Algebra,* Leningr. Ped. Inst., 1978, 118–133 (in Russian).

[6] E.S. Ljapin – A.E. Evseev, Semigroups in which every subsemigroup is unit ideal, *Izv. Vysš. Učebn. Zaved. Matematika,* 10 (1970), 44–48 (in Russian).

[7] L. O'Caroll – B.M. Schein, On exclusive semigroups, *Semigroup Forum,* 3 (1972), 338–348.

[8] L.N. Ševrin, On the general theory of semigroups, *Mat. Sb.*, 53 (1961), 367–386 (in Russian).

[9] E.G. Šutov, Semigroups with ideal subsemigroups, *Mat. Sb.*, 57 (1962), 179–186 (in Russian).

E.S. Ljapin

196 066 Leningrad, Moskovskiĭ prospekt 208, kv. 16, USSR.

COLLOQUIA MATHEMATICA SOCIETATIS JÁNOS BOLYAI
39. SEMIGROUPS, SZEGED (HUNGARY), 1981.

THE REES CONSTRUCTION IN REGULAR SEMIGROUPS

J. MEAKIN

1. INTRODUCTION

During the past decade there has been something of an explosion in the structure theory of regular semigroups. Much of the recent work in this area has occurred as an outgrowth of the rich structure theory for inverse semigroups which has been developed during the past thirty years. Since the introduction of inverse semigroups into the literature in the papers of Vagner [143] and Preston [120] in the early 1950's, there has been a large number of papers which involve the general theme of constructing inverse semigroups from groups and semilattices. Clifford's paper [14] on the structure of unions of groups with commuting idempotents is an early and important example of a paper involving this theme. Reilly's paper [124] describing the structure of bisimple semigroups whose idempotents form an ω-chain was the forerunner of numerous papers dealing with inverse semigroups whose idempotents satisfy restricted conditions: some of these papers are discussed in the survey article of Lallement [53].

In [128], Schein showed how to construct an arbitrary inverse semigroup from its trace and natural partial order: since the structure of the trace of an inverse semigroup is obtained from the Rees–Suschkewitsch theorem, Schein's results provide a method for constructing inverse semigroups from groups and semilattices. His results were later rediscovered and formulated in terms of "structure mappings" by Meakin [76]. A different approach developed as a result of the important work of Munn [89], [90] who introduced the Munn semigroup T_E of a semilattice E and the notion of a "fundamental" inverse semigroup: since every inverse semigroup is an $\mathscr{H}$-coextension of a fundamental inverse semigroup, Munn's theory, together with the standard theories of $\mathscr{H}$-coextensions of inverse semigroups (see, for example, Coudron [20], Leech [61], D'Alarcao [21], Lausch [60], Loganathan [63], Meakin [77], Širjaev [134], Sribala [135], etc.) provide some techniques for constructing inverse semigroups from groups and semilattices. A somewhat related approach is taken by Grillet [30] who provides structure data for constructing an inverse semigroup from its semilattice of idempotents and a family of groups (one for each $\mathscr{D}$-class). A different approach, dual to that of Munn, is taken by McAlister [66], [67] who introduces the notion of a "proper" inverse semigroup (or "E-unitary" inverse semigroup), obtains the structure of E-unitary inverse semigroups in terms of groups and semilattices, and shows that every inverse semigroup is an idempotent-separating homomorphic image of an E-unitary inverse semigroup. An excellent account of the structure theory of inverse semigroups (and of several other topics in inverse semigroup theory) may be found in the forthcoming book of Petrich [119].

Some other important papers which prepared the ground for the recent surge of activity in regular semigroup theory were of course the Rees–Suschkewitsch papers (Rees [123], Suschkewitsch [139]), the Miller–Clifford theory of regular $\mathscr{D}$-classes [74], the Schützenberger–Preston representation theorem [133], [121], Lallement's thesis and survey article [54], [53], Clifford's theorem on unions of groups [14], Petrich's theorem on the structure of bands [116] and the structure theory for orthodox semigroups (see, for example, Fantham [28], Yamada [146], [147], Hall [35], [36], [37] and Szendrei [141]).

There have been at least three successful theories of fundamental regular semigroups which have been published during the past decade. The first is due to Hall [38], who constructs a fundamental Munn semigroup $T_{\langle E \rangle}$ corresponding to each regular idempotent-generated semigroup $\langle E \rangle$; the second is due to Grillet [31], [32], [33], [34] who constructs fundamental regular semigroups via his theory of "cross connections"; the third is due to Nambooripad [91], [93] who constructs a Munn semigroup T_E corresponding to each "biordered set" E (see also the papers of Clifford [15], [16] for related information). The relationship between the Grillet theory of cross connections and the Nambooripad theory of biordered sets is explored in Nambooripad's paper [94]; Nambooripad's paper [93] also contains a construction of idempotent-generated semigroups and develops a theory of "inductive groupoids", inspired to some extent by Schein's paper [128]. In another paper [95], Nambooripad introduces a natural partial order on a regular semigroup and shows how to reduce all products in a regular semigroup to trace products by means of the natural partial order: along these lines, the "structure mapping" approach to regular semigroups has been studied by Nambooripad [92] and Meakin [78], [79]. Recent work has been devoted to the problem of extending the McAlister "P-theory" from inverse semigroups to some class of regular semigroups. The paper of Pastijn [110] and the thesis of Veeramony [144] develop P-theories for the class of pseudo-inverse semigroups, which was introduced by Nambooripad [97] and has been the subject of much recent study.

Pastijn's P-theory of pseudo-inverse semigroups and several other recent papers make use of the Rees construction, which has become an increasingly important tool in the structure theory of regular semigroups. In this paper I shall survey a few areas of regular semigroup theory where the Rees construction has been recently exploited and where further useful exploitation seems plausible.

2. THE REES CONSTRUCTION

Let S be a semigroup (with or without 0), let I and Λ be sets and $P = (p_{\lambda i})$ a $\Lambda \times I$ matrix over S; P is said to be regular if each row and each column of P contains at least one non-zero entry. On the set $M = I \times S \times \Lambda$ define a multiplication by

$$(1) \qquad (i, s, \lambda)(j, t, \mu) = (i, sp_{\lambda j}t, \mu).$$

It is easy to check that this multiplication is associative: the resulting semigroup is called a *Rees matrix semigroup over* S *with sandwich matrix* P and is denoted by $\mathscr{M}(S; I, \Lambda; P)$. If S has a zero, the set $Z = \{(i, 0, \lambda): i \in I, \lambda \in \Lambda\}$ is clearly an ideal of M; the Rees quotient M/Z (identify all elements of Z with 0) is denoted by $\mathscr{M}^\circ(S; I, \Lambda; P)$. If $S = G^\circ$, a group with zero, the resulting semigroup is usually denoted by $\mathscr{M}^\circ(G; I, \Lambda; P)$ rather than $\mathscr{M}^\circ(G^\circ; I, \Lambda; P)$. Of course the Rees construction first appeared in the literature in the famous Rees–Suschkewitsch theorem which characterizes completely simple [completely 0-simple] semigroups as Rees matrix semigroups over a group [group with zero] with regular sandwich matrix P. An account of this theorem may be found in any of the standard books on semigroup theory (for example, Clifford and Preston [19], Howie [44], Lallement [55] or Ljapin [62]). I shall not attempt to survey the vast literature on completely 0-simple semigroups here. My concern is with other classes of regular semigroups in which the Rees construction is used.

The Rees construction provides us with a potentially powerful technique for constructing new classes of semigroups from ”known”, ”simpler” classes. One can place various restrictions on the sandwich matrices P or on the semigroups S and ask for a classification of the resulting class of semigroups.

Some work along these lines was initiated by Steinfeld [136] who studied Rees matrix semigroups of the form $\mathscr{M}^\circ(S; I, \Lambda; P)$ where S is a semigroup with 0 and 1 and P is *locally regular*; that is, every row of P contains a right unit of S, every column of P contains a left unit of S and P contains at least one entry which is a unit of S. Actually, Steinfeld called such Rees matrix semigroups ”locally regular”; in order to avoid possible confusion with terminology which is used later in this paper, I shall call the sandwich matrix P ”locally regular” and the resulting semigroup $\mathscr{M}^\circ(S; I, \Lambda; P)$ a ”Rees matrix semigroup with locally regular sandwich matrix". Steinfeld [136] calls two left ideals L_1 and L_2 of a semigroup S *left similar* if there is a bijection φ of L_1 onto L_2 such that $(sx)\varphi = s(x\varphi)$ for all $s \in S$ and $x \in L_1$; right similarity of right ideals is defined dually. One of the main theorems of Steinfeld [136] is the following interesting generalization of the Rees–Suschkewitsch theorem.

Theorem 2.1 ([136], Theorem 4.1). *A semigroup S with zero admits a decomposition of the form*

$$S = \bigcup_{\lambda \in \Lambda} Se_\lambda = \bigcup_{i \in I} e_i S \quad (e_\lambda^2 = e_\lambda,\ e_i^2 = e_i,\ I \cap \Lambda \neq \phi)$$

where the Se_λ $(\lambda \in \Lambda)$ $[e_i S\ (i \in I)]$ are left similar [right similar] 0-disjoint left [right] ideals of S if and only if S is isomorphic to a Rees matrix semigroup of the form $\mathcal{M}^\circ(T; I, \Lambda; P)$ where T is a semigroup with 0 and 1 and P is locally regular.

In a later paper Márki [65] studied regularity and simplicity properties of such semigroups. Some of his results were extended by Tran Quy Tien [142] who showed that if S has 0 and 1, then $\mathcal{M}^\circ(S; I, \Lambda; P)$ is regular (0-bisimple, completely 0-simple) if and only if S has this property and every row [column] of P has a right [left] unit of S. Steinfeld's Theorem 2.1 has been extended in three different directions in the papers of Hoehnke [42], Hotzel [43] and Szabó [140]. See also Steinfeld's survey article [137] for related information. Some analogous results, making use of the Rees construction in rings, are contained in the paper of Ánh and Márki [3].

In [58] Lallement and Petrich used the Rees construction to study certain matrix and 0-matrix decompositions of semigroups. They were concerned with Rees matrix semigroups of the form $M = \mathcal{M}^\circ(S^\circ; I, \Lambda; P)$ (or $\mathcal{M}(S; I, \Lambda; P)$) in which S is a semigroup with 1, S° means S with zero adjoined even if S has a zero, and P is a regular $\Lambda \times I$ matrix with entries in G°, where G is the group of units of S. Clearly, any such sandwich matrix is locally regular in the sense of Steinfeld [136] and hence M is regular if and only if S is regular. Lallement and Petrich called a semigroup S a "Rees 0-composition" if S has a zero and S admits a congruence ρ such that $\{0\}$ is a ρ-class, S/ρ is a rectangular 0-band $\mathcal{M}^\circ(\{1\}; I, \Lambda; P)$ (with P regular) and the ρ-classes $A_{i\lambda}$ $(i \in I,\ \lambda \in \Lambda)$ satisfy the condition: (A) for each $i, j \in I$ and $\lambda, \mu \in \Lambda$, there is an element $x_{i\lambda} \in A_{i\lambda}$ such that $x_{i\lambda} A_{j\mu} = A_{i\mu}$ or 0 and $A_{j\mu} x_{i\lambda} = A_{j\lambda}$ or 0. One of the main results of Lallement and Petrich [58] is the following.

Theorem 2.2 ([58], Theorem 3.4). *A semigroup S is a Rees 0-composition if and only if S is isomorphic to a Rees matrix semigroup $\mathcal{M}^\circ(D^\circ; I, \Lambda; P)$ where D is a monoid with group of units G and P is a regular $\Lambda \times I$ matrix over G°.*

The above definitions and theorem have an obvious modification in the case without zero. In this case it follows easily from the results of Lallement and Petrich that each ρ-class $A_{i\lambda}$ is a submonoid of S and that $A_{i\lambda} A_{j\mu} = A_{i\mu}$ for each $i, j \in I$ and $\lambda, \mu \in \Lambda$. These results can be sharpened somewhat in case all of the submonoids $A_{i\lambda}$ are inverse monoids. In [109] Pastijn calls a semigroup S a *rectangular band of inverse semigroups* $S_{i\lambda}$ $(i \in I,\ \lambda \in \Lambda)$ if S is the disjoint union of the inverse subsemigroups $S_{i\lambda}$ and if $S_{i\lambda} S_{j\mu} \subseteq S_{i\mu}$ for all $i, j \in I$ and $\lambda, \mu \in \Lambda$: he calls S an *elementary* rectangular band of the $S_{i\lambda}$ if $S_{i\lambda} S_{j\mu} = S_{i\mu}$ for all $i, j \in I$ and $\lambda, \mu \in \Lambda$. He obtains the following version of the Lallement–Petrich result in this case.

Theorem 2.3 ([109], Theorem 4.1). *A semigroup S is an elementary rectangular band of inverse monoids if and only if S is isomorphic to a Rees matrix semigroup $\mathcal{M}(D; I, \Lambda; P)$ where D is an inverse monoid with group of units G and P is a $\Lambda \times I$ matrix over G.*

Pastijn goes on to determine a Rees-type construction of elementary rectangular bands of E-unitary inverse semigroups in [109] and uses these results to obtain a version of McAlister's "P-theorem" for pseudo-inverse semigroups in a later paper [110]; I shall take another look at his results in Section 4 of this paper. His paper [109] provides a detailed examination of rectangular bands of inverse semigroups; other general studies of bands of semigroups and monoids may be found in the papers of Clifford [17], Petrich [114] and Schein [130]; see also the bibliography to Pastijn's paper [109].

It is easy to see (and was observed by Venkatesan [145]) that the regularity of $M = \mathscr{M}^\circ(S; I, \Lambda; P)$ implies the regularity of S; the converse is not true in general. McAlister [70] has observed that the element $(i, s, \lambda) \in \mathscr{M}^\circ(S; I, \Lambda; P)$ is regular if and only if $V(s) \cap p_{\lambda j} S p_{\mu i} \neq \phi$ for some $j \in I$, $\mu \in \Lambda$ (here $V(s)$ denotes the set of inverses of s); he has also shown ([70], Lemma 2.1) that the set of regular elements of $\mathscr{M}^\circ(S; I, \Lambda; P)$ is a regular subsemigroup of $\mathscr{M}^\circ(S; I, \Lambda; P)$. This set of regular elements of $\mathscr{M}^\circ(S; I, \Lambda; P)$, for S a regular semigroup, will be denoted by $\mathscr{R}\mathscr{M}^\circ(S; I, \Lambda; P)$ in later sections of this paper and is called a regular Rees matrix semigroup over S. From the results of Márki [65] or Tien [142] one sees that $\mathscr{R}\mathscr{M}^\circ(S; I, \Lambda; P) = \mathscr{M}^\circ(S; I, \Lambda; P)$ if S is regular and P is locally regular. McAlister [70] provides an example to show that $\mathscr{M}(S; I, \Lambda; P)$ is not necessarily regular even if S is a semilattice and P is regular.

3. PRINCIPAL SUBMONOIDS AND DIVISION THEOREMS

If $e = e^2$ is an idempotent of the semigroup S, then eSe is clearly a submonoid of S with identity e. In this paper I shall refer to the submonoids eSe for $e = e^2 \in S$ as the *principal submonoids* of S. McAlister [70] and [71] refers to principal submonoids as *local submonoids:* his terminology is somewhat more closely related to terminology used in finite semigroup theory and language theory. The principal submonoids of S coincide with the quasi-ideals with identity (see Steinfeld's book [138] for an extensive discussion of the theory of quasi-ideals). The group of units of the principal submonoid eSe is of course H_e, the maximal subgroup ($\mathscr{H}$-class) with identity e; it is clear that the set of idempotents of eSe is $\omega(e) = \{f \in S\colon f = f^2 \text{ and } f \leqslant e\}$, the principal ideal generated by e in the partially ordered set of idempotents of S. (The partial order is the usual partial order: $e \leqslant f$ if and only if $e = ef = fe$.) One may remark that if e and f are $\mathscr{D}$-related idempotents of S, then $H_e \cong H_f$, $eSe \cong fSf$ and $\omega(e) \cong \omega(f)$ (as partially ordered sets). A proof of this for S finite is given in Proposition 1.4 of Tilson's Chapter XI in Eilenberg's book [26]; the same proof works in general, using properties of the regular $\mathscr{D}$-class $D_e = D_f$. Steinfeld [136] observed that if the left-ideals Se_1, Se_2, (or right-ideals e_1S, e_2S)

($e_1^2 = e_1 \neq 0$, $e_2^2 = e_2 \neq 0$) of a semigroup S with zero are left similar [right similar] then the principal submonoids e_1Se_1 and e_2Se_2 are isomorphic.

An important strategy in both finite semigroup theory and regular semigroup theory has been to study semigroups S with the property that every principal submonoid of S belongs to a certain class $\mathscr{C}$ of monoids: in this case one may say that S is *principally in* $\mathscr{C}$. For information about this concept in finite semigroup theory and for additional references, see Eilenberg's discussion of local varieties in Chapter V of his book [26]; see also Lallement [56] and Margolis [64] for additional results and references relating this concept to the theory of prefix codes and the theory of complexity of finite semigroups. From the point of view of regular semigroup theory there have been several important recent developments along these lines. Notice first the rather obvious facts that rectangular bands are principally trivial, completely simple semigroups are principally groups and normal bands (Yamada [146]) are principally semilattices. A generalization of the notion of a normal band was introduced into the literature by Zalcstein [150], who defined a regular semigroup to be *locally testable* if it is periodic and all of its principal submonoids are semilattices: such semigroups arise naturally in connection with his study of locally testable languages [151]. Locally testable regular semigroups have been studied extensively by Nambooripad [97], [98] (who showed that the requirement of periodicity is redundant), Pastijn [111] and Rajan [122]. Locally testable semigroups and languages are discussed in Chapter 7 of Lallement's book [55]. In [97], Nambooripad calls a regular semigroup *pseudo-inverse* if all of its principal submonoids are inverse; McAlister [70] uses the terminology *locally inverse* to describe the same class of semigroups. There is a growing literature concerned with the structure of pseudo-inverse semigroups and I shall provide some more information about this in Section 4 of this paper. In [86] Meakin and Nambooripad defined a regular semigroup S to be *locally orthodox* if, for all $e = e^2 \in S$, the subsets $\omega^r(e) = \{f \in S\colon f = f^2$ and $f = ef\}$ and $\omega^l(e) = \{f \in S\colon f = f^2$ and $f = fe\}$ are subbands of S. Hall [39] showed that S is locally orthodox if and only if S is regular and $\omega(e)$ is a band for each $e = e^2 \in S$; it follows easily that

S is locally orthodox if and only if S is regular and all of its principal submonoids are orthodox. The structure of locally orthodox semigroups has been studied by Meakin and Nambooripad [86] and Veeramony [144].

An important technique which has developed in the past decade or so is an application of the Rees construction to build certain classes of regular semigroups from their principal submonoids. This technique first appeared in Steinfeld's paper [136] (in his proof of Theorem 2.1) and resurfaced in a basic paper of Allen [1] who showed how to construct "max-principal" regular semigroups as particularly nice homomorphic images of max-principal regular semigroups whose distinct maximal principal left (or right) ideals are disjoint. Allen's paper has stimulated a lot of important work in finite semigroup theory and in particular has led to a formulation of the prime decomposition theorem of Krohn and Rhodes [52] in terms of the Rees construction. In order to describe this formulation, due to Rhodes and Allen [126], it is convenient to introduce some notation. If G is a group, identify G with its image under the right regular representation and let $\bar{G}$ be the semigroup obtained by adjoining the constant maps on G. If G is a group and S is a semigroup, let $S + G$ denote any ideal extension of S by G such that G acts as a group of units. If $G_1, G_2, \ldots, G_n$ are groups, define semigroups S_i $(i = 1, \ldots, n)$ by $S_1 = \bar{G}_1$ and for $i \geqslant 1$,

$$S_{i+1} = \mathscr{M}(S_i; I_i, \Lambda_i; P_i) + G_{i+1}$$

where I_i and Λ_i are sets and P_i a $\Lambda_i \times I_i$ matrix over S_i; write $S_n = \mathscr{M}(\{G_i\}; \{I_i\}, \{\Lambda_i\}; \{P_i\})$. One may show that $\mathscr{M}(S; I, \Lambda; P)$ divides the wreath product of S and the rectangular band on $I \times \Lambda$; also, if G is a group, $S + G$ divides the wreath product of S^1 and G. (One says that the semigroup S *divides* the semigroup T if S is a homomorphic image of a subsemigroup of T.) The Rhodes–Allen theorem linking the prime decomposition theorem with the Rees construction may now be stated as follows.

Theorem 3.1 ([126], Theorem 5.2). *If S is a finite semigroup then there exist groups $G_1, \ldots, G_n$ dividing S, an iterated Rees matrix semi-*

group $R = \mathcal{M}(\{G_i\}; \{I_i\}, \{\Lambda_i\}; \{P_i\})$, *a subsemigroup* T *of* R *and an epimorphism* $\theta: T \to S$ *of* T *onto* S *such that* θ *enjoys the following properties:*

(2) θ *is one-one when restricted to* $\mathcal{H}$*-classes of* T,

(3) *if* s_1 *and* s_2 *are regular elements of* T *with* $s_1\theta = s_2\theta$, *then* $s_1 \mathcal{J} s_2$ *in* T.

Furthermore, if S *is regular then* T *may be chosen to be regular.*

For additional information concerning the application of the Rees construction to finite semigroups, see Tilson's Chapters XI and XII in Eilenberg's book [26]. From the point of view of regular semigroup theory, McAlister [70], [71] has recently reworked Allen's paper [1] and has put it into a somewhat more general setting. McAlister [70] calls a homomorphism θ from a regular semigroup S onto a regular semigroup T a *local isomorphism* if θ is an isomorphism when restricted to principal submonoids of S. It is routine to check from the results of Meakin and Nambooripad [84] that θ is a local isomorphism if and only if θ satisfies the following two conditions:

(4) each congruence class $e(\theta \circ \theta^{-1})$ for $e = e^2 \in S$ is a rectangular subband of S;

(5) if $e = e^2 \in S$, $e\theta = g \in T$ and $h = h^2 \in \omega(g)$, then there is a unique idempotent $f \in S$ such that $f\theta = h$ and $f \in \omega(e)$.

Meakin and Nambooripad [84] have called the semigroup S a *coextension of* T *by rectangular bands* if there is an epimorphism $\theta: S \to T$ from S onto T which satisfies (4); they have called S a *normal coextension of* T if θ satisfies (4) and (5). Thus the regular semigroup S is a normal coextension of the (regular) semigroup T if and only is T is a locally isomorphic image of S. Coextensions of regular semigroups by rectangular bands and normal coextensions of regular semigroups have been extensively studied by Meakin and Nambooripad [84], [85] and [86]. In particular it is shown in [86] that a regular semigroup S is locally orthodox if and only if it is a coextension of a pseudo-inverse semigroup by rectangular bands; a coextension of a pseudo-inverse semigroup by rectangular bands is again pseudo-inverse if and only if the coextension is normal. Note that every orthodox semigroup is a coextension of an inverse semigroup by rectangular bands (Schein [128], Hall [35], Meakin [75]). In terms of the notion of "local isomorphism" (or equivalently "normal coextension"), McAlister's version of Allen's theorem may now be stated as follows.

Theorem 3.2 ([70], Theorem 2.4 and [71], Theorem 2.2). *If* S *is a regular semigroup and* e *is an idempotent of* S *then* SeS *is a locally isomorphic image of a regular Rees matrix semigroup of the form* $\mathcal{RM}(eSe; I, \Lambda; P)$ *where* P *is a unital* $\Lambda \times I$ *matrix over* eSe. (P is called *unital* if at least one entry of P is e, the identity of eSe).

McAlister [70] calls this theorem the "local isomorphism theorem": it provides an interesting technique for constructing the principal ideals of a regular semigroup from the principal submonoids; in particular, a regular semigroup which has a maximum $\mathcal{J}$-class (in the usual partial order on $\mathcal{J}$-classes) may be constructed from its principal submonoids this way. McAlister [70] has used a version of the local isomorphism theorem to study certain classes of partially ordered regular semigroups. One expects pleasant structure theorems for regular semigroups whose principal submonoids belong to some restricted class. Recently McAlister and Hall (announced in [71]) and independently Margolis (private communication) have obtained the following result, a proof of which is closely related to proofs developed in Allen's paper [1].

Theorem 3.3. *Let* $\mathcal{C}$ *be a class of regular semigroups with zero with the following properties:*

(i) $\mathcal{C}$ *is closed under* 0-*direct unions;*

(ii) $S \in \mathcal{C}$ *if and only if* $IG(S)$, *the subsemigroup generated by the idempotents of* S, *is in* $\mathcal{C}$.

Then a regular semigroup S *with zero is principally in* $\mathcal{C}$ *if and only if it is a locally isomorphic image of a regular Rees matrix semigroup over a member of* $\mathcal{C}$.

It is not known at present whether the requirement that S should have a zero may be removed. Note that Theorem 3.3 applies in particular to the class of pseudo-inverse semigroups and the class of locally orthodox semigroups (but not to the class of locally testable semigroups). I shall discuss some related theorems for these classes of semigroups in the next section. The theorems discussed in this section are examples of what are usually called "division theorems" for regular semigroups. Loosely speaking, a good "division theorem" for a certain class of semigroups should tell us how to obtain semigroups in this class as "particularly pleasant" divisors of semigroups in a more restricted class whose structure is more clearly understood. McAlister's theorem characterizing every inverse semigroup as an idempotent separating image of an E-unitary inverse semigroup [66], [67] is a typical example of such a theorem. For examples of recent division theorems of this type in the theory of regular semigroups, see the papers of McAlister [68], [72], McAlister and Reilly [73], Veeramony [144], Pastijn [110], [112] and Chen and Hsieh [13].

4. PSEUDO-INVERSE SEMIGROUPS

Recall that a regular semigroup is called *pseudo-inverse* if all of its principal submonoids are inverse. Pseudo-inverse semigroups were introduced by Nambooripad [97]. They form a wide class of regular semigroups which includes, for example, all inverse semigroups, all subdirect products of completely simple and completely 0-simple semigroups (Lallement [54], Lallement and Petrich [59]), all generalized inverse semigroups and normal bands (Yamada [146]), all locally testable regular semigroups (Zalcstein [150]), all normal bands of groups (Petrich [115]) and all rectangular bands of inverse semigroups (Pastijn [109]). Pseudo-inverse semigroups may be characterized in several different ways, some of which are collected in the following theorem, due to Nambooripad [95] and [97].

Theorem 3.1. *The following conditions on a regular semigroup S are equivalent:*

(a) *S is pseudo-inverse;*

(b) *for each* $e = e^2 \in S$, $\omega(e)$ *is a semilattice;*

(c) S *does not contain any subsemigroup isomorphic to* U^1 *or* V^1, *where* $U[V]$ *denotes the* 2*-element right* [*left*] *zero semigroup;*

(d) *the natural partial order on* S *(defined for any regular semigroup* S *by*

$$a \leqslant b \quad \textit{iff} \quad R_a \leqslant R_b \quad \textit{and} \quad a = eb \quad \textit{for some} \quad e = e^2 \in R_a)$$

is compatible with the multiplication in S.

The natural partial order on a regular semigroup, defined in (d) above, was introduced by Nambooripad [95] who showed that the definition given here is equivalent to the left-right dual (using $\mathscr{L}$-classes instead of $\mathscr{R}$-classes); it extends the natural partial order on the idempotents of S and reduces to the usual natural partial order (see, for example, Clifford and Preston [19], Volume 2, Chapter 7) if S is inverse. It has been used by Nambooripad [95] to reduce all products in S to trace products and is of great importance in much recent work in regular semigroup theory. The natural partial order on a regular semigroup S is closely related to the structure mappings on S introduced by Nambooripad [92] and Meakin [78], [79]. If e and f are idempotents of the regular semigroup S, the mappings $\varphi_{e,f}$: $R_e \to R_f$, (for $R_e \geqslant R_f$) and $\psi_{e,f}$: $L_e \to L_f$ (for $L_e \geqslant L_f$) defined by:

$$a\varphi_{e,f} = fa \quad \text{for } a \in R_e \text{ and } R_e \geqslant R_f$$

and

$$a\psi_{e,f} = af \quad \text{for } a \in L_e \text{ and } L_e \geqslant L_f$$

are called the *structure mappings* of S. It is easy to see that $a \leqslant b$ if and only if $a = b\varphi_{e,f}$ (or $a = b\psi_{g,h}$) for some idempotents $e, f, g, h \in S$. McAlister [70] has shown that a regular semigroup S may be turned into a partially ordered semigroup in which the imposed partial order extends the natural partial order on idempotents if and only if S is pseudo-inverse.

A general construction of pseudo-inverse semigroups along the lines of Schein's construction [128] of inverse semigroups from their trace and natural partial order was provided by Nambooripad in [97]. Combinatorial pseudo-inverse semigroups (and indeed combinatorial regular semigroups in general) have been constructed by Rajan [122] (see also Nambooripad and Rajan [100]). Division theorems for pseudo-inverse semigroups in terms of subdirect products of fundamental pseudo-inverse semigroups and completely simple semigroups have been provided by Veeramony [144] (see also Nambooripad and Veeramony [103]). A pseudo-inverse semigroup S is called *E-unitary* if its structure mappings are all injective: this coincides with McAlister's definition of "proper" in the inverse case. Pastijn [110] has shown that every pseudo-inverse semigroup is a strictly compatible image of a suitable E-unitary pseudo-inverse semigroup (a homomorphism φ from a regular semigroup S onto a regular semigroup T is called *strictly compatible* if $e(\varphi \circ \varphi^{-1})$ is a completely simple subsemigroup of S for each $e = e^2 \in S$; equivalently (Nambooripad [95]), φ is strictly compatible if and only if $a \leqslant b$ and $a\varphi = b\varphi$ implies $a = b$). An alternative proof of Pastijn's result is provided by Veeramony [144], making heavy use of Nambooripad's concept [93] of an "inductive groupoid". Veeramony also shows that every E-unitary pseudo-inverse semigroup is a special kind of subdirect product of a fundamental pseudo-inverse semigroup and a completely simple semigroup. His results extend many of the results of Schein [129], McAlister [68], [69] and McAlister and Reilly [73] from the class of inverse semigroups to the classes of regular or pseudo-inverse semigroups. His results also relate to the construction of DeBodt and Pastijn [24].

It is routine to check that if S is an inverse semigroup then every regular Rees matrix semigroup $\mathcal{RM}(S; I, \Lambda; P)$ over S is pseudo-inverse. It follows that if P is a locally regular matrix (see Section 2) and S is an inverse monoid then $\mathcal{M}(S; I, \Lambda; P)$ is a pseudo-inverse semigroup. Theorems 3.2 and 3.3 apply to pseudo-inverse semigroups and enable us to represent pseudo-inverse semigroups with zero or pseudo-inverse semigroups with a maximum $\mathcal{J}$-class as locally isomorphic images of regular Rees matrix semigroups over an inverse monoid. In [110], Pastijn obtained a division theorem along these lines for all pseudo-inverse semigroups. Pastijn's proof makes use of McAlister's representation [66] of an E-unitary inverse semigroup as a "P-semigroup" $P(G; X, Y)$. He first constructs elementary rectangular bands of E-unitary inverse semigroups as Rees matrix semigroups over P-semigroups, then obtains a representation of an arbitrary pseudo-inverse semigroup as a strictly compatible image of a special kind of E-unitary pseudo-inverse semigroup and finally shows how to embed an E-unitary pseudo-inverse semigroup of this kind as a subsemigroup and order-ideal of a suitable elementary rectangular band of E-unitary inverse semigroups. His theorem may be stated as follows.

Theorem 4.2 ([110], Theorem 4.1). *Let X be a partially ordered set which contains the semilattice L as a subsemilattice and as an order-ideal. Let G be a group which acts on X (on the left) as a group of order automorphisms. Let I be an index set, let $\{A_i\colon i \in I\}$ be a family of elements of L, and let $\{p_{ij}\colon (i,j) \in I \times I\}$ be a family of elements of G which have actions on X that induce automorphisms on L. Let M be the set which consists of the elements $(A, g)_{ij}$, $A \in L, g \in G$, $(i,j) \in I \times I$, where $A \leqslant A_i$ and $p_{jj}^{-1} g^{-1} A \leqslant A_j$. Define a multiplication on M by*

$$(6) \qquad (A, g)_{ij}(B, h)_{mn} = (A \wedge gp_{jm}B, gp_{jm}h)_{in}.$$

Then M is a pseudo-inverse semigroup. Every strictly compatible image of M is pseudo-inverse. Conversely, every pseudo-inverse semigroup may be obtained as a strictly compatible image of a semigroup M constructed in this way.

Note that McAlister's theorem [66] describing E-unitary inverse semigroups as P-semigroups $P(G; X, Y)$ may be obtained from the above theorem: the multiplication (6) may be viewed as a Rees matrix multiplication over $P(G; X, L)$ if we assume that L has a 1. Veeramony [144] has obtained a quite different version of McAlister's "P-theorem" for pseudo-inverse semigroups, in terms of a completely simple inductive groupoid acting on both sides of a partially ordered set. A theorem describing locally testable regular semigroups as divisors of Rees matrix semigroup over semilattices is provided by Pastijn in [111]. I shall provide much more information about the idempotents of pseudo-inverse semigroups in Section 6 of this paper.

5. IDEMPOTENT-GENERATED SEMIGROUPS

Idempotent-generated semigroups, and in particular idempotent-generated regular semigroups, have received a lot of attention in the literature in recent years. There is a number of papers which are concerned with the general theme of characterizing the idempotent-generated part of certain classes of semigroups. For example, Howie [45] determines the idempotent-generated part of the full transformation semigroup on a set, Erdos [27] determines the idempotent-generated part of the semigroup of linear transformations on a finite-dimensional vector space and Schein [131] determines the idempotent-generated part of the semigroup of order-preserving transformations of a chain: additional references to papers along these lines may be found in Schein's paper [131].

In [29], Fitz-Gerald showed that the subsemigroup generated by the idempotents of a regular semigroup is itself regular: the result was also obtained by Eberhart, Williams and Kinch [23]. Hall [38] used Fitz-Gerald's ideas to show that any element in the idempotent-generated part of a regular semigroup S can be expressed as a product of $\mathscr{D}$-related idempotents of S and initiated a study of regular semigroups whose idempotent-generated part satisfies certain restricted conditions. Additional results along these lines may be found in Hall's supplement to Yamada's paper [148] (see also Yamada [149]) and in Hall [41], where the idempotent-generated parts of the principal submonoids of the semigroups considered play a basic role. The idea of Fitz-Gerald and Hall was refined by Nambooripad [93] who showed that every element in the idempotent-generated part of a regular semigroup S may be written as a product $x = e_1 e_2 \ldots e_n$ where the e_i are idempotents of S such that $e_i (\mathscr{L} \cup \mathscr{R}) e_{i+1}$ for $i = 1, \ldots, n-1$: Nambooripad [93] (see also [99]) called such a sequence of idempotents an *E-chain* and developed a structure theory for idempotent-generated regular semigroups by using the groupoid of E-chains of a biordered set. This idea is also essential to his notion of "inductive groupoid" and his major structure theory for regular semigroups in terms of inductive groupoids [93]; it is also used heavily by Pastijn [106] who is concerned with locating products of an arbitrary finite number of idempotents in an idempotent-generated semigroup. See also Nambooripad and Pastijn [101] for some related results. Idempotent-generated completely 0-simple semigroups were considered by Kim [51], Howie [47] and Pastijn [105]. Semigroups generated by two idempotents were classified by Benzaken and Mayr [4]. For additional references on idempotent-generated semigroups, see the bibliography in Pastijn's Aggregaatsthesis [108].

In his paper [45], Howie showed that every semigroup may be embedded in an idempotent-generated semigroup (in fact in the idempotent-generated part of a suitable full transformation semigroup). He refined his results somewhat in his paper [46] and in [48] and [49] studied various questions relating to the minimum number of idempotents needed to express elements of the full transformation semigroup as a product of idempotents. Similar results for the semigroup of singular linear transformations of a finite dimensional vector space were obtained by Dawlings in [22] and [23]. Howie's embedding theorem [45] was strengthened by Pastijn [104], who showed that every semigroup may in fact be embedded in a bisimple idempotent-generated semigroup, thus providing a negative answer to a question raised by Eberhart, Williams and Kinch [25]. An even more remarkable theorem was obtained by Howie [50] who constructed a bisimple idempotent-generated congruence-free semigroup S_c^*, corresponding to each infinite regular cardinal c, with the property that S_c^* contains an isomorphic copy of every semigroup having order less than c. Hall [40] had earlier shown (in a less constructive fashion) that any semigroup can be embedded in a bisimple idempotent-generated congruence-free semigroup.

Pastijn's results [104] stimulated considerable additional activity in the study of bisimple idempotent-generated semigroups. In [10], Byleen, Meakin and Pastijn showed that the set E of idempotents of a regular semigroup may occur as the set of idempotents of some bisimple idempotent-generated semigroup if and only if E is *connected* (that is, any two idempotents in E are connected by some E-chain); they also showed that if e and f are distinct idempotents with $e \leqslant f$, then any E-chain linking e and f must have length at least four. This led to a consideration of the "fundamental four-spiral semigroup" Sp_4, which can be described as the semigroup presented as follows:

$$Sp_4 = \langle a, b, c, d\colon\ a = ba,\ ab = b = bc,$$

$$cb = c = dc,\ cd = d = da \rangle.$$

The semigroup Sp_4 was introduced in [10] as a "minimal" example of a bisimple idempotent-generated semigroup which is not completely simple. Its (biordered) set of idempotents may be visualized as in Diagram 1: in this diagram, $\mathscr{R}$-related idempotents are in the same row, $\mathscr{L}$-related idempotents are in the same column and the order relation is indicated by arrows.

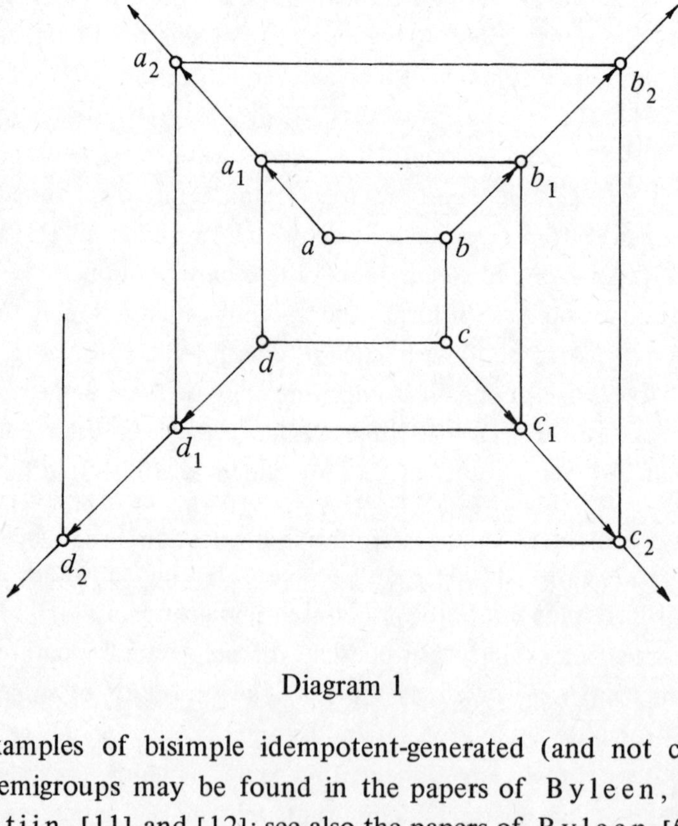

Diagram 1

Other examples of bisimple idempotent-generated (and not completely simple) semigroups may be found in the papers of Byleen, Meakin and Pastijn [11] and [12]; see also the papers of Byleen [5], [6] and Pastijn's Aggregaatsthesis [108] for additional related information along these lines. A semigroup closely related to Sp_4 has been constructed by

L a l l e m e n t in [57]. The fundamental four-spiral semigroup and all of its $\mathscr{H}$-coextensions (M c a k i n [80]) are obviously pseudo-inverse, since all principal ideals of the form $\omega(e)$ for $e = e^2 \in Sp_4$ are ω-chains. The double four-spiral semigroup by B y l e e n, M e a k i n and P a s t i j n [12] is locally orthodox but not pseudo-inverse.

In [7], B y l e e n pointed out that much of the work which has been done on idempotent-generated semigroups can be simplified by use of the Rees construction. He observed first that the four-spiral semigroup Sp_4 may be represented as a Rees matrix semigoup $\mathscr{M}\left(\mathscr{C}(p, q); 2, 2; \left(\begin{smallmatrix} 1 & q \\ 1 & 1 \end{smallmatrix}\right)\right)$ over the bicyclic semigroup $\mathscr{C}(p, q)$ and went on to construct $\mathscr{H}$-coextensions of Sp_4, previously constructed by M e a k i n [80], as Rees matrix semigroups over bisimple ω-semigroups (R e i l l y [124]). He then obtained quick proofs of several of the embedding theorems of H o w i e [45] and P a s t i j n [104], by use of the Rees construction. B y l e e n [7] showed that any semigroup S can be embedded in an idempotent-generated Rees matrix semigroup $\mathscr{M}(S^1; I, I; P)$ over S^1: I and P are chosen so that $P_{1i} = P_{i1} = P_{ii} = 1$ ($i \in I$) and the entries of P generate S. As corollaries to this theorem Byleen easily deduces the results of P a s t i j n [104] that any countable semigroup can be embedded in a semigroup generated by 3 idempotents and any semigroup can be embedded in a bisimple idempotent-generated semigroup. H a l l (verbal communication to B y l e e n) has shown how a slight modification in the proof of Byleen's theorem enables one to establish that every semigroup may be embedded in a semigroup in which every element is a product of at most two idempotents, thus answering in the affirmative a question raised by H o w i e (private communication); it should be obvious that not every semigroup can be embedded in a semigroup generated by two idempotents − in fact B e n z a k e n and M a y r [4] have shown that the only infinite semigroup generated by two idempotents is the free idempotent-generated semigroup on two generators. Other examples of embedding theorems of this type may be found in B y l e e n's paper [8].

In [9], B y l e e n refined the local isomorphism theorem of McAlister (Theorem 3.2) to obtain detailed information about the structure of finitely generated bisimple regular semigroups. He showed that if S is a finitely generated bisimple regular semigroup, then so is every principal submonoid of S and used this, together with the Rees construction, to prove the following theorem.

Theorem 5.1 ([9], Theorem 5). *If S is a finitely generated bisimple regular semigroup and $e = e^2 \in S$, then S is a locally isomorphic image of a finitely generated bisimple Rees matrix semigroup $\mathcal{M}(eSe; I, I; P)$ over the finitely generated bisimple monoid eSe with I finite and with P having diagonal entries all e.*

Byleen [9] was able to improve on this result if S is generated by a finite number of idempotents: in this case the Rees matrix semigroup $\mathcal{M}(eSe; I, I; P)$ in the statement of the above theorem is also generated by a finite number of idempotents and P may be chosen to have entries e on the diagonal and on the two adjacent diagonals. Byleen went on to use the Rees construction to examine certain embedding questions relating to the fundamental four-spiral semigroup Sp_4. He showed that every non-completely simple bisimple semigroup generated by a finite number of idempotents such that $\omega(e)$ is well-ordered for each $e = e^2 \in S$, must contain a copy of Sp_4. He also produced an example of a non-completely simple bisimple semigroup which is generated by 5 idempotents and which does not contain a copy of Sp_4.

6. BIORDERED SETS

Let E be the set of idempotents of a regular semigroup S. On E define relations ω^r and ω^l as follows:

$$e\,\omega^r f \text{ if and only if } e = fe,$$

$$e\,\omega^l f \text{ if and only if } e = ef.$$

Let $\omega = \omega^r \cap \omega^l$ (the natural partial order on E), $\kappa = \omega^r \cup \omega^l$, $\mathcal{R} = \omega^r \cap (\omega^r)^{-1}$ and $\mathcal{L} = \omega^l \cap (\omega^l)^{-1}$. For $\rho \in \{\omega^l, \omega^r, \omega\}$ and $e \in E$, let $\rho(e) = \{f \in E\colon f\,\rho\, e\}$. If $e\,\omega^r f$ it is easy to see that $ef \in E$, $ef\,\mathcal{R}\, e$ and $ef\,\omega\, f$. In particular, the products ef and fe are defined in E. Similarly, if $e\,\omega^l f$, the products ef and fe are defined in E. Thus the product ef of $e, f \in E$ must be defined in E whenever $e(\kappa \cup \kappa^{-1})f$.

Products of this form are referred to as *basic products.* Relative to these basic products, E forms a partial binary algebra with domain $D_E = \kappa \cup \kappa^{-1}$. In [93], Nambooripad characterized the partial binary algebra E (relative to these basic products) axiomatically as a (regular) *biordered set.* His axioms for biordered sets are provided in the following definition.

Definition 6.1. Let E be a partial binary algebra. On E define relations $\omega^r, \omega^l, \omega, \kappa, \mathcal{R}$ and $\mathcal{L}$ as above (where one interprets the product ef for $e, f \in E$ as the product in the partial binary algebra E). Then E is called a *biordered set* if the following axioms and their duals hold. (Here e, f, g, h etc. denote arbitrary elements of E.)

(B1) ω^r and ω^l are quasi-orders on E and $D_E = \kappa \cup \kappa^{-1}$. ($D_E$ is the domain of the partial binary operation on E.)

(B21) $f \in \omega^r(e)$ implies $f \mathcal{R} fe \, \omega \, e$.

(B22) $g \, \omega^l f, \; f, g \in \omega^r(e)$ implies $ge \, \omega^l fe$.

(B31) $g \, \omega^r f \, \omega^r e$ implies $(ge)f = gf$.

(B32) $g \, \omega^l f, \; g, f \in \omega^r(e)$ implies $(fg)e = (fe)(ge)$.

(B4) $f, g \in \omega^r(e)$ implies $S(f, g)e = S(fe, ge)$.

Here $S(e, f)$ is defined for $e, f \in E$ as follows: first let $M(e, f)$ denote the quasi-ordered set

$$M(e, f) = (\omega^l(e) \cap \omega^r(f), \prec)$$

where $\prec$ is defined by

$$g \prec h \text{ (for } g, h \in M(e, f)) \text{ if and only if } eg \, \omega^r \, eh \text{ and } gf \, \omega^l \, hf;$$

then $S(e, f) = \{h \in M(e, f) \colon g \prec h \text{ for all } g \in M(e, f)\}$ is called the *sandwich set* of e and f (in that order).

The biordered set E is said to be *regular* if

(R) $S(e, f) \neq \phi$ for all $e, f \in E$.

I shall be concerned only with regular biordered sets in this article: consequently I shall omit the adjective "regular" and consider a biordered set as a partial algebra satisfying (B1)–(B4) and their duals and (R). A mapping θ: $E \to F$ from the biordered set E to the biordered set F is called a *bimorphism* if it satisfies (a) and (b) below:

(a) if $e, f \in E$ then $(e, f) \in D_E$ implies $(e\theta, f\theta) \in D_F$ and $(ef)\theta = (e\theta)(f\theta)$;

(b) $S(e, f)\theta \subseteq S(e\theta, f\theta)$.

One may consider biordered sets as forming a category in which objects are biordered sets and morphisms are bimorphisms.

The definition of the sandwich set $S(e, f)$ for $e, f \in E$ given above is formulated entirely in terms of the basic products; if E is the biordered set of idempotents of a regular semigroup S, then one may formulate other characterizations of $S(e, f)$ which involve non-basic products: for example,

$$S(e, f) = \{h \in M(e, f):\ h \in V(ef)\},$$

or

$$S(e, f) = \{h \in M(e, f):\ ef = ehf\}.$$

Sandwich sets are very useful in locating products. If S is a regular semigroup and $a, b \in S$, then the product ab may be expressed in the form $ab = ahb$ for any $h \in S(e, f)$ where $e = e^2 \in L_a$ and $f = f^2 \in R_b$. The product $ahb = (ah)(hb)$ lies in the $\mathscr{H}$-class $R_{ah} \cap L_{hb}$ since the idempotent h lies in the $\mathscr{H}$-class $L_{ah} \cap R_{hb}$ (i.e. $(ah)(hb)$ is a "trace product").

Using the concept of a biordered set, Nambooripad [93] has developed a theory of fundamental regular semigroups analogous to Munn's theory [90] of fundamental inverse semigroups; he has also constructed all regular idempotent-generated semigroups determined by a given biordered set. His theory rests solidly on the notion of a biordered set and clearly demonstrates a need to study the structure of biordered sets. The axioms are rather complicated and it is desirable to find ways of building biordered sets from simpler kinds of structures. Some work along these lines has recently been done by Rajan [122], who has constructed combinatorial biordered sets (the biordered sets of combinatorial regular semigroups) from objects which he calls "admissible categories". The biordered sets of strongly regular Baer semigroups (including the multiplicative semigroups of von Neumann regular rings) have been constructed by Pastijn [107] and more generally, the biordered sets of strongly V-regular semigroups have been characterized by Nambooripad and Pastijn [102].

The simplest kinds of biordered sets which readily come to mind are semilattices and rectangular biordered sets. A *semilattice* is a biordered set E in which $\omega^r = \omega^l = \omega$ (equivalently $\mathcal{R} = \mathcal{L} = \iota_E$); a *rectangular biordered set* is a biordered set E in which, for all $e, f \in E$ there exist $g, h \in E$ such that $e \,\mathcal{R}\, g \,\mathcal{L}\, f \,\mathcal{R}\, h \,\mathcal{L}\, e$; alternatively, a rectangular biordered set may be characterized as a biordered set E in which $\omega = \iota_E$. Rectangular biordered sets are the biordered sets of completely simple semigroups; semilattices are the biordered sets of inverse semigroups. From Nambooripad [93] it follows that if E is any biordered set and $e, f \in E$, then $S(e, f)$ is a rectangular biordered subset of E.

It is of interest to determine the extent to which certain classes of biordered sets may be built from semilattices and rectangular biordered sets. Some useful processes for building new biordered sets from "known" ones are the processes of taking coextensions or images by semilattices or rectangular biordered sets. If $\mathcal{C}$ is a class of biordered sets and $\theta\colon E \to F$ is a bimorphism from the biordered set E onto the biordered set F, one says that E is a *coextension* of F by $\mathcal{C}$ (or that F is an *image of E by $\mathcal{C}$*) if $e(\theta \circ \theta^{-1}) \in \mathcal{C}$ for each $e \in E$. Coextensions of rectangular biordered sets by semilattices were constructed in Byleen, Meakin and Pastijn [11] and greatly exploited by Pastijn in [109]. A general theory of coextensions of biordered sets by rectangular biordered sets was developed by Meakin and Nambooripad [84] (see also [85] and [86]). *Solid* biordered sets (biordered sets in which $\mathcal{R} \circ \mathcal{L} = \mathcal{L} \circ \mathcal{R}$) may be characterized as the biordered sets of completely regular semigroups (Clifford [18]); they are the biordered sets which are coextensions of a semilattice by rectangular biordered sets and have been constructed by Meakin and Nambooripad [84]. For additional information about constructing biordered sets from semilattices and rectangular biordered sets, see the paper of Meakin [81]; in particular, an inductive process for constructing all finite biordered sets from semilattices and rectangular biordered sets is discussed in Meakin [81].

The biordered sets of pseudo-inverse semigroups form a large class of biordered sets which have been the object of considerable recent study. They may be characterized in several different ways, as the following theorem of Nambooripad [96], [97] shows.

Theorem 6.2. *The following are equivalent for a biordered set* E:

(a) E *is the biordered set of some pseudo-inverse semigroup;*

(b) *every regular semigroup with biordered set* E *is pseudo-inverse;*

(c) *for all* $e, f \in E$, $|S(e,f)| = 1$;

(d) *for all* $e \in E$, $\omega(e)$ *is a semilattice;*

(e) *for all* $e, f \in E$, *there is an element* $h \in E$ *such that* $M(e,f) = \omega(h)$.

In this paper I shall refer to a biordered set which satisfies any one of the above conditions as a *pseudo-semilattice.* Pseudo-semilattices were first studied by Nambooripad [96], who called them "partially associative pseudo-semilattices"; some subsequent authors have referred to them as "pseudo-semilattices" (e.g. Meakin and Pastijn [87], [88], Pastijn [110]); Schein [132] used the terminology "pseudo-semilattice" in a somewhat broader setting; Meakin referred to them as "local semilattices" in [82]. In [96], Nambooripad showed that there are two alternative methods of viewing a pseudo-semilattice, namely:

(i) as a set E, together with two quasi-orders ω^r and ω^l such that to every pair $e, f \in E$ there exists a unique element $e \wedge f \in E$ for which

$$\omega^r(e) \cap \omega^l(f) = \omega(e \wedge f),$$

and such that the conditions (PA1) and (PA2) (and their duals) of [96] are satisfied;

(ii) as a binary algebra $(E, \wedge)$ in which $\wedge$ satisfies the following identities and their duals:

(a) $x \wedge x = x$;

(b) $(x \wedge y) \wedge (x \wedge z) = (x \wedge y) \wedge z$;

(c) $(x \wedge y) \wedge ((x \wedge z) \wedge (x \wedge u)) = ((x \wedge y) \wedge (x \wedge z)) \wedge (x \wedge u)$.

The binary operation $\wedge$ is related to the sandwich sets in E by the following: $S(e, f) = \{f \wedge e\}$; the operation $\wedge$ is defined in such a way as to extend the basic products.

In [87], Meakin and Pastijn showed that if E is any pseudo-semilattice then E is an image by rectangular biordered sets of a pseudo-semilattice $\widetilde{E}$ which is a coextension of a rectangular biordered set by semilattices. Thus pseudo-semilattices may be constructed from semilattices and rectangular biordered sets. The biordered set of a locally orthodox semigroup (Meakin and Nambooripad [86]) is a coextension of a pseudo-semilattice by rectangular biordered sets. Since pseudo-semilattices may be defined as binary algebras $(E, \wedge)$ which satisfy certain laws ((a), (b) and (c) above), it is clear that they form a variety and so free pseudo-semilattices exist. The free pseudo-semilattice on a set X is a pair (FLS_X, i) where FLS_X is a pseudo-semilattice and i is a mapping from X to FLS_X, with the property that for each pseudo-semilattice E and mapping f: $X \to E$, there is a unique bimorphism φ: $FLS_X \to E$ such that the following diagram commutes:

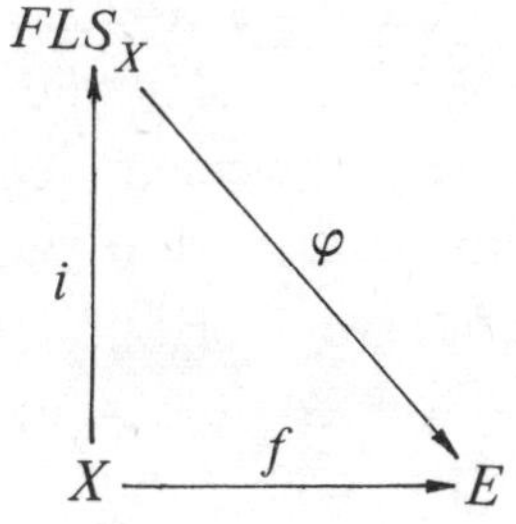

One may view a pseudo-semilattice as being generated by a subset of its elements by successive applications of the $\wedge$ operation: for example, the four-spiral biordered set pictured in Diagram 1 is a pseudo-semilattice generated (as a pseudo-semilattice) by a and c.

The free pseudo-semilattice on two generators was constructed by Meakin and Pastijn in [88] and all of its images (i.e. all pseudo-semilattices on two generators) were described by Meakin in [82]. Recently, [83] I have been able to describe the structure of the free pseudo-semilattice on an arbitrary set X: this structure uses the Rees construction and depends on a knowledge of the free inverse monoid on a set. I shall use Scheiblich's construction [127] of the semilattice of idempotents of the free inverse monoid on a set: additional references to the extensive literature on free inverse semigroups may be found in Reilly's paper [125] and Petrich's book [119]. Some notation from Scheiblich's paper [127] will be needed here.

Let X be a set and X' a set of the same cardinality as X with $X \cap X' = \phi$ and $x \to x'$ a bijection from X onto X'. Identify $(x')'$ with x for each $x \in X$. A word $w = x_1 x_2 \ldots x_n$ of the free semigroup on $X \cup X'$ is called *reduced* if $x_i \neq x'_{i+1}$ for any i; denote the set of all reduced words by R. The multiplication in the free group $R \cup \{1\}$ will be denoted by $\cdot$ if necessary, to distinguish it from the multiplication in the free monoid, which will be denoted by concatenation. A finite subset T of $R \cup \{1\}$ is called *closed* if $1 \in T$ and $w = x_1 x_2 \ldots x_n \in T$ implies $x_1 \ldots x_i \in T$ for all $i \leq n$. Scheiblich [127] showed that, relative to the operation of taking unions of sets, the set of all finite closed subsets of $R \cup \{1\}$ forms a semilattice E_X isomorphic to the semilattice of idempotents of the free inverse monoid on X.

Let $X_0 = X \cup \{x_0\}$, where $x_0 \notin X \cup X'$. Denote the principal ideal of E_X generated by $A \in E_X$ by $\langle A \rangle$: thus $\langle A \rangle = \{B \in E_X : B \supseteq A\}$. Note that the antiatoms of E_X (elements covered by $\{1\}$) are of the form $\{1, x\}$ or $\{1, x'\}$ for $x \in X$. For $x, y \in X$ let $\pi(x, y)$ be the 2-cycle of S_X (the symmetric group on X) which interchanges x and y ($\pi(x, y)$ is the identity map on X if $x = y$) and let $\pi(x, y)$ act on a word $w = x_1 \ldots x_n \in R \cup \{1\}$ by

$$\pi(x, y)w = \pi(x, y)(x_1)\pi(x, y)(x_2) \ldots \pi(x, y)(x_n),$$

where $\pi(x, y)(z)$ denotes the action of $\pi(x, y)$ on z if $z \in X$ and $\pi(x, y)(z') = (\pi(x, y)(z))'$ if $z \in X$. Note that, if $w \in R$, then $\pi(x, y)w \in R$. Extend the action of $\pi(x, y)$ to E_X by defining $\pi(x, y)A = \{\pi(x, y)w \colon w \in A\}$ for $A \in E_X$; note that $\pi(x, y)A \in E_X$ if $A \in E_X$. For $x \in X \cup X'$ define $\bar{x}\colon R \cup \{1\} \to R \cup \{1\}$ by $\bar{x}w = x \cdot w$ for $w \in R \cup \{1\}$ and define $\bar{x}A = \{\bar{x}w \colon w \in A\}$ for $A \in E_X$. Now define a principal ideal isomorphism $\gamma_{x,y}\colon \langle\{1, x\}\rangle \to \langle\{1, y'\}\rangle$ (for $x, y \in X$) by $\gamma_{x,y} = \bar{y}'\pi(x, y) = \pi(x, y)\bar{x}'$; i.e. $\gamma_{x,y}(A) = \bar{y}'\pi(x, y)A = \pi(x, y)\bar{x}'A$, for $A \in \langle\{1, x\}\rangle$. Note that $\gamma_{x,y}$ is a principal ideal isomorphism from $\langle\{1, x\}\rangle$ onto $\langle\{1, y'\}\rangle$ and so $\gamma_{x,y} \in T_{E_X}$ (the Munn semigroup [90] of E_X). Finally, let ι denote the identity automorphism of E_X and for $x, y \in X_0$ let

$$(7) \qquad p_{xy} = \begin{cases} \iota & \text{if } x = y \\ \gamma_{x,x} & \text{if } y = x_0 \text{ and } x \neq x_0 \\ \gamma_{y,y} & \text{if } x = x_0 \text{ and } y \neq x_0 \\ \gamma_{y,x} & \text{if } x \neq y \text{ and } x, y \neq x_0 \end{cases}$$

The theorem of Meakin [83] may now be stated as follows.

Theorem 6.3. *Let X be a non-empty set, $x_0 \notin X \cup X'$ and let $P = (p_{xy})$ be the $X_0 \times X_0$ matrix with (x, y) entry p_{xy} defined by* (7). *Form the $X_0 \times X_0$ Rees matrix semigroup $M = \mathcal{M}(T_{E_X}; X_0, X_0; P)$ over T_{E_X}. Then M is a pseudo-inverse semigroup whose biordered set $FLS_{X_0} = E(M)$ is a pseudo-semilattice. Define a map $i\colon X_0 \to FLS_{X_0}$ by $i(x) = (x, \iota, x)$ for $x \in X_0$. Then the pair (FLS_{X_0}, i) is a free pseudo-semilattice on X_0.*

A diagram of the free pseudo-semilattice on two generators is provided in Meakin and Pastijn [88] and also in Meakin [81]. Diagrams of several other examples of pseudo-semilattices on two generators are provided in Meakin [82]. One may show that every pseudo-semilattice on a set X can be expressed as an image by rectangular biordered sets of an image by semilattices of the free pseudo-semilattice on X.

Acknowledgements. I wish to thank D.B. McAlister, K. Byleen and L. Márki for several useful discussions about this paper and for providing additional references and access to unpublished results. I also wish to thank L. Márki, O. Steinfeld and G. Pollák for the invitation to Szeged, without which this paper would not have been written. The research was supported by the U.S. National Science Foundation, the University of Nebraska–Lincoln Research Council and the János Bolyai Mathematical Society.

ADDENDUM

Since this survey was written I have received preprints of several papers which extend some of the results discussed in the survey in significant ways. In particular, I should like to draw the reader's attention to the three papers [A1], [A2] and [A3] below. In [A1], Byleen develops a construction which is simultaneously a generalization of the Rees construction and of the construction of the Bruck monoid over a semigroup to prove that any countable semigroup can be embedded in a bisimple semigroup which is generated by 3 idempotents. In [A2], McAlister shows that a regular semigroup is pseudo-inverse if and only if it is a locally isomorphic image of a regular Rees matrix semigroup over an inverse semigroup. This theorem strengthens the local isomorphism theorem (Theorem 3.2) in the pseudo-inverse case and also the P-theorem of Pastijn (Theorem 4.2). In [A3], Rhodes provides an extension of the Rhodes–Allen theorem (Theorem 3.1) to infinite semigroups. His papers are deep and powerful and filled with important insights into the structure of semigroups.

[A1] K. Byleen, Embedding any countable semigroup in a 2-generated bisimple monoid, preprint.

[A2] D.B. McAlister, Rees matrix covers for locally inverse semigroups, *Trans. Amer. Math. Soc.*, 277 (1983), 727–738.

[A3] J. Rhodes, Infinite iteration of matrix semigroups. Part I: Structure theorem for torsion semigroups; Part II: Structure theorem for arbitrary semigroups up to aperiodic morphisms, preprints, Center for Pure and Applied Mathematics, University of California, Berkeley.

REFERENCES

[1] D. Allen, Jr., A generalization of the Rees theorem to a class of regular semigroups, *Semigroup Forum,* 2 (1971), 321–331.

[2] D. Allouch, Extensions de demi-groupes inverses, *Semigroup Forum,* 16 (1978), 111–116.

[3] P.N. Ánh – L. Márki, Rees matrix rings, *J. Algebra,* 81 (1983), 340–369.

[4] C. Benzaken – H.C. Mayr, Notion de demi-bande: demi-bandes de type deux, *Semigroup Forum,* 10 (1975), 115–128.

[5] K. Byleen, *The structure of regular and inverse semigroups,* Thesis, University of Nebraska–Lincoln, 1977.

[6] K. Byleen, Spirals of idempotents from *P*-semigroups, *Semigroup Forum,* 17 (1979), 95–99.

[7] K. Byleen, Regular four-spiral semigroups, idempotent-generated semigroups and the Rees construction, *Semigroup Forum,* 22 (1981), 97–100.

[8] K. Byleen, Applications of the Rees construction, *Collected Abstracts, Special Session on Semigroups,* University of California, Davis, April 1980, 1–4.

[9] K. Byleen, On bisimple semigroups generated by a finite number of idempotents, *J. Austral. Math. Soc.,* Ser. A, 33 (1982), 92–101; results announced in Proc. Nebraska Conf. on Semigroups, University of Nebraska, Sept. 1980, 14–24.

[10] K. Byleen – J. Meakin – F. Pastijn, The fundamental four-spiral semigroup, *J. Algebra,* 54 (1978), 6–26.

[11] K. Byleen – J. Meakin – F. Pastijn, Building bisimple idempotent-generated semigroups, *J. Algebra,* 65 (1980), 60–83.

[12] K. Byleen – J. Meakin – F. Pastijn, The double four-spiral semigroup, *Simon Stevin,* 54 (1980), 75–105.

[13] S.Y. Chen – S.C. Hsieh, Factorizable inverse semigroups, *Semigroup Forum,* 8 (1974), 283–297.

[14] A.H. Clifford, Semigroups admitting relative inverses, *Ann. of Math.,* 42 (1941), 1037–1049.

[15] A.H. Clifford, The fundamental representation of a regular semigroup. Department of Mathematics, Tulane Univ., 1974; Announcement: *Semigroup Forum,* 10 (1975), 84–92.

[16] A.H. Clifford, The partial groupoid of idempotents of a regular semigroup, Department of Mathematics, Tulane Univ., 1974; Announcement: *Semigroup Forum,* 10 (1975), 262–268.

[17] A.H. Clifford, Bands of semigroups, *Proc. Amer. Math. Soc.,* 5 (1954), 499–504.

[18] A.H. Clifford, The fundamental representation of a completely regular semigroup, *Semigroup Forum,* 12 (1976), 341–346.

[19] A.H. Clifford – G.B. Preston, *The algebraic theory of semigroups,* Math. Surveys No. 7, Amer. Math. Soc., Providence, R.I., Vol. 1, 1961, Vol. 2, 1967.

[20] A. Coudron, Sur les extensions de demi-groupes réciproques, *Bull. Soc. Roy. Sci. Liège,* 37 (1968), 409–419.

[21] H. D'Alarcao, Idempotent-separating extension of inverse semigroups, *J. Austral. Math. Soc.,* 9 (1969), 211–217.

[22] R.H.J. Dawlings, *Semigroups of singular endomorphisms of vector spaces,* Ph. D. Thesis, University of St. Andrews, 1980.

[23] R.H.J. Dawlings, The semigroup of singular endomorphisms of a finite dimensional vector space, *Semigroups,* Proc. Conf. Monash Univ., 1979, Academic Press, London, 1980, 121–131.

[24] A. DeBodt – F. Pastijn, A class of rectangular bands of inverse semigroups, *Semigroup Forum,* 21 (1980), 9–12.

[25] C. Eberhart – W. Williams – L. Kinch, Idempotent-generated regular semigroups, *J. Austral. Math. Soc.,* 15 (1973), 27--34.

[26] S. Eilenberg, *Automata, Languages and Machines,* Vol. B, Academic Press, 1976.

[27] J.A. Erdos, On products of idempotent matrices, *Glasgow Math. J.,* 8 (1967), 118–122.

[28] P.H.H. Fantham, On the classification of a certain type of semigroup, *Proc. London Math. Soc.,* (3) 10 (1960), 409–427.

[29] D.G. Fitz-Gerald, On inverses of products of idempotents in regular semigroups, *J. Austral. Math. Soc.,* 13 (1972), 335–337.

[30] P.A. Grillet, A construction of inverse semigroups, *Semigroup Forum,* 8 (1974), 169–176.

[31] P.A. Grillet, The structure of regular semigroups I: a representation, *Semigroup Forum,* 8 (1974), 177–183.

[32] P.A. Grillet, The structure of regular semigroups II: cross connections, *Semigroup Forum,* 8 (1974), 254–259.

[33] P.A. Grillet, The structure of regular semigroups III: the reduced case, *Semigroup Forum,* 8 (1974), 260–265.

[34] P.A. Grillet, The structure of regular semigroups IV: the general case, *Semigroup Forum,* 8 (1974), 368–373.

[35] T.E. Hall, On regular semigroups whose idempotents form a subsemigroup, *Bull. Austral. Math. Soc.,* 1 (1969), 195–208; Addenda: *Bull. Austral. Math. Soc.,* 3 (1970), 287–288.

[36] T.E. Hall, On orthodox semigroups and uniform and antiuniform bands, *J. Algebra,* 16 (1970), 204–217.

[37] T.E. Hall, Orthodox semigroups, *Pacific J. Math.,* 39 (1971), 677–686.

[38] T.E. Hall, On regular semigroups, *J. Algebra,* 24 (1973), 1–24.

[39] T.E. Hall, Some properties of local subsemigroups inherited by larger semigroups, *Semigroup Forum,* 25 (1982), 35–48.

[40] T.E. Hall, Inverse and regular semigroups and amalgamation: a brief survey, *Symposium on Regular Semigroups,* Northern Illinois Univ., April 1979, 49–79.

[41] T.E. Hall, The idempotent-generated subsemigroups of 0-bisimple and 0-simple regular semigroups, *Proc. Nebraska Conf. on Semigroups,* Univ. of Nebraska, Sept. 1980, 25–30.

[42] H.-J. Hoehnke, Über Verallgemeinerungen das Scharbegriffs, *Math. Nachr.,* 38 (1968), 365–382.

[43] E. Hotzel, Dual *D*-operands and the Rees theorem, *Algebraic Theory of Semigroups,* Proc. Conf. Szeged, 1976, Coll. Math. Soc. J. Bolyai, Vol. 20, North-Holland, Amsterdam, 1979, 247–275.

[44] J.M. Howie, *An introduction to semigroup theory,* Academic Press, London, 1976.

[45] J.M. Howie, The subsemigroup generated by the idempotents of a full transformation semigroup, *J. London Math. Soc.,* 41 (1966), 707–716.

[46] J.M. Howie, Idempotent generators in finite full transformation semigroups, *Proc. Roy. Soc. Edinburgh Sect. A,* 81 (1978), 317–323.

[47] J.M. Howie, Idempotents in completely 0-simple semigroups, *Glasgow Math. J.,* 19 (1978), 109–113.

[48] J.M. Howie, Gravity, depth and homogeneity in full transformation semigroups, *Semigroups,* Proc. Conf. Monash Univ., 1979, Academic Press, London, 1980, 111–120.

[49] J.M. Howie, Some subsemigroups of infinite transformation semigroups, *Proc. Roy. Soc. Edinburgh Sect. A,* 88 (1981), 159–167.

[50] J.M. Howie, A class of bisimple idempotent-generated congruence-free semigroups, *Proc. Roy. Soc. Edinburgh Sect. A,* 88 (1981), 169–184.

[51] J.B. Kim, Idempotent-generated Rees matrix semigroups, *Kyungpook Math. J.,* 10 (1970), 7–13.

[52] K. Krohn – J. Rhodes, Algebraic theory of machines I. Prime decomposition theorem for finite semigroups and machines, *Trans. Amer. Math. Soc.,* 116 (1965), 450–464.

[53] G. Lallement, Structure theorems for regular semigroups, *Semigroup Forum,* 4 (1972), 95–123.

[54] G. Lallement, Demi-groupes réguliers, Thèse Sc. Mathl, Paris, 1966, *Annali di Matem. pura ed. appl.,* 77 (1967), 43–130.

[55] G. Lallement, *Semigroups and combinatorial applications,* Wiley, New York, 1979.

[56] G. Lallement, Some recent results on languages and codes, *Proc. Nebraska Conf. on Semigroups,* Univ. of Nebraska, Sept. 1980, 49–57.

[57] G. Lallement, Some remarks on the four-spiral semigroup, *Semigroup Forum,* 18 (1979), 341–346.

[58] G. Lallement – M. Petrich, A generalization of the Rees theorem in semigroups, *Acta Sci. Math. (Szeged),* 30 (1969), 113–132.

[59] G. Lallement – M. Petrich, Structure d'une classe de demi-groupes réguliers, *J. Math. Pures Appl.,* 48 (1969), 345–397.

[60] H. Lausch, Cohomology of inverse semigroups, *J. Algebra,* 35 (1975), 273–303.

[61] J. Leech, $\mathscr{H}$-coextensions and the structure of bands of groups, *Mem. Amer. Math. Soc.,* Vol. 1 (2), No. 157 (1975).

[62] E.S. Ljapin, *Semigroups,* Gos. Izdat. Fiz.-Mat. Lit., Moscow, 1960; English translation: Amer. Math. Soc., Providence, R.I., 1963.

[63] M. Loganathan, *Extensions of regular semigroups and cohomology of semigroups,* Ph. D. Thesis, Ramanujan Institute, Univ. of Madras, Dec. 1978.

[64] S. Margolis, An invitation to finite semigroups – some results and open problems, *Proc. Nebraska Conf. on Semigroups,* Univ. of Nebraska, Sept. 1980, 58–81.

[65] L. Márki, On locally regular Rees matrix semigroups, *Acta Sci. Math. (Szeged),* 37 (1975), 95–102.

[66] D.B. McAlister, Groups, semilattices and inverse semigroups, *Trans. Amer. Math. Soc.,* 192 (1974), 227–244.

[67] D.B. McAlister, Groups, semilattices and inverse semigroups II, *Trans. Amer. Math. Soc.,* 196 (1974), 351–370.

[68] D.B. McAlister, Regular semigroups, fundamental semigroups and groups, *J. Austral. Math. Soc.,* 29 (1980), 475–503.

[69] D.B. McAlister, $\vee$-Prehomomorphisms on inverse semigroups, *Pacific J. Math.,* 67 (1976), 215–231.

[70] D.B. McAlister, Regular Rees matrix semigroups and regular Dubreil–Jacotin semigroups, *J. Austral. Math. Soc. Ser. A,* 31 (1981), 325–336.

[71] D.B. McAlister, Regular Rees matrix semigroups and regular Dubreil–Jacotin semigroups, *Proc. Nebraska Conf. on Semigroups,* Univ. of Nebraska, Sept. 1980, 1–13.

[72] D.B. McAlister, Regular semigroups, fundamental semigroups and groups, *Collected Abstracts, Special Session on Semigroups,* Univ. of California, Davis, April, 1980, 41–44.

[73] D. B. McAlister – N.R. Reilly, *E*-unitary covers for inverse semigroups, *Pacific J. Math.*, 68 (1977), 161–174.

[74] D.D. Miller – A.H. Clifford, Regular $\mathcal{D}$-classes in semigroups, *Trans. Amer. Math. Soc.*, 82 (1956), 270–280.

[75] J. Meakin, Congruences on orthodox semigroups, *J. Austral. Math. Soc.*, 12 (1971), 323–341.

[76] J. Meakin, On the structure of inverse semigroups, *Semigroup Forum;* 12 (1976), 6–14.

[77] J. Meakin, Coextensions of inverse semigroups, *J. Algebra,* 46 (1977), 315–333.

[78] J. Meakin, The structure mappings on a regular semigroup, *Proc. Edinburgh Math. Soc.*, 21 (1978), 135–142.

[79] J. Meakin, The structure mapping approach to regular and inverse semigroups, *Algebraic Theory of Semigroups,* Proc. Conf. Szeged, 1976, Coll. Math. Soc. J. Bolyai, Vol. 20, North-Holland, Amsterdam, 1979, 371–383.

[80] J. Meakin, Structure mappings, coextensions and regular four-spiral semigroups, *Trans. Amer. Math. Soc.,* 255 (1979), 111–134.

[81] J. Meakin, Constructing biordered sets, *Semigroups,* Proc. Conf. Monash Univ., 1979, Academic Press, London, 1980, 67–84.

[82] J. Meakin, Local semilattices on two generators, *Semigroup Forum,* 24 (1982), 95–116; results announced in Proc. Nebraska Conf. on Semigroups, Univ. of Nebraska, Sept. 1980, 121–134.

[83] J. Meakin, The free local semilattice on a set, *J. Pure Appl. Algebra,* to appear in 1983; results announced at Halbgruppentheorie Conference, Mathematisches Forschungsinstitut Oberwolfach, May, 1981.

[84] J. Meakin – K.S.S. Nambooripad, Coextensions of regular semigroups by rectangular bands I, *Trans. Amer. Math. Soc.*, 269 (1982), 197–224.

[85] J. Meakin – K.S.S. Nambooripad, Coextensions of regular semigroups by rectangular bands II, *Trans. Amer. Math. Soc.*, 272 (1982), 555–568.

[86] J. Meakin – K.S.S. Nambooripad, Coextensions of pseudo-inverse semigroups by rectangular bands, *J. Austral. Math. Soc.*, 30 (1980), 73–86.

[87] J. Meakin – F. Pastijn, The structure of pseudo-semilattices, *Algebra Universalis*, 13 (1981), 355–372.

[88] J. Meakin – F. Pastijn, The free pseudo-semilattice on two generators, *Algebra Universalis*, 14 (1982), 297–309.

[89] W.D. Munn, Uniform semilattices and bisimple inverse semigroups, *Quart. J. Math.*, 17 (1966), 151–159.

[90] W.D. Munn, Fundamental inverse semigroups, *Quart. J. Math.*, 21 (1970), 157–170.

[91] K.S.S. Nambooripad, Structure of regular semigroups I: fundamental regular semigroups, *Semigroup Forum*, 9 (1975), 354–363.

[92] K.S.S. Nambooripad, Structure of regular semigroups II: the general case, *Semigroup Forum*, 9 (1975), 364–371.

[93] K.S.S. Nambooripad, Structure of regular semigroups I, *Mem. Amer. Math. Soc.*, 224 (1979).

[94] K.S.S. Nambooripad, Relations between cross-connections and biordered sets, *Semigroup Forum*, 16 (1978), 67–82.

[95] K.S.S. Nambooripad, The natural partial order on a regular semigroup, *Proc. Edinburgh Math. Soc.*, 23 (1980), 249–260.

[96] K.S.S. Nambooripad, Pseudo-semilattices and biordered sets I, *Simon Stevin*, 55 (1981), 103–110.

[97] K.S.S. Nambooripad, Pseudo-semilattices and biordered sets II: pseudo-inverse semigroups, *Simon Stevin,* 56 (1982), 143–160.

[98] K.S.S. Nambooripad, Pseudo-semilattices and biordered sets III: locally testable regular semigroups, *Simon Stevin,* 56 (1982), 239–256.

[99] K.S.S. Nambooripad, Regular idempotent-generated semigroups and regular partial bands, *Proc. Conf. on Semigroups in honor of A.H. Clifford,* Tulane Univ., Sept. 1978, 44–84.

[100] K.S.S. Nambooripad – A.R. Rajan, Structure of combinatorial regular semigroups, *Quart. J. Math. Oxford,* 29 (2) (1978), 489–504.

[101] K.S.S. Nambooripad – F. Pastijn, Subgroups of free idempotent-generated regular semigroups, *Semigroup Forum,* 21 (1980), 1–8.

[102] K.S.S. Nambooripad – F. Pastijn, *V*-regular semigroups, *Proc. Royal Soc. Edinburgh,* 88A (1981), 275–291.

[103] K.S.S. Nambooripad – R. Veeramony, Subdirect products of regular semigroups, *Semigroup Forum,* 27 (1983), 265–308.

[104] F. Pastijn, Embedding semigroups in semibands, *Semigroup Forum,* 14 (1977), 247–264.

[105] F. Pastijn, Idempotent-generated completely 0-simple semigroups, *Semigroup Forum,* 15 (1977), 41–50.

[106] F. Pastijn, The biorder on the partial groupoid of idempotents of a semigroup, *J. Algebra,* 65 (1980), 147–187.

[107] F. Pastijn, Biordered sets and complemented modular lattices, *Semigroup Forum,* 21 (1980), 205–220.

[108] F. Pastijn, *Idempotente elementen in reguliere semigroepen,* Aggregaatsthesis, Rijksuniversiteit Gent, 1979.

[109] F. Pastijn, Rectangular bands of inverse semigroups, *Simon Stevin,* 56 (1982), 3–95.

[110] F. Pastijn, The structure of pseudo-inverse semigroups, *Trans. Amer. Math. Soc.,* 273 (1982), 631–656.

[111] F. Pastijn, Regular locally testable semigroups as semigroups of quasi-ideals, *Acta Math. Acad. Sci. Hungar.,* 36 (1980), 161–166.

[112] F. Pastijn, Division theorems for inverse and pseudo-inverse semigroups, *J. Austral. Math. Soc. Ser. A,* 31 (1981), 415–420.

[113] F. Pastijn, Essential normal and conjugate extensions of inverse semigroups, *Glasgow Math. J.,* 23 (1982), 123–130.

[114] M. Petrich, The maximal matrix decomposition of a semigroup, *Portugaliae Math.,* 25 (1) (1966), 15–33.

[115] M. Petrich, The structure of completely regular semigroups, *Trans. Amer. Math. Soc.,* 189 (1974), 211–236.

[116] M. Petrich, A construction and a classification of bands, *Math. Nachr.,* 48 (1971), 263–274.

[117] M. Petrich, Extensions normales de demi-groupes inverses, *Fund. Math.,* 112 (1981), 187–203.

[118] M. Petrich, The conjugate hull of an inverse semigroup, *Glasgow Math. J.,* 21 (1980), 103–124.

[119] M. Petrich, *Inverse Semigroups,* Wiley, New York, 1984.

[120] G.B. Preston, Inverse semigroups, *J. London Math. Soc.,* 29 (1954), 396–403.

[121] G.B. Preston, Matrix representations of semigroups, *Quart. J. Math. Oxford,* 9 (1958), 169–176.

[122] A.R. Rajan, *Structure of combinatorial regular semigroups,* Ph. D. Thesis, University of Kerala, 1981.

[123] D. Rees, On semi-groups, *Proc. Cambridge Phil. Soc.,* 36 (1940), 387–400.

[124] N.R. Reilly, Bisimple ω-semigroups, *Proc. Glasgow Math. Assoc.,* 7 (1966), 160–167.

[125] N.R. Reilly, Free inverse semigroups, *Algebraic Theory of Semigroups,* Proc. Conf. Szeged, 1976, Coll. Math. Soc. J. Bolyai, Vol. 20, North-Holland, Amsterdam, 1979, 479–508.

[126] J. Rhodes – D. Allen, Synthesis of classical and modern theory of finite semigroups, *Adv. in Math.* 11 (1973), 238–266.

[127] H.E. Scheiblich, Free inverse semigroups, *Proc. Amer. Math. Soc.,* 38 (1973), 1–7.

[128] B.M. Schein, On the theory of generalized groups and generalized heaps, *Theory of semigroups and its applications I,* Izdat. Saratov. Univ., Saratov, 1965, 286–324 (in Russian); English translation: Amer. Math. Soc. Transl., (2) 113 (1979), 89–122.

[129] B.M. Schein, Semigroups of strong subsets, *Volž. Mat. Sb.,* 4 (1966), 180–186 (in Russian).

[130] B.M. Schein, Bands of monoids, *Acta Sci. Math. (Szeged),* 36 (1974), 145–154.

[131] B.M. Schein, Products of idempotents of order-preserving transformations of arbitrary chains, *Semigroup Forum,* 11 (1975/76), 297–309.

[132] B.M. Schein, Pseudo-semilattices and pseudo-lattices, *Izv. Vysš. Ucebn. Zaved. Mat.,* 2 (117) (1972), 81–94 (in Russian).

[133] M.P. Schützenberger, $\bar{\mathscr{D}}$-représentations des demi-groupes, *Compt. Rend. Acad. Sci. Paris,* 244 (1957), 1994–1996.

[134] V.M. Širjaev, Inverse semigroups with given *G*-radical, *Dokl. Akad. Nauk BSSR,* 14 (1970), 782–785 (in Russian).

[135] S. Sribala, Cohomology and extensions of inverse semigroups, *J. Algebra,* 47 (1977), 1–17.

[136] O. Steinfeld, On a generalization of completely 0-simple semigroups, *Acta Sci. Math. (Szeged),* 28 (1967), 135–145.

[137] O. Steinfeld, On similarly decomposable semigroups, *Algebraic Theory of Semigroups,* Proc. Conf. Szeged, 1976, Coll. Math. Soc. J. Bolyai, Vol. 20, North-Holland, Amsterdam, 1979, 601–611.

[138] O. Steinfeld, *Quasi-ideals in rings and semigroups,* Akadémiai Kiadó, Budapest, 1978.

[139] A. Suschkewitsch, Über die endlichen Gruppen ohne das Gesetz der eindeutigen Umkehrbarkeit, *Math. Ann.,* 99 (1928), 30–50.

[140] I. Szabó, Rees matrix semigroups with 4-dimensional sandwich matrices, *Acta Math. Acad. Sci. Hungar.,* 35 (1980), 339–350.

[141] M.B. Szendrei, On the structure of orthodox semigroups, *Semigroup Forum,* 13 (1976/77), 271–280.

[142] Tran Quy Tien, *Investigations in the theory of Rees matrix semigroups and Rees matrix rings,* Thesis, Hungarian Acad. of Sciences, Budapest, 1975 (in Hungarian).

[143] V.V. Vagner, Generalized groups, *Dokl. Akad. Nauk SSSR (N. S.),* 84 (1952), 1119–1122 (in Russian).

[144] R. Veeramony, *Subdirect products of regular semigroups,* Ph. D. Thesis, University of Kerala, 1981.

[145] P.S. Venkatesan, Matrix semigroups over a semigroup with zero, *Math. Ann.,* 181 (1969), 60–64.

[146] M. Yamada, Regular semigroups whose idempotents satisfy permutation identities, *Pacific J. Math.*, 21 (1967), 371–392.

[147] M. Yamada, On a regular semigroup whose idempotents form a band, *Pacific J. Math.*, 33 (1970), 261–272.

[148] M. Yamada, On a certain class of regular semigroups, *Symposium on regular semigroups,* Northern Illinois Univ., April, 1979, 146–179.

[149] M. Yamada, The structure of quasi-orthodox semigroups, *Mem. Fac. Sci. Shimane Univ.*, 14 (1980), 1–18.

[150] Y. Zalcstein, Locally testable semigroups, *Semigroup Forum,* 5 (1973), 216–227.

[151] Y. Zalcstein, Locally testable languages, *J. Comp. System Sci.*, 6 (1972), 151–167.

J.C. Meakin

Department of Mathematics and Statistics, University of Nebraska, Lincoln, Nebraska, 68588, USA.

COLLOQUIA MATHEMATICA SOCIETATIS JÁNOS BOLYAI
39. SEMIGROUPS, SZEGED (HUNGARY), 1981.

BASIS CLASS FOR SOME CLASSES OF SEMIGROUPS

S. MILIĆ — S. CRVENKOVIĆ

The elementary language L of semigroup theory may be taken to consist of first order logic in which the sole non-logical constant is a binary function symbol, to be interpreted as denoting semigroup multiplication.

Let $\varphi(x, y)$ be a formula in the language L. Denote by $\mathbf{S}_\varphi$ the class of semigroups satisfying the condition

$$(*) \qquad (\forall x)(\exists y)\varphi(x, y).$$

A semigroup S belongs to the class $\mathbf{QS}_\varphi$ of semigroups if each proper subsemigroup of S belongs to $\mathbf{S}_\varphi$.

In [4] E.S. Ljapin introduced the concept of a basis class for some classes of semigroups.

Definition 1. Let $\mathbf{M}, \mathbf{N}, \mathbf{P}$ be three classes of semigroups such that $\mathbf{M} \subset \mathbf{N} \subset \mathbf{P}$. The class $\mathbf{M}$ is said to be a *basis class* for the class $\mathbf{N}$ relative to the class $\mathbf{P}$ if the following conditions are satisfied.

(a) Each semigroup of $\mathbf{N}$ may be represented in the form of a union of subsemigroups of it belonging to the class $\mathbf{M}$.

(b) Every semigroup of **P** which can be represented in the form of the union of subsemigroups belonging to the class **M** must belong to the class **N**.

(c) If some class of semigroups $\mathbf{M}_1$, all the semigroups of which belong to **M**, satisfies the conditions (a) and (b) formulated above for the class **M**, then $\mathbf{M}_1$ coincides with **M**.

In the present paper we give necessary and sufficient condition for the existence of a basis class of $\mathbf{QS}_\varphi$, where $\varphi \in$ Form (L).

Theorem 1. $\mathbf{QS}_\varphi$ *has a basis class relative to the class of all semigroups if and only if* $\mathbf{QS}_\varphi \subset \mathbf{S}_\varphi$.

Proof. Suppose $\mathbf{QS}_\varphi$ has a basis class relative to the class of all semigroups. If $\mathbf{QS}_\varphi \not\subset \mathbf{S}_\varphi$ then there exists a semigroup S with $S \in \mathbf{QS}_\varphi$ and $S \notin \mathbf{S}_\varphi$. The semigroup $S^0 = S \cup \{0\}$ is a union of semigroups from $\mathbf{QS}_\varphi$ but $S^0 \notin \mathbf{QS}_\varphi$ so it follows that $\mathbf{QS}_\varphi$ does not have a basis class, since the condition (b) of the definition of a basis class does not hold.

Assume that $\mathbf{QS}_\varphi \subset \mathbf{S}_\varphi$. Let **M** be the set of all monogenic semigroups which are from $\mathbf{QS}_\varphi$. Each cyclic group of prime order belongs to $\mathbf{QS}_\varphi$. Let $S \in \mathbf{QS}_\varphi$. Then $S = \bigcup_{x \in S} \langle x \rangle$. Every proper subsemigroup of $\langle x \rangle$ is a proper subsemigroup of S so that $\langle x \rangle \in \mathbf{QS}_\varphi$. Let $S = \bigcup_{\alpha \in I} \langle x_\alpha \rangle$ and $\langle x_\alpha \rangle \in \mathbf{M}$. If S_1 is a proper subsemigroup of S then from $a \in S_1$ it follows that $a \in \langle x_{\alpha_0} \rangle$ for some $\alpha_0 \in I$. If $\langle a \rangle = \langle x_{\alpha_0} \rangle$, as $\mathbf{QS}_\varphi \subset \mathbf{S}_\varphi$, then $\langle a \rangle \in \mathbf{S}_\varphi$ so that there exists $b \in \langle a \rangle$ such that $\varphi(a, b)$ holds. Analogously, if $\langle a \rangle$ is a proper subsemigroup of $\langle x_{\alpha_0} \rangle$. From the previous considerations it follows that $S_1 \in \mathbf{S}_\varphi$ so that $S \in \mathbf{QS}_\varphi$.

Semigroups belonging to **M** can not be represented as a union of proper subsemigroups from $\mathbf{QS}_\varphi$ so **M** is a minimal class.

If $\varphi(x, y)$ is one of the following formulas

(i) $x = xyx$,

(ii) $x = xyx \wedge xy = yx$,

(iii) $x = yx^2$,

(iv) $x = x^2 y$,

then we have

Proposition 1. *Let S be a semigroup. The following conditions are equivalent:*

1° $S \in \mathbf{QS}_\varphi$.

2° *S is a monogenic semigroup of index 2 or* $(\forall x \in S)$ $x^n = x$ *for some integer* $n \geqslant 2$.

Proof. Immediate.

A monogenic semigroup $\langle a \rangle = \{a, a^2, \ldots, a^{r+1}\}$, i.e. the semigroup in which $a^2 = a^{2+r}$ holds belongs to the class $\mathbf{QS}_\varphi$ but $\langle a \rangle \notin \mathbf{S}_\varphi$ and it follows from Theorem 1 that $\mathbf{QS}_\varphi$ does not have a basis class.

If m and r are the index and period, respectively, of the monogenic semigroup $\langle a \rangle$ we denote by $K_a = \{a^m, a^{m+1}, \ldots, a^{m+r-1}\}$ the subgroup of $\langle a \rangle$.

Let $\varphi(x, y)$ be one of the following formulas

(i) $xy = y$,

(ii) $yx = y$,

(iii) $xy = y \wedge yx = y$.

Proposition 2. $S \in \mathbf{QS}_\varphi$ *iff one of the following conditions hold.*

$\overline{1}$ *S is a cyclic group of prime order or S is the trivial group;*

$\overline{2}$ $(\forall x \in S)$ $x^{m+1} = x^m$ *for some positive integer m.*

Proof. The "if" part follows immediately. Conversely, suppose $S \in \mathbf{QS}_\varphi$. If S is a cyclic group it is of prime order or trivial. If $S = \langle a \rangle$ is a monogenic semigroup with an index m, $m \neq 1$, then $K_a = \{a^m\}$ is the trivial group so that $\overline{2}$ holds. If S is not monogenic and $x \in S$ then $\langle x \rangle$ is a proper subsemigroup of S. As $\langle x \rangle \in \mathbf{S}_\varphi$ we have that $x^{t+1} = x^t$

where t is the index of $\langle x \rangle$.

Denote by Π the class of semigroups $\mathbf{S}_\varphi$ where $\varphi(x,y) \equiv xy = y$. The class Π has a basis class (see [4]). If C_p is a cyclic group of a prime order then C_p^0 is a semigroup which is a union of semigroups from $\mathbf{Q}\Pi$. $C_p \notin \Pi$ so that $\mathbf{Q}\Pi$ does not have a basis class relative to the class of all semigroups.

Let $\varphi(x,y)$ be the formula $x = xy$.

Proposition 3. *Let S be a semigroup. The following conditions are equivalent.*

(1) $S \in \mathbf{QS}_\varphi$;

(2) S *is a union of finite cyclic groups.*

Proof. Immediate.

Obviously, $\mathbf{QS}_\varphi \subset \mathbf{S}_\varphi$. The semigroup S given by the table ([3])

	0	e	f	a	b
0	0	0	0	0	0
e	0	e	0	a	0
f	0	0	f	0	b
a	0	0	a	0	e
b	0	b	0	f	0

belongs to $\mathbf{S}_\varphi$ but $S \notin \mathbf{QS}_\varphi$ so that $\mathbf{QS}_\varphi$ is a proper subset of $\mathbf{S}_\varphi$.

Let $\varphi(x,y)$ be the following formula

$$(1') \qquad x^m = y^m \wedge yx = x^{m+1}y \wedge x^n = x,$$

where m, n are positive integers, $n > 1$. $\mathbf{S}_\varphi$ is the class $\mathbf{S}^*_{m,n}$ (see [1]). If $S \in \mathbf{S}^*_{m,n}$ denote by $[\{x,y\}]$ the semigroup generated by $x, y \in S$ such that the formula (1′) holds. $[\{x,y\}]$ is a finite group. We have the following theorem.

Theorem 2 ([1]). *Let S be a semigroup. Then*

$$S \in \mathbf{S}^*_{m,n} \Leftrightarrow (\forall x \in S)\,(\exists y \in S)\,([\{x,y\}] \in \mathbf{S}^*_{m,n}).$$

According to [2], the set $\mathbf{M}$ of all groups $[\{x, y\}] \in \mathbf{S}^*_{m,n}$ which can not be represented as a union of proper subgroups of the same type, is a basis class for $\mathbf{S}^*_{m,n}$. If $m = 6$ and $n = 13$ we have that

$$\mathbf{M} = \{\{e\}, C_2, C_6, G_8, G_{24}\},$$

where G_8 is the quaternion group and G_{24} is a group of 24 elements.

The proof of the following theorem is straightforward and will not be given explicitely.

Theorem 3. *Let S be a semigroup. $S \in \mathbf{QS}^*_{m,n}$ iff exactly one of the following conditions holds.*

1° $(\forall a \in S)\ a = a^{(m,n-1)+1}$ *where* $(m, n-1)$ *is the* GCD *of* m *and* n;

2° S *is a cyclic group,* $|S| = p^\alpha, p^{\alpha-1} \mid (m, n-1)$ *and* $p^\alpha \nmid (m, n-1)$, *where* p *is a prime number and* α *a nonnegative integer;*

3° S *is a monogenic semigroup with index* 2 *and* $r \mid (m, n-1)$, *where* r *is the period of the semigroup* S.

From the definition of the class $\mathbf{S}^*_{m,n}$ it follows that $x^{n-1} = e_x$. Let p be a prime number such that $p > n-1$. Then $C_p \in \mathbf{QS}^*_{m,n}$ and $C_p \notin \mathbf{S}^*_{m,n}$. Therefore, $\mathbf{QS}^*_{m,n}$ does not have a basis class relative to the class of all semigroups.

REFERENCES

[1] S. Bogdanović – S. Crvenković, On some classes of semigroups, *Zbor. rad. PMF Novi Sad,* 8 (1978), 69–77.

[2] S. Crvenković, On some properties of a class of the completely regular semigroups, *Zbor. rad. PMF Novi Sad,* 9 (1979), 153–160.

[3] J.M. Howie, *An Introduction to Semigroup Theory,* Academic Press, London, 1976.

[4] E.S. Ljapin, *Semigroups,* Translations of Mathematical Monographs, Amer. Math. Soc., Providence, R.I., 1963.

S. Milić – S. Crvenković

Institute of Mathematics, University of Novi Sad, 21000 Novi Sad P.O.B. 224, Yugoslavia.

COLLOQUIA MATHEMATICA SOCIETATIS JÁNOS BOLYAI
39. SEMIGROUPS, SZEGED (HUNGARY), 1981.

SEMIGROUPS OF CONTINUOUS LINEAR TRANSFORMATIONS

L. G. MUSTAFAEV

The problem to what extent topological spaces are determined by the semigroup of their continuous transformations was posed long ago. Recently, a lot of papers have been devoted to this question. Taking into account that every linear space of dimension greater than one is determined by the semigroup of its linear transformations [1] it is natural to raise the question: which topological linear spaces are determined by the semigroup of their continuous linear transformations in certain classes? We proved ([2], [3]) that in the class of real normed spaces every space of dimension greater than one is determined up to topological isomorphism by the semigroup of its one-dimensional continuous linear transformations, that is of those continuous linear transformations which map the whole space into subspaces of dimension 1. Wood [4] verified that every real locally convex Hausdorff space with weak or Mackey topology is determined by the semigroup of its continuous linear transformations. Furthermore, every automorphism of this semigroup is an inner automorphism. In the present paper we continue the investigation of the previous determination problem for locally convex spaces. We show that every real locally convex Hausdorff space with weak or Mackey topology is determined by the semigroup of its one-dimensional continuous linear transformations. This implies as a corollary Wood's result [4] mentioned above and a result due to Mackey [5] concerning the rings of continuous linear transformations. We investigate the analogous problem for spaces over the field of complex numbers, too.

1.

Let X be a locally convex space over the field $\mathbf{R}$ of real numbers or over the field $\mathbf{C}$ of complex numbers. We denote by $S(X)$ the multiplicative semigroup of all continuous linear trasnformations of the space X, and by $H(X)$ the subsemigroup of $S(X)$ consisting of all those transformations which map X into subspaces with dimension 1. X^* will denote the topological conjugate of X. Let $\{A_\lambda\}_{\lambda \in \Lambda}$ and $\{B_i\}_{i \in I}$ be the set of all subspaces of X and X^*, resp. with dimension 1. For every $i \in I$, $\lambda \in \Lambda$, we choose and fix non-zero elements $f_i \in B_i$, $e_\lambda \in A_\lambda$. In what follows, we denote by $\mathfrak{M}\mathbf{R}$ and $\mathfrak{M}\mathbf{C}$ the multiplicative semigroup of the fields $\mathbf{R}$ and $\mathbf{C}$, respectively. For $f \in X^*$ and $x \in X$, $[x, f]$ will stand for $f(x)$. We say that the topological linear spaces $\mathscr{A} = (A, \mathbf{C})$ and $\mathscr{B} = (B, \mathbf{C})$ over the field of complex numbers are in a topological semilinear correspondence u (cf. [6]) if u is a topological isomorphism of the group A onto the group B such that one of the relations $(\alpha x)u = \alpha(xu)$ or $(\alpha x)u = \bar{\alpha}(xu)$ holds for all $\alpha \in \mathbf{C}$ and $x \in A$.

2.

Theorem. *Let X be a real locally convex space. Then the semigroup $H(X)$ is isomorphic to the completely simple semigroup*

$$S(\mathfrak{M}\mathbf{R}; I, \Lambda; [e_\lambda, f_i]).$$

Proof. It is clear that, for every $a \in H(X)$, there is a functional $f = f_i \alpha \in X^*$ for which

$$(1) \qquad xa = [x, f_i]\alpha e_\lambda \qquad (x \in X,\ \alpha \in \mathbf{R}).$$

(1) shows that the correspondence φ assigning the triple $(\alpha; i, \lambda)$ to $a \in H(X)$ is one-to-one. If $b\varphi = (\beta; j, \mu)$ then, by (1), we have

$$x(ab) = (xa)b = ([x, f_i]\alpha e_\lambda)b = [[x, f_i]\alpha e_\lambda, f_j]\beta e_\mu =$$

$$= [x, f_i]\alpha[e_\lambda, f_j]\beta e_\mu$$

for every x and hence $(ab)\varphi = (\alpha[e_\lambda, f_j]\beta; i, \mu)$. Thus φ is an isomorphism of $H(X)$ onto $S(\mathfrak{M}\mathbf{R}; I, \Lambda; [e_\lambda, f_i])$ which completes the proof.

It is clear that the same theorem holds when X is a complex locally convex space.

3.

Theorem. *Let X and Y be real locally convex Hausdorff spaces both with weak topology or both with Mackey topology. If $\dim X > 1$ then the semigroups $H(X)$ and $H(Y)$ are isomorphic to each other if and only if X and Y are topologically isomorphic. Every isomorphism of the semigroup $H(X)$ onto $H(Y)$ has the form*

$$a\varphi = u^{-1}au \qquad (a \in H(X))$$

where u is a topological isomorphism of X onto Y.

Proof. First we show the "only if" part. Let φ be an isomorphism of $H(X)$ onto $H(Y)$. By Theorem 2 there are isomorphisms ϑ_1 and ϑ_2 of $H(X)$ and $H(Y)$ onto $S_1 = S(\mathfrak{M}\mathbf{R}; I, \Lambda; [e_\lambda, f_i])$ and $S_2 = S(\mathfrak{M}\mathbf{R}; I', \Lambda'; [e'_{\lambda'}, f'_{i'}])$, respectively. Since $\varphi_1 = \vartheta_1^{-1}\varphi\vartheta_2$ is an isomorphism of S_1 onto S_2 (see [7], [8]) there exists an automorphism ψ of the semigroup $\mathfrak{M}\mathbf{R}$, there are one-to-one mappings $i \mapsto i'$, $\lambda \mapsto \lambda'$ of I and Λ onto I' and Λ' respectively, and there exist mappings $i \mapsto g_i$, $\lambda \mapsto v_\lambda$ of I and Λ into the group $\mathfrak{M}\mathbf{R} \setminus \{0\}$ such that, for all $\alpha \in \mathbf{R}$, $i \in I$ and $\lambda \in \Lambda$, we have

$$(\alpha; i, \lambda)\varphi_1 = \{g_i(\alpha\psi)v_\lambda; i', \lambda'\},$$

$$[e_\lambda, f_i]\psi = v_\lambda [e'_{\lambda'}, f'_{i'}]g_i.$$

We define one-to-one mappings u and v of X and X^* onto Y and Y^*, respectively, by

$$(\alpha e_\lambda)u = \alpha\psi v_\lambda e'_{\lambda'} \quad \text{and} \quad (f_i\beta)v = f'_{i'}g_i\beta\psi.$$

One can easily prove that, for every $x \in X$ and $f \in X^*$,

$$[x, f]\psi = [xu, fv]$$

holds. By making use of the method of [1] it can be shown that ψ is an automorphism of the field **R** and u is the semilinear mapping of X onto Y corresponding to ψ. Since the only automorphism of the field **R** is the identity automorphism we obtain that u is a linear mapping of X onto Y. Hence, for every $x \in X$ and $f \in X^*$, we have

$$[x, f] = [xu, fv].$$

We show that u is continuous in the weak topologies $\sigma(X, X^*)$ and $\sigma(Y, Y^*)$. In fact, if the generalized sequence $\{x_s\}$ from X converges weakly to the zero element of X then $[x_s, f]$ converges to zero for every $f \in X^*$ and so the same is valid for $[x_s u, fv]$. This implies that the generalized sequence $\{x_s u\}$ converges weakly to the zero element of Y by which we infer that u is a continuous mapping of X onto Y with respect to the topologies $\sigma(X, X^*)$ and $\sigma(Y, Y^*)$. Hence, by applying Proposition 14 in [9], p. 96, we obtain that u is continuous also in the Mackey topologies $\tau(X, X^*)$ and $\tau(Y, Y^*)$. One can prove similarly that u^{-1} is continuous both in the weak and in the Mackey topologies. Thus necessity is proved.

The sufficiency of the condition is obvious. Therefore we turn to the last assertion of the theorem. If y is an arbitrary point from Y and $a \in H(X)$, $a\vartheta_1 = (\alpha; i, \lambda)$ then

$$\begin{aligned} y(a\varphi) &= y(a\vartheta_1 \varphi_1 \vartheta_2^{-1}) = y\{(\alpha; i, \lambda)\varphi_1 \vartheta_2^{-1}\} = \\ &= y\{(g_i \alpha v_\lambda; i', \lambda')\vartheta_2^{-1}\} = [y, f'_{i'}] g_i \alpha v_\lambda e'_{\lambda'} = \\ &= [y, f_i v]\alpha v_\lambda e'_{\lambda'} = [yu^{-1}, f_i]\alpha v_\lambda e'_{\lambda'} = [yu^{-1}, f_i \alpha] v_\lambda e'_{\lambda'} = \\ &= \{[yu^{-1}, f_i \alpha] e_\lambda\} u = y(u^{-1} a u). \end{aligned}$$

Hence $a\varphi = u^{-1} a u$ and the proof is complete.

Corollary (W o o d [4]). *Let X and Y satisfy the assumptions of the previous theorem. The semigroups $S(X)$ and $S(Y)$ are isomorphic to each other if and only if X and Y are topologically isomorphic.*

Proof. The proof follows directly from the previous Theorem because $H(X)$ and $H(Y)$ are the kernels of the semigroups $S(X)$ and $S(Y)$, respectively.

4.

Theorem. *Let* X *be a real locally convex Hausdorff space with weak or Mackey topology, and let* $\dim X > 1$. *Every automorphism of* $H(X)$ *is a restriction of an inner automorphism of* $S(X)$ *to* $H(X)$.

The proof is straightforward from Theorem 3.

Corollary (Wood [4]). *Every automorphism of the semigroup* $S(X)$ *is an inner automorphism.*

Proof. Let φ be an automorphism of $S(X)$. Since $H(X)$ is the kernel of $S(X)$ it follows from Theorem 4 that there exists a topological isomorphism u of X onto itself for which $a\varphi = u^{-1}au$ is satisfied for every $a \in H(X)$. Clearly, the mapping φ_u of $S(X)$ onto itself defined by the equality $b\varphi_u = u^{-1}bu$ for every $b \in S(X)$ is an automorphism of the semigroup $S(X)$. Hence $\varphi\varphi_u^{-1} = \varphi_1$ is an automorphism of $S(X)$ which is identical on $H(X)$. If $(\alpha; i, \lambda)$ is an arbitrary element in $H(X)$ and $b \in S(X)$ then

$$(\alpha; i, \lambda)b = [(\alpha; i, \lambda)b]\varphi_1 = (\alpha; i, \lambda)b\varphi_1 .$$

This implies

$$x[(\alpha; i, \lambda)b] = x[(\alpha; i, \lambda)b\varphi_1]$$

or

$$\{[x, f_i]\alpha e_\lambda\}b = \{[x, f_i]\alpha e_\lambda\}b\varphi_1$$

for every $x \in X$. By choosing f_i such that $[x, f_i] \neq 0$ we obtain

$$(\alpha e_\lambda)b = (\alpha e_\lambda)b\varphi_1 ,$$

i.e. $b\varphi_1 = b$. Thus, φ_1 is the identity automorphism of $S(X)$ and hence $\varphi = \varphi_u$. We mention that Theorem 3 implies as a corollary that a space X is determined by the ring of its continuous linear transformations [5].

5.

Theorem. *Let* X *and* Y *be real locally convex Hausdorff spaces both with weak topology or both with Mackey topology, such that* $\dim X > 1$ *and let* C $[D]$ *be a subsemigroup of the semigroup* $S(X)$ $[S(Y)]$ *containing* $H(X)$ $[H(Y)]$. *Every isomorphism* φ *of* C *onto* D *has the form*

$$a\varphi = u^{-1}au \qquad (a \in C)$$

where u *is a topological isomorphism of* X *onto* Y.

Proof. This theorem can be proved by the method used in proving the Corollary to Theorem 4.

6.

Now let X be a complex locally convex space.

Theorem. *For every maximal subgroup* G *of the semigroup* $H(X)$ *there exists a metric on the set* $G^0 = G \cup \{0\}$ *where* 0 *is the zero of* X, *such that the semigroup* G^0 *together with this metric is a metric and hence a topological semigroup.*

Proof. Let G be an arbitrary maximal subgroup of $H(X)$. Then there exists an $\mathscr{H}$-class $H_{i\lambda}$ in the semigroup $S = S(\mathfrak{M}\mathbf{C}; I, \Lambda; [e_\lambda, f_i])$ which is a group and $G\vartheta = H_{i\lambda}$, where ϑ is the isomorphism of the semigroup $H(X)$ onto S, defined in the proof of Theorem 2. Hence, for every $a \in G^0$ we have $a\vartheta = (\alpha; i, \lambda)$ for some $\alpha \in \mathfrak{M}\mathbf{C}$, which implies $xa = [x, f_i]\alpha e_\lambda$ for every $x \in X$. Since the subspace $X_\lambda = [e_\lambda]$ in X is of dimension 1 it is normed (cf. [9], Theorem 5, p. 60). A norm can be defined for every transformation $a \in G^0$ by

$$\|a\| = \sup\{\|xa\| : \|x\| \leqslant 1, \ x \in [e_\lambda]\}.$$

On G^0 we define the metric $\rho(a, b) = \|a - b\|$. Clearly, the semigroup G^0_ρ is metric and so it is also a topological semigroup. Indeed, the validity of this assertion follows from the inequality

$$\rho(ab, a_0 b_0) = \| ab - a_0 b_0 \| \leqslant \| a_0 \| \| b - b_0 \| +$$

$$+ \| b_0 \| \| a - a_0 \| + \| b - b_0 \| \| a - a_0 \|.$$

The theorem is proved.

7.

Theorem. *For every maximal subgroup G in the semigroup $H(X)$ the topological semigroup G^0_ρ is topologically isomorphic to the semigroup $\mathfrak{M}\mathbf{C}$ (with the natural topology of $\mathbf{C}$).*

Proof. Let G be a maximal subgroup of $H(X)$. By Theorem 2, there exists an isomorphism ϑ of the semigroups $H(X)$ onto $S = S(\mathfrak{M}\mathbf{C}; I, \Lambda; [e_\lambda, f_i])$. Clearly, $G^0 \vartheta = H^0_{i\lambda} = H_{i\lambda} \cup \{0\}$ where $H_{i\lambda}$ is an $\mathscr{H}$-class of the semigroup S. It is easy to see that the mapping of $\mathfrak{M}\mathbf{C}$ onto $H^0_{i\lambda}$ defined by

$$\alpha\varphi = ([e_\lambda, f_i]^{-1} \alpha; i, \lambda)$$

is an isomorphism of the semigroup $\mathfrak{M}\mathbf{C}$ onto $H^0_{i\lambda}$.

Hence the isomorphism $\vartheta\varphi^{-1}$ is a homomorphism of the space G^0_ρ onto $\mathfrak{M}\mathbf{C}$. In fact for every $a, b \in G^0_\rho$ we have

$$\rho(a, b) = \| a - b \| = \sup \{ \| x(a - b) \| : \| x \| \leqslant 1, \; x \in [e_\lambda] \} =$$

$$= \sup \{ \| xa - xb \| : \| x \| \leqslant 1, \; x \in [e_\lambda] \} =$$

$$= \sup \{ \| [x, f_i] \alpha e_\lambda - [x, f_i] \beta e_\lambda \| : \| x \| \leqslant 1, \; x \in [e_\lambda] \} =$$

$$= \sup \{ \| [x, f_i] (\alpha - \beta) e_\lambda \| : \| x \| \leqslant 1, \; x \in [e_\lambda] \} =$$

$$= \sup \{ \| [x, f_i] \| : \| x \| \leqslant 1, \; x \in [e_\lambda] \} \cdot | \alpha - \beta | =$$

$$= \| f_i \| \cdot | \alpha - \beta | =$$

$$= \| f_i \| \cdot | a\vartheta\varphi^{-1} [e_\lambda, f_i]^{-1} - b\vartheta\varphi^{-1} [e_\lambda, f_i]^{-1} | =$$

$$= \| f_i \| \cdot | [e_\lambda, f_i]^{-1} | \cdot | a\vartheta\varphi^{-1} - b\vartheta\varphi^{-1} |.$$

This implies that $\vartheta\varphi^{-1}$ is a homeomorphism of the space G^0_ρ onto $\mathfrak{M}\mathbf{C}$.

8.

Definition. Let X and Y be complex locally convex spaces. An isomorphism of the semigroup $H(X)$ onto $H(Y)$ is said to be partially topological if φ is a topological isomorphism on the semigroup G^0_ρ for some maximal subgroup G of the semigroup $H(X)$.

Theorem. *Let X and Y be complex locally convex Hausdorff spaces both with weak topology or both with Mackey topology and let* $\dim X > 1$. *The semigroups $H(X)$ and $H(Y)$ are partially topologically isomorphic to each other if and only if the spaces X and Y are in a topological semilinear correspondence. Every partially topological isomorphism φ of $H(X)$ onto $H(Y)$ has the form*

$$a\varphi = u^{-1}au \qquad (a \in H(X))$$

where u is a topological semilinear correspondence between X and Y.

Proof. The sufficiency of the condition of the theorem can be checked easily. We prove its necessity. By Theorem 2, there are isomorphisms ϑ_1 and ϑ_2 of $H(X)$ and $H(Y)$ onto $S_1 = S(\mathfrak{M}\mathbf{C}; I, \Lambda; [e_\lambda, f_i])$ and $S_2 = S(\mathfrak{M}\mathbf{C}; I', \Lambda'; [e'_{\lambda'}, f'_{i'}])$, respectively. Let φ be a partially topological isomorphism of the semigroup $H(X)$ onto $H(Y)$. Let G be the maximal subgroup of $H(X)$ on which φ is a topological isomorphism. Then $G\varphi$ is a maximal subgroup of $H(Y)$. Clearly, there are indices $i \in I$, $\lambda \in \Lambda$, $i' \in I'$, $\lambda' \in \Lambda'$ with the property that $G^0\vartheta_1 = H^0_{i\lambda}$, $G'^0\vartheta_2 = H'^0_{i'\lambda'}$ where $G' = G\varphi$ and $H_{i\lambda}$ and $H'_{i'\lambda'}$ are $\mathscr{H}$-classes of the semigroups S_1 and S_2, respectively. Let us denote by ψ_1 and ψ_2 the isomorphisms of $\mathfrak{M}\mathbf{C}$ onto $H^0_{i\lambda}$ and $H'^0_{i'\lambda'}$, respectively, defined in the proof of Theorem 7. By Theorem 7, $\vartheta_1\psi_1^{-1}$ is a topological isomorphism of the topological semigroup G^0_ρ onto $\mathfrak{M}\mathbf{C}$ and $\vartheta_2\psi_2^{-1}$ is a topological isomorphism of G'^0_ρ onto $\mathfrak{M}\mathbf{C}$. Thus $\psi = \psi_1\vartheta_1^{-1}\varphi\vartheta_2\psi_2^{-1}$ is a topological automorphism of the semigroup $\mathfrak{M}\mathbf{C}$. The isomorphism $\varphi_1 = \vartheta_1^{-1}\varphi\vartheta_2$ of S_1 onto S_2 induces (see [7], [8]) one-to-one mappings $i \mapsto i'$, $\lambda \mapsto \lambda'$ of the sets I and Λ onto I' and Λ', respectively, and mappings $i \mapsto u_i$, $\lambda \mapsto v_\lambda$ of the sets I and Λ into the group $\mathfrak{M}\mathbf{C} \setminus \{0\}$ such that, for every $\alpha \in \mathbf{C}$, $i \in I$ and $\lambda \in \Lambda$, we have

$$(\alpha; i, \lambda)\varphi_1 = \{u_i(\alpha\psi)v_\lambda ; i', \lambda'\},$$

$$[e_\lambda, f_i]\psi = v_\lambda[e'_{\lambda'}, f'_{i'}]u_i.$$

Let the mapping u $[v]$ of the space X $[X^*]$ onto Y $[Y^*]$ be defined by $(\alpha e_\lambda)u = \alpha\psi v_\lambda e'_{\lambda'}$ $[(f_i\beta)v = f'_{i'}\beta\psi u_i]$. Then, for every $x \in X$, $f \in X^*$, the equality

$$(2) \qquad [x, f]\psi = [xu, fv]$$

holds. Using the method of [1] it can be shown that ψ is an automorphism of the field $\mathbf{C}$ and that if u is the semilinear transformation between X and Y corresponding to ψ then u is an isomorphism of the group X onto Y. We rewrite equality (2) in the form

$$(3) \qquad [x, f'v^{-1}]\psi = [xu, f'].$$

Since ψ is a topological automorphism of the space $\mathbf{C}$ we have, for every $\alpha \in \mathbf{C}$, either $\alpha\psi = \alpha$ or $\alpha\psi = \bar{\alpha}$. Thus, for every $x \in X$ and $f' \in Y^*$, we have either $[x, f'v^{-1}] = [xu, f']$ or $\overline{[x, f'v^{-1}]} = [xu, f']$. By [9], Proposition 12, p. 61 and by its corollary this implies that u is continuous in the weak topologies $\sigma(X, X^*)$ and $\sigma(Y, Y^*)$ and v^{-1} is continuous in the weak topologies $\sigma(Y^*, Y)$ and $\sigma(X^*, X)$. Using this and [9], Proposition 14, p. 96 we obtain that u is continuous in the Mackey topologies $\tau(X, X^*)$, $\tau(Y, Y^*)$. Similarly, it can be proved that u^{-1} is continuous in both the weak and the Mackey topologies. Thus, u is a topological semilinear correspondence between the spaces X and Y.

Finally, we prove the last statement of the theorem. If $x \in Y$ and $a \in H(X)$, $a\vartheta_1 = (\alpha; i, \lambda)$ then

$$\begin{aligned} x(a\varphi) &= x(a\vartheta_1\varphi_1\vartheta_2^{-1}) = x\{(\alpha; i, \lambda)\varphi_1\vartheta_2^{-1}\} = \\ &= x[u_i(\alpha\psi)v_\lambda; i', \lambda']\vartheta_2^{-1} = [x, f'_{i'}]u_i(\alpha\psi)v_\lambda e'_{\lambda'} = \\ &= [x, f_i v](\alpha\psi)v_\lambda e'_{\lambda'} = [xu^{-1}, f_i]\psi\alpha\psi v_\lambda e'_{\lambda'} = \\ &= \{[xu^{-1}, f_i\alpha]e_\lambda\}u = x(u^{-1}au) \end{aligned}$$

by which $a\varphi = u^{-1}au$. The proof is complete.

9.

Definition. The isomorphism ψ of the semigroup $S(X)$ onto $S(Y)$ is called k-partially topological if ψ is a partially topological isomorphism on the semigroup $H(X)$.

Theorem. *Let X and Y be complex locally convex Hausdorff spaces both with the weak topology or both with the Mackey topology, and let* $\dim X > 1$. *The semigroups $S(X)$ and $S(Y)$ are k-partially topologically isomorphic if and only if the spaces X and Y are in a topological semilinear correspondence. Every k-partially topological isomorphism φ of the semigroup $S(X)$ onto $S(Y)$ has the form*

$$a\varphi = u^{-1}au \qquad (a \in S(X))$$

where u is a topological semilinear mapping of X onto Y.

Proof. The first part of the theorem follows from Theorem 8 because $H(X)$ and $H(Y)$ are the kernels of the semigroups $S(X)$ and $S(Y)$, respectively. Let φ be a k-partially topological isomorphism of the semigroup $S(X)$ onto $S(Y)$. Since $H(X)$ and $H(Y)$ are the kernels of the semigroups $S(X)$ and $S(Y)$, respectively, Theorem 8 guarantees the existence of a topological semilinear mapping u between X and Y such that $a\varphi = u^{-1}au$ holds for every $a \in H(X)$. Clearly the mapping φ_u between the semigroups $S(X)$ and $S(Y)$ defined by $b\varphi_u = u^{-1}bu$ $(b \in S(X))$ is an isomorphism. Hence $\varphi_1 = \varphi\varphi_u^{-1}$ is an automorphism of the semigroup $S(X)$ which is identical on $H(X)$. If $(\alpha; i, \lambda) \in H(X)$ and $b \in S(X)$ then

$$(\alpha; i, \lambda)b = [(\alpha; i, \lambda)b]\varphi_1 = (\alpha; i, \lambda)b\varphi_1 \,.$$

Thus we have

$$x[(\alpha; i, \lambda)b] = x[(\alpha; i, \lambda)b\varphi_1]$$

or

$$\{[x, f_i]\alpha e_\lambda\}b = \{[x, f_i]\alpha e_\lambda\}b\varphi_1$$

for every $x \in X$. By choosing f_i such that $[x, f_i] \neq 0$ we obtain that

$$(\alpha e_\lambda)b = (\alpha e_\lambda)b\varphi_1$$

whence $b\varphi_1 = b$. Thus φ_1 is the identity automorphism of the semigroup $S(X)$, and so $\varphi = \varphi_u$. The proof is complete.

REFERENCES

[1] L.M. Gluskin, Semigroups and rings of endomorphisms of linear spaces, *Izv. Akad. Nauk SSSR,* 23 (1959), 841–870 (in Russian).

[2] L.G. Mustafaev, Semigroups and rings of continuous linear operators, *Theses of the II. All-Union Symposium on the Theory of Semigroups,* Sverdlovsk, 1978 (in Russian).

[3] L.G. Mustafaev, Semigroups of continuous linear operators, *Soobšč. Akad. Nauk Gruzin.* SSR, 95 (1979) (in Russian).

[4] G.R. Wood, A note on automorphisms of semigroups and near-rings of mappings, *Studia Math.,* 62 (1978), 209–218.

[5] G.W. Mackey, On convex topological linear spaces, *Trans. Amer. Math. Soc.,* 60 (1946), 519–537.

[6] A.V. Mihalev, On the isomorphism of rings of continuous endomorphisms, *Sibirsk. Mat. Ž.,* 4 (1963), 177–186 (in Russian).

[7] E.S. Ljapin, *Semigroups,* Fizmatgiz, Moskow, 1959 (in Russian); English translation: Amer. Math. Soc., Providence, R. I., 1963.

[8] A.H. Clifford – G.B. Preston, *The Algebraic Theory of Semigroups* I, II, Amer. Math. Soc., Providence, R. I., 1961, 1967.

[9] A.P. Robertson – W.J. Robertson, *Topological Vector Spaces,* Cambridge Tracts in Mathematics, 53, Cambridge, 1964.

L.G. Mustafaev

Institute of Mathematics and Mechanics of the Azerbaijan Academy, Baku, USSR.

COLLOQUIA MATHEMATICA SOCIETATIS JÁNOS BOLYAI
39. SEMIGROUPS, SZEGED (HUNGARY), 1981.

TOPICS IN THE STRUCTURE THEORY OF REGULAR SEMIGROUPS

K.S.S. NAMBOORIPAD

1. INTRODUCTION

A semigroup S is said to be regular if $x \in xSx$ for all $x \in S$. This is an adaptation to semigroups of a concept, that of a regular ring, introduced by von Neumann [40] in order to coordinatize certain lattices of projections of a Hilbert space. An important class of examples for regular semigroups is therefore the class of multiplicative semigroups of regular rings. Other examples include the semigroup of all full [partial] transformations of a set, the endomorphism semigroup of a vector space, the semigroup of all projectivities of a projective space etc. These examples show that regular semigroups form an extensive class of semigroups that arise naturally in various contexts. It is not surprising therefore that a large number of structure theorems in the algebraic theory of semigroups, including some of the basic results of the theory, concern regular semigroups. The Suschkewitsch – Rees theorem [36], [39] describing the structure of completely 0-simple semigroups and Clifford's theorem [2] describing the structure of inverse semigroups that are unions of groups are examples of such fundamental results which had definite influence on the later developments in the whole semigroup theory.

The crucial observation regarding regular semigroups underlying implicitly or explicitly all the structure theorems obtained so far is that the structure of a regular semigroup S is determined to a large extent by the set $E(S)$ of its idempotents. The principal technique used prior to 1970 to produce structure theorems was to impose various types of restrictions on $E(S)$. An excellent survey of such results are available in G. Lallement's review paper [12]. Rees construction provides another powerful technique. Results obtained using this method are discussed in J. Meakin's paper [19]. A more detailed bibliography of papers on the structure theory of regular semigroups is available in these papers.

Even though interest in various special classes of regular semigroups started very early in the development of semigroup theory, a systematic investigation of regular semigroups in general, started only in 1960. In fact G. Lallement's thesis [11] was the first paper to be published containing such an investigation. In this survey I shall confine myself to a few results obtained after 1970 to illustrate various techniques that were useful in analysing the structure of regular semigroups.

2. THE FUNDAMENTAL REGULAR SEMIGROUPS

A semigroup S is said to be fundamental if the maximum congruence μ contained in the Green's relation $\mathscr{H}$ is the identity relation. For any semigroup S, it is known that S/μ is fundamental. This fact indicates the importance of the class of fundamental semigroups, since it shows that every semigroup can be obtained as an $\mathscr{H}$-coextension of a suitable fundamental semigroup. General methods for constructing $\mathscr{H}$-coextension are known from Grillet [6] or Leech [13].

It was W.D. Munn [25] who first determined the structure of a non-trivial class of fundamental semigroups. He showed in [25] that the set of all principal ideal isomorphisms T_E of a semilattice E is a fundamental inverse semigroup whose semilattice of idempotents is isomorphic to E and having the property that if S is any fundamental inverse semigroup whose semilattice is isomorphic to E, then S is isomorphic to a full subsemigroup of T_E. T.E. Hall generalized Munn's construction by considering a band B instead of a semilattice and constructed the fundamental orthodox semigroup W_B having properties analogous to the Munn semigroup T_E [8], [9]. Later he generalized this further and gave a construction of fundamental regular semigroups [10]. Let S be a regular idempotent generated semigroup and $e \in E(S)$. Define $\lambda_e: S/\mathscr{R} \to S/\mathscr{R}$ by $\lambda_e R_g = R_{eg}$ and $\rho_e: S/\mathscr{L} \to S/\mathscr{L}$ by $L_h \rho_e = L_{he}$. Let $\alpha: \langle e \rangle \to \langle f \rangle$, where $e, f \in E(S)$ and $\langle e \rangle$ denotes the idempotent generated subsemigroup of eSe, be an isomorphism. Define $\bar{\alpha}$ and $\bar{\bar{\alpha}}$ by $\bar{\alpha} R_g = R_{g\alpha}$ and $L_g \bar{\bar{\alpha}} = L_{g\alpha}$ for all $g \in E(eSe)$. Hall shows that the set of all pairs of mappings of the form $(\bar{\alpha}\lambda_e, \rho_e \bar{\bar{\alpha}})$ is a fundamental regular subsemigroup T_S of $\mathscr{T}^{\circ}_{S/\mathscr{R}} \times \mathscr{T}_{S/\mathscr{L}}$ (where $\mathscr{T}^{\circ}_{S/\mathscr{R}}$ denotes the left-right dual of $\mathscr{T}_{S/\mathscr{R}}$ in which maps are written as left operators) such that there exists a homomorphism $\varphi: S \to T_S$ with $\ker \varphi = \mu$; that is, S/μ is isomorphic to the idempotent generated part of T_S.

Grillet [7] refined Hall's construction of fundamental regular semigroups. It is well-known that for any semigroup S, $S/\mathscr{R}$ $[S/\mathscr{L}]$ is a partially ordered set with respect to the relation defined by:

$$R_x \leqslant R_y \Leftrightarrow xS^1 \subseteq yS^1 \quad [L_x \leqslant L_y \Leftrightarrow S^1 x \subseteq S^1 y].$$

Grillet [7] gave an abstract characterization of the partially ordered sets $I = S/\mathscr{R}$ and $\Lambda = S/\mathscr{L}$ when S is regular. To formulate this characterization, we need some preliminary definitions.

Let $\varphi: I \to \Lambda$ be an order preserving map of partially ordered sets. We say that φ is *normal* if $\operatorname{im} \varphi = \{y' \in \Lambda: y' \leqslant y\} = \Lambda(y)$ for some $y \in \Lambda$ and for all $x \in I$ there exists $x' \leqslant x$ such that $\varphi \mid I(x')$ is an isomorphism onto $\Lambda(x\varphi)$. In particular, since $y \in \Lambda(y) = \operatorname{im} \varphi$ there exists $x \in I$ such that $\varphi | I(x)$ is an isomorphism onto $\Lambda(y) = \operatorname{im} \varphi$. We denote the set of all such elements, x, by $M(\varphi)$. We say that a principal order ideal $I(x)$ of I is a *normal retract* if there exists an idempotent normal mapping ϵ of I into I such that $\operatorname{im} \epsilon = I(x)$; ϵ is called a *projection* upon $I(x)$. If every principal order ideal of a partially ordered set I is a normal retract of I then I is said to be *regular*. If $I = S/\mathscr{R}$ for some regular semigroup S, then for every $R_e \in I$ $(e \in E(S))$ it can be checked that $\lambda_e: R_g \mapsto R_{eg}$ is a projection upon $I(R_e)$ and so I is a regular partially ordered set (cf. [27]).

It is easy to see that normal mappings compose and so the subset $S(I)$ of $\mathscr{T}_I$ consisting of normal mappings, is a subsemigroup of $\mathscr{T}_I$. We denote by $S^\circ(I)$ the left-right dual of $S(I)$ (the corresponding subsemigroup of $\mathscr{T}_I^\circ$). Suppose that $f \in S(I)$, $x \in M(f)$ and let ϵ be a projection upon $\operatorname{im} f = I(y)$. Then $g = \epsilon(f \mid I(x))^{-1}$ is an inverse of f in $S(I)$ and so f is a regular element of $S(I)$. Conversely if f is regular, then L_f contains an idempotent in $S(I)$ and so, by Lemma 2.5 of [5], $\operatorname{im} f$ is a normal retract of I. Therefore, it follows that $S(I)$ is a regular semigroup such that $S(I)/\mathscr{L} \approx I$ if and only if I is a regular partially ordered set. We thus have the following result due to Grillet [7].

Theorem 2.1. *A partially ordered set I is isomorphic to the partially ordered set of $\mathscr{R}$ [$\mathscr{L}$]-classes of a regular semigroup if and only if I is regular.*

By a *normal equivalence relation* on a partially ordered set I we mean an equivalence relation σ such that $\sigma = \ker f$ for some $f \in S(I)$. The partially orderd set of all normal equivalence relations on I ordered by reverse inclusion, is denoted by I°. It follows from Lemma 2.6 of [5] that I° is isomorphic to $S(I)/\mathscr{R}$. Consequently, if I is regular, so is I°.

Let I and Λ be two regular partially ordered sets. A cross-connection between I and Λ is a pair (Γ, Δ) of mappings $\Gamma\colon \Lambda \to I^\circ$ and $\Delta\colon I \to \Lambda^\circ$ satisfying the following conditions [27]:

(C1) For $x \in I$, $y \in \Lambda$, $x \in M(\Gamma(y)) \Rightarrow y \in M(\Delta(x))$.

Here, for $y \in \Lambda$, $M(\Gamma(y))$ denotes the set $M(f)$ for some $f \in S(I)$ with $\ker f = \Gamma(y)$. It is easy to see that $\ker f = \ker g$ (or $f \mathscr{R} g$ in $S(I)$) implies $M(f) = M(g)$ and so our notation $M(\Gamma(y))$ is unambiguous.

(C2) Let $x \in M(\Gamma(y))$ and let ϵ_1 [ϵ_2] be the projection upon $I(x)$ [$\Lambda(y)$] with $\ker \epsilon_1 = \Gamma(y)$ [$\ker \epsilon_2 = \Delta(x)$]. Then

$$\Delta(\epsilon_1 u) = (\Delta(u))\epsilon_2^{-1}, \quad \Gamma(v\epsilon_2) = \epsilon_1^{-1}(\Gamma(v))$$

for all $u \in I$ and $v \in \Lambda$.

In what follows we shall refer to the quadruple $[I, \Lambda; \Gamma, \Delta]$ as a *cross-connection.*

Let $\alpha: I \to I'$ be an order-isomorphism of regular partially ordered sets. For each $\sigma \in I^\circ$, define

$$\sigma\alpha^\circ = \{(u', v') \in I' \times I' : (u'\alpha^{-1}, v'\alpha^{-1}) \in \sigma\}.$$

It is easy to see that $\alpha^\circ: I^\circ \to I'^\circ$ is also an order-isomorphism. Given two cross-connections $[I, \Lambda; \Gamma, \Delta]$ and $[I', \Lambda'; \Gamma', \Delta']$ an isomorphism from $[I, \Lambda; \Gamma, \Delta]$ to $[I', \Lambda'; \Gamma', \Delta']$ is a pair (α, β) of order-isomorphisms $\alpha: I \to I'$, $\beta: \Lambda \to \Lambda'$ such that the following diagrams commute:

$$\begin{array}{ccc} I & \xrightarrow{\ \alpha\ } & I' \\ \Big\downarrow{\scriptstyle \Delta} & & \Big\downarrow{\scriptstyle \Delta'} \\ \Lambda^\circ & \xrightarrow[\ \beta^\circ\ ]{} & \Lambda'^\circ \end{array} \qquad \begin{array}{ccc} \Lambda & \xrightarrow{\ \beta\ } & \Lambda' \\ \Big\downarrow{\scriptstyle \Gamma} & & \Big\downarrow{\scriptstyle \Gamma'} \\ I^\circ & \xrightarrow[\ \alpha^\circ\ ]{} & I'^\circ \end{array}$$

We are now ready to state the result describing Grillet's construction of fundamental regular semigroups [7].

Theorem 2.2. *Let* $[I, \Lambda; \Gamma, \Delta]$ *be a cross-connection and let* $U = U(I, \Lambda; \Gamma, \Delta)$ *denote the subset of* $S^\circ(I) \times S(\Lambda)$ *consisting of all pairs* (f, g) *satisfying the following conditions:*

(C3) $\operatorname{im} f = I(x)$, $\operatorname{im} g = \Lambda(y) \Rightarrow \ker f = \Gamma(y)$, $\ker g = \Delta(x)$.

(C4) *For all* $u \in I$ *and* $v \in \Lambda$, $\Delta(fu) = (\Delta(u))g^{-1}$, $\Gamma(vg) = f^{-1}(\Gamma(v))$.

Then U *is a fundamental regular subsemigroup of* $S^\circ(I) \times S(\Lambda)$.

Conversely if S *is any regular semigroup, then* S *induces a cross-connection* (Γ_S, Δ_S) *between regular partially ordered sets* $I_S = S/\mathscr{R}$ *and* $\Lambda_S = S/\mathscr{L}$ *defined as follows: For all* $L_g \in \Lambda_S$ *and* $R_h \in I_S$

$$\Gamma(L_g) = \ker \lambda_g, \quad \Delta(R_h) = \ker \rho_h .$$

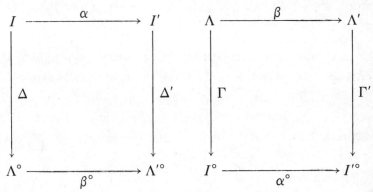

Further, there exists a natural homomorphism φ: $S \to U(I_S, \Lambda_S; \Gamma_S, \Delta_S)$ *such that* $\ker\varphi = \mu$. S *is isomorphic to a full subsemigroup of* $U(I, \Lambda; \Gamma, \Delta)$ *if and only if* S *is fundamental and the cross-connection* $[I_S, \Lambda_S; \Gamma_S, \Delta_S]$ *induced by* S *is isomorphic to* $[I, \Lambda; \Gamma, \Delta]$.

If S is any regular semigroup, it is easy to see that the semigroup $U = U(I_S, \Lambda_S; \Gamma_S, \Delta_S)$ is the same as the semigroup T_{S_1} constructed by Hall [10], where S_1 is the idempotent generated part of S/μ. Also, the homomorphism φ: $S \to U(I_S, \Lambda_S; \Gamma_S, \Delta_S)$ given in the theorem above is the same as Hall's representation (λ, ρ) (cf. [10], §4).

The method of studying semigroups and rings by means of their one-sided ideals is classical. Since I $[\Lambda]$ is isomorphic to the partially ordered set of principal right [left] ideals, Grillet's theory of cross-connections may be viewed as a formalization of this classical method and cross-connections as representing the ideal structure of regular semigroups.

3. INDUCTIVE GROUPOIDS

Let S be a semigroup. By an *invariant* of S we mean an object A_S (such as a group, a partially ordered set, a topological space etc.) associated with S in some natural way such that the assignment $S \mapsto A_S$ is functorial from some category of semigroups containing S. For example, the largest group homomorphic image G_S the regular partially ordered sets I_S, Λ_S etc. are all invariants of a regular semigroup S. Many existing structure theorems give an explicit procedure for constructing the required semigroup from a given set of its invariants. The semigroup so obtained, depends not only on the invariants themselves, but also certain given relations between them. For example, McAlister's construction of E-unitary inverse semigroups describes the construction of the E-unitary inverse semigroup $P = P(G, \mathscr{X}, \mathscr{Y})$ from the invariants G (the maximum group homomorphic image of P), the semilattice $\mathscr{Y}$ and the partially ordered set $\mathscr{X}$ [15], [16]. The relation between them consists of the given embedding (inclusion) of $\mathscr{Y}$ in $\mathscr{X}$ and the action of G on $\mathscr{X}$. Similarly Theorem 2.2 constructs the semigroup $U(I, \Lambda; \Gamma, \Delta)$ from I and Λ; the relation between I and Λ is specified by the cross-connection (Γ, Δ). In this section we shall describe another construction of regular semigroups that uses two other invariants, viz., a biordered set and an ordered groupoid.

For definitions of biordered sets and related concepts like bimorphisms, sandwich sets etc. the reader may refer [4] and [28]. The paper by J. Meakin [19] (which appears elsewhere in this volume) also contains an adequate account of these. We observe that the class of all regular biordered sets forms a category **RB** with morphisms as bimorphisms and that there exists a functor, denoted by E from the category **RS** of regular semigroups to **RB** which sends $S \in$ **RS** to its biordered sets $E(S)$. In this paper all biordered sets considered are regular.

By a *groupoid,* we mean a small category in which every morphism is an isomorphism. As in [28], if G is a groupoid, G also denotes its morphism set and $V(G)$ denotes the set of vertices of G. We shall always identify identities of G with the corresponding vertices (objects). If $x \in G$, e_x and f_x denote the domain [left identity] and codomain [right identity] of x respectively.

Let G be a groupoid and $\leqslant$ be a partial order on G. We shall say that G is an ordered groupoid with respect to $\leqslant$ if the following axioms hold:

(OG1) Let $x \leqslant u$, $y \leqslant v$. If xy and uv exist in G then $xy \leqslant uv$.

(OG2) $x \leqslant y$ implies $x^{-1} \leqslant y^{-1}$.

(OG3) Let $x \in G$, $e \in V(G)$ and $e \leqslant e_x$. Then there exists a unique $e * x \in G$ (called the *restriction* of x to e) such that $e * x \leqslant x$ and $e_{e * x} = e$.

If $x \in G$, $f \in V(G)$ and $f \leqslant f_x$ then the element

(3.1) $\qquad x * f = (f * x^{-1})^{-1}$

may be called the *corestriction* of x to f; $x * f$ is the unique element such that $x * f \leqslant x$ and $f_{x * f} = f$. Axioms for ordered groupoids are self dual in the sense that we may define ordered groupoids equivalently by replacing the one-sided axiom (OG3) by its dual which asserts the existence of corestriction. It is easy to see that the class of all ordered groupoids forms a category **OG** in which morphisms are order-preserving functors and that V defined by the assignments $G \mapsto V(G)$, $\varphi \mapsto V(\varphi)$ is a functor from **OG** to the category of ordered sets.

Examples of ordered groupoids are many. If X is a set then the set of all one-to-one partial transformations with usual composition but restricted to the pairs (α, β) such that $\operatorname{codom} \alpha = \operatorname{dom} \beta$ and with inclusion as partial order, is an ordered groupoid G_X. More generally, if S is an inverse semigroup, Schein [37] showed that $G(S) = (S(*), \leqslant)$ where $S(*)$ denotes the trace of S and $\leqslant$ denotes the natural partial order on S (cf. [5], Vol. II) is an ordered groupoid. Recall (cf. [28]) that the trace of a regular semigroup S is the partial algebra $S(*)$ where the partial binary operation $(*)$ is defined as follows:

$$x * y = \begin{cases} xy & \text{if } xy \in R_x \cap L_y \\ \text{undefined otherwise.} \end{cases} \tag{3.2}$$

Schein showed that $G(S)$ is an ordered groupoid in which $V(G(S))$ is a semilattice and that every ordered groupoid whose vertex set is a semilattice arises as $G(S)$ for some inverse semigroup S [37]. Note that the ordered groupoid G_X is in fact the ordered groupoid $G(\mathscr{I}_X)$ of the symmetric inverse semigroup $\mathscr{I}_X$ on the set X.

Let E be a biordered set. Since $R = \omega^r \cap (\omega^r)^{-1}$ is an equivalence relation on E we may regard R as a groupoid with $V(R) = E$; R becomes an ordered groupoid when we define partial order on R by

$$(g, h) \leqslant (e, f) \Leftrightarrow g\omega e \quad \text{and} \quad h = gf.$$

Similarly $L = \omega^l \cap (\omega^l)^{-1}$ is also an ordered groupoid. The identity relation 1_E is similarly an ordered groupoid which is an ordered subgroupoid of both R and L. As in [28], it can be shown that there exists an ordered groupoid $\mathscr{G}(E)$ which is the "free product of R and L amalgamating 1_E"; that is, the following diagram is a push-out in **OG**:

$$\begin{array}{ccc} R & \longrightarrow & \mathscr{G}(E) \\ \uparrow & & \uparrow \\ 1_E & \longrightarrow & L \end{array}$$

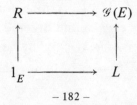

Here $1_E \to R$ and $1_E \to L$ are inclusions. It follows that morphisms in $\mathscr{G}(E)$ may be represented as

$$c = c(e_0, \ldots, e_n) = (e_0, e_1)(e_1, e_2) \ldots (e_{n-1}, e_n)$$

where for $i = 1, 2, \ldots, n$, $(e_{i-1}, e_i) \in R \cup L$. If $c = c(e_0, \ldots, e_n)$, $c' = c(f_0, \ldots, f_m)$ are morphisms in $\mathscr{G}(E)$ the composition cc' exists in $\mathscr{G}(E)$ if and only if $e_n = f_0$ and then $cc' = c(e_0, \ldots, e_n, f_1, \ldots, f_m)$. Identities in $\mathscr{G}(E)$ are morphisms of the form $c(e, e)$ which we identify with e. The partial order in $\mathscr{G}(E)$ is the natural extension of the partial orders in R and L; in fact if $c \in \mathscr{G}(E)$ and $e \omega e_c$, then the restriction of $c = c(e_0, \ldots, e_n)$ to e is $e * c = c(h_0, h_1, \ldots, h_n)$ where $h_0 = e$ and for $i = 1, \ldots, n$, $h_i = e_i h_{i-1} e_i$. Morphisms of $\mathscr{G}(E)$ are called E-chains and the groupoid $\mathscr{G}(E)$ is called the groupoid of E-chains. An E-cycle is an E-chain c with $e_c = f_c$; that is, an automorphism of some vertex of $\mathscr{G}(E)$. E-cycles of the form $c(e, f, fg, eg, e)$ with $e, f \in \omega^r(g)$, eLf or $c(e, f, gf, ge, e)$ with $e, f \in \omega^l(g)$, eRf are called singular E-cycles.

The construction of $\mathscr{G}(E)$ can be extended to a functor $\mathscr{G}$: **RB** $\to$ $\to$ **OG**. If $\theta\colon E \to E'$ is a bimorphism, $\mathscr{G}(\theta)$ is defined by the assignment $c \mapsto c\theta$ where $c\theta = c(e_0\theta, e_1\theta, \ldots, e_n\theta)$ and setting $V(\mathscr{G}(\theta)) = \theta$.

We are now ready to define the basic concept of this section.

Definition. Let E be a biordered set, G an ordered groupoid and $\epsilon\colon \mathscr{G}(E) \to G$ an order preserving functor such that $V(\epsilon)$ is an order-isomorphism. We say that the pair (G, ϵ) is an *inductive groupoid* if the following axioms and their duals (shown in brackets) hold.

(IG1) Let $x \in G$ and $e \in E$ with $\epsilon(e) = e_x$. Suppose that $e_i \in \omega(e)$ for $i = 1, 2$ and for each i let f_i be the unique element in E such that $\epsilon(f_i) = f_{\epsilon(e_i) * x}$. If $e_1 \omega^r e_2$ $[e_1 \omega^l e_2]$ then $f_1 \omega^r f_2$ $[f_1 \omega^l f_2]$ and

$$\epsilon(e_1, g)(g * x) = (x * f_1)\epsilon(f_1, h)$$

where $g = e_2 e_1 e_2$ and $h = f_2 f_1 f_2$.

Note that $e_2 e_1 e_2 = e_1 e_2$ if $e_1 \omega^r e_2$ and $e_2 e_1 e_2 = e_2 e_1$ if $e_1 \omega^l e_2$.

(IG2) If γ is any singular E-cycle in E then $\epsilon(\gamma) = \epsilon(e_\gamma)$.

We refer to the functor ϵ in the above definition as the *evaluation* of $\mathcal{G}(E)$ in G.

Let (G, ϵ), (G', ϵ') be inductive groupoids. An *inductive functor* from (G, ϵ) to (G', ϵ') is a pair (θ, φ) where $\theta\colon E \to E'$ is a bimorphism and $\varphi\colon G \to G'$ is an order preserving functor such that the following commutativity condition holds:

$$\mathcal{G}(\theta)\epsilon' = \epsilon\varphi.$$

To simplify notations, we shall denote inductive groupoids by G, G' etc. whose evaluations are ϵ, ϵ' etc. (or ϵ_G, $\epsilon_{G'}$ etc.). Since $V(\epsilon)$ is an isomorphism, we may identify $E = V(\mathcal{G}(E))$ with $V(G)$ by the isomorphism $V(\epsilon)$; that is, we may assume that $V(\epsilon) = 1_E$. If (θ, φ) is an inductive functor, this identification will also identify θ with $V(\varphi)$. Thus we may denote inductive functors by φ, φ' etc. where $V(\varphi)$, $V(\varphi')$ etc. are bimophisms of the associated biordered sets.

The class of all inductive groupoids clearly forms a category **IG** with morphisms as inductive functors and $V\colon G \mapsto V(G)$, $\varphi \mapsto V(\varphi)$ is a functor of **IG** to the category **RB** of biordered sets. Also there exists a natural forgetful functor from **IG** to **OG**.

Let S be a regular semigroup and let

$$G(S) = \{(x, x')\colon x \in S,\ x' \in V(x)\}$$

where $V(x)$ denotes the set of inverses of x. Define a partial binary operation in $G(S)$ by

$$(3.3) \qquad (x, x')(y, y') = \begin{cases} (xy, y'x') & \text{if } x'x = yy', \\ \text{undefined otherwise.} \end{cases}$$

When $x'x = yy'$, it is easy to see that $y'x' \in V(xy)$ and so (3.3) define a single-valued partial binary operation in $G(S)$. $G(S)$ is a groupoid with respect to this composition such that identities in $G(S)$ are (e, e), $e \in \in E(S)$. Hence setting $V(G(S)) = E(S)$ we obtain a groupoid whose vertex set is $E(S)$. It can be seen that $G(S)$ becomes an inductive groupoid when we define partial order and evaluation of $\mathscr{G}(E(S))$ in $G(S)$ as follows (cf. [28], p. 50–55):

$$(3.4) \qquad (x,x') \leqslant (y,y') \Leftrightarrow (xx')y = x,\ y'(xx') = x'$$

and if $c = c(e_0, e_1, \ldots, e_n) \in \mathscr{G}(E(S))$,

$$(3.5) \qquad \epsilon(c) = (e_0\, e_1 \ldots e_n,\ e_n\, e_{n-1} \ldots e_0).$$

We have the following result (cf. [28], Theorems 3.8 and 4.12).

Theorem 3.1. *Let* S *be a regular semigroup. Then the set* $\{(x,x'): x \in S,\ x' \in V(x)\}$ *becomes an inductive groupoid* $G(S)$ *with* $V(G(S)) = E(S)$ *when composition, partial order and evaluation are defined by* (3.3), (3.4) *and* (3.5), *respectively.*

Conversely, let G *be an inductive groupoid. Define the relation* p *on* G *by*

$$(3.6) \qquad xpy \Leftrightarrow e_x R e_y,\ f_x L f_y \quad \text{and} \quad \epsilon(e_x, e_y)y = x\epsilon(f_x, f_y).$$

Then p *is an equivalence relation on* G. *For* $\bar{x}, \bar{y} \in G/p$ *let*

$$(3.7) \qquad \bar{x}\bar{y} = \overline{(x \circ y)_h}$$

where $h \in S(f_x, e_y)$ *and* $(x \circ y)_h = (x * f_x h)\epsilon(f_x h, h)\epsilon(h, he_y)(he_y * y)$. *Then* (3.7) *defines a single-valued binary operation in* $G/p = S(G)$ *and* $S(G)$ *is a regular semigroup such that* $E(S(G)) \approx V(G)$.

Furthermore, we have natural isomorphisms

$$S \approx S(G(S)) \quad \text{and} \quad G \approx G(S(G))$$

for all regular semigroups S *and all inductive groupoids* G.

Observe that an ordered groupoid in which the vertex set is a semilattice is an inductive groupoid in the sense of our definition above in which the relation p defined by (3.6) is the identity relation. Further when S is an inverse semigroup, the uniqueness of inverse enables us to identify the set $G(S)$ with the set S itself and the composition defined by (3.3) coincides with that defined by (3.2). We thus obtain Schein's construction of inverse semigroups [37] as a particular case of Theorem 3.1 above.

The constructions of $G(S)$ and $S(G)$ described in the theorem above can be extended to functors G: **RS** → **IG** and S: **IG** → **RS**, respectively. If $\varphi: S \to S'$ is a homomorphism of regular semigroups, it can be shown that

$$(3.8) \qquad G(\varphi)(x, x') = (x\varphi, x'\varphi)$$

for all $(x, x') \in G(S)$ defines an inductive functor $G(\varphi): G(S) \to G(S')$ (cf. [28], p. 54) and if $\psi: G \to G'$ is an inductive functor, then

$$(3.9) \qquad \bar{x}S(\psi) = \overline{\psi(x)}$$

defines a homomorphism $S(\psi): S(G) \to S(G')$. We have the following (cf. [28], Theorems 4.13 and 4.14):

Theorem 3.2. *G and S defined by assignments:*

$$G\colon S \mapsto G(S), \qquad \varphi \mapsto G(\varphi)$$

and

$$S\colon G \mapsto S(G), \qquad \psi \mapsto S(\psi)$$

are functors G: **RS** → **IG** *and* S: **IG** → **RS**, *respectively. Moreover, these functors determine an adjoint equivalence of the categories* **IG** *and* **RS**.

To illustrate the application of the construction of regular semigroups using inductive groupoids, consider the set $T^*(E)$ of all ω-isomorphisms of a biordered set E. Recall [28] that an ω-isomorphism α is a biorder isomorphism of an ω-ideal $\omega(e_\alpha)$ onto an ω-ideal $\omega(f_\alpha)$ of E. It is clear that $T^*(E)$ is an ordered subgroupoid of the symmetric ordered groupoid $G_E = G(\mathscr{I}_E)$ of one-to-one partial transformations of E such that $V(T^*(E))$ may be identified with E; we set $V(T^*(E)) = E$. Define $\tau: \mathscr{G}(E) \to T^*(E)$ by

$$(3.10) \qquad g\tau(c) = f_{g*c}$$

for all $c \in \mathscr{G}(E)$ and $g \in \omega(e_c)$. It can be shown that $\tau(c)$ is an ω-isomorphism and that the assignment $\tau: c \mapsto \tau(c)$ defines an order preserving functor such that $V(\tau) = 1_E$. We have the following result (cf. [28], Proposition 3.6 and Theorem 5.2).

Theorem 3.3. *Let E be a biordered set and $T^*(E)$ be the ordered groupoid of all ω-isomorphisms of E and τ be the V-isomorphism defined by* (3.10). *Then $T^*(E)$ is an inductive groupoid with evaluation τ such that $T(E) = S(T^*(E))$ is a fundamental regular semigroup. Moreover, if S is any fundamental regular semigroup, then S is isomorphic to a full subsemigroup of $T(E(S))$.*

This gives an alternate construction of fundamental regular semigroups (see also [3]). It can be shown that every biordered set E determines a unique cross-connection $[I, \Lambda; \Gamma, \Delta]$ such that $T(E) \approx U(I, \Lambda; \Gamma, \Delta)$ (cf. [27]) so that the two constructions are equivalent.

As another application of the construction described in Theorem 3.1, we can construct all idempotent generated regular semigroups of a given biordered set E as semigroups of inductive groupoids obtained as certain quotients of the groupoid $\mathscr{G}(E)$ (cf. [28], §6). The technique is also useful in constructing an important class of subdirect products of regular semigroups called S^*-direct products [32] which generalizes the construction of subdirect products of inverse semigroups and groups in [17]. Let $\theta: E \to E(S)$ be a bimorphism of a biordered set E to the biordered set of a regular semigroup S. By an extension of θ by S, we mean a pair (T, φ) where T is a regular semigroup and $\varphi: T \to S$ is a homomorphism onto S such that $E(\varphi) = \theta$. An important application of S^*-direct products is that every extension of θ by S can be expressed as an $\mathscr{H}$-coextension of a suitable S^*-direct product. Since the structure of all strictly compatible bimorphisms are known from the construction given in [20], it follows that every coextension of a regular semigroup S by completely simple semigroups can be constructed as an $\mathscr{H}$-coextension of an S^*-direct product of a fundamental regular semigroup and S. S^*-direct products are also useful in generalizing covering theorems for inverse semigroups given in [17] to pseudo-inverse semigroups and obtaining an alternate proof of Pastijn's division theorem [33]. These results together with the principal result of [21] yield covering theorem for locally orthodox semigroups [22] as well (cf. [32]).

The applications of inductive groupoids discussed above are as technical devices for constructing certain semigroups. There are other important conceptual applications as well. In view of the equivalence of categories **RS** and **IG** it is clear that any problem about regular semigroups has a suitable formulation in terms of inductive groupoids and vice versa. In view of the symmetry of inductive groupoids, many problems that have been studied for inverse semigroups have a simple and natural analogue for inductive groupoids which may not be so obvious for regular semigroups. For example, it is known that if S is an inverse semigroup with idempotent metacentre, then the semigroup $\Phi(S)$ of all isomorphisms of principal submonoids (subsemigroups of the form eSe with $e \in E(S)$) is the maximum essential normal extension of S [34]. For regular semigroups, it is not immediately clear as to what the analogue of $\Phi(S)$ is. It is not even clear that such an analogue exists. However, it turns out that the set of all isomorphisms of principal submonoids of a regular semigroup is an inductive groupoid whose semigroup (constructed as in Theorem 3.1) satisfies the requirements [30].

4. COMBINATORIAL PSEUDO-INVERSE SEMIGROUPS

A *pseudo-inverse semigroup* S is a regular semigroup whose biordered set is a pseudo-semilattice or equivalently every principal submonoid of S is inverse. Pseudo-semilattice was introduced as an abstract order structure in [38] and its relation to regular semigroups was determined in [29] by introducing the class of pseudo-inverse semigroups. A regular semigroup S is *combinatorial* if the Green's relation $\mathscr{H}$ of S is the identity relation. In this concluding section of the paper we shall discuss a theorem due to A.R. Rajan [35] which shows the usefulness of certain general constructions and results of abstract category theory in studying regular semigroups.

If S is a regular semigroup, we have seen that the set of $\mathscr{R}$-classes I_S [$\mathscr{L}$-classes Λ_S] may be characterized as a regular partially ordered set (cf. §2). It is possible to regard I_S [Λ_S] as a small category. For if $e, f \in E(S)$ and $f \omega e$ then $\varphi_{e,f}\colon u \mapsto fu$ is a mapping of R_e into R_f. These mappings are called *$\mathscr{R}$-structure mappings.* $\mathscr{L}$-structure mappings $\psi_{e,f}\colon L_e \to L_f$ are defined dually. $S/\mathscr{R}$ [$S/\mathscr{L}$] together with $\mathscr{R}$ [$\mathscr{L}$]-structure mappings form a small category, say **R** [**L**]. By a preorder, we mean a small category in which between any two objects, there exists atmost one morphism [14]. It is clear that any preorder uniquely determines a quasiordered set (on the set of objects of the preorder) and vice versa. Also any small category determines a preorder (obtained by identifying morphisms belonging to the same hom-set). Note that **R** and **L** are small categories whose associated preorder is a regular partially ordered set. Also it follows from Result 1 of [33] that S is pseudo-inverse if and only if **R** and **L** are preorders (so that they are practically the same as regular partially ordered sets I_S and Λ_S, respectively). It is possible to develop a structure theory similar to the theory of cross-connections replacing I_S and Λ_S by **R** and **L**, respectively. In fact it is shown in [18] and [26] that the structure of a regular semigroup may be determined in terms of its trace, biordered set (or regular partial band as in [18]) and the structure mappings. The situation simplifies considerably for combinatorial regular semigroups. In this case the structure of S may be described in terms of the biordered set $E(S)$ and a functor $\varphi(S)$ from the preorder $(E(S), \omega)$ to the category of combinatorial Rees groupoids [31] defined in terms of the structure mappings of S. A Rees groupoid is the partial algebra of nonzero elements of a completely 0-simple semigroup; it is combinatorial if the associated completely 0-simple semigroup is combinatorial (cf. [31]). The functor $\varphi(S)$ is an invariant of the combinatorial regular semigroup S.

Let D be a $\mathscr{D}$-class of a regular semigroup. By the trace $T(D)$ of D we mean the partial algebra on D with the partial binary operation obtained by restricting the partial binary operation defined by (3.2) to D. It is known that $T(D)$ is a Rees groupoid (cf. [5], Theorem 3.4) and so by Rees Theorem [3] (see also [5]) $T(D)$ may be written as a partial algebra of Rees-matrices in the form

$$M(G; I, \Lambda; P) = M^{\circ}(G; I, \Lambda; \mathrm{P}) \setminus \{0\}$$

for some group G, sets $I = D/\mathscr{R}$, $\Lambda = D/\mathscr{L}$ and a regular matrix P over G°. If S is combinatorial, G must be the trivial group and P must be a Boolean matrix. Hence P is completely determined by the subset $\Delta = \{(i, \lambda) \colon p_{\lambda i} \neq 0\}$ of $I \times \Lambda$. Regularity of P is equivalent to the fact that projections of Δ to I and Λ are surjective; that is, Δ is a subdirect product of I and Λ. We may thus represent $T(D)$ as the partial algebra $G(I, \Lambda; \Delta)$ on the set $I \times \Lambda$ with product defined by:

$$(4.1) \qquad (i, \lambda)(j, \mu) = \begin{cases} (i, \mu) & \text{if } (j, \lambda) \in \Delta; \\ \text{undefined otherwise.} \end{cases}$$

Also a (partial algebra) homomorphism $\varphi\colon G(I, \Lambda; \Delta) \to G(I', \Lambda'; \Delta')$ is completely determined by a pair of mappings $\varphi_1\colon I \to I'$, $\varphi_2\colon \Lambda \to \Lambda'$ such that $\varphi = \varphi_1 \times \varphi_2$ and $\varphi_1 \times \varphi_2 \mid_\Delta$ is a mapping of Δ into Δ'.

Suppose that $D, D' \in S/\mathscr{D}$, $e \in E(D)$ and $f \in E(D')$ with $f\omega e$. Here $E(D)$ denotes the set of idempotents in D. Since S is combinatorial, every $x \in D$ may be uniquely written as $x = uv$ with $u \in L_e$ and $v \in R_e$. It is easy to see that

$$x\theta_{e,f} = (u\psi_{e,f})(v\varphi_{e,f})$$

is a homomorphism of $T(D)$ into $T(D')$. Homomorphisms of this type are called *structure mappings of the corresponding* $\mathscr{D}$*-classes.* It can be shown that composition of two structure mappings is again a structure mapping (cf. [31]) so that there is a small category $\mathbf{D}$ whose object set is $S/\mathscr{D}$ and whose morphisms are structure mappings. It is easy to see that $\mathbf{D}$ does not have non-trivial isomorphisms; that is, the only isomorphisms in category $\mathbf{D}$ are identity morphisms. We regard $\mathbf{D}$ as an abstract small category and T as a functor from $\mathbf{D}$ to the category of combinatorial Rees groupoids which sends the object $D \in \mathbf{D}$ to its trace $T(D)$ and the morphism θ in $\mathbf{D}$ to the corresponding homomorphism of Rees groupoids. Note that $\mathbf{D}$ may be considered as a small category of combinatorial Rees groupoids and T as the inclusion of $\mathbf{D}$ in the large category of all combinatorial Rees groupoids.

Now for each object $D \in \mathbf{D}$ write

$$T(D) = G(I_D, \Lambda_D; \Delta_D)$$

where as above, $I_D = D/\mathscr{R}$, $\Lambda_D = D/\mathscr{L}$ and Δ_D is a subdirect product of I_D and Λ_D and, for each morphism $\theta\colon D \to D'$, let

$$T(\theta) = \theta_1 \times \theta_2$$

where $\theta_1\colon I_D \to I_{D'}$, $\theta_2\colon \Lambda_D \to \Lambda_{D'}$ are maps induced by the structure mapping $T(\theta)$ of D into D'. Then $\Delta(\theta) = \theta_1 \times \theta_2 \mid_{\Delta_D}$ is a map of Δ_D into $\Delta_{D'}$. It follows that the assignments

$$P_1 : D \mapsto I_D, \quad \theta \mapsto \theta_1;$$

$$P_2 : D \mapsto \Lambda_D, \quad \theta \mapsto \theta_2;$$

$$\Delta : D \mapsto \Delta_D, \quad \theta \mapsto \Delta(\theta),$$

define functors P_1, P_2 and Δ from $\mathbf{D}$ to the category **Set**. Also since $\Delta(\theta) = \theta_1 \times \theta_2 \mid_{\Delta(D)}$ the inclusion $\Delta(D) \subseteq P_1(D) \times P_2(D)$ is a natural transformation of the functor Δ to the direct product $P_1 * P_2$ of P_1 and P_2 in the functor category $\mathbf{Set}^{\mathbf{D}}$. Thus Δ is a subfunctor of $P_1 * P_2$. Also the map p_i: $D \mapsto p_i(D)$ where $p_i(D)$ is the projection of $P_1(D) \times P_2(D)$ to $P_i(D)$ is a natural transformation of $P_1 * P_2$ to P_i (the projection of the product $P_1 * P_2$ to P_i) and, consequently, the composition $\pi_i = \subseteq \circ\, p_i$ where $\subseteq$ is the inclusion of Δ in $P_1 * P_2$ (the natural transformation where the components are the respective inclusions) is a natural transformation of Δ to p_i. Since the components of π_i are surjective, we may say that Δ is a subdirect product of P_1 and P_2.

Thus a combinatorial regular semigroup S determines a small category $\mathbf{D}$ without non-trivial isomorphisms and three set-valued functors P_1, P_2 and Δ such that Δ is a subdirect product of P_1 and P_2. These are invariants for S. To determine the relation between these invariants, we use the standard category theoretic construction of comma categories [14]. By Yoneda embedding theorem (cf. [14]), there exists an embedding (called Yoneda embedding) Y: $\mathbf{D}^{op} \to \mathbf{Set}^{\mathbf{D}}$ which sends each object D to the hom-functor $\mathbf{D}(D, -)$ and each morphism θ: $D \to D'$ to the natural transformation $\mathbf{D}(\theta, -)$: $\mathbf{D}(D, -) \to \mathbf{D}(D', -)$. So the comma category $(Y \downarrow P_1)$ has, for objects, natural transformations m_1: $Y(D) = \mathbf{D}(D, -) \to P_1$ and if m_1: $Y(D) \to P_1$, m_1': $Y(D') \to P_1$ are objects in $(Y \downarrow P_1)$, a morphism from m_1 to m_1' is a morphism θ: $D' \to D$ in $\mathbf{D}$ such that $Y(\theta) m_1' = m_1$. Further, the inclusion $\Delta \subseteq P_1 * P_2$ induces a functor δ: $(Y \downarrow \Delta) \to (Y \downarrow P_1) \times (Y \downarrow P_2)$ defined as follows:

$$\delta : m \mapsto (m\pi_1, m\pi_2), \quad \theta \mapsto (\theta, \theta)$$

for each object $m \in (Y \downarrow \Delta)$ and each morphism θ in $(Y \downarrow \Delta)$. The following theorem, due to A. R. Rajan [35], describes combinatorial pseudo-inverse semigroups in terms of the invariants $\mathbf{D}, P_1, P_2$ and Δ.

Theorem 4.1. *Let* $\mathbf{D}$ *be a small category without non-trivial isomorphisms and* $P_1, P_2, \Delta \in \mathbf{Set}^{\mathbf{D}}$ *be such that* Δ *is a subdirect product of* P_1 *and* P_2. *Assume further that these satisfy the following axioms:*

(I) $(Y \downarrow P_i)$ $(i = 1, 2)$ *are preorders;*

(II) *the functor* $\delta\colon (Y \downarrow \Delta) \to (Y \downarrow P_1) \times (Y \downarrow P_2)$ *has a right-adjoint* $\bar{\delta}$.

Let

$$S = \bigcup \{\mathrm{Nat}\,(Y(D), P_1) \times \mathrm{Nat}\,(Y(D), P_2)\colon D \in \mathrm{Obj}\,\mathbf{D}\}$$

and define product in S *by:*

$$(x, u)(y, v) = (Y(\varphi)x, Y(\theta)v)$$

where $(\theta, \varphi) = \epsilon_{(y,u)}\colon \delta(\bar{\delta}(y, u)) \to (y, u)$ *is the component of the co-unit of the adjunction given by axiom* (II) *at* (y, u). S *with this product is a combinatorial pseudo-inverse semigroup. Conversely, every combinatorial pseudo-inverse semigroup is isomorphic to one obtained in this way.*

Simplifications of this procedure occur in special cases. If S is inverse it is clear that for each $D \in \mathrm{Obj}\,\mathbf{D}$, $\Delta(D)$ is a bijection of $P_1(D)$ to $P_2(D)$ and that the mapping $D \mapsto \Delta(D)$ is natural in D. Identifying P_1 and P_2 by this natural isomorphism, we may take $P_1 = P_2 = \Delta = P$ in the foregoing theorem. Then δ becomes the usual diagonal functor $\delta\colon (Y \downarrow P) \to$ $\to (Y \downarrow P) \times (Y \downarrow P)$ and so axiom (II) is equivalent to the statement that the category $(Y \downarrow P)$ has products. Also the hypothesis that $\mathbf{D}$ does not have non-trivial isomorphisms implies, by axiom (I), that $(Y \downarrow P)$ is a partial order. Thus, in this case, axioms (I) and (II) are equivalent to the statement that $(Y \downarrow P)$ is a partial order with products; that is, a semilattice. Thus we have the following corollary to Theorem 4.1 which describes combinatorial inverse semigroups [35]:

Corollary 4.2. *Let* $\mathbf{D}$ *be a small category without non-trivial isomorphisms and let* $P\colon \mathbf{D} \to \mathbf{Set}$ *be a functor such that* $(Y \downarrow P)$ *is a semilattice.*

Let

$$S = \bigcup \{\text{Nat}\,(Y(D), P) \times \text{Nat}\,(Y(D), P)\colon D \in \text{Obj}\,\mathbf{D}\}$$

and define product in S *by:*

$$(x, u)(y, v) = (Y(\varphi)x, Y(\theta)v)$$

when $(\theta, \varphi) = \epsilon_{(y,u)}$, ϵ *being the unit of the adjunction given by the product in* $(Y \downarrow P)$. *Then,* S *with this product is a combinatorial inverse semigroup. Conversely, every combinatorial inverse semigroup is isomorphic to one constructed as above.*

Similarly, taking $\mathbf{D}$ to be a semilattice and $\Delta = P_1 * P_2$ one recovers from Theorem 4.1, the usual description of normal bands as 'strong semilattices' of rectangular bands [35].

Even though there has been significant advance in the theory of regular semigroups during the past decade, the theory still remains very new and much remains to be done in this area. The theory has not achieved the level of maturity comparable to the theory of inverse semigroups or groups. Even more work remains to be done in the theory of specific semigroups that arise in other branches of algebra such as multiplicative semigroups of regular rings, matrix semigroups, etc. and the applications of this theory. Recent works by Anh and Márki [1] and Mohan Putcha [23], [24] are significant steps in this direction.

Acknowledgements. I wish to thank John C. Meakin and R. Veeramony for discussion about this paper and A.R. Rajan for discussion and for permission to include some of his unpublished results in this paper. I also wish to thank L. Márki, O. Steinfeld, G. Pollák and other members of the organizing committee of the conference "Semigroups, structure theory and universal algebraic problems" for their invitation and University of Kerala for officially deputing me to attend the conference.

REFERENCES

[1] P.N. Ánh – L. Márki, Rees matrix rings, *J. Algebra,* 81 (1983), 340–369.

[2] A.H. Clifford, Semigroups admitting relative inverses, *Ann. of Math.,* 42 (1941), 1037–1049.

[3] A.H. Clifford, *The fundamental representation of regular semigroup,* Department of Mathematics, Tulane University (1974); Announcement: Semigroup Forum, 10 (1975), 84–92.

[4] A.H. Clifford, *The partial groupoid of idempotents of regular semigroups,* Department of Mathematics, Tulane University (1974); Announcement: Semigroup Forum, 10 (1975) 262–268.

[5] A.H. Clifford – G.B. Preston, *The Algebraic Theory of Semigroups,* Math. Surveys No. 7, Amer. Math. Soc., Providence, R.I., Vol. I, 1961, Vol. II, 1967.

[6] P.A. Grillet, Left coset extensions, *Semigroup Forum,* 7 (1974), 260–263.

[7] P.A. Grillet, Structure of regular semigroups – I: a representation; II: cross-connections; III: reduced case; IV: the general case, *Semigroup Forum,* 8 (1974), 177–183 (I), 254–259 (II), 260–265 (III), 368–373 (IV).

[8] T.E. Hall, On regular semigroups whose idempotents form a subsemigroup, *Bull. Austral. Math. Soc.,* 1 (1969), 195–208.

[9] T.E. Hall, On orthodox semigroup and uniform and antiuniform bands, *J. Algebra,* 16 (1970), 204–217.

[10] T.E. Hall, On regular semigroups, *J. Algebra,* 24 (1973), 1–24.

[11] G. Lallement, Demi-groupes réguliers, *Annali di Matem. pura ed. appl.,* 7 (1967), 43–130.

[12] G. Lallement, Structure theorems for regular semigroups, *Semigroup Forum,* 4 (1972), 95–123.

[13] J. Leech, $\mathscr{H}$-coextensions of monoids, *Mem. Amer. Math. Soc.*, 157 (1975), 1–66.

[14] S. MacLane, *Categories for the Working Mathematician,* Springer-Verlag, New York, 1971.

[15] D.B. McAlister, Groups, semilattices and inverse semigroups I, *Trans. Amer. Math. Soc.*, 192 (1974), 213–233.

[16] D.B. McAlister, Groups, semilattices and inverse semigroups II, *Trans. Amer. Math. Soc.*, 196 (1974), 351–369.

[17] D.B. McAlister – N.R. Reilly, *E*-unitary covers for inverse semigroups, *Pacific J. Math.*, 68 (1977), 161–174.

[18] J. Meakin, The structure mappings on a regular semigroup, *Proc. Edinburgh Math. Soc.*, 21 (1978) 135–142.

[19] J. Meakin, The Rees construction in regular semigroups, this volume.

[20] J. Meakin – K.S.S. Nambooripad, Co-extensions of regular semigroups by rectangular bands I, *Trans. Amer. Math. Soc.*, 269 (1982), 197–224.

[21] J. Meakin – K.S.S. Nambooripad, Co-extensions of regular semigroups by rectangular bands II, *Trans. Amer. Math. Soc.*, 272 (1982), 555–568.

[22] J. Meakin – K.S.S. Nambooripad, Co-extensions of pseudo-inverse semigroups by rectangular bands, *J. Austral. Math. Soc.*, 30 (1980), 73–86.

[23] Mohan S. Putcha, Linear algebraic semigroups, *Semigroup Forum,* 22 (1981), 287–309.

[24] Mohan S. Putcha, A semigroup approach to linear algebraic groups, submitted for publication.

[25] W.D. Munn, Fundamental inverse semigroups, *Quart. J. Math.*, 21 (1970), 157–170.

[26] K.S.S. Nambooripad, Structure of regular semigroup II. The general case, *Semigroup Forum,* 9 (1975), 364–371.

[27] K.S.S. Nambooripad, Relations between biordered sets and cross-connections, *Semigroup Forum,* 16 (1978), 67–81.

[28] K.S.S. Nambooripad, Structure of regular semigroups I, *Mem. Amer. Math. Soc.,* 224 (1979).

[29] K.S.S. Nambooripad, Pseudo-semilattices and biordered sets I, *Simon Stevin,* 55 (1981), 103–110.

[30] K.S.S. Nambooripad – Radhakrishnan Chettiyar, Essential and normal extensions of regular semigroups, preprint.

[31] K.S.S. Nambooripad – A.R. Rajan, Structure of combinatorial regular semigroups, *Quart. J. Math. Oxford,* (2) 29 (1978), 489–504.

[32] K.S.S. Nambooripad – R. Veeramony, *Subdirect products of regular semigroup,* to appear in *Semigroup Forum.*

[33] F. Pastijn, The structure of pseudo-inverse semigroups, *Trans. Amer. Math. Soc.,* 273 (1982), 631–656.

[34] M. Petrich, *Extensions normales de demi-groupes inverses,* preprint.

[35] A.R. Rajan, A structure theorem for combinatorial pseudo-inverse semigroups, this volume.

[36] D. Rees, On semigroups, *Proc. Cambridge Phil. Soc.,* 36 (1940), 387–400.

[37] B.M. Schein, On the theory of generalized groups and generalized heaps, *Theory of semigroups and its applications* I, Izv. dat. Saratov Univ., Saratov, 1966, 286–324 (in Russian); Amer. Math. Soc. Transl., (2) 113 (1979), 89–122.

[38] B.M. Schein, Pseudo-semilattices and pseudo-lattices, *Izv. Vysš, Učebn. Zaved. Mat.,* 2 (117) (1972), 81–94 (in Russian).

[39] A.K. Suschkewitsh, Über die endlichen Gruppen ohne das Gesetz der eindeutigen Umkehrbarkeit, *Math. Ann.*, 99 (1928), 30–50.

[40] J. von Neumann, On regular rings, *Proc. Nat. Acad. Sci. USA,* 22 (1936), 707–713.

K.S.S. Nambooripad

Department of Mathematics, University of Kerala, Kariavattom 695 581, Trivandrum, India.

COLLOQUIA MATHEMATICA SOCIETATIS JÁNOS BOLYAI
39. SEMIGROUPS, SZEGED (HUNGARY), 1981.

REGULAR INVOLUTION SEMIGROUPS

K.S.S. NAMBOORIPAD – F.J.C.M. PASTIJN

The fundamental representation of regular involution semigroups is studied, and the different approaches which are suggested in [13], [14], [15], [17], [30] are presented in an integrated view. We show how some of the results by Foulis [9], [10], [11], Imaoka [22], [23], [24], [25] and Yamada [45] fit in the general framework which is outlined here.

1. PRELIMINARIES

We shall use the standard notation and terminology, as established in [4], [7], [20], [30], [39].

Let S be a semigroup. A transformation $^\star : S \to S,\ x \to x^\star$ will be called an *involution* on S if for all $x, y \in S$

(1.1) $$(xy)^\star = y^\star x^\star$$

and

(1.2) $$(x^\star)^\star = x.$$

The purpose of this note is to sketch a general scheme for describing the structure of regular semigroups with involution. This first section will introduce some definitions which lead to a preliminary classification, a list of "natural" examples, and a first investigation on the behaviour of congruences.

It follows at once that the involution * on S gives rise to a permutation on S which induces a permutation on the set $E(S)$ of idempotents of S. Note that * maps $\mathscr{L}$-classes onto $\mathscr{R}$-classes, $\mathscr{R}$-classes onto $\mathscr{L}$-classes, and $\mathscr{H}$-classes onto $\mathscr{H}$-classes. If an element is fixed by *, then its $\mathscr{H}$-class is left invariant (setwise). This is e.g. the case with the $\mathscr{H}$-classes which contain elements of the form $aa^\star$, $a \in S$. An idempotent of S which is fixed by the involution * will be called a *projection* of S. Evidently an $\mathscr{L}$-class or an $\mathscr{R}$-class of S can contain at most one projection.

Theorem 1.1. *Let S be a regular semigroup with involution *. Then the following statements and their duals are equivalent.*

(i) *Every $\mathscr{L}$-class of S contains a projection,*

(ii) *for every $x \in S$ we have $x^\star x \mathrel{\mathscr{L}} x$,*

(iii) *for every $x \in S$, $x^\star$ is $\mathscr{L}$-related to some inverse of x.*

Proof. Obviously (iii) implies (ii). Let us suppose that (ii) holds, and let $x \in S$. From (ii) it follows that $xx^\star \mathrel{\mathscr{L}} x^\star$, and hence also $xx^\star = (xx^\star)^\star \mathrel{\mathscr{R}} (x^\star)^\star = x$ for all $x \in S$. Thus $x \mathrel{\mathscr{R}} xx^\star \mathrel{\mathscr{L}} x^\star$ and consequently also $x^\star \mathrel{\mathscr{R}} x^\star x \mathrel{\mathscr{L}} x$. It follows that the $\mathscr{H}$-classes $H_{xx^\star}$ and $H_{x^\star x}$ contain idempotents. Since $xx^\star$ and $x^\star x$ are fixed by the involution, we have that * induces an involution on the groups $H_{xx^\star}$ and $H_{x^\star x}$ respectively. Thus the identities of these groups are projections of S. We proved (i) and its dual.

Let us now assume that (i) holds, and let $x \in S$. Let e be the projection in the $\mathscr{L}$-class of $x^\star$. From $e \mathrel{\mathscr{L}} x^\star$, it follows that $e = e^\star \mathrel{\mathscr{R}} (x^\star)^\star = x$. Thus $x \mathrel{\mathscr{R}} e \mathrel{\mathscr{L}} x^\star$, and we conclude that (iii) holds. It is now easy to show that (ii) and (iii) are equivalent to their duals.

Corollary 1.2. *The equivalent statements of Theorem 1.1 are equivalent to*

(iv) *for every* $x \in S$, $x^\star$ *is* $\mathscr{H}$*-equivalent to some inverse of* x,

(v) *for every* $x \in S$ *there exists an element* $x^\dagger$ *which satisfies*

$$(1.3) \qquad x^\dagger x x^\dagger = x^\dagger, \quad x x^\dagger x = x, \quad (x x^\dagger)^\star = x x^\dagger, \quad (x^\dagger x)^\star = x^\dagger x.$$

Proof. If $x \in S$, then the element $x^\dagger$ is the inverse of x which is $\mathscr{L}$-related to $x^\star$, $x^\dagger x$ is the projection which is $\mathscr{L}$-related to x, and $x x^\dagger$ is the projection which is $\mathscr{R}$-related to x.

A regular semigroup S with involution * which satisfies the equivalent conditions of Theorem 1.1 and Corollary 1.2 will be called a *$\star$-regular semigroup.* Our definition accords with the definition of $\star$-regularity which has been given in [8], [10]. As in [8] we may call the inverse $x^\dagger$ of x which is defined by (1.3) the MP- *(Moore–Penrose) inverse* of x (see also [11], [19]). We may consider the class of $\star$-regular semigroups as the variety of algebras of type $\langle 2, 1, 1 \rangle$ defined by the identities (1.1), (1.2), (1.3), together with the identity which guarantees the associativity of the binary operation. In $\star$-regular semigroups each $\mathscr{L}$-class and each $\mathscr{R}$-class contains exactly one projection. The elements which are fixed by * must belong to a maximal subgroup which contains a projection.

Let R be a regular ring, and let * be a transformation of R. Then R is called a *$\star$-regular ring* if (1.1), (1.2),

$$(1.4) \qquad (x + y)^\star = x^\star + y^\star$$

and

$$(1.5) \qquad x x^\star = 0 \Rightarrow x = 0$$

are satisfied [44]. One can easily show that the multiplicative reduct of a $\star$-regular ring yields a $\star$-regular semigroup ([44], Proposition 88). Further, every $\star$-regular semigroup which has a zero 0 satisfies (1.5). Thus $\star$-regular rings yield the first examples of $\star$-regular semigroups, and in fact they formed much of the motivation for introducing $\star$-regularity for semigroups in the way described above.

The following may be recorded for later use.

Theorem 1.3. *Let* S *be a* $\star$*-regular semigroup. Then for every* $x \in S$, *we have*

$$(1.6) \qquad (x^\star)^\dagger = (x^\dagger)^\star .$$

Proof. Clearly $(x^\star)^\dagger$ and $(x^\dagger)^\star$ are $\mathscr{H}$-related to x. Therefore it suffices to show that $(x^\dagger)^\star$ is an inverse of $x^\star$. And indeed,

$$x^\star(x^\dagger)^\star x^\star = (xx^\dagger x)^\star = x^\star,$$

$$(x^\dagger)^\star x^\star (x^\dagger)^\star = (x^\dagger x x^\dagger)^\star = (x^\dagger)^\star .$$

A semigroup S with involution * will be called a *special* $\star$*-semigroup* if (1.1), (1.2) and

$$(1.7) \qquad xx^\star x = x, \quad x^\star x x^\star = x^\star$$

are satisfied. A special $\star$-semigroup is a $\star$-regular semigroup, where

$$(1.8) \qquad x^\dagger = x^\star .$$

That (1.7) or (1.8) is a strong restriction is suggested by the following.

Lemma 1.4. *Let* S *be a* $\star$*-regular semigroup which satisfies*

$$(1.9) \qquad x^\star x = x^\star y = y^\star x = y^\star y \Rightarrow x = y.$$

Then S *is reductive.* S *is a special* $\star$*-semigroup if and only if* S *is an inverse semigroup where* $x^\star = x^\dagger = x^{-1}$ *for all* $x \in S$.

Proof. Let us suppose that S is a $\star$-regular semigroup which satisfies (1.9). It was remarked in [8] that S must be reductive. We give a short proof. Let $a, b \in S$ such that $ta = tb$ for every $t \in S$. Then $a^\star a = a^\star b$ and $b^\star a = b^\star b$. Thus $a^\star b = (b^\star a)^\star = (b^\star b)^\star = b^\star a$, and we may put $a^\star a = a^\star b = b^\star a = b^\star b$. By (1.9) we have $a = b$, and so S is left reductive. From the fact that * is an involution we may immediately conclude that S is also right reductive.

Let us now suppose that S is a special $\star$-semigroup which satisfies (1.9). Let e be any idempotent of S. Then $e^\star$ is an idempotent which is an inverse of e, and $e, ee^\star, e^\star, e^\star e$ form an E-square. Clearly, $ee^\star$ and $e^\star e$ are projections. Putting $x = e^\star$ and $y = ee^\star$, we see that the antecedent of (1.9) is satisfied. Thus $e^\star = ee^\star$, and we see that the

E-square considered above reduces to one element. Thus every idempotent is a projection. Since in a $\star$-regular semigroup every $\mathscr{L}$-class and every $\mathscr{R}$-class contains exactly one projection, we may conclude that S is an inverse semigroup where $x^\star = x^\dagger = x^{-1}$ for all $x \in S$. The converse is obvious.

Corollary 1.5. *A $\star$-regular ring R has a reduct which is a special $\star$-semigroup if and only if R is an abelian regular ring where $x^\star = x^\dagger$.*

Proof. A $\star$-regular ring satisfies (1.9) (see e.g. [8], [11]).

Note that we use the terminology "abelian regular ring" in the sense of [12].

In [22], [23], [24], [25], [35], [40], [45], special $\star$-semigroups are called regular $\star$-semigroups. In order to avoid confusion with the situation for $\star$-regular rings, and also in view of Corollary 1.5, we do not want to adopt this terminology any longer, and we rather prefer the terminology of [8], [10]. The name SIR (Special Involution Regular) semigroup was used in [3] for what we here will call a special $\star$-semigroup which is completely simple.

We illustrate the foregoing with three major classes of examples.

Example 1.6. Let S be an orthodox semigroup, and let $\mathbf{S}$ be the set which consists of the pairs (x, x'), where $x' \in V(x)$, $x \in S$. On $\mathbf{S}$ we define a multiplication and an involution by

$$(x, x')(y, y') = (xy, y'x'),$$

$$(x, x')^\star = (x', x),$$

for all $(x, x'), (y, y') \in \mathbf{S}$. It follows from [41], [42] that the above product is well defined, and that * yields an involution. Obviously $\mathbf{S}$ becomes a special $\star$-semigroup which is orthodox. The projections are of the form (e, e), $e \in E(S)$.

One easily checks that $(e, e)\, \omega\, (f, f)$ in $E(\mathbf{S})$ if and only if $e\, \omega\, f$ in $E(S)$. Thus, the poset of projections of $\mathbf{S}$ is order-isomorphic to the poset $(E(S), \omega)$. This also leads to the surprising conclusion that bands, endowed with the natural partial order, constitute regular partially ordered sets (in the sense of [13]).

If S is a band, then $\mathbf{S}$ is a $\star$-regular band. $\star$-regular bands were studied extensively in [1], [2].

Example 1.7. A rectangular band $I \times \Lambda$ of inverse semigroups $S_{i\lambda}$, $(i, \lambda) \in I \times \Lambda$, is called elementary if $S_{i\lambda} S_{j\mu} = S_{i\mu}$ for all $(i, \lambda), (j, \mu) \in \in I \times \Lambda$ [38]. We shall describe here how to introduce an involution on an elementary rectangular band of E-unitary inverse semigroups. The corresponding results for completely simple semigroups and E-unitary inverse semigroups will come as special cases.

Let G be a group which acts (from the left) on the partially ordered set $\mathscr{X}$ as a group of order automorphisms; let $\mathscr{Y}$ be a subsemilattice and an ideal of $\mathscr{X}$, such that $\mathscr{X} = G\mathscr{Y}$ and such that $a\mathscr{Y} \cap \mathscr{Y} \neq \square$ for every $a \in G$ (see also [27]). Let α_1 be an automorphism of order 2 of $\mathscr{X}$ which leaves $\mathscr{Y}$ (setwise) invariant, and let α_2 be an automorphism of order 2 of G, such that $(gA)^{\alpha_1} = g^{\alpha_2} A^{\alpha_1}$ for all $g \in G$ and $A \in \mathscr{Y}$. Let I be an index set, and let $P = (p_{ij})$ be an $I \times I$-matrix with entries in G, such that

(i) for each $(i, j) \in I \times I$, p_{ij} induces an automorphism on $\mathscr{Y}$,

(ii) for each $(i, j) \in I \times I$, $p_{ji} = (p_{ij}^{-1})^{\alpha_2}$.

Let $\mathscr{M} = \mathscr{M}(P(G, \mathscr{X}, \mathscr{Y}); I; P; \alpha_1, \alpha_2)$ be the set which consists of the elements $(A, g)_{ij}$, where $(A, g) \in \mathscr{Y} \times G$, $i, j \in I$, and $g^{-1}A \in \mathscr{Y}$. On $\mathscr{M}$ we define a multiplication and a $\star$-operation by

$$(1.10) \qquad (A, g)_{ij}(B, h)_{mn} = (A \wedge gp_{jm}B, gp_{jm}h)_{in}$$

and

$$(1.11) \qquad (A, g)_{ij}^{\star} = ((g^{-1})^{\alpha_2} A^{\alpha_1}, (g^{-1})^{\alpha_2})_{ji}.$$

Then $\mathscr{M}$ is an elementary rectangular band of E-unitary inverse semigroups, and $\star$ is an involution. Conversely, every elementary rectangular band of E-unitary inverse semigroups which has an involution, can be so constructed. The proof of this statement is routine from [36], [38]. We obtain the corresponding result for E-unitary inverse semigroups by putting $|I| = 1$ and $P = (1)$, and we obtain the corresponding result for completely simple semigroups by putting $\mathscr{X} = \mathscr{Y} = \{1\}$ in which case we may identify $P(G, \mathscr{X}, \mathscr{Y})$ with G.

Though the involution semigroup $\mathscr{M}$ which is defined above is a regular semigroup with involution, it is not in general a $\star$-regular semigroup, even not if $\mathscr{M}$ reduces to an E-unitary inverse semigroup. One can check that $\mathscr{M}$ is $\star$-regular if and only if $\alpha_1 \mid \mathscr{Y}$ is the identity on $\mathscr{Y}$. This condition is obviously satisfied if $\mathscr{M}$ reduces to a completely simple semigroup. One sees that in this case for every $(A, g)_{ij} \in \mathscr{M}$, $(A, g)^{\dagger}_{ij} = (p_{jj}^{-1} g^{-1} A, p_{jj}^{-1} g^{-1} p_{ii}^{-1})_{ji}$.

The semigroup with involution which is defined above is a special $\star$-semigroup if and only if α_1 is the identity on $\mathscr{X}$ and α_2 is the identity on G. If $\mathscr{M}$ reduces to an inverse semigroup, we have $^{\star} = {}^{-1} = {}^{\dagger}$, and in case $\mathscr{M}$ reduces to a completely simple semigroup we obtain a result from [40].

Example 1.8. Let X be a set, and B_X the semigroup of all binary relations on the set X. For any relation $\alpha \in B_X$, let α^{-1} be the inverse relation: $(a, b) \in \alpha \Leftrightarrow (b, a) \in \alpha^{-1}$, for all $a, b \in X$. It is well known that B_X, endowed with the unary operation $^{-1}$ becomes an involution semigroup. Since $^{-1}$ leaves the greatest regular ideal M_X of B_X setwise invariant, we can state that M_X becomes a regular involution semigroup [18]. Yet, M_X is not a $\star$-regular semigroup. To see this, we take $X = \{0, 1\}$, and we consider the $\mathscr{D}$-class consisting of the four three-element binary relations: in this $\mathscr{D}$-class $^{-1}$ interchanges the two idempotents, and fixes the two non-idempotents. B_X has a smallest non-zero ideal which consists of the so-called rectangular binary relations [43], [46]. The involution $^{-1}$ turns this ideal into a special $\star$-semigroup: the projections are the rectangular binary relations which are of the form $A \times A$, for $A \subseteq X$.

Example 1.9. The following example is taken from the work of Foulis. We refer to [4], [9], [10], [11], [26] for more details and for more references to the relevant literature in this connection.

Let $(L, \wedge, \vee, 0, 1, ^{\perp})$ be an orthocomplemented lattice, and let $\mathrm{Res}\,(L)$ be the semigroup of residuated mappings of L into L. If $f \in \mathrm{Res}\,(L)$, and if $\overset{+}{f}$ is the residual of f, then

$$^{\perp}\overset{+}{f}{}^{\perp} = f^{\star}: L \to L, \quad a \mapsto (a^{\perp}\overset{+}{f})^{\perp}$$

is a residuated mapping, and

$$^{\star}: \mathrm{Res}\,(L) \to \mathrm{Res}\,(L), \quad f \to f^{\star}$$

is an involution of $\mathrm{Res}\,(L)$. If L is an orthocomplemented modular lattice, then the strongly range-closed residuated mappings form a subsemigroup $B(L)$ which is closed for the involution *. The semigroup $B(L)$ is a strongly regular Baer semigroup, which is at the same time a $\star$-regular semigroup. Here the projections are the so-called Sasaki projections $L \to L$, $a \mapsto n \wedge (n^{\perp} \vee a)$, $n \in L$.

Remark that a $\star$-regular ring R yields a strongly regular Baer semigroup with involution *, which is at the same time a $\star$-regular semigroup. Further, if L is the orthocomplemented modular lattice which is coordinatized by R, then there exists a canonical idempotent-separating (multiplicative) homomorphism of R into $B(L)$ which respects the involution (see also [26]). Corollary 1.5 states that if R yields a special $\star$-semigroup, then L must be a Boolean algebra, and in this case $B(L)$ will be an inverse semigroup.

It could be interesting to characterize the orthocomplemented modular lattices L which can be coordinatized by some strongly regular Baer $\star$-semigroup which is also a special $\star$-semigroup. The answer to this question will be given in Example 2.14. If L is the lattice which is depicted below, then the idempotent-generated part of $B(L)$ is a special $\star$-semigroup.

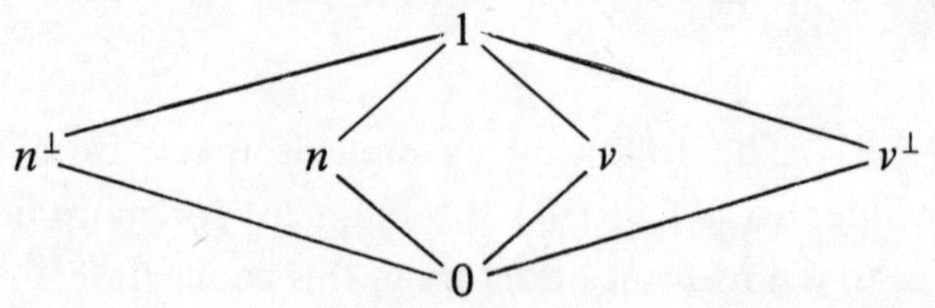

We conclude this section with some observations concerning congruences. A semigroup congruence ρ on an involution semigroup S will be called a $\star$-congruence if $x \rho y$ implies $x^\star \rho y^\star$ for all $x, y \in S$. A $\star$-homomorphism between involution semigroups will be defined accordingly (see also [24], [35]).

In the following, S will be a regular involution semigroup, and $\mathcal{L}^\star(S)$ the complete lattice of $\star$-congruences on S. If X is any set, then we denote the partition lattice on X by $\Pi(X)$. It was already observed in [35] that $\mathcal{L}^\star(S)$ is a complete sublattice of $\Pi(S)$. In particular, $\mathcal{L}^\star(S)$ is a complete sublattice of $\mathcal{L}(S)$, i.e. the complete lattice of all semigroup congruences on S. Let θ be the relation on $\mathcal{L}(S)$ which is defined by

(1.12) $\quad \theta = \{(\rho, \sigma) \in \mathcal{L}(S) \times \mathcal{L}(S)\colon\ \rho \mid E(S) = \sigma \mid E(S)\}.$

Using [16], we can state that θ induces a congruence on $\mathcal{L}^\star(S)$ in such a way that $\theta^\natural$ induces a complete lattice homomorphism of $\mathcal{L}^\star(S)$ into the partition lattice $\Pi(E(S))$. Since the θ-classes form complete modular sublattices of $\mathcal{L}(S)$ [41], the $\theta \mid \mathcal{L}^\star(S)$-classes will form complete modular sublattices of $\mathcal{L}^\star(S)$.

The semigroup congruence κ^c which is generated by a relation κ will be a $\star$-congruence if

(1.13) $\quad (x, y) \in \kappa \Rightarrow (x^\star, y^\star) \in \kappa^c$

is satisfied. If S is an involution semigroup, then it follows from [17] and the foregoing that the greatest idempotent-separating congruence μ_S of S is also a $\star$-congruence. Let ρ be a $\star$-congruence, and let $\rho_{\min}$ [$\rho_{\max}$] be the least [greatest] element in the θ-class of ρ. Putting $\kappa = \rho \mid E(S)$, we have that $\rho_{\min} = \kappa^c$, and that (1.13) is satisfied. Thus $\rho_{\min}$ is a $\star$-congruence. Further, $\rho_{\max}$ is the congruence which is induced by the canonical $\star$-homomorphism $\rho^\natural \mu^\natural_{S/\rho}$, and so $\rho_{\max}$ is also a $\star$-congruence. We conclude that, whenever a θ-class contains a $\star$-congruence, then its least and its greatest elements are also $\star$-congruences. If S is a special $\star$-semigroup, then the behaviour of the $\star$-congruences is even nicer. We refer to [35] for more details concerning this case. One can also directly verify that the least group congruence (generated by $E(S) \times E(S)$) and the least completely simple congruence (generated by the natural partial order on $E(S)$) are $\star$-congruences. The least quasi-orthodox congruence, the least completely regular congruence, the least inverse congruence, the least band congruence, the least semilattice congruence, . . . , are all $\star$-congruences: it is routine to show this from the foregoing, and from the results in [21].

Theorem 1.10. *Let S be a $\star$-regular semigroup. The semigroup congruence which is generated by*

$$\kappa = \{(x^\star, x^\dagger) \mid x \in S\}$$

is a $\star$-congruence. This $\star$-congruence is the least among all $\star$-congruences ρ on S for which the quotient is a special $\star$-semigroup.

Proof. From Theorem 1.3 it follows that the congruence κ^c satisfies (1.13). Hence κ^c is a $\star$-congruence. Now $\star$-congruences on $\star$-regular semigroups are always compatible with the †-operation; thus S/κ^c is a special $\star$-semigroup since it satisfies (1.8). Further, every $\star$-congruence ρ on S which yields a special $\star$-semigroup S/ρ filters through κ^c. Hence the statement of our theorem holds.

2. BIORDERED SETS OF REGULAR INVOLUTION SEMIGROUPS

Let us first characterize the biordered set of regular involution semigroups. If $^\star: S \to S$ is an involution of the regular semigroup S, then * gives rise to an isomorphism of S onto S^{op} (S^{op} is the left-right dual of S). Therefore the restriction * of * to $E(S)$ is an isomorphism of the biordered set $E(S)$ onto the biordered set $E(S^{op})$. It is clear that * is a mapping of $E(S)$ onto itself which satisfies the following conditions:

(i) $e^{\star\star} = e$ for all $e \in E(S)$,

(ii) $(ef)^\star = f^\star e^\star$ for all $e, f \in E(S)$ for which the basic product ef exists in the biordered set.

In particular, for all $e, f \in E(S)$, we have

(2.1) $\quad e\, \omega^r f \Leftrightarrow e^\star\, \omega^l f^\star,$

(2.2) $\quad S(e,f)^{\star} = S(f^{\star}, e^{\star}).$

Let E be a biordered set, and let ${}^{\star}\colon E \to E$ be a mapping of E onto itself which satisfies the above conditions (i) and (ii). Then * will be called an involution on E.

Theorem 2.1. *Let S be a regular semigroup with an involution *. Let * be the restriction of * to $E(S)$. Then * is an involution of the biordered set $E(S)$.*

Conversely, if E is a biordered set with an involution *, then there exists an involution * of $T(E)$ which extends the involution * of $E =$ $= E(T(E))$.

Proof. It is sufficient to prove the converse. Let us consider the set $T^{*}(E)$ of ω-isomorphisms of the biordered set E, and let $\alpha\colon \omega(e_{\alpha}) \to$ $\to \omega(f_{\alpha})$ be any element of $T^{*}(E)$. We define ${}^{\star}(\alpha^{-1})^{\star}$ to be the ω-isomorphism

(2.3) $\quad {}^{\star}(\alpha^{-1})^{\star}\colon \omega(f_{\alpha}) \to \omega(e_{\alpha}), \quad g \mapsto ((g^{\star})\alpha^{-1})^{\star}.$

The mapping

$$\phi\colon T^{*}(E) \to T^{*}(E), \quad \alpha \mapsto {}^{\star}(\alpha^{-1})^{\star} = \phi\alpha$$

is obviously a permutation of order 2 of the set $T^{*}(E)$. Let $e, f \in E$ and let $e \,\mathcal{R}\, f$. Then $e^{\star} \,\mathcal{L}\, f^{\star}$, and

$$\phi(\tau^{r}(e,f)) = {}^{\star}\tau^{r}(f,e)^{\star} = \tau^{l}(f^{\star}, e^{\star}).$$

Let us now suppose that α and β are p-related in $T^{*}(E)$, i.e.

(2.4) $\quad e_{\alpha} \,\mathcal{R}\, e_{\beta}, \quad f_{\alpha} \,\mathcal{L}\, f_{\beta}, \quad \text{and} \quad \alpha\tau^{l}(f_{\alpha}, f_{\beta}) = \tau^{r}(e_{\alpha}, e_{\beta})\beta.$

We then have

$$(2.4) \Leftrightarrow e_{\alpha} \,\mathcal{R}\, e_{\beta}, \quad f_{\alpha} \,\mathcal{L}\, f_{\beta} \quad \text{and} \quad ({}^{\star}\tau^{l}(f_{\beta}, f_{\alpha})^{\star})({}^{\star}(\alpha^{-1})^{\star}) =$$
$$= ({}^{\star}(\beta^{-1})^{\star})({}^{\star}\tau^{r}(e_{\beta}, e_{\alpha})^{\star}) \Leftrightarrow$$
$$\Leftrightarrow e_{\alpha}^{\star} \,\mathcal{L}\, e_{\beta}^{\star}, \quad f_{\alpha}^{\star} \,\mathcal{R}\, f_{\beta}^{\star} \quad \text{and} \quad \tau^{r}(f_{\beta}^{\star}, f_{\alpha}^{\star})(\phi\alpha) =$$
$$= (\phi\beta)\tau^{l}(e_{\beta}^{\star}, e_{\alpha}^{\star}) \Leftrightarrow$$

$$\Leftrightarrow e_{\phi\alpha}\ \mathcal{R}\ e_{\phi\beta},\ f_{\phi\alpha}\ \mathcal{L}\ f_{\phi\beta} \quad \text{and} \quad \tau^r(e_{\phi\beta}, e_{\phi\alpha})(\phi\alpha) =$$

$$= (\phi\beta)\tau^l(f_{\phi\beta}, f_{\phi\alpha}) \Leftrightarrow$$

$$\Leftrightarrow \phi\alpha\ p\ \phi\beta.$$

We see that $\alpha\, p\, \beta$ in $T^\star(E)$ if and only if $\phi\alpha\, p\, \phi\beta$ in $T^\star(E)$. For any $\alpha \in T^\star(E)$, $\bar{\alpha}$ denotes the p-class of α. The above reasoning shows that

(2.5) $\quad {}^\star\colon T(E) \to T(E), \quad \bar{\alpha} \mapsto \bar{\alpha}^\star = \overline{\phi\alpha}$

is a permutation of order 2 of $T(E)$.

Let $\alpha, \beta \in T(E)$, and let $h \in S(f_\alpha, e_\beta)$. We then have ([30], Theorem 4.12)

$$\begin{aligned}
(\bar{\alpha}\bar{\beta})^\star &= \overline{\alpha\tau^l(f_\alpha h, h)\tau^r(h, he_\beta)\beta}^{\,\star} = \\
&= \overline{\phi(\alpha\tau^l(f_\alpha h, h)\tau^r(h, he_\beta)\beta)} = \\
&= \overline{{}^{\star}\beta^{-1}\tau^r(he_\beta, h)\tau^l(h, f_\alpha h)\alpha^{-1\star}} = \\
&= \overline{({}^{\star}\beta^{-1\star})({}^{\star}\tau^r(he_\beta, h)^{\star})({}^{\star}\tau^l(h, f_\alpha h)^{\star})({}^{\star}\alpha^{-1\star})} = \\
&= \overline{(\phi\beta)\tau^l((e_\beta^{\star})(h^{\star}), h^{\star})\tau^r(h^{\star}, (h^{\star})(f_\alpha^{\star}))(\phi\alpha)} = \\
&= \overline{(\phi\beta)\tau^l(f_{\phi\beta}(h^{\star}), h^{\star})\tau^r(h^{\star}, (h^{\star})e_{\phi\alpha})(\phi\alpha)} = \\
&= \overline{\phi\beta}\,\overline{\phi\alpha} = \bar{\beta}^\star\bar{\alpha}^\star
\end{aligned}$$

since $h^{\star} \in S(f_\alpha, e_\beta)^{\star} = S(e_\beta^{\star}, f_\alpha^{\star}) = S(f_{\phi\beta}, e_{\phi\alpha})$. Thus $\star$ is an involution of $T(E)$. For any $e \in E$, ϵ_e denotes the identity transformation on $\omega(e)$. Then $\bar{\epsilon}^\star = \overline{\phi\epsilon_e} = \bar{\epsilon}_{e^{\star}}$. If we identify E with $E(T(E))$, by the canonical biorder isomorphism $E \to E(T(E))$, $e \mapsto \bar{\epsilon}_e$, then the above shows that $\star$ extends the involution * of $E = E(T(E))$.

Let S be a regular semigroup. If $x \in S$, and if x' is an inverse of x in S, then $\theta_{x,x'}\colon \omega(xx') \to \omega(x'x)$, $g \mapsto x'gx$ belongs to $T^\star(E(S))$, and $\overline{\theta_{x,x'}}$ is independent of the choice of x' in $V(x)$, i.e. the set of inverses of x. The mapping $\theta\colon S \to T(E(S))$, $x \mapsto \overline{\theta_{x,x'}}$ is an idempotent separating homomorphism of S onto a full regular subsemigroup of $T(E(S))$, and θ is called the canonical homomorphism of S onto $T(E(S))$ [30].

Theorem 2.2. *Let S be a regular semigroup with involution *, and let * be the involution of $E = E(S)$ which is induced by *. Let * be the involution of $T(E)$ which is defined by (2.3) and (2.4). Then the canonical homomorphism $\theta\colon S \to T(E)$ is a *-homomorphism, i.e. $\overline{\theta_{x,x'}}^{\star} = \overline{\theta_{x^\star,x'^\star}}$ for all $x \in S$, $x' \in V(x)$.*

If in particular S is a full regular subsemigroup of $T(E)$, then the involution * on S must be the restriction to S of the involution * on $T(E)$.

Proof. Let $x \in S$ and $x' \in V(x)$. Then

$$\overline{\theta_{x,x'}}^{\star} = \overline{\phi(\theta_{x,x'})} = \overline{{}^{\star}\theta_{x',x}{}^{\star}}.$$

Here

$${}^{\star}\theta_{x',x}{}^{\star}\colon \omega((x'x)^{\star}) \to \omega((xx')^{\star}), \quad h \mapsto (x(h^{\star})x')^{\star}.$$

Since $(x'x)^{\star} = x^\star x'^\star$, we see that $\theta_{x^\star,x'^\star}$ and ${}^{\star}\theta_{x',x}{}^{\star}$ have the same domain. Further, if $h \in \omega((x'x)^{\star}) = \omega(x^\star x'^\star)$, then

$$h({}^{\star}\theta_{x',x}{}^{\star}) = (x(h^{\star})x')^{\star} = x'^\star h x^\star = h\theta_{x^\star,x'^\star}$$

and so ${}^{\star}\theta_{x',x}{}^{\star} = \theta_{x^\star,x'^\star}$. Consequently $\overline{\theta_{x,x'}}^{\star} = \overline{\theta_{x^\star,x'^\star}}$. Thus θ is a $\star$-homomorphism.

If S is a full regular subsemigroup of $T(E)$, then the canonical homomorphism $\theta\colon S \to T(E)$ is the inclusion mapping and so the last assertion in the statement of the theorem is clear.

Corollary 2.3. *Let E, * and * be as in Theorem 2.1. Then * is the unique involution on $T(E)$ which extends the involution * of $E = E(T(E))$.*

Remark that from Theorem 2.2 it again follows that the greatest idempotent separating congruence on the regular involution semigroup S must be compatible with the involution. Indeed, the canonical homomorphism θ of S into $T(E(S))$ induces the greatest idempotent separating congruence on S [30].

Theorem 2.4. *Let S be a regular semigroup with involution $\star$. Then S is $\star$-regular if and only if the involution induced on $E(S)$ satisfies the following condition:*

(A) *for all $e \in E(S)$, there exist $h, k \in E(S)$, such that*

$$\begin{pmatrix} e & h \\ k & e^{\star} \end{pmatrix} \tag{2.6}$$

is an E-square.

Proof. If S is a $\star$-regular semigroup, and if $e \in E(S)$, *then it is easy* to show from Corollary 1.2 that $h = ee^{\dagger}$ and $k = e^{\dagger}e$ are projections such that (2.6) is an E-square.

Conversely, if for $e \in E(S)$ there exist $h, k \in E(S)$ such that (2.6) is an E-square, then k must be a projection, and k must be in the $\mathcal{L}$-class of e. Thus, if the regular involution semigroup S satisfies the above mentioned condition, then it immediately follows from Theorem 1.1 that S is $\star$-regular.

Combining Theorems 2.1 and 2.4 we see that a biordered set E is the biordered set of some $\star$-regular semigroup if and only if E admits an involution such that the condition (A) is satisfied. In what follows we shall call such a biordered set a *$\star$-regular biordered set.* We shall also use the notation $P(E)$ to denote the set of projections of E, that is the set of elements of the biordered set E which are fixed by the involution.

$\star$-regular biordered sets have the following important property.

Proposition 2.5. *Let E be a $\star$-regular biordered set. Then for all $e, f \in P(E)$, there exists a unique $h \in S(e, f)$ such that eh and hf are projections.*

Proof. In view of Theorem 2.4, we may assume that E is the biordered set of some $\star$-regular semigroup S. Let $e, f \in P(E)$, and put $h = (ef)^{\dagger}$. Then $(ef)(ef)^{\dagger}$ and $(ef)^{\dagger}(ef)$ are projections. It follows from $ef(ef)^\dagger \ \omega^r \ e$ that $ef(ef)^\dagger \ \omega \ e$, and from $(ef)^\dagger (ef) \ \omega^l \ f$ that $(ef)^\dagger (ef) \ \omega \ f$. Hence $f(ef)^\dagger e = (ef)^\dagger$, and so $h = (ef)^\dagger \in S(e,f)$ ([30], Theorem 1.1).

Let k be any element of $S(e,f)$ such that ek and kf are projections. Then $(ef)^\dagger (ef)$ and kf are both projections in the $\mathscr{L}$-class of ef, and thus $(ef)^\dagger (ef) = kf$. Analogously $(ef)(ef)^\dagger = ek$. We conclude that $(ef)^\dagger \ \mathscr{H} k$, thus $(ef)^\dagger = k$.

Given the $\star$-regular biordered set E, and $e, f \in P(E)$, we denote the element $h \in S(e,f)$ which satisfies the condition of the above proposition by $h = (ef)^\dagger$. The proof given shows that h is in fact equal to $(ef)^\dagger$ whenever E appears as the biordered set of a $\star$-regular semigroup. Also, since * and † commute (see (1.6)), we also have

(2.7) $$(fe)^\dagger = ((ef)^\dagger)^\star \in S(f, e)$$

for all $e, f \in P(E)$.

We now turn to the characterization of regular involution semigroups that are special $\star$-semigroups in terms of their regular partial bands. We refer to [25], Section 2, for another approach to the converse of the following theorem.

Theorem 2.6. *Let S be a special $\star$-semigroup. Then the regular partial band of S satisfies the following condition:*

(B) *for all $e \in E(S)$*

$$\begin{pmatrix} e & ee^\star \\ e^\star e & e^\star \end{pmatrix}$$

is a 2×2 rectangular subband of $E(S)$.

Conversely, if the regular partial band of a regular involution semigroup S satisfies (B), *then*

(2.8) $$T = \{x \in S \mid xx^\star \in E(R_x)\}$$

is the greatest full subsemigroup of S which is a special $\star$-semigroup.

Proof. If S is a special $\star$-semigroup, and if $e \in E(S)$, then $e^\star = e^\dagger$ is an idempotent inverse of e, and so (B) is satisfied.

Let us conversely suppose that S is a regular involution semigroup whose regular partial band $E(S)$ satisfies (B). In particular S satisfies condition (A) of Theorem 2.4, and so S is $\star$-regular. Thus, if $e, f \in \in P(E(S))$, then by Proposition 2.5, $(ef)^\dagger$ is an idempotent, and so $((ef)^\dagger)^\star$ is an idempotent inverse of $(ef)^\dagger$ by the condition (B). Now ef is also an inverse of $(ef)^\dagger$, and we know that ef must be $\mathscr{H}$-equivalent to $((ef)^\dagger)^\star$. Thus $ef = ((ef)^\dagger)^\star$, and so ef is an idempotent. From this it follows that $fe = (ef)^\star = ((ef)^\dagger)^{\star\star} = (ef)^\dagger$ is also an idempotent.

It is clear that $E(S) \subseteq T$. If $x \in T$, then from (2.8) is follows that $xx^\star \in P(E(S))$, thus $xx^\star = xx^\dagger$. Also, since $x^\star \mathscr{H} x^\dagger$, we must have $x^\star = x^\dagger$. Hence $x^\star x = x^\dagger x \in E(R_{x^\star})$, thus $x^\star = x^\dagger \in T$. Let $x, y \in T$. We have that $x^\star x$ and $yy^\star$ are projections, and that $x^\dagger = x^\star$ and $y^\dagger = y^\star$ belong to T. Therefore $(x^\star x)(yy^\star)$ is an idempotent, so that

$$(xy)(y^\star x^\star)(xy) = x(yy^\star)(x^\star x)y = x(x^\star x)(yy^\star)(x^\star x)(yy^\star)y =$$

$$= x((x^\star x)(yy^\star))^2 y = x(x^\star x)(yy^\star)y = xy$$

and by duality, $(y^\star x^\star)(xy)(y^\star x^\star) = y^\star x^\star$. We conclude that $(xy)^\star = = y^\star x^\star$ is an inverse of xy. Therefore $(xy)(xy)^\star \in E(R_{xy})$, and we may conclude that $xy \in T$. Thus T is a full subsemigroup of S. Further, since for every $x \in T$, we have $x^\star = x^\dagger \in T$, T is a full regular subsemigroup of S which must be $\star$-special. One readily verifies that every full regular subsemigroup of S which is $\star$-special must be contained in T.

Remark that in the statement of the above theorem, the semigroup T may also be characterized by

(2.9) $\qquad T = \{x \in S \mid x^\dagger = x^\star\}.$

Recall that an E-square

$$\begin{pmatrix} e & f \\ g & h \end{pmatrix}$$

in a biordered set E is a 2×2 rectangular band in some regular semigroup S with $E(S) = E$ if and only if this E-square is τ-commutative [5], [30]. If this is the case, then this E-square forms a rectangular band in $T(E)$. We thus have the following characterization of biordered sets of special $\star$-semigroups.

Corollary 2.7. *A biordered set E is the biordered set of a special $\star$-semigroup if and only if E admits an involution * satisfying the following condition:*

(B′) *if $e \in E$, then there exist $h, k \in E$ such that*

$$\begin{pmatrix} e & h \\ k & e^{\star} \end{pmatrix}$$

is a τ-commutative E-square in E.

We shall see in Example 2.14 that not every $\star$-regular biordered set needs to be the biordered set of some special $\star$-semigroup.

A bimorphism $\alpha \colon E \to E'$ of involution biordered sets is called a *$\star$-bimorphism* if for all $e \in E$, $(e\alpha)^{\star} = e^{\star}\alpha$. Similarly a biorder congruence is called a *$\star$-congruence* if the associated canonical bimorphism is a $\star$-bimorphism. A $\star$-bimorphism which is a biorder isomorphism is called a *$\star$-isomorphism.*

Lemma 2.8. *Let E be a $\star$-regular biordered set. Then for $e \in E$, $\omega(e)$ is closed under * if and only if $e \in P(E)$.*

Proof. Let $\omega(e)$ be closed under * for $e \in E$. Then $e^{\star} \in \omega(e)$. From condition (A) follows that $e = e^{\star}$, and thus $e \in P(E)$. Let us conversely suppose that $e \in P(E)$, and let $f \in \omega(e)$. From condition (A) follows that there exist $g, j \in E$ such that

$$\begin{pmatrix} f & g \\ j & f^{\star} \end{pmatrix}$$

is an E-square. Clearly $g, j \in P(E)$. But then $g\, \omega^{r}\, e$ and $j\, \omega^{l}\, e$ imply $g, j \in \omega(e)$. Consequently $f^{\star} \in \omega(e)$, and we see that $\omega(e)$ is closed under *.

Theorem 2.9. *Let* E *be a special* $\star$*-biordered set. The set of all* $\alpha \in T(E)$ *for which* $\alpha\colon \omega(e_\alpha) \to \omega(f_\alpha)$, $e_\alpha, f_\alpha \in P(E)$, *is a* $\star$*-isomorphism, forms the greatest full regular subsemigroup* $\bar{T}(E)$ *of* $T(E)$ *which is a special* $\star$*-semigroup.*

Proof. Let $\alpha\colon \omega(e_\alpha) \to \omega(f_\alpha)$ be any element of $T^\star(E)$, and let h, k be the projections of E, with $h \mathcal{R} e_\alpha$ and $k \mathcal{L} f_\alpha$. Then

$$\beta = \tau^r(h, e_\alpha)\alpha\tau^l(f_\alpha, k)\colon \omega(h) \to \omega(k)$$

belongs to $T^\star(E)$ and $\alpha \, p \, \beta$ in $T^\star(E)$. Thus every p-class of $T^\star(E)$ contains a (unique) $\star$-isomorphism which maps a principal ω-ideal which is generated by a projection onto a principal ω-ideal generated by a projection.

Let $\bar{\alpha}$ be any element of the greatest full regular subsemigroup of $T(E)$ which is a special $\star$-semigroup. We may suppose that $\alpha\colon \omega(e_\alpha) \to \omega(f_\alpha)$ is an element of $T^\star(E)$ such that $e_\alpha, f_\alpha \in P(E)$. We denote the involution on E by *. From Theorem 2.6 we know that $\bar{\alpha}^\star$ must be equal to $\bar{\alpha}^\dagger$ in $T(E)$. Thus $\bar{\alpha}\bar{\alpha}^\star$ must be the projection $\bar{\epsilon}_{e_\alpha}$ of $T(E)$. From Theorem 2.1, we know that $\bar{\alpha}^\star = \overline{{}^\star\alpha^{-1\star}}$. One verifies that ${}^\star\alpha^{-1\star}\colon \omega(f_\alpha) \to \omega(e_\alpha)$, since $e_\alpha, f_\alpha \in P(E)$. Thus $\alpha^\star\alpha^{-1\star}\colon \omega(e_\alpha) \to \omega(e_\alpha)$. Since $\alpha^\star\alpha^{-1\star} \, p \, \epsilon_{e_\alpha}$, we must have $\alpha^\star\alpha^{-1\star} = \epsilon_{e_\alpha}$ and thus ${}^\star\alpha^{-1\star} = \alpha^{-1}$. Consequently $\alpha^\star = {}^\star\alpha$, and we may conclude that α is a $\star$-isomorphism. Thus $\bar{T}(E)$ contains the greatest full regular subsemigroup of $T(E)$ which is a special $\star$-semigroup.

Let α be any element of $T(E)$, where $\alpha\colon \omega(e_\alpha) \to \omega(f_\alpha)$, $e_\alpha, f_\alpha \in P(E)$, is a $\star$-isomorphism. If $e \in \omega(e_\alpha)$, then $e\alpha = (e^{\star\star})\alpha = (e^\star\alpha)^\star$. Hence $\alpha = {}^\star\alpha^\star$ and thus also $\alpha^{-1} = {}^\star\alpha^{-1\star}$. By Theorem 2.1, we have that $\bar{\alpha}^\star = \overline{{}^\star\alpha^{-1\star}} = \overline{\alpha^{-1}}$, and we see that $\bar{\alpha}^\star$ is an inverse of α. In other words, $\bar{\alpha}^\star = \bar{\alpha}^\dagger$, and $\bar{\alpha}$ belongs to the greatest full regular subsemigroup of $T(E)$ which is a special $\star$-semigroup. This completes the proof of the theorem.

We refer to [22], [23], [45] for other constructions of fundamental special $\star$-semigroups.

Example 2.10. It is clear that if B is a band which is $\star$-regular, then B must be a special $\star$-band: obviously all E-squares in B are τ-commutative. The orthodox $\star$-regular semigroup $T(B)$ then contains a greatest full regular subsemigroup $\bar{T}(B)$ which is a special $\star$-semigroup. The following shows that $\bar{T}(B)$ does not coincide with $T(B)$ in general. We suppose that B is the $\star$-regular band which consists of the E-square

$$\begin{pmatrix} e & h \\ k & e^\star \end{pmatrix}$$

with an identity 1 adjoined. The projections are h, k and 1. The permutation

$$\alpha = \begin{pmatrix} 1 & e & h & k & e^\star \\ 1 & h & e & e^\star & k \end{pmatrix}$$

of B is an ω-isomorphism, but not a $\star$-isomorphism. Thus, $\bar{\alpha}$ belongs to $T(B)$ but not to $\bar{T}(B)$.

However, for pseudo-semilattices we have the following (see [29], [32], [33], [34] for definitions).

Theorem 2.11. *Let E be a $\star$-regular pseudo-semilattice. Then E is a special $\star$-biordered set, and $T(E)$ is a special $\star$-semigroup.*

Proof. Let g be any projection of the $\star$-regular pseudo-semilattice E, and let $e \in \omega(g)$. Then there exist projections h and k such that

$$\begin{pmatrix} e & h \\ k & e^\star \end{pmatrix}$$

is an E-square in E. From $h\,\omega^r\,g$ and $k\,\omega^l\,g$ now follows that $h, k \in$ $\in \omega(g)$. Yet, $\omega(g)$ is a semilattice, and so the above E-square reduces to one single element. Thus, if g is a projection, then $\omega(g)$ is contained in the set of projections of E.

Let us now suppose that e is any element of E. Then

$$\begin{pmatrix} e & h \\ k & e^\star \end{pmatrix}$$

is an E-square in E for some projections h, k of E, since E is $\star$-regular. Let j be any element of $\omega(h)$. The above argument demonstrates that j must be a projection. Thus,

$$j\tau(h, e)\tau(e, k) = k(je) \in \omega(k)$$

and

$$j\tau(h, e)\tau(e, k) = (k(je))^{\star} = (je)^{\star} k = (e^{\star} j)k = j\tau(h, e^{\star})\tau(e^{\star}, k)$$

since $k(je)$ is then a projection. Hence the above considered E-square is τ-commutative, and E must be a special $\star$-biordered set.

Let α be any element of $T^{\star}(E)$. Then there exists a unique element $\beta \in T^{\star}(E)$ such that $e_{\beta}, f_{\beta} \in P(E)$ and $\alpha\, p\, \beta$. Since we know that $\omega(e_{\beta})$ and $\omega(f_{\beta})$ are subsets of $P(E)$, the ω-isomorphism β must be a $\star$-isomorphism. Thus $\bar{\alpha} = \bar{\beta}$ belongs to the greatest full regular subsemigroup of $T(E)$ which is a special $\star$-semigroup. We conclude that $T(E)$ itself must be a special $\star$-semigroup.

Example 2.12. The pseudo-semilattice which is determined by an elementary rectangular band of inverse semigroups must be of the form $E = (L; M_{i\lambda}; \phi_{i\lambda}, \psi_{i\lambda}; I, \Lambda)$ [38]. Let us suppose that $I = \Lambda$. We can always take $M_{i_0 i_0} = L$ and $\phi_{i_0 i_0} = \psi_{i_0 i_0} = \iota_L$ for some fixed $i_0 \in I$. Let us put

(2.10) $$\theta_{ij} = \psi_{i_0 i}^{-1} \phi_{i_0 i} \phi_{ji}^{-1} \psi_{ji} \psi_{ji_0}^{-1} \phi_{ji_0}$$

for all $i, j \in I$. Then θ_{ij}, $i, j \in I$, is an automorphism of L. Let $^{\circ}$ be an automorphism of order 2 of L, such that

(2.11) $$^{\circ}\theta_{ij}{}^{\circ} = \theta_{ij}^{-1}$$

for all $i, j \in I$. Then the mapping

(2.12) $$^{\star}: e_{ij} \to e_{ij}^{\star} = (e_{ij} \psi_{ij} \psi_{ii_0}^{-1} \phi_{ii_0})^{\circ} \psi_{i_0 i}^{-1} \phi_{i_0 i} \phi_{ji}^{-1}$$

is an involution on $E = (L; M_{ij}; \phi_{ij}, \psi_{ij}; I, I)$. Conversely, if the pseudo-semilattice of an elementary rectangular band of inverse semigroups admits an involution, then this pseudo-semilattice and the involution on it must be constructed in this way.

Let $E = (L; M_{ij}; \phi_{ij}, \psi_{ij}; I, I)$, let $M_{i_0 i_0} = L$ for some fixed i_0 in I, let θ_{ij} be defined by (2.10) for all $i, j \in I$, and let $^\circ$ be an automorphism of order 2 of L such that (2.11) is satisfied. Let * be the involution on E which is defined by (2.12). Let $T(L)$ be the Munn semigroup of L, and let $(\theta_{ij})_{i,j \in I} = Q$ be the $I \times I$-matrix which has θ_{ij} on the (i,j)-position. $\mathfrak{M} = \mathcal{M}(T(L); I; Q)$ is the set which consists of the elements α_{ij}, with $\alpha \in T(L)$, $i, j \in I$. The multiplication on $\mathfrak{M}$ is defined as follows. For all $\alpha_{ij}, \beta_{mn} \in \mathfrak{M}$

$$\alpha_{ij}\beta_{mn} = (\alpha\theta_{jm}\beta)_{in}.$$

One can show that $\mathfrak{M}$ is isomorphic to $T(E)$ [38]. Remark that $\mathfrak{M}$ is an elementary rectangular band of fundamental inverse semigroups which are all isomorphic with $T(L)$. On $\mathfrak{M}$ one defines an involution * by

(2.13) $\quad \alpha^\star_{ij} = ({}^\circ(\alpha^{-1})^\circ)_{ji}.$

The idempotents of $\mathfrak{M}$ are of the form $(\theta_{ji}^{-1}\iota_e)_{ij}$, where ι_e is the identity transformation on the principal ideal of L which is generated by e. One shows that

(2.14) $\quad E \to E(\mathfrak{M}), \; e\psi_{i_0 j}^{-1}\phi_{i_0 j}\phi_{ij}^{-1} \to (\theta_{ji}^{-1}\iota_e)_{ij}, \quad e \in L,$

is a biorder isomorphism. Further, if we identify E with $E(\mathfrak{M})$ by the biorder isomorphism (2.14), then the given pseudo-semilattice E with involution * is exactly the involution biordered set which is determined by the regular semigroup $\mathfrak{M}$ with involution *.

The above constructed pseudo-semilattice E is $\star$-regular if and only if the involution $^\circ$ on L, and the automorphisms θ_{ii}, $i \in I$, of L are equal to the identity transformation of L. Then the set of projections is given by $P(E) = \bigcup_{i \in I} M_{ii}$. Remark that in this case $\mathfrak{M}$ ($\cong T(E)$) becomes a special $\star$-semigroup: for all $\alpha_{ij} \in \mathfrak{M}$, we then have $\alpha^\star_{ij} = (\alpha^{-1})_{ji} = \alpha^\dagger_{ij}$. This illustrates Theorem 2.11.

Let us again consider the regular involution semigroup $\mathcal{M} = \mathcal{M}(P(G, \mathcal{X}, \mathcal{Y}); I; P; \alpha_1, \alpha_2)$ of Example 1.7. Here we can always suppose that the matrix P is normalized [38]: for some fixed $i_0 \in I$, and all $i \in I$, $p_{ii_0} = p_{i_0 i} = 1$ is the identity of G. For $i, j \in I$, put

$$M_{ij} = \{(A, p_{ji}^{-1})_{ij} \mid A \in \mathscr{Y}\},$$

$$M_{i_0 i_0} = L,$$

$$\theta_{ij}\colon L \to L, \qquad (A, 1)_{i_0 i_0} \mapsto (p_{ij}^{-1}A, 1)_{i_0 i_0},$$

$$\phi_{ij}\colon M_{ij} \to L, \quad (A, p_{ji}^{-1})_{ij} \mapsto (p_{ji}A, 1)_{i_0 i_0},$$

$$\psi_{ij}\colon M_{ij} \to L, \quad (A, p_{ji}^{-1})_{ij} \mapsto (A, 1)_{i_0 i_0}.$$

Then $E = (L; M_{ij}; \phi_{ij}, \psi_{ij}; I, I)$ is the pseudo-semilattice which is determined by $\mathscr{M}$ and (2.10) is satisfied. On E we may consider the involution * which is induced by the involution * on $\mathscr{M}$ (compare with (1.11)):

$$(A, p_{ji}^{-1})_{ij}^{\star} = (p_{ji}^{\alpha_2} A^{\alpha_1}, p_{ji}^{\alpha_2})_{ji} = (p_{ij}^{-1} A^{\alpha_1}, p_{ij}^{-1})_{ji}.$$

Consider the automorphism $^{\circ}$ of order 2 on L, which is given by

$$(A, 1)_{i_0 i_0}^{\circ} = (A^{\alpha_1}, 1)_{i_0 i_0}.$$

It is routine to check that (2.11) and (2.12) are satisfied. This exemplifies the fact that the pseudo-semilattice E with involution * which is determined by the regular involution semigroup $\mathscr{M}$, is constructed in the way described above.

If $(A, g) \in P(G, \mathscr{X}, \mathscr{Y})$, then $\alpha_{(A,g)}$ which is defined by

$$\operatorname{dom} \alpha_{(A,g)} = \{(B, 1)_{i_0 i_0} \mid B \leqslant A\},$$

and

$$(B, 1)_{i_0 i_0} \alpha_{(A,g)} = (g^{-1}B, 1)_{i_0 i_0} \quad \text{for all } B \leqslant A,$$

belongs to $T(L)$. The mapping

$$\text{(2.15)} \qquad \mathscr{M} \to \mathfrak{M}, \quad (A, g)_{ij} \to (\alpha_{(A,g)})_{ij}$$

is an idempotent separating homomorphism of $\mathscr{M}$ onto a full regular subsemigroup of $\mathfrak{M}$. One may verify that this homomorphism is in fact a $*$-homomorphism. Since $\mathfrak{M}$ is, up to isomorphism, the fundamental involution semigroup $T(E)$, the above homomorphism may considered to be the fundamental representation of $\mathcal{M}$. This illustrates Theorem 2.2.

Let E be a special $\star$-biordered set. Then the special $\star$-semigroup $\bar{T}(E)$ of Theorem 2.9 may be constructed in an easier way. The construction in the following theorem immediately entails Munn's construction of $T(E)$ in case E is a semilattice. The ideas of the following theorem resemble Yamada's procedure in [45]. However, Yamada's starting point is a warp (in the sense of [6]) and not a biordered set.

Theorem 2.13. *Let E be a special $\star$-biordered set, and let P be the set of projections of E. Let $T(P)$ be the set of ω-isomorphisms of $T^\star(E)$, $\alpha: \omega(e_\alpha) \to \omega(f_\alpha)$, with $e_\alpha, f_\alpha \in P$, which are also $\star$-isomorphisms. For all $e, f \in P$, let*

$$\gamma_{e,f} = \tau^l(e(ef)^\dagger, (ef)^\dagger)\tau^r((ef)^\dagger, (ef)^\dagger f) \tag{2.16}$$

where $(ef)^\dagger$ is the unique element in $S(e,f)$ for which $e(ef)^\dagger$ and $(ef)^\dagger f$ are projections. On $T(P)$ we introduce a multiplication and an involution $\star$ by the following: for all $\alpha, \beta \in T(P)$,

$$\begin{aligned} \alpha\beta &= \alpha \circ \gamma_{f_\alpha, e_\beta} \circ \beta, \\ \alpha^\star &= \alpha^{-1}, \end{aligned} \tag{2.17}$$

where the product on the right is a composition of one-to-one partial transformations. Then $T(P)$ is the greatest fundamental special $\star$-semigroup with biordered set E.

Proof. Let us again consider the semigroup $\bar{T}(E)$ of Theorem 2.9. An element of $\bar{T}(E)$ is of the form α, where $\alpha \in T^\star(E)$, $e_\alpha, f_\alpha \in P$, such that α is also a $\star$-isomorphism. The mapping $\phi: \bar{\alpha} \to \alpha$ is then well defined, and is a bijection of $\bar{T}(E)$ onto $T(P)$.

For all $e, f \in P$, $(ef)^\dagger$ is well defined by Proposition 2.5. Since $\bar{T}(E)$ is a full regular subsemigroup of $T(E)$, $\bar{T}(E)$ contains the idempotents

$$\overline{\tau^l(e(ef)^\dagger, (ef)^\dagger)} \quad \text{and} \quad \overline{\tau^r((ef)^\dagger, (ef)^\dagger f)}$$

and consequently also $\overline{\gamma_{e,f}}$. Since $\gamma_{e,f}: \omega(e(ef)^\dagger) \to \omega((ef)^\dagger f)$, where $e(ef)^\dagger, (ef)^\dagger f \in P$, we have $\overline{\gamma_{e,f}}\phi = \gamma_{e,f} \in T(P)$. From this we have

$\alpha \circ \gamma_{f_\alpha, e_\beta} \circ \beta \in T(P)$ for all $\alpha, \beta \in T(P)$. One readily verifies that

$$\overline{\alpha\beta} = \overline{\alpha \circ \gamma_{f_\alpha, e_\beta} \circ \beta} = \bar{\alpha}\bar{\beta}$$

in $T(E)$, since $(f_\alpha e_\beta)^\dagger \in S(f_\alpha, e_\beta)$. Therefore the above considered mapping ϕ is a multiplicative morphism.

If $\bar{\alpha} \in \bar{T}(E)$, where $e_\alpha, f_\alpha \in P$, then $\bar{\alpha}^\star = \overline{\alpha^{-1}}$, from which $\bar{\alpha}^\star \phi = = \alpha^{-1}$. Thus ϕ is a $\star$-isomorphism of $\bar{T}(E)$ onto $T(P)$.

Example 2.14 (see also Example 1.9). Let $(L, \wedge, \vee, 0, 1, {}^\perp)$ be an orthocomplemented modular lattice. If n and v are any pair of complementary elements in L, then

$$(n; v)\colon L \to L,\ x \to v \wedge (n \vee x) = x(n; v)$$

is an idempotent order preserving mapping of L onto the principal ideal of L which is generated by v. We denote by $P(L)$ the subsemigroup of the full transformation semigroup on the set L which is generated by the above considered mappings $(n; v)$. From [37] we have that

$$E(L) = \{(n; v) \mid n, v \in L,\ n \text{ and } v \text{ complementary in } L\}$$

is the set of idempotents of $P(L)$. In $(E(L), \omega^l, \omega^r, \tau^l, \tau^r)$ we have [37] $(n_1; v_1)\, \omega^l\, (n_2; v_2)$ if and only if $v_1 \leqslant v_2$ in L and then

$$(n_1; v_1)\, \tau^l\, (n_2; v_2) = (n_2 \vee (v_2 \wedge n_1); v_1),$$

$(n_1; v_1)\, \omega^r\, (n_2; v_2)$ if and only if $n_1 \geqslant n_2$ in L and then

$$(n_1; v_1)\, \tau^r\, (n_2; v_2) = (n_1; v_2 \wedge (n_2 \vee v_1)).$$

On $E(L)$ we can define an involution * by

$$(n; v)^\star = (v^\perp; n^\perp),\quad (n; v) \in E(L).$$

This involution * on the biordered set $E(L)$ extends in a unique way to an involution * on $P(L)$. The set of projections consists of the idempotents which are of the form $(n^\perp; n)$, $n \in P$. Obviously this set of so-called "Sasaki projections" forms for the natural partial order on the idempotents a lattice which is isomorphic to the given lattice L. With this involution *, the semigroup $P(L)$ becomes a Baer $\star$-semigroup in the sense of [9]. Since $P(L)$ is a fundamental regular semigroup [37], $P(L)$ may be identified with the idempotent generated part of $T(E(L))$.

Remark that $P(L)$ (and thus also $T(E(L))$) is then $\star$-regular, since for all $(n; \nu) \in E(L)$,

$$(2.18) \qquad \begin{pmatrix} (n;\nu) & (n;n^\perp) \\ (\nu^\perp;\nu) & (\nu^\perp;n^\perp) \end{pmatrix}$$

is an E-square. Given two projections $(n_1; n_1^\perp)$ and $(n_2; n_2^\perp)$, we see that $(n_2 \vee (n_2^\perp \wedge n_1); n_1^\perp \wedge (n_1 \vee n_2^\perp))$ is the unique element in the sandwich set of $(n_1; n_1^\perp)$ and $(n_2; n_2^\perp)$ which satisfies the condition expressed in the statement of Proposition 2.5:

$$(n_2 \vee (n_2^\perp \wedge n_1); n_1^\perp \wedge (n_1 \vee n_2^\perp))\, \tau^l\, (n_1; n_1^\perp) =$$
$$= (n_1 \vee (n_1^\perp \wedge n_2); n_1^\perp \wedge (n_1 \vee n_2^\perp)),$$

$$(n_2 \vee (n_2^\perp \wedge n_1); n_1^\perp \wedge (n_1 \vee n_2^\perp))\, \tau^r\, (n_2; n_2^\perp) =$$
$$= (n_2 \vee (n_2^\perp \wedge n_1); n_2^\perp \wedge (n_2 \vee n_1^\perp))$$

are projections, and in $P(L)$ we have

$$(n_2 \vee (n_2^\perp \wedge n_1); n_1^\perp \wedge (n_1 \vee n_2^\perp)) = ((n_1; n_1^\perp)(n_2; n_2^\perp))^\dagger .$$

In order to see that the $\star$-regular biordered set $E(L)$ is not always the biordered set of some special $\star$-semigroup, we take L to be the orthocomplemented modular lattice of subspaces of an Euclidean vector space. Let ν be any plane and n any line such that $n \neq \nu$ and $n \nsubseteq \nu$. Then $(n; \nu) \in E(L)$ and the E-square (2.18) consists of four different elements. If x is any line such that $x \subseteq \nu$, $x \neq \nu \wedge n^\perp$, $x \nsubseteq (\nu \wedge n^\perp)^\perp$, then $x(n;\nu)(\nu^\perp; n^\perp) \neq x(n; n^\perp)$. Thus the E-square (2.18) cannot be τ-commutative, and $E(L)$ does not satisfy the condition (B$'$) of Corollary 2.7.

The orthocomplemented modular lattice L of length 2 which is depicted in Example 1.9 yields a 14-element biordered set $E(L)$. The special $\star$-semigroup $\bar{T}(E(L))$ is properly contained in $T(E(L))$, and $\bar{T}(E(L))$ is a combinatorial completely 0-simple semigroup with a unit group adjoined.

Let L be any orthocomplemented lattice. Then L can be coordinatized by some strongly regular Baer $\star$-semigroup which is also a special $\star$-semigroup if and only if $P(L)$ is a special $\star$-semigroup. In view of Proposition 2.5 and Theorem 2.6, this is the case if and only if the product of any two projections in $P(L)$ yields an idempotent. Thus, L can be coordinatized by a strongly regular Baer $\star$-semigroup which is a special $\star$-semigroup if and only if the composition of any two Sasaki projections on L is an idempotent transformation on L. This latter condition can be expressed in the form of an identity in three variables which must be satisfied in L:

$$z \wedge (z^{\perp} \vee (y \wedge (y^{\perp} \vee (z \wedge (z^{\perp} \vee (y \wedge (y^{\perp} \vee x))))))) =$$

$$= z \wedge (z^{\perp} \vee (y \wedge (y^{\perp} \vee x))).$$

The lattices under consideration thus form a variety of orthocomplemented modular lattices (which contains the variety of Boolean algebras properly).

3. CROSS-CONNECTIONS OF REGULAR INVOLUTION SEMIGROUPS

In this section we assume that the reader is familiar with the concepts of regular partially ordered sets, cross-connections and related notions. For definitions of these concepts and their basic properties, we refer the reader to [13], [14], [15], [31].

Recall from [13] or from [31] that if I is a regular partially ordered set, then the set of all normal equivalences on I is again a regular partially ordered set under the reverse of inclusion. We denote this partially ordered set by I°.

Lemma 3.1. *Let* $\alpha\colon I \to I'$ *be an isomorphism of regular partially ordered sets. For each* $\nu \in I^{\circ}$, *define*

$$\nu\alpha^{\circ} = \{(x, y) \in I' \times I' \mid (x\alpha^{-1}, y\alpha^{-1}) \in \nu\}.$$

Then $\alpha^{\circ}\colon I^{\circ} \to I'^{\circ}$ *is an isomorphism such that for all* $\nu \in I^{\circ}$,

$$A(\nu)\alpha = A(\nu\alpha^{\circ}) \quad \textit{and} \quad M(\nu)\alpha = M(\nu\alpha^{\circ}).$$

Furthermore, the mapping $\bar{\alpha}: S(I) \to S(I')$ *defined by*

$$f\bar{\alpha} = \alpha^{-1} f \alpha$$

for all $f \in S(I)$ *is an isomorphism of the semigroup* $S(I)$ *onto* $S(I')$, *such that for all* $f \in S(I)$,

$$\operatorname{im} f\bar{\alpha} = (\operatorname{im} f)\alpha \quad \textit{and} \quad \ker f\bar{\alpha} = (\ker f)\alpha^{\circ}.$$

Proof. Recall that $S(I)$ denotes the semigroup of all normal mappings of I written as right operators. Since α is an isomorphism, it is easy to see that $f: I \to I$ is a normal mapping if and only if $f\bar{\alpha} = \alpha^{-1} f \alpha$ is a normal mapping of I'. $\bar{\alpha}: S(I) \to S(I')$ is clearly an isomorphism. Since $I(x)\alpha = I'(x\alpha)$ for all $x \in I$, it follows that $\operatorname{im} f\bar{\alpha} = I(x)\alpha = (\operatorname{im} f)\alpha$ if $\operatorname{im} f = I(x)$. Also

$$\begin{aligned}(u, v) \in \ker f\bar{\alpha} &\Leftrightarrow u\alpha^{-1} f\alpha = v\alpha^{-1} f\alpha \Leftrightarrow \\ &\Leftrightarrow (u\alpha^{-1})f = (v\alpha^{-1})f \Leftrightarrow \\ &\Leftrightarrow (u\alpha^{-1}, v\alpha^{-1}) \in \ker f \Leftrightarrow \\ &\Leftrightarrow (u, v) \in (\ker f)\alpha^{\circ}.\end{aligned}$$

Since I is a regular partially ordered set, for all $v \in I^{\circ}$, there exists at least one $f \in S(I)$ such that $\ker f = v$; the result proved above then shows that $v\alpha^{\circ} \in I'^{\circ}$. Since the partial order on I° is the reverse of inclusion, α° is an order isomorphism. Now, if $v \in I^{\circ}$, $u \in A(v)$ if and only if for some $f \in S(I)$ with $\ker f = v$, $f \mid I(u)$ is an isomorphism onto $I(uf)$. Since α is an isomorphism, this is true if and only if $f\bar{\alpha} \mid I'(u\alpha) = \alpha^{-1} f\alpha \mid I'(u\alpha)$ is an isomorphism onto $I'((u\alpha)f\bar{\alpha}) = I'((uf)\alpha)$, that is if and only if $u\alpha \in A(f\bar{\alpha})$. The equality $M(v)\alpha = M(v\alpha^{\circ})$ is proved similarly.

Recall from [14] (see also [31]) that a cross-connection $[I, \Lambda; \Gamma, \Delta]$ consists of two regular partially ordered sets I and Λ and two mappings $\Gamma: \Lambda \to I^{\circ}$ and $\Delta: I \to \Lambda^{\circ}$ satisfying the conditions (C1) and (C2) of [31]. If S is a regular semigroup then $I_S = S/\mathscr{R}$ and $\Lambda_S = S/\mathscr{L}$ are regular partially ordered sets and there exist mappings $\Gamma_S: \Lambda_S \to I_S^{\circ}$ and $\Delta_S: I_S \to \Lambda_S^{\circ}$ such that $[I_S, \Lambda_S; \Gamma_S, \Delta_S]$ is a cross-connection. We refer to this cross-connection as the cross-connection which is induced by S.

It follows from [31] that the mappings Γ_S and Δ_S are defined as follows. For each $e \in E(S)$, (that is, for each $L_e \in \Lambda_S$), we have

$$(3.1) \qquad \Gamma_S(L_e) = \ker \lambda_e$$

where λ_e is the normal retraction of I_S which sends $R_g \in I_S$ to R_{ek} with $k \in S(e, g)$ (see equations (2.2) and (2.3) of [31]). Now in S, $eg \mathcal{R} ek$ and so λ_e is the mapping which sends R_g to R_{eg} so that λ_e is the image of e under the representation λ defined by Hall [17]). Similarly for each $R_f \in I_S$ (with $f \in E(S)$),

$$(3.2) \qquad \Delta_S(R_f) = \ker \rho_f$$

where ρ_f is the image of f under the representation ρ of [17], that is, the mapping sending $L_h \in \Lambda_S$ to L_{hf}.

Conversely, let $[I, \Lambda; \Gamma, \Delta]$ be a cross-connection. Then the subset $U = U(I, \Lambda; \Gamma, \Delta)$ of $S^{op}(I) \times S(\Lambda)$ (where $S^{op}(I)$ denotes the left-right dual of $S(I)$ in which mappings are written as left operators) consisting of all pairs (f, g) satisfying the conditions (C3) and (C4) of [31], is a fundamental regular subsemigroup of $S^{op}(I) \times S(\Lambda)$ such that U is isomorphic to $T(E(U))$. Also

$$E(U) = \{(e^1_{x,y}, e^2_{x,y}) \mid x \in M(\Gamma(y))\}$$

where $e^1_{x,y}$ $[e^2_{x,y}]$ is the canonical projection along $\Gamma(y)$ $[\Delta(x)]$ upon $I(x)$ $[\Lambda(y)]$. Further, there exist natural isomorphisms $\eta_R : I \to I_U$ defined by $x\eta_R = R_{(e^1_{x,y}, e^2_{x,y})}$ and $\eta_L : \Lambda \to \Lambda_U$ defined by $y\eta_L = L_{(e^1_{x,y}, e^2_{x,y})}$ (where $x \in M(\Gamma(y))$) such that up to these isomorphisms the cross-connections $[I, \Lambda; \Gamma, \Delta]$ and $[I_U, \Lambda_U; \Gamma_U, \Delta_U]$ are the same. This means that if we identify I with I_U and Λ with Λ_U by the isomorphisms η_R and η_L respectively, then Γ becomes equal to Γ_U and Δ to Δ_U. Equivalently, the following diagrams commute:

$$\begin{array}{ccc} I & \xrightarrow{\eta_R} & I_U \\ \downarrow{\scriptstyle \Delta} & & \downarrow{\scriptstyle \Delta_U} \\ \Lambda^\circ & \xrightarrow[\eta_L^\circ]{} & \Lambda_U^\circ \end{array} \qquad\qquad \begin{array}{ccc} \Lambda & \xrightarrow{\eta_L} & \Lambda_U \\ \downarrow{\scriptstyle \Gamma} & & \downarrow{\scriptstyle \Gamma_U} \\ I^\circ & \xrightarrow[\eta_R^\circ]{} & I_U^\circ \end{array}$$

By an isomorphism from a cross-connection $[I, \Lambda; \Gamma, \Delta]$ to a cross-connection $[I', \Lambda'; \Gamma', \Delta']$ we mean a pair of order isomorphisms $\phi_1 \colon I \to I'$ and $\phi_2 \colon \Lambda \to \Lambda'$ making the diagrams (D1) and (D2) commute. In this case we write $(\phi_1, \phi_2) \colon [I, \Lambda; \Gamma, \Delta] \approx [I', \Lambda'; \Gamma', \Delta']$.

$$\begin{array}{ccc} I & \xrightarrow{\phi_1} & I' \\ \downarrow{\scriptstyle \Delta} & & \downarrow{\scriptstyle \Delta'} \\ \Lambda^\circ & \xrightarrow{\phi_2^\circ} & \Lambda'^\circ \end{array} \qquad\qquad \begin{array}{ccc} \Lambda & \xrightarrow{\phi_2} & \Lambda' \\ \downarrow{\scriptstyle \Gamma} & & \downarrow{\scriptstyle \Gamma'} \\ I^\circ & \xrightarrow{\phi_1^\circ} & I'^\circ \end{array}$$

(D1) (D2)

Thus $(\eta_R, \eta_L) \colon [I, \Lambda; \Gamma, \Delta] \approx [I_U, \Lambda_U; \Gamma_U, \Delta_U]$.

If

$$(\phi_1, \phi_2) \colon [I, \Lambda; \Gamma, \Delta] \approx [I', \Lambda'; \Gamma', \Delta']$$

and

$$(\phi_1', \phi_2') \colon [I', \Lambda'; \Gamma', \Delta'] \approx [I'', \Lambda''; \Gamma'', \Delta'']$$

are isomorphisms of cross-connections, then it is clear that

$$(\phi_1\phi_1', \phi_2\phi_2') \colon [I, \Lambda; \Gamma, \Delta] \approx [I'', \Lambda''; \Gamma'', \Delta'']$$

is also an isomorphism.

Theorem 3.2. *Let* $(\phi_1, \phi_2) \colon [I, \Lambda; \Gamma, \Delta] \approx [I', \Lambda'; \Gamma', \Delta']$ *be an isomorphism of cross-connections. Then the mapping*

$$(\bar{\phi}_1, \bar{\phi}_2) \colon (f, g) \mapsto (f\bar{\phi}_1, g\bar{\phi}_2)$$

is an isomorphism of $U = U(I, \Lambda; \Gamma, \Delta)$ *onto* $U' = U(I', \Lambda'; \Gamma', \Delta')$. *Conversely, if* $\psi \colon U \to U'$ *is an isomorphism, then there exists an isomorphism of cross-connections* $(\psi_R, \psi_L) \colon [I, \Lambda; \Gamma, \Delta] \approx [I', \Lambda'; \Gamma', \Delta']$ *such that* $\psi = (\bar{\psi}_R, \bar{\psi}_L)$.

Proof. By Lemma 3.1 it is clear that $(\bar{\phi}_1, \bar{\phi}_2)$ is an isomorphism of U onto a subsemigroup of $S^{op}(I') \times S(\Lambda')$. It is therefore sufficient to show that $\operatorname{im}(\bar{\phi}_1, \bar{\phi}_2) = U'$, that is, $(f, g) \in U$ if and only if $(f\bar{\phi}_1, g\bar{\phi}_2) \in$ $\in U'$. So, assume $(f, g) \in U$, $\operatorname{im} f\bar{\phi}_1 = I'(x')$ and $\operatorname{im} g\bar{\phi}_2 = \Lambda'(y')$. Then there exist $x \in I$ and $y \in \Lambda$ such that $x\phi_1 = x'$, $y\phi_2 = y'$, $\operatorname{im} f = I(x)$ and $\operatorname{im} g = \Lambda(y)$. By condition (C3) of [31] and Lemma 3.1, $\ker f\bar{\phi}_1 = (\ker f)\phi_1^\circ = \Gamma(y)\phi_1^\circ = \Gamma'(y\phi_2) = \Gamma'(y')$, since diagram (D2) commutes. Similarly $\ker g\bar{\phi}_2 = \Delta'(x')$, and so the pair $(f\bar{\phi}_1, g\bar{\phi}_2)$ satisfies the condition (C3). To prove (C4), consider $u' \in I'$. Since (f, g) satisfies (C4), by the commutativity of the diagram (D1) we have

$$\Delta'((f\bar{\phi}_1)(u')) = \Delta'((f(u'\phi_1^{-1}))\phi_1) = (\Delta((f(u'\phi_1^{-1}))))\phi_2^\circ =$$

$$= ((\Delta(u'\phi_1^{-1}))g^{-1})\phi_2^\circ = (((\Delta'(u'))(\phi_2^\circ)^{-1})g^{-1})\phi_2^\circ .$$

Now,

$$(s', t') \in (((\Delta'(u'))(\phi_2^\circ)^{-1})g^{-1})\phi_2^\circ \Leftrightarrow$$

$$\Leftrightarrow (s'\phi_2^{-1}g\phi_2, t'\phi_2^{-1}g\phi_2) = (s'g\bar{\phi}_2, t'g\bar{\phi}_2) \in \Delta'(u').$$

Thus for all $u' \in I'$, $\Delta'((f\bar{\phi}_1)(u')) = (\Delta'(u'))(g\bar{\phi}_2)^{-1}$. Similarly for all $v' \in \Lambda'$, $\Gamma'((v')(g\bar{\phi}_2)) = (f\bar{\phi}_1)^{-1}(\Gamma'(v'))$. Hence $(f\bar{\phi}_1, g\bar{\phi}_2) \in U'$. It can be shown similarly that for all $(f', g') \in U'$, $(f'\bar{\phi}_1^{-1}, g'\bar{\phi}_2^{-1}) = (f'\overline{\phi_1^{-1}}, g'\overline{\phi_2^{-1}}) \in U$ and so $(\bar{\phi}_1, \bar{\phi}_2)$ is an isomorphism of U onto U'.

Suppose now that $\psi\colon U \to U'$ is an isomorphism. Define $\phi_1'\colon I_U \to I_{U'}$ and $\phi_2'\colon \Lambda_U \to \Lambda_{U'}$ by

$$R_e\phi_1' = R_{e\psi}, \quad L_f\phi_2' = L_{f\psi}$$

for all $R_e \in I_U$ and $L_f \in \Lambda_U$. Clearly ϕ_1' and ϕ_2' are order isomorphisms. Let $e \in E = E(U)$ and $g' \in E' = E(U')$. Then

$$\lambda_{e\psi} R_{g'} = R_{(e(g'\psi^{-1}))\psi} = R_{e(g'\psi^{-1})}\phi_1' =$$

$$= (\lambda_e(R_{g'}\phi_1'^{-1}))\phi_1' = (\lambda_e\overline{\phi_1'})(R_{g'})$$

and hence $\lambda_{e\psi} = \lambda_e\overline{\phi_1'}$. Similarly $\rho_{e\psi} = \rho_e\overline{\phi_2'}$. Therefore

$$\Gamma_{U'}(L_e\phi_2') = \Gamma_{U'}(L_{e\psi}) = \ker \lambda_{e\psi} = (\ker \lambda_e)\phi_1'^\circ = \Gamma_U(L_e)\phi_1'^\circ$$

by Lemma 3.1. This proves that diagram (D2) commutes. Similarly diagram (D1) also commutes and so

$$(\phi_1', \phi_2')\colon [I_U, \Lambda_U; \Gamma_U, \Delta_U] \approx [I_{U'}, \Lambda_{U'}; \Gamma_{U'}, \Delta_{U'}]$$

is an isomorphism. If

$$(\eta_R, \eta_L)\colon [I, \Lambda; \Gamma, \Delta] \approx [I_U, \Lambda_U; \Gamma_U, \Delta_U]$$

and

$$(\eta'_R, \eta'_L)\colon [I', \Lambda'; \Gamma', \Delta'] \approx [I_{U'}, \Lambda_{U'}; \Gamma_{U'}, \Delta_{U'}]$$

are natural isomorphisms, it follows that

$$(\psi_R, \psi_L)\colon (\eta_R \phi'_1 \eta_R'^{-1}, \eta_L \phi'_2 \eta_L'^{-1})$$

is an isomorphism of $[I, \Lambda; \Gamma, \Delta]$ to $[I', \Lambda'; \Gamma', \Delta']$. From the direct part it follows that $(\bar{\psi}_R, \bar{\psi}_L)$ is an isomorphism of U onto U'. Now an automorphism of a fundamental regular semigroup coincides with the identity transformation if and only if it fixes the idempotents. Therefore, to show that $\psi = (\bar{\psi}_R, \bar{\psi}_L)$, it is sufficient to show that the two isomorphisms coincide on E. Accordingly, choose $e \in E$. Then for some $x \in I$, $y \in \Lambda$ with $x \in M(\Gamma(y))$, $e = (e^1_{x,y}, e^2_{x,y})$. By Lemma 3.1, $e^1_{x,y}\bar{\eta}_R$ is the projection along $\Gamma(y)\eta_R^\circ = \Gamma_U(y\eta_L) = \Gamma_U(L_{(e^1_{x,y}, e^2_{x,y})}) = \Gamma_U(L_e)$ upon $I_U(x\eta_R) = I_U(R_e)$ and so $e^1_{x,y}\bar{\eta}_R = \lambda_e$. Similarly $e^2_{x,y}\bar{\eta}_L = \rho_e$. Hence

$$e(\bar{\psi}_R, \bar{\psi}_L) = e(\bar{\eta}_R, \bar{\eta}_L)(\bar{\phi}_1, \bar{\phi}_2)(\bar{\eta}_R'^{-1}, \bar{\eta}_L'^{-1}) =$$

$$= (\lambda_e \bar{\phi}'_1, \rho_e \bar{\phi}'_1)(\bar{\eta}_R'^{-1}, \bar{\eta}_L'^{-1}) = (\lambda_{e\psi}, \rho_{e\psi})(\bar{\eta}_R'^{-1}, \bar{\eta}_L'^{-1}) = e\psi.$$

Remark. It may be noted that the isomorphism $(\bar{\eta}_R, \bar{\eta}_L)\colon U = U(I, \Lambda; \Gamma, \Delta) \to U(I_U, \Lambda_U; \Gamma_U, \Delta_U)$ corresponding to the natural isomorphism $(\eta_R, \eta_L)\colon [I, \Lambda; \Gamma, \Delta] \approx [I_U, \Lambda_U; \Gamma_U, \Delta_U]$ is the same as the representation (λ, ρ) of Hall [17].

If S^{op} denotes the left-right dual of the regular semigroup S, then it is clear that the cross-connection $[I_{S^{op}}, \Lambda_{S^{op}}; \Gamma_{S^{op}}, \Delta_{S^{op}}]$ is isomorphic to $[\Lambda_S, I_S; \Delta_S, \Gamma_S]$. In particular if $U = U(I, \Lambda; \Gamma, \Delta)$, then the mapping

(3.3) $\zeta\colon (f, g) \to (g, f)$

is an isomorphism of $U(I, \Lambda; \Gamma, \Delta)$ onto $U(\Lambda, I; \Delta, \Gamma)$. We use these observations in the following characterization of cross-connections of a regular involution semigroup.

Theorem 3.3. *Let* $U = U(I, \Lambda; \Gamma, \Delta)$ *where* $[I, \Lambda; \Gamma, \Delta]$ *is a cross-connection. Suppose that* $\alpha \colon I \to \Lambda$ *is an isomorphism making the following diagram commute.*

$$\begin{array}{ccc} I & \xrightarrow{\alpha} & \Lambda \\ \downarrow{\scriptstyle \Delta} & & \downarrow{\scriptstyle \Gamma} \\ \Lambda^{\circ} & \xrightarrow{\alpha^{\circ}} & I^{\circ} \end{array}$$

(D3)

Then α *induces an involution* $^{\star} \colon U \to U$ *defined by:*

$$(f, g)^{\star} = (g\bar{\alpha}^{-1}, f\bar{\alpha}) \tag{3.4}$$

for all $(f, g) \in U$.

Conversely if * *is an involution of* U, *then there exists an isomorphism* $\alpha \colon I \to \Lambda$ *making the diagram* (D3) *commute and such that the involution on* U *induced by* α *coincides with* *.

Proof. Let $\alpha \colon I \to \Lambda$ be an isomorphism making the diagram (D3) commute. Then $(\alpha, \alpha^{-1}) \colon [I, \Lambda; \Gamma, \Delta] \approx [\Lambda, I; \Delta, \Gamma]$ is an isomorphism and so, if ζ is the anti-isomorphism defined by (3.3), by Theorem 3.2 $(\alpha, \alpha^{-1})\zeta^{-1}$ is an anti-isomorphism of U onto itself. Since the mapping defined by (3.4) is the same as $(\bar{\alpha}, \bar{\alpha}^{-1})\zeta^{-1}$, it follows that (3.4) defines an anti-isomorphism of U onto U. Since

$$(f, g)^{\star\star} = (f\bar{\alpha}\bar{\alpha}^{-1}, g\bar{\alpha}^{-1}\bar{\alpha}) = (f, g),$$

we conclude that (3.4) defines an involution on U.

Conversely, let * be an involution on U. Then $\psi = {}^{\star}\zeta$ is an isomorphism of U onto $U(\Lambda, I; \Delta, \Gamma)$. By Theorem 3.2 there exists an isomorphism (ψ_R, ψ_L) of $[I, \Lambda; \Gamma, \Delta]$ to $[\Lambda, I; \Delta, \Gamma]$ such that $\psi = (\bar{\psi}_R, \bar{\psi}_L)$. Now for all $(f, g) \in U$, $(f, g)^{\star\star} = (f, g)$ and since $^{\star} = \psi\zeta^{-1}$, we get

$$(f, g) = (f, g)^{\star\star} = (f\bar{\psi}_R\bar{\psi}_L, g\bar{\psi}_L\bar{\psi}_R).$$

This implies that $\psi_L = \psi_R^{-1}$ and so the diagram (D3) commutes (with $\alpha = \psi_R$). Since $^\star = \psi\zeta^{-1}$, the involution induced by ψ_R coincides with the given involution.

The existence of the isomorphism $\alpha: I \to \Lambda$ suggests that the structure of a fundamental regular involution semigroup may be expressed in terms of a regular partially ordered set I and a mapping $\eta: I \to I^\circ$. We have the following.

Theorem 3.4. *Let I be a regular partially ordered set and $\eta: I \to I^\circ$ a mapping satisfying the following conditions.*

(I1) *For all $x, y \in I$, $x \in M(\eta(y))$ implies $y \in M(\eta(x))$.*

(I2) *Let $x \in M(\eta(y))$ and let $p(x, y)$ denote the projection along $\eta(x)$ upon $I(y)$. Then for all $z \in I$,*

$$\eta(zp(x, y)) = (\eta(z))(p(y, x))^{-1}.$$

Let $U = U(I, \eta)$ denote the subset of $S^{op}(I) \times S(I)$ consisting of all pairs (f, g) satisfying the following conditions.

(I3) $\operatorname{im} f = I(x)$, $\operatorname{im} g = I(y)$ *implies* $\ker f = \eta(y)$, $\ker g = \eta(x)$.

(I4) *For all $z \in I$,*

$$\eta(fz) = (\eta(z))g^{-1}, \quad \eta(zg) = f^{-1}(\eta(z)).$$

Then U is a fundamental regular subsemigroup of $S^{op}(I) \times S(I)$ and the mapping defined by

$$(f, g)^\star = (g, f) \tag{3.5}$$

for all $(f, g) \in U$ is an involution in U.

Conversely every fundamental regular involution semigroup U determines a unique mapping $\eta_U: I_U \to I_U^\circ$ satisfying (I1) *and* (I2), *and U is $\star$-isomorphic to a full subsemigroup of $U(I_U, \eta_U)$.*

The fundamental regular involution semigroups $U(I, \eta)$ and $U(I', \eta')$ are $\star$-isomorphic if and only if there exists an isomorphism $\phi: I \to I'$ such that the following diagram commutes.

$$\begin{array}{ccc} I & \xrightarrow{\ \phi\ } & I' \\ \downarrow{\scriptstyle \eta} & & \downarrow{\scriptstyle \eta'} \\ I^{\circ} & \xrightarrow{\ \phi^{\circ}\ } & I'^{\circ} \end{array}$$

(D4)

The unique $\star$*-isomorphism determined by* ϕ *is* $(\bar{\phi}, \bar{\phi})$.

Proof. If the mapping $\eta\colon I \to I^{\circ}$ satisfies (I1) and (I2), it is clear that $[I, I; \eta, \eta]$ is a cross-connection. Further, $(f, g) \in S^{op}(I) \times S(I)$ satisfies conditions (C3) and (C4) of [31] if and only if (f, g) satisfies (I3) and (I4). Hence $U(I, \eta) = U(I, I; \eta, \eta)$ and so $U(I, \eta)$ is a fundamental regular subsemigroup of $S^{op}(I) \times S(I)$. Further the isomorphism $\alpha = \iota_I\colon I \to I$ satisfies the condition of Theorem 3.3 and so $(f, g) \to (g\bar{\alpha}^{-1}, f\bar{\alpha}) = (g, f)$ is an involution of $U(I, \eta)$.

Conversely let U be a fundamental regular involution semigroup and let $[I, \Lambda; \Gamma, \Delta]$ be the cross-connection induced by U. It follows from Theorems 2.1 and 2.2 that $U(I, \Lambda; \Gamma, \Delta)$ is an involution semigroup and U is $\star$-isomorphic to a full subsemigroup of $U(I, \Lambda; \Gamma, \Delta)$. By Theorem 3.3, there exists an isomorphism $\alpha\colon I \to \Lambda$ making diagram (D3) commute. We shall show that the mapping

(3.6) $\qquad \eta = \Delta(\alpha^{\circ})^{-1} = \alpha\Gamma$

satisfies conditions (I1) and (I2). To prove (I1) suppose that $x \in M(\eta(y))$. Then by Lemma 3.1, $x\alpha \in M(\eta(y))\alpha = M((\eta(y))\alpha^{\circ}) = M(\Delta(y))$ and so by condition (C1) of [31], $y \in M(\Gamma(x\alpha))$. Since diagram (D3) commutes, we have $M(\Gamma(x\alpha)) = M(\Delta(x)\alpha^{\circ -1}) = M(\eta(x))$ and so the condition (I1) holds. Now by Lemma 3.1, $p(x, y)\bar{\alpha}$ is the projection along $(\eta(x))\alpha^{\circ} = \Delta(x)$ upon $I(y)\alpha = I(y\alpha)$ and so $p(x, y)\bar{\alpha} = e^2_{x, y\alpha}$. Similarly $e^1_{x, y\alpha}$ is the projection along $\Gamma(y\alpha) = \Delta(y)\alpha^{\circ -1} = \eta(y)$ upon $I(x)$, and so $e^1_{x, y\alpha} = p(y, x)$. Hence for all $z \in I$, since (D3) commutes, we have

$$\eta(zp(x, y)) = \eta(((z\alpha)e^2_{x, y\alpha})\alpha^{-1}) = \Gamma((z\alpha)e^2_{x, y\alpha}) =$$

$$= (e^1_{x, y\alpha})^{-1}(\Gamma(z\alpha)) = \qquad \text{(by (C2) of [31])}$$

$$= (\eta(z))p(y, x)^{-1}.$$

Hence (I2) also holds and so $[I, I; \eta, \eta]$ is a cross-connection and $(1, \alpha)$: $[I, I; \eta, \eta] \approx [I, \Lambda; \Gamma, \Delta]$ is an isomorphism of cross-connections. Therefore, by Theorem 3.2, $(1, \bar{\alpha})$ is an isomorphism of $U(I, \eta)$ onto $U(I, \Lambda; \Gamma, \Delta)$. If $(f, g) \in U(I, \eta)$, then $((f, g)^\star)(1, \bar{\alpha}) = (g, f)(1, \bar{\alpha}) =$ $= (g, f\bar{\alpha})$ and $((f, g)(1, \bar{\alpha}))^\star = (f, g\bar{\alpha})^\star = (g\overline{\alpha\alpha^{-1}}, f\bar{\alpha}) = (g, f\bar{\alpha})$ by (3.4) and (3.5). This proves that $(1, \bar{\alpha})$ is a $\star$-isomorphism and we conclude that U is $\star$-isomorphic to a full subsemigroup of $U(I, \eta)$.

Let ψ: $U(I, \eta) \to U(I', \eta')$ be a $\star$-isomorphism. By Theorem 3.2, there exists a unique isomorphism (ψ_R, ψ_L): $[I, I; \eta, \eta] \approx [I', I'; \eta', \eta']$ such that $\psi = (\bar{\psi}_R, \bar{\psi}_L)$. We shall show that ψ_R makes diagram (D4) commute. Since (D1) and (D2) commute, it is sufficient to show that $\psi_R = \psi_L$. Since ψ is a $\star$-isomorphism, for all $(f, g) \in U(I, \eta)$, we have

$$(g\bar{\psi}_R, f\bar{\psi}_L) = ((f, g)^\star)\psi = ((f, g)\psi)^\star = (g\bar{\psi}_L, f\bar{\psi}_R)$$

and so $g\bar{\psi}_R = g\bar{\psi}_L$, $f\bar{\psi}_R = f\bar{\psi}_L$. Now for all $x \in I$, we can find $(f, g) \in$ $\in U(I, \eta)$ such that $\operatorname{im} g = I(x)$, and from Lemma 3.1 and the result proved above we obtain $I(x\psi_R) = \operatorname{im} g\bar{\psi}_R = \operatorname{im} g\bar{\psi}_L = I(x\psi_L)$ which implies $x\psi_R = x\psi_L$. Hence we conclude that $\psi_R = \psi_L$. Conversely, if ϕ: $I \to I'$ is an isomorphism making (D4) commute, then (ϕ, ϕ): $[I, I; \eta, \eta] \approx [I', I'; \eta', \eta']$ is an isomorphism. It is easy to see that $(\bar{\phi}, \bar{\phi})$ is a $\star$-isomorphism.

Corollary 3.5. *A regular partially ordered set* I *is the poset of principal right ideals of some regular involution semigroup if and only if there exists a mapping* η: $I \to I^\circ$ *such that* (I1) *and* (I2) *are satisfied.*

Let P be a regular partially ordered set and σ: $P \to P^\circ$ be a mapping. We say that the pair (P, σ) is a *P-set* if it satisfies axioms (I1), (I2) and the following.

(I0) For all $e \in P$, $e \in M(\sigma(e))$.

By an isomorphism of a P-set (P, σ) to a P-set (P', σ') we mean an order isomorphism α: $P \to P'$ making the following diagram commute:

$$\begin{array}{ccc} P & \xrightarrow{\alpha} & P' \\ \downarrow{\scriptstyle\sigma} & & \downarrow{\scriptstyle\sigma'} \\ P^{\circ} & \xrightarrow{\alpha^{\circ}} & P'^{\circ} \end{array}$$

(D5)

It may be noted that our definition of a P-set given above is more general than the concept of a P-set introduced by Yamada in [45] in order to characterize the set of projections of a special $\star$-semigroup. Indeed, it follows from Yamada's result and Theorem 3.6 below that a P-set in the sense of Yamada is also a P-set in the sense above, but not conversely. Imaoka [23] gave an alternate characterization of the set of projections of a special $\star$-semigroup using the concept of a P-groupoid. A P-groupoid is a pair (P, θ) consisting of a set P and a mapping $\theta\colon P \to \mathscr{I}_P$ satisfying some axioms [23]. The following proposition, whose proof is straightforward, shows that P-sets may also be defined in a similar fashion.

Proposition 3.6. *Let (P, σ) be a P-set. Define $\theta\colon P \to S(P)$ as follows:*

(3.7) $\qquad \theta(e) = p(e, e), \qquad e \in P$

where, for $x, y \in P$, $p(x, y)$ denotes the projection along $\sigma(x)$ upon $P(y)$. Then θ satisfies the following conditions.

(P0) *For all $e \in P$, $\theta(e)$ is a normal retraction upon $P(e)$.*

(P1) *$e \in M(\theta(f))$ implies $f \in M(\theta(e))$.*

(P2) *Let $e \in M(\theta(f))$ and let $p(e, f)$ denote the projection along $\ker \theta(e)$ upon $P(f)$. Then for all $g \in P$,*

$$\ker(\theta(gp(e, f))) = \ker(p(f, e)\theta(g)).$$

Conversely, if P is a regular partially ordered set, and $\theta\colon P \to S(P)$ is a mapping satisfying (P0), (P1) *and* (P2), *then the mapping σ defined by*

(3.8) $\qquad \sigma(e) = \ker \theta(e), \qquad e \in P$

satisfies axioms (I0), (I1) *and* (I2) *so that (P, σ) is a P-set.*

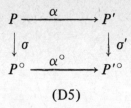

If (P, σ) *and* (P', σ') *are P-sets, an order isomorphism* $\phi: P \to P'$ *is an isomorphism of P-sets if and only if the following diagram commutes.*

$$\begin{array}{ccc} P & \xrightarrow{\ \phi\ } & P' \\ \downarrow{\scriptstyle \theta} & & \downarrow{\scriptstyle \theta'} \\ S(P) & \xrightarrow{\ \bar{\phi}\ } & S(P') \end{array}$$

(D6)

Here θ *and* θ' *denote mappings defined by* (3.7).

In the following it will be convenient to denote P-sets by P, P' etc. If P is a P-set, the underlying regular partially ordered set will also be denoted by P and the associated mappings of P into P° and into $S(P)$ will be denoted by σ and θ respectively (or by σ_P and θ_P respectively, if it is necessary to specify P).

Theorem 3.7. *Let* P *be the partially ordered set of projections of a* $\star$*-regular semigroup* S. *Then there exists a mapping* $\sigma: P \to P^\circ$ *such that* $P(S) = (P, \sigma)$ *is a P-set.*

Conversely if P *is a P-set then there exists a fundamental* $\star$*-regular semigroup* $T = T(P)$ *such that* $P(T)$ *is isomorphic to* P. *Furthermore a fundamental* $\star$*-regular semigroup* S *is* $\star$*-isomorphic to a full subsemigroup of* T *if and only if* $P(S)$ *is isomorphic to* P *as a P-set.*

Proof. Let P be the partially ordered set of projections and $[I, \Lambda; \Gamma, \Delta]$ be the cross-connection of a $\star$-regular semigroup S. By the remarks made at the end of Section 1 and by Theorems 2.1 and 2.2, $U = U(I, \Lambda; \Gamma, \Delta)$ is $\star$-regular and the representation $(\lambda, \rho): x \mapsto (\lambda_x, \rho_x)$ is a $\star$-homomorphism of S into U. Hence for all $e \in E(S)$, $(\lambda_e, \rho_e)^\star = (\lambda_{e^\star}, \rho_{e^\star})$ and by (3.3), $(\lambda_e, \rho_e)^\star = (\rho_e \bar{\alpha}^{-1}, \lambda_e \bar{\alpha})$, where $\alpha: I \to \Lambda$ is the isomorphism making (D3) commute. Hence by Lemma 3.1, $\Lambda(R_e \alpha) = \operatorname{im} \lambda_e \bar{\alpha} = \operatorname{im} \rho_{e^\star} = \Lambda(L_{e^\star})$ and so $R_e \alpha = L_{e^\star}$ for all $e \in E(S)$. Now if $\eta = \Delta(\alpha^\circ)^{-1} = \alpha\Gamma$ is the map defined by (3.6), then for all $g \in P$, λ_g is the projection along $\Gamma(L_g) = \Gamma(R_g \alpha) = \eta(R_g)$ upon $I(R_g)$ and so $R_g \in M(\eta(R_g))$. Thus η satisfies axiom (I0). Axioms (I1) and (I2) hold by Theorem 3.4. Therefore (I, η) is a P-set. Now the mapping

$$\nu\colon R_g \to g, \quad g \in P$$

is clearly an order isomorphism and so, $P(S) = (P, \sigma)$ is a P-set isomorphic to (I, η) if we define σ as $\sigma = \nu^{-1}\eta\nu^{\circ}$.

Conversely let $P = (P, \sigma)$ be a P-set. Then by Theorem 3.4, $T = T(P) = U(P, \sigma)$ is a fundamental regular involution semigroup. Now, idempotents of T are of the form $(p(x,y), p(y,x))$ with $x \in M(\sigma(y))$. Since axiom (I0) holds, $(p(x,x), p(x,x))$ is an idempotent in T which, by (3.5), is a projection. Also if $x \in M(\sigma(y))$, $(p(x,x), p(x,x))$ is clearly $\mathscr{L}$-equivalent to $(p(x,y), p(y,x))$. Hence T is $\star$-regular.

Now let $\eta_T = \Delta_T \alpha^{\circ -1}$ where $\alpha\colon I_T \to \Lambda_T$ is the isomorphism induced by the involution in T. Then by the first part of the proof, (I_T, η_T) is a P-set. Also by the fundamental result on cross-connections [14] we have isomorphisms $T \cong U(I_T, \Lambda_T; \Gamma_T, \Delta_T) \cong U(I_T, \eta_T)$ and so there exists an isomorphism $\beta\colon P \to I_T$ making (D4) commute (by Theorem 3.4). Then β is an isomorphism of the P-set P to (I_T, η_T). Since $\nu\colon R_g \to g$ (g is a projection in T) is an isomorphism of the P-set (I_T, η_T) to $P(T)$, it follows that $\beta\nu$ is an isomorphism of P to $P(T)$.

Finally let S be a fundamental $\star$-regular semigroup. Since S is isomorphic to a full subsemigroup of $T(P(S))$, S is isomorphic to a full subsemigroup of T if and only if $T(P(S))$ is isomorphic to T. By Theorem 3.4, this is true if and only if $P(S)$ is isomorphic to P.

Theorem 3.8. *Let P be a P-set and $\theta = \theta_P$. Then P is the P-set of a special $\star$-semigroup if and only if the following condition holds.*

(P3) *For all $e, f \in P$ with $e \in M(\theta(f))$, $\theta(e)\theta(f)\theta(e) = \theta(e)$.*

Proof. We note that the condition (P3) is equivalent to the statement that for $e, f \in P$ with $e \in M(\theta(f))$, the projections $(p(e,e), p(e,e)) = (\theta(e), \theta(e))$ and $(\theta(f), \theta(f))$ are mutually inverse, that is, the E-square

$$\begin{pmatrix} (p(e,f), p(f,e)) & (p(f,f), p(f,f)) \\ (p(e,e), p(e,e)) & (p(f,e), p(e,f)) \end{pmatrix}$$

is τ-commutative. Now P is the P-set of a special $\star$-semigroup if and only if $T(P)$ contains a full special $\star$-subsemigroup. By Corollary 2.7 and the remark above, this is true if and only if P satisfies (P3).

Remark. The theorem above shows that P-sets satisfying axiom (P3) characterize the sets of projections of special $\star$-semigroups. Since the concept of P-groupoids [23] also does the same, it is of interest to find the explicit relation between the two. Indeed, let S be a special $\star$-semigroup, $\sigma_S = \sigma_{P(S)}$ and $\theta_S = \theta_{P(S)}$. We have seen that the map $\nu\colon R_g \to g$ is an isomorphism of the P-set $I_S = (I_S, \eta_S)$ to $P(S)$. Hence if $\theta' = \theta_{I_S}$, we have $\nu\theta_S = \theta'\bar{\nu}$. Now for each $R_g \in I_S$ with $g \in P(S)$, it is easy to see that $\theta'(g) = \lambda_g$ and hence for all $h \in P(S)$, $\nu(\lambda_g(R_h)) = \nu(R_{gh}) = h\theta_S(g)$. Since S is a special $\star$-semigroup, the projection in R_{gh} is ghg and so for all $h \in P(S)$, $h\theta_S(g) = ghg$. Hence it follows from the construction of the P-groupoid of S in the proof of Theorem 3.2 of [23] that $\theta_S(g)$ is the same as the map θ_g of Imaoka. Thus the map θ_S is the same as the map θ constructed in the proof of Theorem 3.2 of [23]. Thus $(P(S), \theta_S)$ is a P-groupoid. It is easy to see that if two P-sets P and P' are isomorphic (as P-sets) and if (P, θ_P) is a P-groupoid, so is $(P', \theta_{P'})$. Similarly if the P-groupoids (P, θ) and (P', θ') are isomorphic (as P-groupoids) and if θ satisfies axioms (P0), (P1), (P2) and (P3) then so does θ'. Hence it follows from the above and Theorem 3.2 of [23] that every P-set satisfying (P3) is a P-groupoid and conversely.

We have seen that fundamental regular involution semigroups may be constructed from a regular partially ordered set I and a mapping $\eta\colon I \to I^{\circ}$ satisfying axioms (I1) and (I2). In general, the semigroup $U(I, \eta)$ constructed from the pair (I, η) does not admit a faithful representation into $S(I)$. However in several interesting cases, such a representation is possible. It is therefore of interest to characterize those regular involution semigroups that can be so represented.

Lemma 3.9. *Every congruence on a regular semigroup S contained in $\mathscr{L}$ is idempotent separating if and only if Δ_S is injective. When S satisfies this condition, S is left reductive.*

Proof. Recall from [17] that the kernel congruence of the representation ρ is the maximum congruence $\mathscr{L}_c$ contained in $\mathscr{L}$. Therefore it is sufficient to prove that $\mathscr{L}_c \subseteq \mathscr{H}$ if and only if Δ_S is one-to-one.

So first assume that $\mathscr{L}_c \subseteq \mathscr{H}$ and $\Delta_S(R_e) = \Delta_S(R_f)$ for $e, f \in E(S)$. Then $M(\Delta_S(R_e)) = M(\Delta_S(R_f))$ and it follows from equations (2.2), (2.3) and Theorem C of [31] that for every $\mathscr{L}$-class L of S, $L \cap R_e$ is a subgroup of S if and only if $L \cap R_f$ is also a subgroup. In particular there exists an idempotent f' such that $e \mathscr{L} f' \mathscr{R} f$. Then $\ker \rho_e = \Delta_S(R_e) = \Delta_S(R_{f'}) = \ker \rho_{f'}$ and $\operatorname{im} \rho_e = \Lambda_S(L_e) = \Lambda_S(L_{f'}) = \operatorname{im} \rho_{f'}$. Since ρ_e and $\rho_{f'}$ are normal retractions of Λ_S, we conclude that $\rho_e = \rho_{f'}$ and so $(e, f') \in \mathscr{L}_c$. Since $\mathscr{L}_c \subseteq \mathscr{H}$, we have $e = f'$ and hence $R_e = R_{f'} = R_f$. Thus Δ_S is one-to-one. Conversely, if Δ_S is one-to-one and if $(e, f) \in \mathscr{L}_c \cap (E(S) \times E(S))$, then $\rho_e = \rho_f$ and so $\Delta_S(R_e) = \ker \rho_e = \ker \rho_f = \Delta_S(R_f)$. Hence $R_e = R_f$ and so $e \mathscr{R} f$. Since $e \mathscr{L} f$, we conclude that $e = f$. Hence $\mathscr{L}_c \subseteq \mathscr{H}$. The last statement of the lemma follows from the fact that the kernel of the right regular representation of S is contained in $\mathscr{L}$ and it intersects every $\mathscr{H}$-class in a single element.

Theorem 3.10. *For a regular semigroup S, the following statements are equivalent.*

(i) *The representation ρ is faithful.*

(ii) *S is fundamental and Δ_S is injective.*

(iii) *S can be isomorphically embedded in $S(\Lambda_S)$ in such a way that S intersects every $\mathscr{L}$-class of $S(\Lambda_S)$.*

Proof. Obviously (i) ⇒ (iii), and by Lemma 3.9 (i) ⇔ (ii). To prove (iii) ⇒ (i), we may assume that $S \subseteq S(\Lambda_S)$. Consider $f, g \in S$ such that $\rho_f = \rho_g$. Then for all $h \in S$, $hf \mathscr{L} hg$ and so (since $S \subseteq \mathscr{I}_{\Lambda_S}$), $\Lambda_S(xf) = \operatorname{im} hf = \operatorname{im} hg = \Lambda_S(xg)$, where $\operatorname{im} h = \Lambda_S(x)$, that is, $xf = xg$. Now for every $x \in \Lambda_S$, there exists $h \in S$ such that $\operatorname{im} h = \Lambda_S(x)$ and so $xf = xg$ for all $x \in \Lambda_S$. Thus $f = g$.

The theorem above implies that every subsemigroup of $S(\Lambda)$ (for some regular partially ordered set Λ) that intersects every $\mathscr{L}$-class of $S(\Lambda)$ is fundamental. In particular $S(\Lambda)$ is fundamental.

Theorems 3.4 and 3.10 yield the following.

Theorem 3.11. *Let S be a fundamental regular involution semi-*

group. Then ρ *is a faithful representation of* S *as a subsemigroup of* $S(I_S)$ *if and only if* η_S *(the map defined by* (3.6)*) is injective.*

Remark that, if the fundamental regular involution semigroup S satisfies the condition of Theorem 3.11, then η_S must be an order embedding of I_S into the poset (under the reverse of inclusion) of the normal equivalences on I_S.

Before ending this section we shall briefly consider fundamental $\star$-regular semigroups satisfying $\star$-cancellation (that is, condition (1.9)). We first show that condition (1.9) is a regular partial band condition.

Lemma 3.12. *Let* S *be a* $\star$*-regular semigroup. Then* S *satisfies condition* (1.9) *if and only if it satisfies the following.*

(C) *If an E-square*

$$\begin{pmatrix} e & g \\ h & e^\star \end{pmatrix}$$

is a rectangular band, then $g = h$.

Proof. First, let S satisfy (1.9). If

$$\begin{pmatrix} e & g \\ h & e^\star \end{pmatrix}$$

is a rectangular band, then g and h are projections and $e^\star h = e^\star e = h^\star h = h^\star e = h$ and so $e = h$ by (1.9). Hence $g \mathscr{R} h$ and since g and h are projections, we have $g = h$. Conversely let S satisfy (C). Consider $a, b \in S$ such that $a^\star a = a^\star b = b^\star a = b^\star b$. Then $R_a \cap L_{b^\star}$ contains an idempotent, say e. Since $R_a \cap L_{a^\star}$ and $R_b \cap L_{b^\star}$ contain projections, say g and h, the idempotent in $R_b \cap L_{a^\star}$ is $e^\star$. Let a' and b' be the inverses of a and b in the $\mathscr{H}$-classes $H_{b^\star}$ and $H_{a^\star}$ respectively. From $a^\star a = a^\star b$, we obtain $e = ge = (a^\star)^\dagger a^\star aa' = (a^\star)^\dagger a^\star ba' = gba'$ and similarly $e^\star = hab'$. Since $a'a = k$ is the projection in $H_{a^\star a}$, we have

$$ee^\star = gba'hab' = gba'ab' = gbb' = ge^\star = g.$$

This proves that

$$\begin{pmatrix} e & g \\ h & e^\star \end{pmatrix}$$

is a rectangular band and so $g = h$ by (3.9). Therefore $a \mathscr{H} b$. From $a^\star a = a^\star b$ we conclude that $a = b$. Thus S satisfies (1.9).

The result proved above shows that $\star$-cancellation is a regular partial band condition. Hence, if a $\star$-regular semigroup S contains a full subsemigroup satisfying (1.9), then every $\mathscr{H}$-coextension of S also satisfies (1.9) (if the coextension is $\star$-regular). But an idempotent separating homomorphic image of S need not satisfy (1.9).

Since (1.9) is not a biordered set condition, it cannot be characterized in terms of the P-set of projections. In particular, there is no relation between injectivity of σ_S and condition (1.9).

Indeed, let $Z_2 = \{0, 1\}$ be a group of order 2, $I = \Lambda = \{1, 2\}$, $P = \begin{pmatrix} 0 & 0 \\ 0 & 1 \end{pmatrix}$. On $S = \mathscr{M}(Z_2; P; I, \Lambda)$ we define $\star$ by $g_{ij}^\star = (-g)_{ji}$. Then S becomes a completely simple $\star$-regular semigroup which satisfies (1.9). Yet, σ_S is a constant map.

However, for fundamental $\star$-regular semigroups satisfying (1.9), we have the following result.

Theorem 3.13. *Every fundamental $\star$-regular semigroup S satisfying (1.9) admits a faithful representation into $S(P(S))$.*

Proof. In view of Theorem 3.9 it is sufficient to show that the congruence $\mathscr{L}_c$ is idempotent separating. So consider $(e, f) \in \mathscr{L}_c \cap \cap (E(S) \times E(S))$. Since the congruence classes of $\mathscr{L}_c$ containing idempotents are completely simple subsemigroups of S, it follows from [28] that e and f have inverses in the same $\mathscr{H}$-classes. Let g be the projection in R_e. Since $g \in V(e)$, H_g contains an inverse of f, and so $R_f \cap L_g$ contains an idempotent, say f_1. It is easy to see that $(g, f_1) \in \mathscr{L}_c$ and so $\rho_g = \rho_{f_1}$. Hence $\Delta(R_{f_1^\star}) = \Delta(R_g) = \Delta(R_{f_1})$, so that $\rho_{f_1} \mathscr{R} \rho_{f_1^\star}$ in $S(\Lambda_S)$. Also, since $(g, f_1) \in \mathscr{L}_c$, $(g, f_1^\star) \in \mathscr{R}_c$ and so $\Gamma(L_{f_1}) = \Gamma(L_{f_1^\star})$, that is

$\lambda_{f_1} \mathscr{L} \lambda_{f_1^\star}$ in $S^{op}(I_S)$. Hence

$$(\lambda_{f_1^\star}, \rho_{f_1^\star})(\lambda_{f_1}, \rho_{f_1}) = (\lambda_{f_1^\star}\lambda_{f_1}, \rho_{f_1^\star}\rho_{f_1}) = (\lambda_{f_1^\star}, \rho_{f_1}) = (\lambda_g, \rho_g).$$

Since the representation (λ, ρ) is faithful, we have $f_1^\star f_1 = g$. Hence the E-square

$$\begin{pmatrix} f_1 & h \\ g & f_1^\star \end{pmatrix}$$

is a rectangular band and so by Lemma 2.12, $g = h$. It follows that $e \mathscr{R} f$ and so we conclude that $e = f$. Thus $\mathscr{L}_c$ is idempotent separating.

Example 3.14 (see also Examples 1.9 and 2.14). Let $(L, \wedge, \vee, 0, 1, ^\perp)$ be an orthocomplemented modular lattice. Obviously L is a regular partially ordered set. Let $\eta: L \to L^\circ$ be defined by the following. For $\nu \in L$, $\eta(\nu)$ is the normal equivalence on L which is given by $x\, \eta(\nu)\, y$ if and only if $\nu^\perp \vee x = \nu^\perp \vee y$. Obviously η is injective, and η is an order embedding of L into the poset of normal equivalences on L (under the reverse of inclusion).

For all $\nu \in L$, we have that $M(\eta(\nu))$ consists of the complements of $\nu^\perp$ in L. Therefore (I1) is obviously satisfied for the pair (L, η). Remark that for $n, \nu \in L$, we have that n and ν are complementary in L if and only if $n^\perp \in M(\eta(\nu))$, and if this is the case, then

$$p(n^\perp, \nu) = (n; \nu): L \to L, \quad x \to \nu \wedge (n \vee x)$$

is the projection along $\eta(n^\perp)$ upon $L(\nu)$. But then, for $z \in L$,

$$(x, y) \in \eta(zp(n^\perp, \nu)) \Leftrightarrow x \vee (\nu \wedge (n \vee z))^\perp = y \vee (\nu \wedge (n \vee z))^\perp \Leftrightarrow$$

$$\Leftrightarrow x \vee (\nu^\perp \vee (n^\perp \wedge z^\perp)) = y \vee (\nu^\perp \vee (n^\perp \wedge z^\perp)) \Leftrightarrow$$

$$\Leftrightarrow (n^\perp \wedge z^\perp) \vee (n^\perp \wedge (\nu^\perp \vee x)) = (n^\perp \wedge z^\perp) \vee (n^\perp \wedge (\nu^\perp \vee y)) \Leftrightarrow$$

$$\Leftrightarrow z^\perp \vee (n^\perp \wedge (\nu^\perp \vee x)) = z^\perp \vee (n^\perp \wedge (\nu^\perp \vee y)) \Leftrightarrow$$

$$\Leftrightarrow (n^\perp \wedge (\nu^\perp \vee x), n^\perp \wedge (\nu^\perp \vee y)) \in \eta(z) \Leftrightarrow$$

$$\Leftrightarrow (x, y) \in \eta(z)p(\nu, n^\perp)^{-1}.$$

Thus (L, η) satisfies (I2). Further, for all $\nu \in L$, $\nu \in M(\eta(\nu))$, and so (L, η) satisfies (I0). Thus (L, η) is a P-set. The mapping

$$\theta\colon L \to S(L), \quad \nu \to p(\nu, \nu) = (\nu^{\perp}; \nu)$$

satisfies (P0), (P1), (P2) by Proposition 3.6, and by Theorem 3.7, $U(L, \eta)$ is a fundamental $\star$-regular semigroup.

The set $E(U(L, \eta))$ of idempotents of $U(L, \eta)$ consists of the elements $(p(\nu, n^{\perp}), p(n^{\perp}, \nu)) = ((\nu^{\perp}; n^{\perp}), (n; \nu)) \in S^{op}(L) \times S(L)$. It follows that the biordered set $E(L)$ which was considered in Example 2.14 is biorder isomorphic to the biordered set of idempotents of $U(L, \eta)$, and so the fundamental $\star$-regular semigroup $U(L, \eta)$ is $\star$-isomorphic to the fundamental $\star$-regular semigroup $T(E(L))$ [31].

Using Theorem 3.10 (and Theorem 3.4) we can say that the mapping $\pi\colon U(L, \eta) \to S(L)$, $(f, g) \mapsto g$ is a monomorphism. Obviously π maps $E(U(L, \eta))$ onto the set $E(L)$ which was considered in 2.14, and π maps $\langle E(U(L, \eta))\rangle$ isomorphically onto $P(L)$, i.e. the subsemigroup of $S(L)$ which is generated by the mappings $(n; \nu)$, n, ν complementary in L.

Let us consider $(g, f) \in U(L, \eta)$. We shall put $g1 = a$ and $1f = b$. Then in $U(L, \eta)$ we must have

$$(g, f) \mathscr{L} ((b^{\perp}; b), (b^{\perp}; b)) \mathscr{R} (f, g) =$$

$$= (g, f)^{\star} \mathscr{L} ((a^{\perp}; a), (a^{\perp}; a)) \mathscr{R} (g, f),$$

and so in $\pi(U(L))$ we must have

$$f \mathscr{L} (b^{\perp}, b) \mathscr{R} g \mathscr{L} (a^{\perp}; a) \mathscr{R} f.$$

If $y \in L$, then $b \wedge y \in \operatorname{im} f$ since f is a normal mapping. Thus there exists an element $x \in L$ such that $xf = b \wedge y$. From $(a^{\perp}; a) \mathscr{R} f$ follows $\ker (a^{\perp}; a) = \ker f$. Hence $x \vee a^{\perp}$ is the greatest element in the ff^{-1}-class of x. Thus the mapping

$$\overset{+}{f}\colon L \to L, \quad y \mapsto y\overset{+}{f} = \bigvee \{x \mid xf = y \wedge b\}$$

is well defined, and for all $x, y \in L$ we then have

$$xf\overset{+}{f} = x \vee a^{\perp} \geqslant x, \quad y\overset{+}{f}f = y \wedge b \leqslant y.$$

It is easy to see that $\overset{+}{f}$ is order preserving. This means that f is a residuated mapping, and that $\overset{+}{f}$ is its residual [4]. It follows immediately from the definition of a normal mapping that f maps principal ideals onto principal ideals. Hence f is a totally range-closed residuated mapping [4]. Let y be any element of L. Then, since $\overset{+}{f}$ is order preserving, we see that $\overset{+}{f}$ maps the principal filter of L which is generated by y into the principal filter which is generated by $y\overset{+}{f}$. Let z be any element of the latter principal filter: $z \geqslant y\overset{+}{f}$. Therefore $y \wedge b \leqslant zf \leqslant b$, from which $(y \vee zf) \wedge b = zf = zf \wedge b$, by the modularity of L. Hence $(y \vee zf)\overset{+}{f} = z$. We conclude that $\overset{+}{f}$ maps principal filters onto principal filters. Thus the residuated mapping f is strongly range-closed [4].

Let us now suppose that f is a strongly range-closed residuated mapping of the lattice L, and let $\overset{+}{f}$ be the residual of f. From Theorem 13.5 of [4] we know that for all $x, y \in L$

(3.9) $\qquad (x\overset{+}{f} \wedge y)f = x \wedge yf,$

(3.10) $\qquad (xf \vee y)\overset{+}{f} = x \vee y\overset{+}{f}.$

Let us put $a = 0\overset{+}{f}$ and $b = 1f$. Then it follows from (3.10) that $xf\overset{+}{f} = x \vee a^{\perp}$ for all $x \in L$. Hence $xf = xf\overset{+}{f}f = (x \vee a^{\perp})f$ for all $x \in L$, and so

$$(x, y) \in \ker f \Leftrightarrow x \vee a^{\perp} = y \vee a^{\perp} \Leftrightarrow (x, y) \in \ker (a^{\perp}; a).$$

Thus $\ker f$ is a normal equivalence on L. If $x \in L$, then $(x \wedge (x^{\perp} \vee a, x) \in \ker f$, and the principal ideal of L which is generated by $x \wedge (x^{\perp} \vee a)$ intersects the ff^{-1}-classes in at most one element. Since f maps principal ideals onto principal ideals, we see that f maps the principal ideal $L(x \wedge (x^{\perp} \vee a))$ isomorphically onto the principal ideal $L(xf)$. Hence, f is a normal mapping. In an analogous way one shows that ${}^{\perp}\overset{+}{f}{}^{\perp}$ is a normal mapping.

Again, let f be the strongly range-closed residuated mapping of the lattice L which was considered above, and take $({}^{\perp}\overset{+}{f}{}^{\perp}, f) \in S^{op}(L) \times S(L)$.

Then $\operatorname{im} f = L(1f) = L(b)$, $\operatorname{im}({}^{\perp}\overset{+}{f}{}^{\perp}) = L((\overset{+}{f}0)^{\perp}) = L(a)$. Then

$$\ker f = \{(x, y) \in L \times L \mid x \vee a^{\perp} = y \vee a^{\perp}\} = \eta(a),$$

and

$$\begin{aligned}\ker {}^{\perp}\overset{+}{f}{}^{\perp} &= \{(x, y) \in L \times L \mid (x^{\perp}, y^{\perp}) \in \ker \overset{+}{f}\} = \\ &= \{(x, y) \in L \times L \mid x^{\perp} \wedge b = y^{\perp} \wedge b\} = \qquad \text{(by (3.9))} \\ &= \eta(b).\end{aligned}$$

Thus the pair $({}^{\perp}\overset{+}{f}{}^{\perp}, f)$ satisfies (I3). Further, if $z \in L$,

$$\begin{aligned}&(x, y) \in \eta(z)f^{-1} \Leftrightarrow xf \vee z^{\perp} = yf \vee z^{\perp} \Rightarrow \\ &\Rightarrow x \vee \overset{+}{f}(z^{\perp}) = \overset{+}{f}(xf \vee z^{\perp}) = \overset{+}{f}(yf \vee z^{\perp}) = y \vee \overset{+}{f}(z^{\perp}) \Leftrightarrow \qquad \text{(by (3.10)} \\ &\Leftrightarrow (x, y) \in \eta({}^{\perp}\overset{+}{f}{}^{\perp}(z)).\end{aligned}$$

Conversely, if $(x, y) \in \eta({}^{\perp}\overset{+}{f}{}^{\perp}(z))$, then $(xf \vee z^{\perp}, yf \vee z^{\perp}) \in \ker \overset{+}{f}$. Consequently,

$$xf \vee (z^{\perp} \wedge b) = (xf \vee z^{\perp}) \wedge b = (yf \vee z^{\perp}) \wedge b = yf \vee (z^{\perp} \wedge b),$$

from which it follows that $xf \vee z^{\perp} = yf \vee z^{\perp}$, and thus $(x, y) \in \eta(z)f^{-1}$. We conclude that $\eta(z)f^{-1} = \eta({}^{\perp}\overset{+}{f}{}^{\perp}(z))$ for all $z \in L$. Analogously, $\eta(zf) = ({}^{\perp}\overset{+}{f}{}^{\perp})^{-1}\eta(z)$ for all $z \in L$. Thus also (I4) is satisfied, and we may conclude that $({}^{\perp}\overset{+}{f}{}^{\perp}, f) \in U(L, \eta)$.

From the foregoing it follows that the map $\pi \colon U(L, \eta) \to B(L)$, $(g, f) \longmapsto f$ is an isomorphism of $U(L, \eta)$ onto the semigroup of strongly range-closed residuated mappings on the lattice L. Further, if $(g, f) \in U(L, \eta)$, then g must be well defined by f, and is in fact $g = {}^{\perp}\overset{+}{f}{}^{\perp}$. Since $^{\star} \colon U(L, \eta) \to U(L, \eta)$, $({}^{\perp}\overset{+}{f}{}^{\perp}, f) \longmapsto ({}^{\perp}\overset{+}{f}{}^{\perp}, f)^{\star} = (f, {}^{\perp}\overset{+}{f}{}^{\perp})$ is the appropriate involution on $U(L, \eta)$, the mapping $^{\star} \colon B(L) \to B(L)$, $f \longmapsto {}^{\perp}\overset{+}{f}{}^{\perp}$ is an involution on $B(L)$. The latter involution on $B(L)$ is exactly the one which was considered by F o u l i s [9], [10], [11] (see Example 1.9), and π is a $\star$-isomorphism. Therefore also the fundamental $\star$-regular semigroup $T(E(L))$ will be $\star$-isomorphic to $B(L)$ [31]. Further, $E(L)$ is the set of idempotents of $B(L)$, and $P(L)$ is the idempotent generated part of $B(L)$.

Let S be any strongly regular Baer ⋆-semigroup which coordinatizes the orthocomplemented modular lattice L. Then the fundamental representation of S which was considered in [30] yields a representation of S by a full ⋆-subsemigroup of $T(E(S))$ (see Theorem 2.2). Also, following Grillet's approach [13], [14], [15], S may be represented by a full ⋆-subsemigroup of $U(L, \eta)$. Finally, S may be represented by a semigroup of strongly range-closed residuated mappings of L [9], [10], [11]. By the above, and also by [31], one can show that the three representations which are considered here are equivalent.

The orthocomplemented modular lattice L of length 2 which is depicted in Example 1.9 yields a fundamental strongly regular Baer ⋆-semigroup $B(L)$ which does not satisfy the ⋆-cancellation (1.9). Therefore the converse of Theorem 3.13 does not hold. For a discussion of the ⋆-cancellation law in Baer ⋆-semigroups, we refer to Section 6 of [11].

REFERENCES

[1] C.L. Adair, Varieties of ⋆-Orthodox Semigroups, Ph. D. Thesis, University of South Carolina, 1979.

[2] C.L. Adair, Bands which admit an involution, *Symposium on Regular Semigroups,* Northern Illinois University, April 1979, 1–10.

[3] B. Banaschewski, Brandt groupoids and semigroups, *Semigroup Forum,* 18 (1979), 307–312.

[4] T.S. Blyth – M.F. Janowitz, *Residuation Theory,* Pergamon Press, London, 1972.

[5] A.H. Clifford, *The partial groupoid of idempotents of a regular semigroup,* Tulane University, 1974.

[6] A.H. Clifford, *The fundamental representation of a regular semigroup,* Tulane University, 1974.

[7] A.H. Clifford – G.B. Preston, *The Algebraic Theory of Semigroups,* Vol. I, Amer. Math. Soc., Providence, R. I., 1961.

[8] M.P. Drazin, Regular semigroups with involution, *Symposium on Regular Semigroups,* Northern Illinois University, April 1979, 29–46.

[9] D.J. Foulis, Baer ⋆-semigroups, *Proc. Amer. Math. Soc.,* 11 (1960), 648–654.

[10] D.J. Foulis, Conditions for the modularity of an orthomodular lattice, *Pacific J. Math.,* 11 (1961), 889–895.

[11] D.J. Foulis, Relative inverses in Baer ⋆-semigroups, *Michigan Math. J.,* 10 (1963), 65–84.

[12] K.R. Goodearl, *von Neumann Regular Rings,* Pitman, London, 1979.

[13] P.A. Grillet, The structure of regular semigroups, I: a representation, *Semigroup Forum,* 8 (1974), 177–183.

[14] P.A. Grillet, The structure of regular semigroups, II: cross-connections, *Semigroup Forum,* 8 (1974), 254–259.

[15] P.A. Grillet, The structure of regular semigroups, III: the reduced case, *Semigroup Forum,* 8 (1974), 260–265.

[16] T.E. Hall, On the lattice of congruences on a regular semigroup, *Bull. Austral. Math. Soc.,* 1 (1969), 231–235.

[17] T.E. Hall, On regular semigroups, *J. Algebra,* 24 (1973), 1–24.

[18] D. Hardy – F. Pastijn, The maximal regular ideal of the semigroup of binary relations, *Czechoslovak Math. J.,* 31 (1981), 194–197.

[19] R.E. Hartwig – I.J. Katz, Products of EP elements in reflexive semigroups, *Linear Algebra Appl.,* 14 (1976), 11–19.

[20] J.M. Howie, *An Introduction to Semigroup Theory,* Academic Press, London, 1976.

[21] J.M. Howie – G. Lallement, Certain fundamental congruences on a regular semigroup, *Proc. Glasgow Math. Assoc.*, 7 (1966), 145–159.

[22] T. Imaoka, On fundamental regular ⋆-semigroups, *Mem. Fac. Sci. Shimane Univ.*, 14 (1980), 19–23.

[23] T. Imaoka, On regular ⋆-semigroups, to appear.

[24] T. Imaoka, ⋆-Congruences on regular ⋆-semigroups, to appear.

[25] T. Imaoka, Some remarks on fundamental regular ⋆-semigroups, *Semigroups and related topics,* Proc. Symp. Kyoto, Japan, 1980, 82–84.

[26] G. Kalmbach, *Orthomodular lattices,* London Math. Soc., Monograph series 18, London, 1983.

[27] D.B. McAlister, Groups, semilattices and inverse semigroups, II, *Trans. Amer. Math. Soc.*, 196 (1974), 351–370.

[28] J. Meakin – K.S.S. Nambooripad, Coextensions of regular semigroups by rectangular bands, *Trans. Amer. Math. Soc.*, 269 (1982), 197–224.

[29] J. Meakin – F. Pastijn, The structure of pseudo-semilattices, *Algebra Universalis,* 13 (1981), 355–372.

[30] K.S.S. Nambooripad, Structure of regular semigroups, I, *Mem. Amer. Math. Soc.*, 224 (1979).

[31] K.S.S. Nambooripad, Relations between biordered sets and cross-connections, *Semigroup Forum,* 16 (1978), 67–81.

[32] K.S.S. Nambooripad, Pseudo-semilattices and biordered sets, I, *Simon Stevin,* 55 (1981), 103–110.

[33] K.S.S. Nambooripad, Pseudo-semilattices and biordered sets, II, Pseudo-inverse semigroups, *Simon Stevin,* 56 (1982) 143–159.

[34] K.S.S. Nambooripad, Pseudo-semilattices and biordered sets, III, Regular locally testable semigroups, *Simon Stevin,* 56 (1982), 239–256.

[35] T.E. Nordahl – H.E. Scheiblich, Regular ⋆-semigroups, *Semigroup Forum,* 16 (1978), 369–378.

[36] F. Pastijn, Structure theorems for pseudo-inverse semigroups, *Symposium on Regular Semigroups,* Northern Illinois University, April 1979, 128–138.

[37] F. Pastijn, Biordered sets and complemented modular lattices, *Semigroup Forum,* 21 (1980), 205–220.

[38] F. Pastijn, Rectangular bands of inverse semigroups, *Simon Stevin,* 56 (1982), 4–92.

[39] M. Petrich, *Lectures in Semigroups,* John Wiley and Sons, London, 1977.

[40] N.R. Reilly, A class of regular ⋆-semigroups, *Semigroup Forum,* 18 (1979), 385–386.

[41] N.R. Reilly – H.E. Scheiblich, Congruences on regular semigroups, *Pacific J. Math.,* 23 (1967), 349–360.

[42] B.M. Schein, On the theory of generalized groups and generalized heaps, *Theory of Semigroups and its Applications* I, Izv. dat. Saratov. Univ., Saratov, 1966, 286–324 (in Russian).

[43] B.M. Schein, Semigroups of rectangular binary relations, *Dokl. Akad. Nauk SSSR,* 165 (1965), 1563–1566.

[44] L.A. Skornjakov, *Complemented Modular Lattices and Regular Rings,* Oliver and Boyd Ltd, London, 1964.

[45] M. Yamada, On the structure of fundamental regular ⋆-semigroups, to appear.

[46] K.A. Zareckiĭ, Abstract characterization of the semigroup of all binary relations, *Proc. Leningrad Pedagogical Inst.*, 183 (1958), 251–263 (in Russian).

K.S.S. Nambooripad

Department of Mathematics, University of Kerala, Kariavattom 695581, India.

F.J.C.M. Pastijn

Dienst Hogere Meetkunde, Rijksuniversiteit te Gent, Krijgslaan 281, B-9000 Gent, Belgium.

COLLOQUIA MATHEMATICA SOCIETATIS JÁNOS BOLYAI
39. SEMIGROUPS, SZEGED (HUNGARY), 1981.

UNIVERSAL GROUPS ON REVERSIBLE SEMIGROUPS

K.E. OSONDU

ABSTRACT

Let S be a reversible semigroup and ν a homomorphism of S into a group G such that (G, ν) is a universal group on S. This paper studies the structure of G. It is shown that every element of G is expressible in the form $\nu(a)\nu(b)^{-1}$, where $a, b \in S$. This extends the structure theory of universal groups on the subclass of inverse semigroups to the class of arbitrary reversible semigroups.

1. INTRODUCTION

Following ([1], p. 164), a congruence μ on a semigroup S is said to be left or right cancellative if S/μ is respectively left or right cancellative. Thus μ is cancellative if S/μ is both left and right cancellative. That is, $(ab, ac) \in \mu$ and $(ba, ca) \in \mu$, each implies $(b, c) \in \mu$.

Starting with the minimum cancellative congruence on a reversible semigroup S, one obtains a cancellative and reversible quotient semigroup S' which, by a theorem of Ore [3], can be embedded in a group G, constructed from the homogeneous quotients of S'. The basic undefined terms are those of [1], while the construction of homogeneous quotients is given in [4].

It is shown that G together with a suitably defined homomorphism ν forms a universal group (G, ν) on S, such that every element of G is expressible in the form $\nu(a)\nu(b)^{-1}$, where $a, b \in S$. Furthermore, any other universal group on S has this property. This characterization of universal groups on inverse semigroup exhibited in [4] and [6] is thus shown to apply to those of arbitrary reversible semigroups of which the class of inverse semigroups is a subclass.

2. MINIMUM CANCELLATIVE CONGRUENCE

Let S be a left reversible semigroup. If $a, b \in S$, define the relation μ on S by the rule:

$$(a, b) \in \mu \text{ if and only if } ax = bx \text{ for some } x \in S.$$

Lemma 2.1. *μ is a congruence on S such that S/μ is right cancellative.*

Proof. The relation is reflexive and symmetric. To show that it is transitive, let $a, b, c \in S$ such that $ax = bx$ and $by = cy$. Since S is left reversible, there also exist $s, t \in S$ such that $xs = yt = x^* \in S$. It now follows that

$$ax^* = axs = bxs = byt = cyt = cx^*.$$

Thus $(a, c) \in \mu$ and hence μ is transitive.

If $(a, b) \in \mu$, then for any $c \in S$, $cax = cbx$ for some $x \in S$ and hence $(ca, cb) \in \mu$. Furthermore, since S is left reversible, there exist $u, v \in S$ such that $cu = xv$. We now have

$$acu = axv = bxv = bcu.$$

That is, $(ac, bc) \in \mu$. It follows that μ is a congruence on S.

Let $\bar{a}$ denote the congruence class of a (modulo μ). Suppose $\bar{a}\bar{b} = \bar{c}\bar{b}$. Then $\overline{ab} = \overline{cb}$ and therefore there exists $x \in S$ such that

$abx = cbx$. Hence $\bar{a} = \bar{c}$. This completes the proof of Lemma 2.1.

Corollary 2.2. *If S is left cancellative and left reversible then μ is the minimum cancellative congruence on S.*

Proof. Suppose $\bar{a}\bar{b} = \bar{a}\bar{c}$. Then for some $x \in S$, $abx = acx$. Since S is left cancellative $bx = cx$ and hence $\bar{b} = \bar{c}$. It now follows that S/μ is cancellative.

To show the minimality of μ, suppose ρ is another cancellative congruence on S. Let $(a, b) \in \mu$. Then $ax = bx$ for some $x \in S$. Since ρ is reflexive, $(ax, bx) \in \rho$, and by the cancellativity of ρ, we have $(a, b) \in \rho$. Hence $\mu \subseteq \rho$.

Corollary 2.3. *If S is an inverse semigroup, μ coincides with the minimum group congruence on S.*

Proof. Let ρ denote the minimum group congruence on S. If $(a, b) \in \rho$, then by Munn [2], there exists an idempotent $e \in S$ such that $ae = be$. Hence $\rho \subseteq \mu$.

Conversely suppose $(a, b) \in \mu$, then there exists $x \in S$ such that $ax = bx$. Since $axx^{-1} = bxx^{-1}$ and xx^{-1} is an idempotent, we have $\mu \subseteq \rho$ and therefore the congruences are equal.

Remark 2.4. Dually, if S is right reversible and one defines on S the relation μ^* by the rule, $(a, b) \in \mu^*$ if and only if $xa = xb$ for some $x \in S$, one obtains a congruence on S such that S/μ^* is left cancellative. If S is also right cancellative, then μ^* is the minimum cancellative congruence on S.

Corollary 2.5. *Let S be a reversible semigroup. Then the minimum cancellative congruence on S is the congruence $\alpha = \mu \vee \mu^*$ generated by μ and μ^*. Furthermore $(a, b) \in \alpha$, $a, b \in S$, if and only if $xay = xby$ for some $x, y \in S$.*

Proof. This follows from the preceding analysis or may be established directly as in Lemma 2.1.

3. THE QUOTIENT SEMIGROUP

Throughout this section, S will be assumed to be a semigroup which is left reversible and left cancellative.

Lemma 3.1. *If* μ *is the minimum cancellative congruence on* S, *then* S/μ *is embeddable in a group of right quotients.*

Proof. Since S/μ is a homomorphic image of a left reversible semigroup, it follows from Lemma 2.2 of [5] that S/μ is left reversible. By the definition of μ, it follows also that S/μ is cancellative. Hence by Ore [3], (see also [1], Theorem 1.24), S/μ is embeddable in a group G of right quotients of S/μ.

The elements of G are of the form $\bar{a}\bar{b}^{-1}$, or simply equivalence classes $[\bar{a}/\bar{b}]$ of $(\bar{a}, \bar{b})$, $a, b \in S$. It is known from [3] that $[\bar{a}/\bar{b}] = [\bar{c}/\bar{d}]$ if and only if there exist $x, y \in S$ such that $\bar{a}\bar{x} = \bar{c}\bar{y}$ and $\bar{b}\bar{x} = \bar{d}\bar{y}$. Furthermore, the binary operation in G is given by $[\bar{a}/\bar{b}][\bar{c}/\bar{d}] = [\bar{a}\bar{x}/\bar{d}\bar{y}]$, where $\bar{x}, \bar{y} \in S/\mu$ such that $\bar{b}\bar{x} = \bar{c}\bar{y}$.

Let π: $S \to S/\mu$ be the canonical homomorphism and α: $S/\mu \to G$ the embedding. Set $\nu = \alpha \cdot \pi$. With these notations, we shall prove

Lemma 3.2. *Every element of* G *is expressible in the form* $\nu(a)\nu(b)^{-1}$, $a, b \in S$.

Proof. From [3], if $\bar{a} \in S/\mu$, then $\alpha(\bar{a}) = [\bar{a}\bar{s}/\bar{s}]$, $\bar{s} \in S/\mu$. Also $[\bar{a}/\bar{b}] = [\bar{a}\bar{s}/\bar{b}\bar{s}] = [\bar{a}\bar{s}/\bar{s}][\bar{s}/\bar{b}\bar{s}] = \nu(a)\nu(b)^{-1}$. It also follows that $\nu(S)$ is a set of group generators for G.

Now, let (H, η) be any group on the semigroup S or simply an S-group ([1], p. 289). That is, H is a group and η is a homomorphism of S into H such that $\eta(S)$ is a set of group generators of H. Define θ: $G \to H$ by $\theta([\bar{a}/\bar{b}]) = \eta(a)\eta(b)^{-1}$, where $\eta(b)^{-1} = (\eta(b))^{-1}$.

Lemma 3.3. θ *is a homomorphism.*

Proof. We shall first show that θ is well defined. Supposse $[\bar{a}/\bar{b}] =$ $= [\bar{c}/\bar{d}]$, $a, b, c, d \in S$. Then there exist $x, y \in S$ such that

$$\bar{a}\bar{x} = \bar{c}\bar{y}, \quad \bar{b}\bar{x} = \bar{d}\bar{y}.$$

Hence $(ax, cy), (bx, dy) \in \mu$, and therefore there exist $s, t \in S$ such that

$$axs = cys, \quad bxt = dyt.$$

Applying the homomorphism η to each equation, we have

$$\eta(a)\eta(x)\eta(s) = \eta(c)\eta(y)\eta(s),$$

$$\eta(b)\eta(x)\eta(t) = \eta(d)\eta(y)\eta(t),$$

whence

$$\eta(a)^{-1}\eta(c) = \eta(x)\eta(y)^{-1} = \eta(b)^{-1}\eta(d).$$

Thus $\eta(a)\eta(b)^{-1} = \eta(c)\eta(d)^{-1}$, showing that θ is well defined.

To show that it is a homomorphism, we note from [3] that multiplication in G is given by $[\bar{a}/\bar{b}][\bar{c}/\bar{d}] = [\overline{ax}/\overline{dy}]$, where $\bar{x}, \bar{y}$ are such that $\overline{bx} = \overline{cy}$. Hence from the definition of μ, there exists $s \in S$ such that $bxs = cys$. Thus, we have

$$\eta(x)\eta(y)^{-1} = \eta(b)^{-1}\eta(c).$$

Now,

$$\theta([\bar{a}/\bar{b}][\bar{c}/\bar{d}]) = \theta([\overline{ax}/\overline{dy}]) = \eta(ax)\eta(dy)^{-1} =$$

$$= \eta(a)\eta(x)\eta(y)^{-1}\eta(d)^{-1} = \eta(a)\eta(b)^{-1}\eta(c)\eta(d)^{-1} =$$

$$= \theta([\bar{a}/\bar{b}])\theta([\bar{c}/\bar{d}]).$$

This completes the proof of Lemma 3.3.

Lemma 3.4. (G, ν) *is a universal group on* S.

Proof.

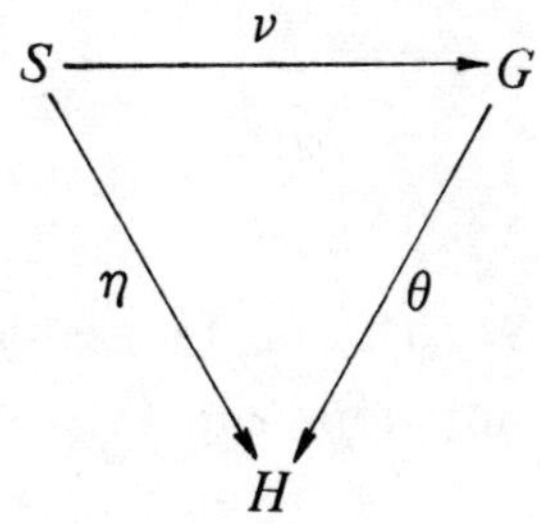

Let (H, η) be another S-group or group on S. One verifies that for $a \in S$,

$$\theta(\nu(a)) = \theta([\overline{as}/\overline{s}]) = \eta(as)\eta(s)^{-1} = \eta(a).$$

Hence $\eta = \theta \cdot \nu$ and the diagram commutes.

4. REVERSIBLE SEMIGROUPS

In this section, S^* will denote a reversible semigroup which is not necessarily left or right cancellative. As indicated in Remark 2.4, if one defines the relation μ^* on S^* by requiring that for $a^*, b^* \in S^*$, $(a^*, b^*) \in \mu^*$ if and only if $x^*a^* = x^*b^*$ for some $x^* \in S^*$, then μ^* is the minimum congruence on S^* such that S^*/μ^* is left cancellative.

Set $S' = S^*/\mu^*$ throughout this section. We shall now prove

Lemma 4.1. *S' has a universal group (G^*, ν') with every element of G^* expressible in the form $\nu'(a')\nu'(b')^{-1}$, $a', b' \in S'$.*

Proof. Noting that $S' = S^*/\mu^*$, it is clear that S' is a homomorphic image of a left reversible semigroup. Therefore, S' is both left reversible and left cancellative in view of Lemma 2.2 of [5]. Proceeding exactly as in Section 3, one constructs a universal group (G^*, ν') on S' in which every element of G^* is expressible in the form $\nu'(a')\nu'(b')^{-1}$, $a', b' \in S'$.

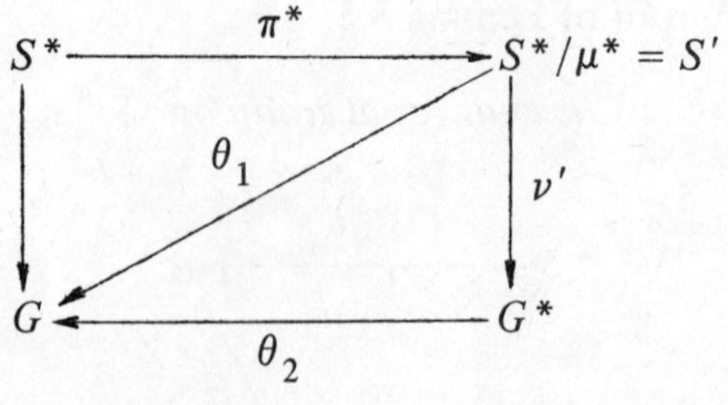

Now, let $\pi^*: S^* \to S^*/\mu^*$ be the canonical homomorphism. Set $\nu^* = \nu' \cdot \pi^*$. With those notations and those of the diagram above, we shall now prove

Theorem 4.2. *(G^*, ν^*) is a universal group on S^* such that every element of G^* is expressible in the form $\nu^*(a^*)\nu(b^*)^{-1}$, $a^*, b^* \in S^*$.*

Proof. Suppose (G, ν) is another group on S^*. If $a^*, b^* \in S^*$ such that $(a^*, b^*) \in \mu^*$, then there exists $x^* \in S^*$ such that $x^*a^* = x^*b^*$. Hence $\nu(x^*)\nu(a^*) = \nu(x^*)\nu(b^*)$; so $\nu(a^*) = \nu(b^*)$. We have thus proved that $\mu^* = \mu^*\mu^{*-1} \subseteq \nu\nu^{-1}$. It now follows from [1], Theorem 1.6 (Induced Homomorphism Theorem), that there exists a unique homomorphism θ_1 of S^*/μ^* onto $\nu(S^*)$ and hence into G such that $\nu = \theta_1\pi^*$.

By definition, $\nu(S^*)$ and hence $\theta_1(\pi^*(S^*)) = \theta_1(S')$ is a set of group generators of G. That is, (G, θ_1) is a group on S'. Since (G^*, ν') is by Lemma 4.1 a universal group on S', there exists a homomorphism $\theta_2 \colon G^* \to G$ such that $\theta_1 = \theta_2 \cdot \nu'$. We now have $\nu = \theta_1\pi^* = \theta_2(\nu'\pi^*) = \theta_2\nu^*$. This shows that (G^*, ν^*) is a universal group on S^*.

Furthermore, by Lemma 4.1, every element of G^* is expressible in the form $\nu'(a')\nu'(b')^{-1}$, $a', b' \in S'$. It is therefore of the form $\nu'(\pi^*(a^*))\nu'(\pi^*(b^*)^{-1})$, $a^*, b^* \in S^*$ and thus of the form $\nu^*(a^*)\nu^*(b^*)^{-1}$, $a^*, b^* \in S^*$. This completes the proof of Theorem 4.2.

Theorem 4.3. *Let S be a reversible semigroup and $\nu \colon S \to G$ a homomorphism of S into a group G such that (G, ν) is a universal group on S. Then every element of G is expressible in the form $\nu(a)\nu(b)^{-1}$, $a, b \in S$.*

Proof. We only need to prove that given any other universal group (G', ν') on S, then the elements of G' are of the required form. Starting with the universal group (G^*, ν^*) as in Theorem 4.2, since (G', ν') is another universal group on S, it follows from [1], Lemma 12.1 that there exists an isomorphism θ of G^* onto G' such that $\nu' = \theta \cdot \nu^*$. Since every element of G^* is of the form $\nu^*(a)\nu^*(b)^{-1}$ and since θ is surjective, every element of G' is of the form $\theta(\nu^*(a)\nu^*(b)^{-1})$ and hence of the form $\nu'(a)\nu'(b)^{-1}$.

REFERENCES

[1] A.H. Clifford – G.B. Preston, *The Algebraic Theory of Semigroups,* I and II, Amer. Math. Soc., Providence, R.I., 1961, 1967.

[2] W.D. Munn, A class of irreducible matrix representations of an arbitrary inverse semigroup, *Proc. Glasgow Math. Soc.,* 5 (1961), 41–48.

[3] O. Ore, Linear equations in non-commutative fields, *Ann. of Math.,* 32 (1931), 463–477.

[4] K.E. Osondu, Homogeneous quotients of semigroups, *Semigroup Forum,* 16 (1978), 455–462.

[5] K.E. Osondu, Semilattices of left reversible semigroups, *Semigroup Forum,* 17 (1979), 139–152.

[6] K.E. Osondu, The universal group of homogeneous quotients, *Semigroup Forum,* 21 (1980), 143–152.

K.E. Osondu

Department of Mathematical Sciences, College of Technology, P.M.B. 1036, Owerri, Nigeria.

COLLOQUIA MATHEMATICA SOCIETATIS JÁNOS BOLYAI
39. SEMIGROUPS, SZEGED (HUNGARY), 1981.

MONOIDS OF UPPER TRIANGULAR MATRICES

J.-E. PIN — H. STRAUBING

The subject of the present paper belongs to the theory of varieties of finite monoids and recognizable languages. We refer the reader to the books by Eilenberg [2] and Lallement [3] for the elements of this theory. Here we study the variety generated by monoids of upper triangular boolean matrices. This is a subvariety of the variety of finite aperiodic monoids, and, as we show in this paper, it appears in a number of different contexts.

In the first section we present the variety **W** generated by monoids of upper triangular boolean matrices as a natural extension of the variety **J** of finite $\mathcal{J}$-trivial monoids (which was studied in detail by Simon [12], [13]). We give a description of the family of recognizable languages whose syntactic monoids are in **W**.

In the second section we show that the variety **W** can be described in terms of the generalized Schützenberger product of finite monoids (introduced in [16]). More exactly, we prove that a monoid M belongs to **W** if and only if M divides an n-fold Schützenberger product $\Diamond_n(M_1, \dots, M_n)$ where $M_1, \dots, M_n$ are idempotent and commutative finite monoids.

In the third section we show that **W** can be described in terms of power sets: If **V** is a variety of finite monoids then **PV** denotes the variety generated by $\{\mathscr{P}(M) \mid M \in \mathbf{V}\}$, where $\mathscr{P}(M)$ is the power set of M. The operation $\mathbf{V} \to \mathbf{PV}$ has been studied by several authors [4], [5], [6], [7], [9], [14]. Here we use these earlier results to prove that $\mathbf{W} = \mathbf{PJ}$.

In the final section we consider the membership problem for **W**. That is, we would like an algorithm for determining, given the multiplication table of a finite monoid M, whether or not $M \in \mathbf{W}$. We are able to give an effective necessary condition for membership in **W**, but we do not yet know if our condition is sufficient. In the same section we cite an unpublished result if Straubing [17] which connects the variety **W** to the dot-depth hierarchy of Brzozowski.

Although the majority of our results are purely "semigroup-theoretic" – in the sense that they make no reference to recognizable languages or the theory of automata – we use recognizable languages constantly in the proofs. In essence we are exploiting the correspondence between varieties of monoids and varieties of languages, as described by Eilenberg [2]. This provides us with a powerful tool for proving theorems about varieties of finite monoids.

1. THE VARIETY GENERATED BY MONOIDS OF UPPER TRIANGULAR MATRICES

Let $n \geqslant 1$. We denote by M_n the set of all $n \times n$ matrices over the boolean semiring $B = \{0, 1\}$ (in which $1 + 1 = 1$) and by K_n the set of all upper triangular matrices in M_n all of whose diagonal entries equal 1. That is,

$$K_n = \{m \in M_n \mid m_{ij} = 0 \text{ for } 1 \leqslant j < i \leqslant n \text{ and}$$
$$m_{ii} = 1 \text{ for } 1 \leqslant i \leqslant n\}.$$

K_n is closed under multiplication of matrices and is thus a submonoid of the multiplicative monoid M_n. We define

$$\mathbf{U} = \{M \mid M \prec K_n \text{ for some } n\}.$$

That is, $\mathbf{U}$ consists of all monoids (necessarily finite) which are divisors of K_n for a certain n. (We say that a monoid M divides a monoid M' if M is a quotient of a submonoid of M' – see [2].)

The family $\mathbf{U}$ is evidently closed under division. It is also closed under direct product: To see this observe that there is an injective morphism $\varphi\colon M_m \times M_n \to M_{m+n}$ defined by

$$(p, q)\varphi_{ij} = \begin{cases} 0 & \text{if} \quad i \leqslant m \text{ and } j > m, \\ 0 & \text{if} \quad i > m \text{ and } j \leqslant m, \\ p_{ij} & \text{if} \quad i \leqslant m \text{ and } j \leqslant m, \\ q_{i-m, j-m} & \text{if} \quad i > m \text{ and } j > m, \end{cases}$$

and that φ embeds $K_m \times K_n$ into K_{m+n}. Since $N_1 \prec M_1$ and $N_2 \prec M_2$ implies $N_1 \times N_2 \prec M_1 \times M_2$, it follows that $\mathbf{U}$ contains the direct product of any two of its members. Thus $\mathbf{U}$ is a *variety of finite monoids* in the sense used by Eilenberg [2], Ch. V.

If $\mathbf{V}$ is a variety of finite monoids and A is a finite alphabet, then we denote by $A^*\mathscr{V}$ the family of all recognizable languages in A^* whose syntactic monoids belong to $\mathbf{V}$. Eilenberg has shown [2], Ch. VII that every variety of finite monoids is generated by the syntactic monoids it contains. Thus if $\mathbf{V}_1$ and $\mathbf{V}_2$ are varieties of finite monoids, $\mathbf{V}_1 \subset \mathbf{V}_2$ if and only if $A^*\mathscr{V}_1 \subset A^*\mathscr{V}_2$ for every finite alphabet A. This enables us to show that two varieties are equal by showing that the corresponding families of recognizable languages are equal.

In the case of the variety $\mathbf{U}$ defined above, Straubing [15] showed that $A^*\mathscr{U}$ is the boolean closure of the family of languages of the form

$$A^*a_1A^* \ldots a_nA^*,$$

where $a_1, \ldots, a_k \in A$.

A deep result of Simon [13] asserts that this is precisely the family of languages whose syntactic monoids belong to the variety $\mathbf{J}$ of finite $\mathscr{J}$-trivial monoids. (A monoid is $\mathscr{J}$*-trivial* if and only if the $\mathscr{J}$-relation is the identity.) We thus have

Theorem 1. $\mathbf{U} = \mathbf{J}$. *(That is, a finite monoid M is $\mathcal{J}$-trivial if and only if it divides K_n for some n.)*

We now consider the family T_n of all $n \times n$ matrices over the semiring B which are *upper triangular.* T_n is a submonoid of M_n which contains K_n. We define

$$\mathbf{W} = \{M \mid M \prec T_n \quad \text{for some} \quad n\}.$$

Once again it is evident that $\mathbf{W}$ is closed under division, and that the morphism $\varphi\colon M_m \times M_n \to M_{m+n}$ maps $T_m \times T_n$ into T_{m+n}. Thus $\mathbf{W}$ is a variety of finite monoids. This variety is the principal concern of the present paper.

We begin with describing the family of recognizable languages corresponding to $\mathbf{W}$:

Theorem 2. *$A^*\mathcal{W}$ is the boolean closure of the family of languages of the form*

$$A_0^* a_1 A_1^* \ldots a_k A_k^*$$

where $k \geqslant 0$, $a_1, \ldots, a_k \in A$, and $A_0, \ldots, A_k$ are (possibly empty) subsets of A.

Remark. If $A_i = \phi$ then $A_i^* = \{1\}$.

Proof. Let $\mathcal{F}$ denote the boolean closure of the family of languages of the form $A_0^* a_1 A_1^* \ldots a_k A_k^*$, where $a_1, \ldots, a_k \in A$ and $A_0, \ldots, A_k \subset A$. We first show that $\mathcal{F} \subset A^*\mathcal{W}$.

For this it suffices to show that the syntactic monoid of any language of the form

$$L = A_0^* a_1 A_1^* \ldots a_k A_k^*$$

is in $\mathbf{W}$, since it is known [2], Ch. VII that $A^*\mathcal{V}$ is closed under boolean operations for any variety $\mathbf{V}$. We will show that L is *recognized* by the monoid T_{k+1}: that is, there exists a morphism $\psi\colon A^* \to T_{k+1}$ and a set $X \subset T_{k+1}$ such that $X\psi^{-1} = L$. This implies by [2], Ch. VII that $M(L) \prec T_{k+1}$ and thus that $M(L) \in \mathbf{W}$. The morphism ψ is defined by

$$(a\psi)_{ij} = \begin{cases} 1 & \text{if} \quad i=j \ \text{ and } \ a \in A_{i-1}, \\ 1 & \text{if} \quad j = i+1 \ \text{ and } \ a = a_i, \\ 0 & \text{otherwise} \end{cases}$$

for all $a \in A$, $i,j \in \{1, \ldots, k+1\}$. It is easy to verify that if $w \in A^*$ then $(w\psi)_{ij} = 1$ if and only if there is a path labelled w from state i to state j in the nondeterministic automaton pictured below.

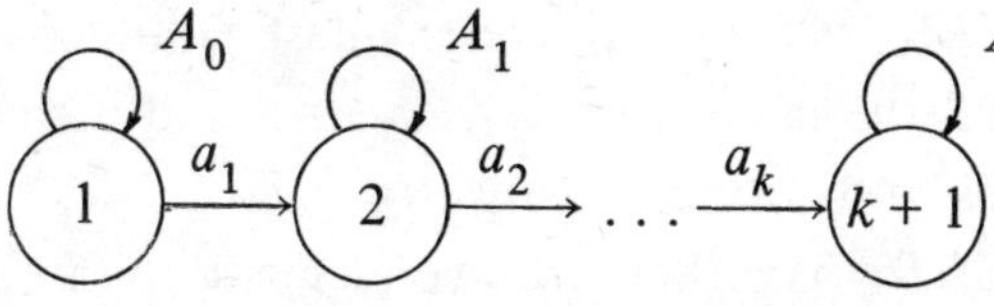

In particular, $(w\psi)_{1,k+1} = 1$ if and only if $w \in A_0^* a_1 A_1^* \ldots a_k A_k^* = L$. Thus $L = X\psi^{-1}$, where $X = \{m \in T_{k+1} \mid m_{1,k+1} = 1\}$. This proves that $\mathscr{F} \subset A^*\mathscr{W}$.

To prove the opposite inclusion, suppose that $L \in A^*\mathscr{W}$. Then $M(L) \in \mathbf{W}$ and consequently L is recognized by T_n for some $n \geqslant 1$. Thus there exists a morphism η: $A^* \to T_n$ and a subset X of T_n such that $L = X\eta^{-1}$. We need to show that $X\eta^{-1} \in \mathscr{F}$. Since $X\eta^{-1} = \bigcup_{x \in X} x\eta^{-1}$ and since $\mathscr{F}$ is closed under boolean operations, it suffices to show that $x\eta^{-1} \in \mathscr{F}$ for each $x \in T_n$. Now

$$\begin{aligned} x\eta^{-1} &= \bigcap_{1 \leqslant i,j \leqslant n} \{w \mid (w\eta)_{ij} = x_{ij}\} = \\ &= \bigcap_{\{(i,j) \mid x_{ij} = 1\}} \{w \mid (w\eta)_{ij} = 1\} - \\ &\quad - \bigcup_{\{(i,j) \mid x_{ij} = 0\}} \{w \mid (w\eta)_{ij} = 1\}. \end{aligned}$$

Thus it suffices to show that each set of the form $\{w \mid (w\eta)_{ij} = 1\}$ belongs to $\mathscr{F}$. Let $A_{k,l} = \{a \in A \mid (a\eta)_{kl} = 1\}$ and let Q_{ij} be the set of all *strictly increasing* sequences $(i_0, \ldots, i_t)$ such that $i_0 = i$ and $i_t = j$. (If $i > j$ then Q_{ij} is empty. If $i = j$ then Q_{ij} consists of the single sequence (i)). Then

$$\{w \mid (w\eta)_{ij} = 1\} = \bigcup_{(i_0, \ldots, i_t) \in Q_{ij}} A^*_{i_0 i_0} A_{i_0 i_1} A^*_{i_1 i_1} \ldots A_{i_{k-1} i_k} A^*_{i_k i_k}.$$

Since each language $A^*_{i_0 i_0} A_{i_0 i_1} A^*_{i_1 i_1} \ldots A_{i_{k-1} i_k} A^*_{i_k i_k}$ is a finite union of the form $A^*_{i_0 i_0} a_1 A^*_{i_1 i_1} \ldots a_k A^*_{i_k i_k}$ it follows that $\{w \mid (w\eta)_{ij} = 1\} \in \in \mathscr{F}$. This completes the proof of Theorem 2.

From Theorem 2 we can deduce that the monoids T_n are *aperiodic*. That is, they contain no nontrivial groups: Let **Ap** denote the variety of finite aperiodic monoids, and $A^* \mathscr{A}p$ the family of recognizable languages in A^* whose syntactic monoids are in **Ap**. According to a theorem of Schützenberger [10], $A^* \mathscr{A}p$ is the smallest family of languages in A^* which contains all the languages $\{a\}$, $a \in A$, and which is closed under boolean operations and product. Now for each subset B of A, $B^* \in A^* \mathscr{A}p$ because $M(B)$ is either the two-element aperiodic monoid $U_1 = \{0, 1\}$ (if $\phi \neq B \neq A$) or $M(B)$ is trivial (if $B = \phi$ or $B = A$). Since $A^* \mathscr{A}p$ contains the letters and is closed under products, it follows that every language of the form $A_0^* a_1 A_1^* \ldots a_k A_k^*$ is in $A^* \mathscr{A}p$. Since $A^* \mathscr{A}p$ is closed under boolean operations, it follows from Theorem 2 that $A^* \mathscr{W} \subset A^* \mathscr{A}p$. Thus $\mathbf{W} \subset \mathbf{Ap}$; in particular the monoids T_n are aperiodic.

2. CONNECTION WITH THE SCHÜTZENBERGER PRODUCT

In [10] Schützenberger introduced a binary product on finite monoids to study the product operation on recognizable languages. This was generalized to an n-fold product by Straubing [16]. Here we recall the definition of this product and some of its basic properties.

Let $M_1, \ldots, M_n$ be finite monoids and consider the set $\mathscr{P}(M_1 \times \ldots \times M_n)$ of all subsets of $M_1 \times \ldots \times M_n$. Multiplication in the direct product $M_1 \times \ldots \times M_n$ is extended to $\mathscr{P}(M_1 \times \ldots \times M_n)$ by the formula $XY = \{xy \mid x \in X,\ y \in Y\}$ for all $X, Y \in \mathscr{P}(M_1 \times \ldots \times M_n)$. Addition in $\mathscr{P}(M_1 \times \ldots \times M_n)$ is defined by $X + Y = X \cup Y$. With these operation $\mathscr{P}(M_1 \times \ldots \times M_n)$ is a semiring (with $\{(1, \ldots, 1)\}$ as the multiplicative identity and ϕ as the additive identity) and we can thus consider the monoid $\mathscr{M}$ of all $n \times n$ matrices over $\mathscr{P}(M_1 \times \ldots \times M_n)$. The Schützenberger product $\Diamond_n(M_1, \ldots, M_n)$ is the submonoid of $\mathscr{M}$ consisting of all matrices P such that

$$P_{ij} = \phi \text{ if } i > j,$$

$$P_{ii} = \{(1, \ldots, s_i, \ldots, 1)\} \text{ for some } s_i \in M_i,$$

$$P_{ij} \subset \{(s_1, \ldots, s_n) \in M_1 \times \ldots \times M_n \mid s_k = 1$$
$$\text{if } k < i \text{ or if } k > j\} \text{ if } i < j.$$

The following property of the Schützenberger product was proved by Reutenauer [9] in the case $n = 2$ and by Pin [8] in general:

If $L \subset A^*$ is recognized by $\diamond_n(M_1, \ldots, M_n)$ then L belongs to the boolean closure of the family of languages of the form

$$L_i a_i L_{i+1} \ldots a_{j-1} L_j$$

where $1 \leqslant i \leqslant j \leqslant n$, $a_i, \ldots, a_{j-1} \in A$, and $M(L_k) \prec M_k$ for $i \leqslant k \leqslant j$.

If $\mathbf{V}$ is a variety of finite monoids we denote by $\diamond\mathbf{V}$ the smallest variety which contains all the Schützenberger products of the form $\diamond_n(M_1, \ldots, M_n)$ where $M_1, \ldots, M_n \in \mathbf{V}$.

In Section 1 we showed that $T_m \times T_n$ is a submonoid of T_{m+n}. An identical argument shows that $\diamond_m(M_1, \ldots, M_m) \times \diamond_n(M_1', \ldots, M_n')$ is a submonoid of $\diamond_{m+n}(M_1, \ldots, M_m, M_1', \ldots, M_n')$. It follows that $M \in \diamond\mathbf{V}$ if and only if M divides a Schützenberger product $\diamond_k(M_1, \ldots, M_k)$ where $M_1, \ldots, M_k$ all belong to $\mathbf{V}$.

As above, we denote by $\mathbf{J}$ the variety of $\mathscr{J}$-trivial monoids. $\mathbf{J}_1$ denotes the variety of idempotent and commutative monoids, $\mathbf{R}$ the variety of $\mathscr{R}$-trivial monoids and $\mathbf{R}^r$ the variety of $\mathscr{L}$-trivial monoids. Finally, $\mathbf{DA}$ denotes the variety of aperiodic monoids with the property that every regular $\mathscr{J}$-class is closed under multiplication. We have the inclusions $\mathbf{J}_1 \subset \mathbf{J} \subset \mathbf{R} \subset \mathbf{DA}$ and $\mathbf{J}_1 \subset \mathbf{J} \subset \mathbf{R}^r \subset \mathbf{DA}$.

Theorem 3. $\mathbf{W} = \diamond\mathbf{J}_1 = \diamond\mathbf{J} = \diamond\mathbf{R} = \diamond\mathbf{R}^r = \diamond\mathbf{DA}$.

Proof. In light of the inclusions cited above it suffices to prove that $\mathbf{W} \subset \diamond\mathbf{J}_1$ and that $\diamond\mathbf{DA} \subset \mathbf{W}$. We prove the first of these inclusions by showing that T_n is a submonoid of $\diamond_n(U_1, \ldots, U_1)$, where U_1 is the two-element monoid $\{0, 1\}$. (Since $U_1 \in \mathbf{J}_1$, this implies that $\mathbf{W} \subset \diamond\mathbf{J}_1$.)

Indeed, if $m \in T_n$ let $m\varphi$ be the element of $\Diamond_n(U_1, \ldots, U_1)$ defined by

$$(m\varphi)_{ij} = \begin{cases} \phi & \text{if } m_{ij} = 0, \\ \{(1, \ldots, 1)\} & \text{if } m_{ij} = 1. \end{cases}$$

(For example, if $n = 3$ and $m = \begin{pmatrix} 1 & 1 & 0 \\ 0 & 0 & 1 \\ 0 & 0 & 1 \end{pmatrix}$ then

$$m\varphi = \begin{pmatrix} \{(1,1,1)\} & \{(1,1,1)\} & \phi \\ \phi & \phi & \{(1,1,1)\} \\ \phi & \phi & \{(1,1,1)\} \end{pmatrix}).$$

It is easy to verify that $\varphi\colon T_n \to \Diamond_n(U_1, \ldots, U_1)$ is an injective morphism.

To prove the inclusion $\Diamond\mathbf{DA} \subset \mathbf{W}$ we make use of the following result of Schützenberger [11]. If $L \subset A^*$ is a recognizable language such that $M(L) \in \mathbf{DA}$, then L is a finite disjoint *union* of languages of the form:

$$A_0^* a_1 A_1^* \ldots a_k A_k^*$$

where $A_0, \ldots, A_k \subset A$, $a_1, \ldots, a_k \in A$, and where the product $A_0^* a_1 A_1^* \ldots a_k A_k^*$ is unambiguous. From this result and the property of the Schützenberger product cited above, we conclude that every language whose syntactic monoid is in $\Diamond\mathbf{DA}$ is in the boolean closure of the family of languages of the form $B_0^* b_1 B_1^* \ldots b_m B_m^*$ where $B_0, \ldots, B_m \subset$ $\subset A$ and $b_1, \ldots, b_m \in A$. (We have used neither the unambiguity of the product $A_0^* a_1 \ldots a_k A_k^*$ nor the fact that the union is disjoint – only the fact that the product of languages distributes over union.) It follows from Theorem 2 that $M(L) \in \mathbf{W}$. Thus $\Diamond\mathbf{DA} \subset \mathbf{W}$. This completes the proof of Theorem 3.

Corollary. *Let M be a finite monoid. The following are equivalent:*

(i) $M \prec T_n$ *for some* $n \geqslant 1$.

(ii) $M \prec \Diamond_m(U_1, \ldots, U_1)$ *for some* $m \geqslant 1$.

Proof. We showed in the proof of Theorem 3 that T_n is a submonoid of $\Diamond_n(U_1, \ldots, U_1)$ and thus (i) implies (ii). Conversely, if $M \prec \Diamond_m(U_1, \ldots, U_1)$, then $M \in \Diamond \mathbf{J}_1$, and, by Theorem 3, $M \in \mathbf{W}$. Thus $M \prec T_n$ for some $n \geqslant 1$ and consequently (ii) implies (i).

3. CONNECTION WITH POWER SETS

If M is a finite monoid then $\mathscr{P}(M)$, the set of subsets of M, is a finite monoid with respect to the operation

$$XY = \{xy \mid x \in X,\ y \in Y\} \quad \text{for all} \quad X, Y \in \mathscr{P}(M).$$

If $\mathbf{V}$ is a variety of finite monoids then we denote by $\mathbf{PV}$ the smallest variety which contains $\{\mathscr{P}(M) \mid M \in \mathbf{V}\}$. Thus $M \in \mathbf{PV}$ if and only if there exists $M_1, \ldots, M_k \in \mathbf{V}$ such that $M \prec \mathscr{P}(M_1) \times \ldots \times \mathscr{P}(M_k)$. The operation $\mathbf{V} \to \mathbf{PV}$ has been studied by several authors [4], [5], [6], [7], [9], [14]. We cite the following result, which appears in [14].

Let A be a finite alphabet, and let $A^*\mathscr{P}\mathscr{V}$ be the family of recognizable languages whose syntactic monoids belong to $\mathbf{PV}$. $A^*\mathscr{P}\mathscr{V}$ is the boolean closure of the family of languages of the form $L\varphi$, where $L \subset B^*\mathscr{V}$ for some finite alphabet B, and where $\varphi: B^* \to A^*$ is a length-preserving morphism (that is, $B\varphi \subset A$).

Theorem 4. $\mathbf{W} = \mathbf{PJ} = \mathbf{PR} = \mathbf{PR}^r = \mathbf{PDA}$.

Proof. As in the proof of Theorem 3, it suffices to show that $\mathbf{W} \subset \mathbf{PJ}$ and that $\mathbf{PDA} \subset \mathbf{W}$.

We begin by showing that $\mathbf{PDA}$ is contained in $\mathbf{W}$. Let A and B be finite alphabets and let $\varphi: B^* \to A^*$ be a length-preserving morphism. If $L \subset B^*$ is a recognizable language such that $M(L) \in \mathbf{DA}$, then, by the result of Schützenberger cited in Section 2, L is a union of languages of the form

$$B_0^* b_1 B_1^* \ldots b_k B_k^*$$

where $B_0, \ldots, B_k \subset B$ and $b_1, \ldots, b_k \in B$. It follows that L is a union of languages of the form

(*) $A_0^*a_1A_1^* \dots a_kA_k^*$

where $A_0, \dots, A_k \subset A$ and $a_1, \dots, a_k \in A$. The result cited above on the variety **PV** implies that if $L \subset A^*$ is a language such that $M(L) \in$ $\in$ **PDA**, then L belongs to the boolean closure of the family of languages of the form (*). It follows from Theorem 2 that $M(L) \in$ **W**. Thus **PDA** $\subset$ **W**.

To prove that **W** $\subset$ **PJ** it suffices (by Theorem 2 and the fact that $A^*\mathscr{P}\mathscr{J}$ is closed under boolean operations) to show that each language of the form $L = A_0^*a_1A_1^* \dots a_kA_k^*$ belongs to $A^*\mathscr{P}\mathscr{J}$. For each $i =$ $= 0, \dots, k$ we consider a copy A_i' of A_i, and for each $j = 1, \dots, k$ a copy a_j' of a_j such that the sets $A_0', \dots, A_k'$, $\{a_1'\}, \dots, \{a_k'\}$ are pairwise disjoint. Let B be the union of these sets. $\varphi: B \to A$ is the map which sends each $b \in A_i'$ to the corresponding letter in A_i and each a_j' to a_j. φ extends to a length-preserving morphism $\varphi: B^* \to A^*$, and we have

$$L = (A_0')^*a_1'(A_1')^* \dots a_k'(A_k')^*.$$

In light of the result on the operation **V** $\to$ **PV** cited above, it remains to show that $L' = (A_0')^*a_1'(A_1')^* \dots a_k'(A_k')^*$ belongs to $B^*\mathscr{J}$. L' is recognized by the automaton

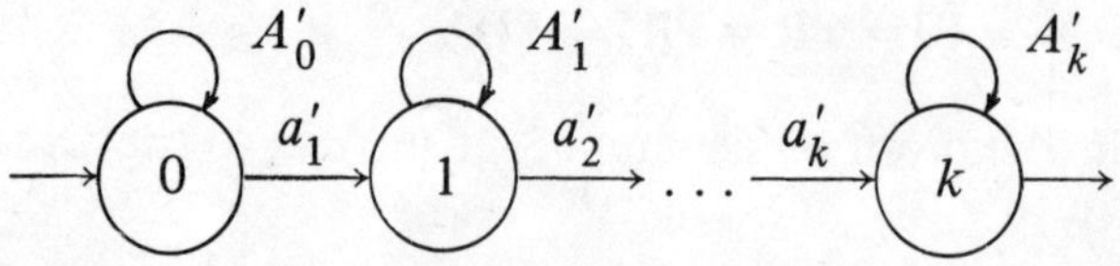

Since $a_i' \notin A_{i-1}'$ for $i = 1, \dots, k$, this automaton is deterministic (though not complete) and reduced, and consequently $M(L')$, the syntactic monoid of L', is the monoid of partial transformations on the states induced by the words of B^*. Now if i is a state of the automaton and $x, y, z \in B^*$ are such that $ixyz = ix = j$, then $(yz) \in (A_j')^*$, consequently $y \in (A_j')^*$, and thus $ixy = ix$. It follows that in $M(L')$ $m_1m_2m_3 = m_1$ implies $m_1m_2 = m_1$ and thus $M(L')$ is $\mathscr{R}$-trivial. The identical argument shows that the syntactic monoid of $(A_k')^*a_k \dots a_1(A_1')^*$, which is the reversal of $M(L')$, is $\mathscr{R}$-trivial. Thus $M(L')$ is $\mathscr{L}$-trivial as well, and thus $\mathscr{J}$-trivial. Hence $L' \in B^*\mathscr{J}$. This completes the proof.

Corollary. $M \in \mathbf{W}$ *if and only if* $M \prec \mathcal{P}(K_n)$ *for some* $n \geqslant 1$.

Proof. By Theorem 1, $\mathcal{P}(K_n) \in \mathbf{PJ}$ and by Theorem 3, $\mathbf{PJ} = \mathbf{W}$; thus $M \prec \mathcal{P}(K_n)$ implies $M \in \mathbf{W}$.

In [11] it is proved that if $\mathbf{V}$ is a nontrivial variety of finite monoids, then $\mathbf{PV}$ is generated by the monoids $\{\mathcal{P}'(M) \mid M \in \mathbf{V}\}$, where $\mathcal{P}'(M)$ denotes the monoid of *nonempty* subsets of M. Thus if $M \in \mathbf{PJ}$ there exist $M_1, \ldots, M_r \in \mathbf{J}$ such that

$$M \prec \mathcal{P}'(M_1) \times \ldots \times \mathcal{P}'(M_r).$$

Now it is easy to see that the map $(X_1, \ldots, X_r) \mapsto X_1 \times \ldots \times X_r$ is an injective morphism embedding $\mathcal{P}'(M_1) \times \ldots \times \mathcal{P}'(M_r)$ into $\mathcal{P}(M_1 \times \ldots \times M_r)$. Thus $M \prec \mathcal{P}(M_1 \times \ldots \times M_r)$. Now $M_1 \times \ldots \times M_r \in$ $\in \mathbf{J}$, and, by Theorem 1, $M_1 \times \ldots \times M_r \prec K_n$ for some $n \geqslant 1$. Since $M' \prec M''$ implies $\mathcal{P}(M') \prec \mathcal{P}(M'')$ we obtained $M \prec \mathcal{P}(K_n)$.

4. FURTHER RESULTS AND OPEN PROBLEMS

The varieties $\mathbf{J}$ and $\mathbf{W} = \mathbf{PJ}$ play a role in the dot-depth hierarchy of Brzozowski (see [1] and [2], Ch. 9 for the definitions). Let $\mathbf{V}_k$ be the variety generated by the syntactic semigroups of languages of dot-depth less than or equal to k.

In [12] it is shown that

$$\mathbf{V}_1 = \mathbf{J} * \mathbf{D}$$

that is, the variety generated by semidirect products of the form $M * S$ where $M \in \mathbf{J}$ and S is a definite semigroup (see [2], Ch. 5). More generally, it is shown in [17] that for all $k \geqslant 1$ the variety $\mathbf{V}_k$ is of the form $\mathbf{V}'_k * \mathbf{D}$ where $\mathbf{V}'_k$ is a variety of finite monoids. Furthermore $\mathbf{V}'_2 = \mathbf{PJ}$, hence

$$\mathbf{V}_2 = \mathbf{PJ} * \mathbf{D}.$$

Thus $\mathbf{J}$ and $\mathbf{PJ}$ are the first two levels in an infinite hierarchy of varieties of finite monoids, whose union is the variety of all *aperiodic* monoids.

The most important open problem concerning the variety $\mathbf{W} = \mathbf{PJ}$ is the decision problem: Is there an algorithm to determine whether or not a finite monoid M, given by its multiplication table, belongs to $\mathbf{W}$? (Such an algorithm exists for the variety $\mathbf{J}$, because we can write down the $\mathcal{J}$-classes of M once we possess the multiplication table.) We have not found such an algorithm – however, we do have an effective necessary condition for membership in $\mathbf{W}$: If M is a finite monoid and $e \in M$ is an idempotent, then we denote by M_e the subsemigroup of M generated by the elements of M which are greater than or equal to e in the $\mathcal{J}$-ordering on M. We can then form the subsemigroup $eM_e e$, which is a monoid whose identity element is e. Our necessary condition is:

Theorem 5. *If* $M \in \mathbf{W}$, *then* $eM_e e \in \mathbf{J}$ *for every idempotent* e *in* M.

Proof. It is easy to show that for any variety $\mathbf{V}$ of finite monoids, the family $\{M \mid eM_e e \in \mathbf{V}\}$ is also a variety (see [8]). Thus, by the corollary to Theorem 3, it suffices to show that if $M = \Diamond_n(U_1, \ldots, U_1)$ then $eM_e e \in \mathbf{J}$. In [16] it is shown that the Schützenberger product $\Diamond_n(M_1, \ldots, M_n)$ has the following property:

There exists a surjective morphism $\pi\colon \Diamond_n(M_1, \ldots, M_n) \to$ $\to M_1 \times \ldots \times M_n$ such that for each idempotent $f \in M_1 \times \ldots \times M_n$, the semigroup $S = f\pi^{-1}$ is locally $\mathcal{J}$-trivial – that is, $eSe \in \mathbf{J}$ for each idempotent $e \in S$.

Now let us consider an idempotent $e \in M = \Diamond_n(U_1, \ldots, U_1)$ and an element a of M_e. Then $a = a_1 \ldots a_k$ where $e \leqslant_{\mathcal{J}} a_i$ for $i = 1, \ldots, k$. Since $U_1 \times \ldots \times U_1$ is idempotent and commutative $(eae)\pi = (e\pi)(a_1\pi)(e\pi) \ldots (e\pi)(a_k\pi)(e\pi)$. In an idempotent and commutative semigroup, $s \leqslant_{\mathcal{J}} t$ implies $st = s$ thus $(e\pi)(a_i\pi) = e\pi$ for $i = 1, \ldots, k$, and consequently $eae \in (e\pi)\pi^{-1}$. Thus $eM_e e =$ $= e[eMe]e \subset e[(e\pi)\pi^{-1}]e$. By the property of the Schützenberger product cited above, $e[(e\pi)\pi^{-1}]e \in \mathbf{J}$. Thus $eM_e e \in \mathbf{J}$.

We do not know if the converse to Theorem 5 is true – if it were it would provide an effective method for testing membership in $\mathbf{W}$.

REFERENCES

[1] J.A. Brzozowski – I. Simon, Characterizations of locally testable events, *Discrete Mathematics,* 4 (1973), 243–271.

[2] S. Eilenberg, *Automata, Languages and Machines,* vol. B, Academic Press, New York, 1976.

[3] G. Lallement, *Semigroups and Combinatorial Applications,* Wiley, New York, 1979.

[4] S.W. Margolis, On *M*-varieties generated by power monoids, *Semigroup Forum,* 22 (1981), 339–353.

[5] S. Margolis – J.-E. Pin, Power semigroups and $\mathcal{J}$-trivial monoids, to appear in *Semigroup Forum.*

[6] S. Margolis – J.-E. Pin, Minimal noncommutative varieties and power varieties, to appear in *Pacific Journal of Mathematics.*

[7] J.-E. Pin, Variétés de langages et monoide des parties, *Semigroup Forum,* 20 (1980), 11–47.

[8] J.-E. Pin, *Variétés de langages et variétés de semigroupes,* These, Paris, 1981.

[9] C. Reutenauer, Sur les variétés de langages et de monoides, *Lecture Notes in Computer Science,* N° 67, Springer-Verlag, Berlin, 1979, 260–265.

[10] M.P. Schützenberger, On finite monoids having only trivial subgroups, *Information and Control,* 8 (1965), 190–194.

[11] M.P. Schützenberger, Sur le produit de concatenation non ambigu, *Semigroup Forum,* 18 (1979), 331–340.

[12] I. Simon, *Hierarchies of events with dot-depth one,* Thesis, University of Waterloo, 1972.

[13] I. Simon, Piecewise testable events, *Lecture Notes in Computer Science,* N° 33, Springer-Verlag, Berlin, 1975, 214–222.

[14] H. Straubing, Recognizable sets and power sets of finite semigroups, *Semigroup Forum,* 18 (1979), 331–340.

[15] H. Straubing, On finite $\mathcal{J}$-trivial monoids, *Semigroup Forum,* 19 (1980), 107–110.

[16] H. Straubing, A generalization of the Schützenberger product of finite monoids, *Theoret. Comput. Sci.,* 13 (1981), 137–150.

[17] H. Straubing, A study of the dot-depth hierarchy, to appear.

J.-E. Pin

Université Paris VI et CNRS (Paris), Laboratoire d'Informatique Théorique, 4 Place Jussieu 75230 Paris Cedex 05, France.

H. Straubing

Université Paris VI, France et Reed College Portland, Oregon, USA.

COLLOQUIA MATHEMATICA SOCIETATIS JÁNOS BOLYAI
39. SEMIGROUPS, SZEGED (HUNGARY), 1981.

ON SEMIPRIMITIVE AND SEMIPRIME FINITE SEMIGROUP RINGS

I.S. PONIZOVSKIĬ

In the present paper a criterion for a contracted semigroup ring to have properties given in the title is considered, and some corollaries from this criterion are formulated. Corollary 1 (the corresponding criterion for finite group ring) is probably known to specialists, though, as far as the author knows, neither this criterion nor its proof has ever been published. The main result (Theorem) generalizes the theorem of Munn – Ponizovskiĭ [1], [2] which gives a criterion for the semisimplicity of a finite semigroup algebra over a field (it should be noted that the theorem in question was the main result of the author's candidate thesis; its public defence was held in 1953 at the Leningrad Herzen Pedagogical Institute, the Dissertation Abstract was published in Leningrad, 1953).

One of the obstacles which arise by the investigation of semigroup rings over a general ring R is the fact that the group $(R,+)$ can be a mixed one, e.g. $TR \neq 0$ and $R \neq TR$. We avoid this difficulty by showing that in our case R is semiprime which gives us the representation of R as a subdirect product of prime rings (which are not mixed by 12). Proposition 10 permits us to exclude in this representation those prime rings which are in a sense alien to R. The proof of the Theorem depends greatly on Weissglass's theorem [3] which gives conditions for the semisimplicity of semigroup rings of completely 0-simple semigroups under certain radical properties.

All rings considered are associative, though not necessary with identity. All modules over rings are right ones.

Let π be a property of rings. A ring R which possesses the property π is called a π-ring. An ideal I of R is a π-ideal if I is itself a π-ring. A property π is a *radical property* if it satisfies the conditions:

(A) A homomorphic image of a π-ring is a π-ring.

(B) Every ring R contains a π-ideal, $\pi(R)$, which contains every π-ideal of R.

(C) $\pi(R/\pi(R)) = 0$.

The ideal $\pi(R)$ of R is called the *π-radical* of R. A radical property π is *hereditary,* if for any ring R with ideal I, the equality $\pi(I) = I \cap \pi(R)$ holds. A hereditary radical property π is *supernilpotent* iff $\pi(R)$ contains the Baer lower radical of R for every ring R.

A class M of rings is a *special class* of rings if it satisfies the conditions:

(D) Every ring in M is a prime ring.

(E) Every nonzero ideal of a ring in M is a ring in M.

(F) If A is a ring in M, and A is an ideal of a ring L, then A/A^* is a ring in M, where $A^* = \{x \in L \mid xA = Ax = 0\}$.

For a ring R, define $\pi(R)$ as the intersection of all ideals A of R such that R/A is in M. Then π is a radical property which is called the *special radical property determined by* M. A radical property is special if it is a special radical property determined by some special class of rings. All special radical properties are supernilpotent. It is well known that such radical properties as Baer lower, Jacobson, Levitzky, Brown–McCoy are special. For details see [5].

Throughout the paper R means a ring, S means a finite semigroup. To avoid triviality, we require $|S| > 1$. We use also the following notations:

$|M|$ cardinality of the set M;

$\mathbf{Z}, \mathbf{Q}$ the ring of integers, the field of rationals respectively;

TM the torsion part of a module M (e.g., the torsion part of a group $(M, +)$; in particular, if L is a ring, then TL is the torsion part of L,

R_n the ring of all $n \times n$-matrices over a ring R;

F_p the prime field of characteristic $p > 0$;

Sq the class of all finite regular semigroups with principal factors isomorphic to completely 0-simple (completely simple) semigroups having square structural matrices;

RS the contracted semigroup ring of S over R;

$O(L)$ $= \{m \in \mathbf{Z} \mid \exists x \in L \ \ (x \neq 0, \ \ xm = 0)\}$, where L is a ring;

$J(L), B(L)$ the Jacobson, the Baer lower radical of a ring L.

For any $S \in Sq$ we define $D(S) \in \mathbf{Z}$ in the following manner: Let $S \cong M^0(G; n, n; P)$ (or $S \cong M(G; n, n; P)$), and let Δ be a right regular matrix representation of $\mathbf{Q}G$. Denote by $P\Delta$ the matrix over $\mathbf{Z}$ obtained from P by substituting all entries of P by their Δ-images. We then define $D(S) = |\det(P\Delta)||G|$.

If S is an arbitrary semigroup in Sq, suppose that $S_1, \ldots, S_n$ are the principal factors of S, and define $D(S) = \prod_{i=1}^{n} D(S_i)$. It is easy to check that $D(S)$ is an invariant of S. Obviously $D(S) \in \mathbf{Z}$.

All semigroup concepts are used in the sense of the book [4].

If π is a radical property, then we write π is J or π is B as abbreviations for "π is the Jacobson radical property" or "π is the Baer lower radical property".

The main result of the paper is:

Theorem. *Let* π *be either* J *or* B. *Then the following conditions are equivalent.*

(i) $\pi(RS) = 0$,

(ii) (a) $\pi(R) = 0$,

(b) *if* $TR = 0$ *then* $J(QS) = 0$,

(c) *if* p *is a positive prime integer,* $p \in O(R)$, *then* $J(F_p S) = 0$.

(iii) $\pi(R) = 0$, $S \in Sq$, $D(S) \notin O(R)$.

Corollary 1. *Let* G *be a finite group. Then* $J(RG) = 0$ *iff* $J(R) = 0$ *and* $|G| \notin O(R)$.

Corollary 2. *Let* π *be either* J *or* B. *Then* $J(F_p S) = 0$ *implies* $\pi(RS) = 0$ *for any ring* R *of characteristic* p *such that* $\pi(R) = 0$.

Corollary 3. *Let* π *be either* J *or* B. *Let* $\pi(RS) = 0$ *for some ring* R. *Then*

(i) $\pi(LS) = 0$ *for any ring* L *such that* $\pi(L) = 0$ *and* $TL = 0$,

(ii) *there exists only a finite number of positive prime integers* p *such that* $J(F_p S) \neq 0$ *(all of them are divisors of* $D(S)$*).*

The proof of the Theorem is given in 15. In 1–14 we establish some auxiliary facts. In 16 we discuss possible generalizations.

1. *Let* K *be a field. Then the following conditions are equivalent.*

(i) $J(KS) = 0$,

(ii) (a) S *is regular,*

(b) *if a principal factor of* S *is isomorphic to* $M^0(G; m, n; P)$ *(to* $M(G; m, n, P)$*), then* $m = n$, P *is invertible in* $(KG)_n$, $|G|$ *is invertible in* K.

(iii) $S \in Sq$ *and* $D(S)$ *is invertible in* K.

Proof. (i) ⇔ (ii). This is merely another form of the Munn–Ponizovskiĭ theorem [1], [2]. The proof of (ii) ⇔ (iii) is straightforward.

2. *Let* $S = M^0(G; n, n; P)$. *Let* π *be either* B *or* J. *Then the following conditions are equivalent.*

(i) $\pi(RS) = 0$,

(ii) (a) $\pi(RG) = 0$,

(b) $xP = Py = 0$ *implies* $x = y = 0$ *for all* $x, y \in (RG)_n$.

Proof. This is just a special case of [3], Theorem 3.8.

3.

(i) *Let* m *be the number of nonzero elements of* S^1. *Then*

$$[J(RS)]^m \subset J(R) \cdot S \subset J(RS).$$

(ii) *Let the* RS*-module* $M \otimes RS^1$ *be fully reducible for every irreducible* R*-module* M. *Then* $J(RS) \subset J(R) \cdot S$.

Proof. This is similar to the proof of Theorems 6.20, 6.11 in [6] (where G is a group instead of a semigroup, R is a ring with identity and RG is a skew group ring instead of a semigroup ring).

Suppose that $B \subset RS$, and assume that $(M \otimes RS^1)B = 0$ for every irreducible R-module M. We state that $B \subset J(R) \cdot S$. In fact, let $y = \Sigma r_i s_i \in B$ $(r_i \in R, \ s_i \in S)$. Then $(M \otimes RS^1)y = 0$. In particular, $(M \otimes R1)y = 0$. Therefore $MR \cdot r_i = 0$, so that $Mr_i = 0$. Since M is an arbitrary irreducible R-module, $r_i \in J(R)$. Hence $y \in J(R) \cdot S$.

We now turn to the proof of 3(i).

Case 1. Suppose $R \neq J(R)$. Then there exist irreducible R-modules; suppose M is one of them. Consider the RS-module $P = M \otimes RS^1$. As an R-module, M is the direct sum of m irreducible constituents. Therefore the length of the composition chain of P as an RS-module is less than or equal to m. Then $P \cdot [J(RS)]^m = 0$, hence from the foregoing $[J(RS)]^m \subset J(R) \cdot S$.

If $J(RS) = RS$ then the inclusion $J(R) \cdot S \subset J(RS)$ is trivial. So we now assume that $J(RS) \neq RS$. Suppose that N is an irreducible RS-module, and let L be any ring with identity which has R as an ideal. Then RS is an ideal of LS^1, and we may consider N as an irreducible LS^1-module [8], Lemma 3. In particular, N is an L-module. As an L-module, N is finitely generated. Therefore there exists a maximal L-submodule N_0 of N (see [6], the commentary at the bottom of p. 46). Consider $N \cdot J(L)$. It is an LS-submodule of N. Since $N \cdot J(L) \subseteq N_0$, therefore $N \cdot J(L) \neq N$. Hence $N \cdot J(L) = 0$. But R is an ideal of L, so that $J(R) \subset J(L)$. Thus $N \cdot J(R) = 0$, whence $N \cdot J(R)S = 0$. Since N is an arbitrary irreducible RS-module, we get $J(R) \cdot S \subset J(RS)$.

Case 2. On the other hand, suppose $R = J(R)$. Then $J(R) \cdot S = R \cdot S = RS$, and the inclusion $[J(RS)]^m \subset J(R) \cdot S$ is trivial. As for the inclusion $J(R) \cdot S \subset J(RS)$, its proof, as the reader could see, does not depend on the condition $J(R) = R$ at all.

Suppose that the conditions of (ii) hold. Then $(M \otimes RS^1) \cdot J(RS) = 0$, and from the foregoing $J(RS) \subset J(R) \cdot S$.

4.

(i) *Let* π *be either* B *or* J. *Then* $\pi(RS) = 0$ *implies* $\pi(R) = 0$.

(ii) *If* m *is a number of nonzero elements of* S^1 *and* $J(R) = 0$, *then* $[J(RS)]^m = 0$.

Proof. Evidently $B(R)S \subset B(RS)$. Hence $B(RS) = 0$ implies $B(R) = 0$. The other assertions in 4 are immediate from 3(i).

5. *Let* $TR = 0$.

(i) *If* R *is primitive, then the ring* $R \otimes \mathbf{Q}$ *is primitive.*

(ii) *If* $J(R) = 0$ *and* $J[(R \otimes \mathbf{Q})S] = 0$, *then* $J(RS) = 0$.

Proof. We note at first two simple facts.

(a) Let L be a ring, and let A be an L-module. Then the kernel of the mapping

$$A \to A \otimes \mathbf{Q}, \quad x \mapsto x \otimes 1$$

is equal to TA (see, e.g., [7], Ch. VII, Proposition 2.1).

(b) Let L be a ring, and let T be a senigroup. Then $LT \otimes \mathbf{Q} \cong$ $\cong (L \otimes \mathbf{Q})T$.

(i) The proof is straightforward.

(ii) Let $J(R) = 0$, $J[(R \otimes \mathbf{Q})S] = 0$. Then by (b) $J(RS \otimes \mathbf{Q}) = 0$. Consider $A = J(RS) \otimes \mathbf{Q}$. Obviously A is an ideal of $RS \otimes \mathbf{Q}$, and A is nilpotent by 4(ii). Thus $J(RS) \otimes \mathbf{Q} \subset J(RS \otimes \mathbf{Q})$. Hence $J(RS) \otimes 1 = 0$, therefore $J(RS) \subset T(RS)$ by (a). But $TR = 0$ implies $T(RS) = 0$. Thus $J(RS) = 0$.

6. *Let L be the ring of right and left operators of R, and let $xy = yx$ for all $x \in L$, $y \in R$. Let $xR \neq 0$ for every $x \in L$, $x \neq 0$. Let π be any supernilpotent radical property. Then $\pi(RS) = 0$ implies $B(LS) = 0$.*

Proof. Assume that all the hypotheses of 6 hold. We may consider LS as the ring of right and left operators of RS. Assume that A is a nilpotent ideal of LS. Then RA is a nilpotent ideal of RS. By the assumption, $A \neq 0$ implies $RA \neq 0$. Thus $\pi(RS) = 0$ implies $B(LS) = 0$.

7. *Let R be an algebra over a field K.*

(i) *If G is a finite group such that $|G|$ is invertible in K, and $J(R) = 0$, then $J(RG) = 0$.*

(ii) *If $S = M^0(G; n, n; P)$ such that $J(KS) = 0$ $[B(KS) = 0]$, and $J(R) = 0$ $[B(R) = 0]$, then $J(RS) = 0$ $[B(RS) = 0]$.*

Proof.

(i) Assume that the hypotheses of (i) hold. If char $K > 0$, then (i) is immediate from [6], Theorem 6.11. Suppose now that char $K = 0$. The following statement is analogous to [6], Lemma 6.12: *Let R be a semiprimitive algebra over a field K, and let M be an RG-module. If M is fully reducible as an R-module, then M is fully reducible as an*

RG-module. The proof is quite similar to the proof of the above mentioned Lemma 6.12 in [6], and we omit it. Assume now that M is an irreducible R-module, and consider an RG-module $P = M \otimes RG$. Evidently R is fully reducible as an R-module. Hence P is fully reducible as an R-module. Then $J(RG) \subset J(R) \cdot G$ by 3(ii). Hence $J(R) = 0$ implies $J(RG) = 0$.

(ii) Let $J(R) = 0$, $J(KS) = 0$. The latter means by 1 that $|G|$ is invertible in K, and P is invertible in $(KG)_n$. Therefore $J(KG) = 0$ by (i). We may consider $(KG)_n$ as a ring of left and right operators of RS. Assume that $x, y \in (RG)_n$, and $xP = Py = 0$. Take $P^{-1} \in (KG)_n$. Then $0 = (xP)P^{-1} = x$, $0 = P^{-1}(Py) = y$. By 2 we obtain $J(RS) = 0$.

The analogous assertion for the Baer lower radical may be proved similarly.

8. *Let* π *be any supernilpotent radical property. Let* $TR \neq 0$, *and let* $\pi(R) = 0$. *Then* $TR = \Sigma \oplus R^{(p)}$ $(p \in O(R))$, *where* $R^{(p)} =$ $= \{x \in R \mid xp = 0\}$.

Proof. If $TR \neq 0$, then there exists a prime $p \in O(R)$. Every $x \in R$ such that $p^2x = 0$, $px \neq 0$ belongs to $\pi(R)$. Thus $\pi(R) = 0$ implies that a p-component of $(R, +)$ coincides with $R^{(p)}$, whence the assertion follows.

9. *Let* π *be any supernilpotent radical property.*

(i) *If* $TR = 0$, *and* $\pi(RS) = 0$, *then* $J(\mathbf{Q}S) = 0$.

(ii) *If* p *is a positive prime integer such that* $p \in O(R)$, *and* $\pi(RS) = 0$, *then* $J(F_pS) = 0$.

Proof.

(i) Assume that $TR = 0$. Then $\mathbf{Z}$ is a ring of left and right operators of R satisfying the hypotheses of 6. Thus $\pi(RS) = 0$ implies $B(\mathbf{Z}S) = 0$. The latter implies $B(\mathbf{Q}S) = 0$. Since $\mathbf{Q}S$ is artinian, $B(\mathbf{Q}S) = J(\mathbf{Q}S)$, whence the assertion follows.

(ii) Assume that the hypotheses of (ii) hold. Since $p \in O(R)$, therefore $R^{(p)} \neq 0$. Thus $R^{(p)}S$ is a nonzero ideal of RS. Since π is hereditary, $\pi(RS) = 0$ implies $\pi(R^{(p)}S) = 0$. But F_p is the ring of left and right operators of $R^{(p)}$ satisfying the hypotheses of 6. Therefore $B(F_pS) = 0$, or equivalently, $J(F_pS) = 0$.

10. *Let* R *be a subdirect product of rings* $\{R_\lambda \mid \lambda \in \Lambda\}$. *Let* $m \notin O(R)$, *and let* $\Lambda_0 = \{\lambda \in \Lambda \mid \text{char}\, R$ *divides* $m\}$. *Then* R *is a subdirect product of rings* $\{R_\lambda \mid \lambda \in \Lambda \setminus \Lambda_0\}$.

Proof. Let $\pi_\lambda : R \to R_\lambda$ be the corresponding projection. Define $x_\lambda = x\pi_\lambda$ for all $x \in R$, $\lambda \in \Lambda$. Then the projections π_λ $(\lambda \in \Lambda \setminus \Lambda_0)$ define a homomorphism of R into the direct product of rings $\{R_\lambda \mid \lambda \in \Lambda \setminus \Lambda_0\}$, and the only thing to prove is that the kernel of this homomorphism is zero. Assume $x \in R$, and suppose that $x_\lambda = 0$ for all $\lambda \in \Lambda \setminus \Lambda_0$. If $\lambda \in \Lambda_0$ then char R_λ divides m according to the definition of Λ_0, so that $x_\lambda m = 0$. Thus $x_\lambda m = 0$ for all $\lambda \in \Lambda$, whence $xm = 0$. Since $m \notin O(R)$, we obtain $x = 0$.

11. *Let* R *be a subdirect product of rings* $\{R_\lambda \mid \lambda \in \Lambda\}$, *and let* T *be a semigroup. Let* π *be any special radical property.*

(i) *The ring* RT *is a subdirect product of rings* $\{R_\lambda T \mid \lambda \in \Lambda\}$.

(ii) *If* $\pi(R_\lambda T) = 0$ *for all* $\lambda \in \Lambda$, *then* $\pi(RT) = 0$.

Proof.

(i) is straightforward.

(ii) Assume that π is a special radical property determined by a special class M of rings. Then, for any ring L, $\pi(L) = 0$ iff L is a subdirect product of rings from M, [5], Lemma 80, p. 139. The assertion of (ii) follows immediately from this.

12. *Let* R *be a prime ring. Then either* $TR = 0$ *or* char $R = p$ *is a positive prime integer.*

Proof. Assume that $TR \neq 0$. Then there exists a positive prime integer $p \in O(R)$. Let $A = \{x \in R \mid xp = 0\}$, and let $B = Rp$. By the assumption, $A \neq 0$. But A, B are ideals of R such that $AB = 0$. Since R is a prime ring, we obtain $B = 0$. Thus $Rp = 0$, and char $R = p$.

13. *Let $S \in Sq$, and let $\{S_i \mid i = 1, \ldots, m\}$ be the set of all principal factors of S.*

(i) *If K is a field such that $J(KS) = 0$, then $J(KS_i) = 0$ for all $i = 1, \ldots, m$.*

(ii) *Let π be any hereditary radical property. If $\pi(RS_i) = 0$ for all $i = 1, \ldots, m$, then $\pi(RS) = 0$.*

Proof. Assume that L is a ring with an ideal A. If L is artinian, then $J(L) = 0$ implies $J(A) = 0$ and $J(L/A) = 0$. If π is any hereditary radical property, then $\pi(A) = 0$ and $\pi(L/A) = 0$ implies $\pi(L) = 0$. The assertions of 13 are immediate from this.

14. *Let R be a prime ring, and let $S \in Sq$.*

(i) *If* char $R = p > 0$, *and p does not divide $D(S)$, then $B(RS) = 0$; if, in addition, R is primitive, then $J(RS) = 0$.*

(ii) *If $TR = 0$, and $D(S) \neq 0$, then $B(RS) = 0$; if, in addition, R is primitive, then $J(RS) = 0$.*

Proof. By 13(ii) it is enough to prove 14 in the case when $S = M^0(G; n, n; P)$.

(i) Assume that the hypotheses of (i) hold. Then $J(F_p S) = 0$ by 1. We may consider R as an algebra over F_p. Now the assertion is immediate from 7(ii).

(ii) Assume that the hypotheses of (ii) hold. Then $B(R) = 0$ and $B(RG) = 0$ by [6], Theorem 13.2. Furthermore, $J(\mathbf{Q}S) = 0$ by 1. The latter means that there exists $P^{-1} \in (\mathbf{Q}G)_n$. Then there exists an integer m such that $mP^{-1} \in (\mathbf{Z}G)_n$. Assume now that $xP = Py = 0$ for $x, y \in (RG)_n$. We may regard $(\mathbf{Z}G)_n$ as the ring of left and right operators of $(RG)_n$. Hence $xP = Py = 0$ implies $(xP)mP^{-1} = mx = 0$. Since

$TR = 0$, therefore $T(RS) = 0$, too. Thus $mx = 0$ means $x = 0$. Similarly we obtain that $y = 0$. Now $B(RS) = 0$ by 2.

Assume that R is primitive. Then $J(R) = 0$. Consider the ring $(R \otimes \mathbf{Q})S$. By 5(i), $R \otimes \mathbf{Q}$ is primitive. Now we have that $R \otimes \mathbf{Q}$ is an algebra over $\mathbf{Q}$ such that $J(R \otimes \mathbf{Q}) = 0$ and $J(\mathbf{Q}S) = 0$. Then $J[(R \otimes \mathbf{Q})S] = 0$, by 7(ii). The latter, together with the condition $J(R) = 0$ and 5(ii), implies $J(RS) = 0$.

15. Proof of Theorem. Recall that B and J are special radical properties. Thus we may use all propositions 1–14 without commentary.

Let π be either B or J.

(i) ⇒ (ii). Assume $\pi(RS) = 0$. Then $\pi(R) = 0$ by 4(i). If $TR = 0$ then $J(\mathbf{Q}S) = 0$ by 9(i). If $TR \neq 0$ then there exists a prime $p \in O(R)$. By 9(ii) we obtain $J(F_p S) = 0$. Thus (ii) holds.

(ii) ⇒ (iii). Assume that (ii) is true. Then (b), (c) show that there exists a field K such that $J(KS) = 0$. By 1 we obtain that $S \in Sq$ and $D(S)$ is invertible in K, in particular $D(S) \neq 0$. Suppose that $D(S) \in O(R)$. Then there exists a prime divisor p of $D(S)$. Then $p \in O(R)$. By (ii) we have $J(F_p S) = 0$. But this is impossible since p divides $D(S)$ and by 1 $J(F_p S) \neq 0$, a contradiction. Thus $D(S) \notin O(R)$, and (iii) holds.

(iii) ⇒ (i). Assume that (iii) is true. Then $\pi(R) = 0$, therefore $B(R) = 0$. Hence R is a subdirect product of prime rings $\{R_\lambda \mid \lambda \in \Lambda\}$. Since $D(S) \notin O(R)$, we may assume by 10 that $\operatorname{char} R_\lambda$ does not divide $D(S)$ for all $\lambda \in \Lambda$. Since $S \in Sq$, we may use 14. The latter shows that in all cases $\pi(R_\lambda S) = 0$ for every $\lambda \in \Lambda$. Then $\pi(RS) = 0$ by 11(ii).

This completes the proof of the Theorem.

16. Let us say that the Theorem is true for a radical property π, iff (i), (ii), (iii) of the Theorem are equivalent for π. A scrupulous analysis shows that the Theorem is true for any special radical property π which satisfies the conditions:

(α) if $S = M^0(G; n, n; P)$ is finite, then conditions 2(i), 2(ii) are equivalent (a special version of Weissglass's theorem [3], Theorem 3.8);

(β) for any finite group G, $\pi(RG) = 0$ implies $\pi(R) = 0$.

It is easy to prove that for any hereditary (thus for any special) radical property π, (β) is equivalent to

(γ) for any finite group G, $\pi(R) = R$ implies $\pi(RG) \neq 0$.

And we conclude with the following

Problem. For which special radical properties do (α) and (γ) hold?

The only example of such a radical property known to the author, except B and J, is the Levitzky radical property. The Brown–McCoy radical property satisfies (γ), but we do not know whether it satisfies (α). Which other special radical properties satisfy (α) and (γ)?

REFERENCES

[1] W.D. Munn, On semigroup algebras, *Proc. Cambridge Phil. Soc.*, 51 (1955), 1–15.

[2] I.S. Ponizovskiĭ, On matrix representations of associative systems, *Mat. Sbornik,* 38 (1956), 241–260 (in Russian).

[3] J. Weissglass, Radicals of semigroup rings, *Glasgow Math. J.*, 10 (1969), 85–93.

[4] A.H. Clifford – G.B. Preston, *The algebraic theory of semigroups,* Amer. Math. Soc., Providence, R. I., 1961, 1967.

[5] N.J. Divinsky, *Rings and radicals,* George Allen and Unwin LTD, London, 1965.

[6] A.E. Zalesskiĭ – A.B. Mihalev, Group rings, *Sovrem. Probl. Matem.*, vol. 2, Moscow, 1973, 5–118 (in Russian).

[7] H. Cartan – S. Eilenberg, *Homological algebra,* Princeton University Press, Princeton, New Jersey, 1956.

[8] I.S. Ponizovskiĭ, On modules over semigroup rings of completely 0-simple semigroups, *Sovrem. Algebra (polugr. constr.)*, Herzen Inst., Leningrad, 1981, 85–94 (in Russian).

I.S. Ponizovskiĭ

194 100 Leningrad, Lesnoĭ pr. 61, kv. 120, USSR.

COLLOQUIA MATHEMATICA SOCIETATIS JÁNOS BOLYAI
39. SEMIGROUPS, SZEGED (HUNGARY), 1981.

ON ALMOST SIMPLE SEMIGROUP IDENTITIES

G. POLLÁK – M.V. VOLKOV

The faith in a pre-ordained harmony mostly leads to disappointment. Mathematical objects studied in our age reveal over and over discouraging properties while we are often reluctant to give up our 19th century expectations. For example, few people could have supposed in the early days of the theory of varieties that finitely based varieties were such rare exceptions as they turned out to be. This lamentable fact was first discovered by E.S. Ljapin in [4] where he proved that there is only a finite number of balanced identities over a fixed finite alphabet which define hereditarily finitely based (h.f.b.) varieties (we shall term such identities, too, h.f.b.). Later the first of the authors of the present paper rendered this result more concrete by pointing out 12 types of identities which contained all h.f.b. identities [8]. In [9] the number of these types has been reduced to 4. These latter ones had the property that at least one of the sides of their identities was either a *simple word* (i.e. contained every letter at most once) or an *almost simple word* (i.e. one containing a single letter twice and the rest at most once). An identity is said to be *simple (almost simple)* if both of its sides are simple (almost simple) words, respectively. A. Ja. Aĭzenstat [1] has proved that a simple identity is h.f.b. iff either it is non-balanced (and then, of course, even non-normal) or its left and right sides start or end with different letters. Our aim here is to describe h.f.b. almost simple identities. Namely, we shall prove the following

Main Theorem. *An almost simple identity* $u = v$ *is h.f.b. if and only if one of the following conditions or their duals is fulfilled (up to the notation of the variables):*

(a) $u \equiv x_1 x_2 x_1$, $v \not\equiv x_1 x_2 x_1 v'$, $v \not\equiv v'' x_1 x_2 x_1$ *and* $v \not\equiv x_2 x_1 x_2$;

(b) $u \equiv x_1 x_2 x_3 x_1$, $v \equiv x_j^2$;

(c) $u \equiv x_1 \ldots x_k x_{k+1} x_k \ldots x_n$,
$v \equiv x_{j(1)} \ldots x_{j(l-1)} x_{j(l)}^2 x_{j(l+1)} \ldots x_{j(m)}$,

and either the identity $u = v$ *is non-balanced or* $j(1) \neq 1$ *or* $j(m) \neq n$.

Corollary. *A balanced identity is h.f.b. iff it is (up to the notation of the variables) of one of both forms*

(0) $\quad x_1 \ldots x_n = x_{1\sigma} \ldots x_{n\sigma}$,

(1) $\quad x_1 \ldots x_{k-1} y x_k y x_{k+1} \ldots x_n = x_{1\sigma} \ldots x_{(l-1)\sigma} y^2 x_{l\sigma} \ldots x_{n\sigma}$,

where σ *is a permutation, and the left and right sides of* (1) *start or end with different letters.*

This paper consists of 8 sections. In Section 1 we prove the necessity of the condition of our theorem, and in cases (a) and (b) also its sufficiency. The treatment of the remaining case constitutes the essential part of the work. This treatment is decomposed into the following four propositions:

Proposition A. *Every variety defined by an identity of the form* (1) *with non-trivial permutation* σ *and an identity of the form*

(2) $\quad y^p x y^q = y^{p+q} x$

is h.f.b.

Proposition B. *Every variety defined by an identity of the form* (2) *and an identity of the form*

(3) $\quad x_1 \ldots x_{k-1} y x_k y x_{k+1} \ldots x_n = x_1 \ldots x_{l-1} y^2 x_l \ldots x_n \quad (l < k)$

or of the form, dual to (3), *is h.f.b.*

Proposition C. *Every variety defined by an identity of the form*

(4) $$x_1 \ldots x_{n-1} y x_n y = x_1 \ldots x_{n-1} y^2 x_n$$

is h.f.b.

Proposition D. *Every variety defined by a non-balanced identity of the form*

(5) $$x_1 \ldots x_k x_{k+1} x_k \ldots x_n = x_{j(1)} \ldots x_{j(l-1)} x_{j(l)}^2 x_{j(l+1)} \ldots x_{j(m)}$$

is h.f.b.

The role of these statements in the proof of the Main Theorem becomes clear if one notes, that every identity of the form (1) with $n\sigma \neq n$ implies some identity (2). Thus, Proposition A, B, C, their duals, and Proposition D imply that an almost simple identity of case (c) is h.f.b.

The proofs of these four propositions run, in first approximation, by the same scheme. This scheme, which seems to be of some interest in itself, is dealt with in Section 2. In Section 3 we construct some quasi-ordered sets which are necessary in the further considerations, and we show that they are well-quasi-ordered (w.q.o.) in G. Higman's sense [3]. In Section 4 some properties of the identities (1) and (5) are established. After these preliminaries, Sections 5–8 deal with the proof of Propositions A–D, resp. In the course of the proofs, new large classes of h.f.b. varieties are pointed out, which contain several classes investigated earlier.

To finish with, let us mention a problem which naturally arises in connection with the Main Theorem.

Problem. Which ones are h.f.b. of the identities with one side simple, and the other one almost simple? In particular, is $xy = x^2y$ h.f.b?

1. REDUCTION TO PROPOSITIONS A, B, C, D

We are going to prove the Main Theorem modulo Propositions A–D.

Necessity. The main result of [8] says that a h.f.b. identity $u = v$ satisfies – up to the notation of the variables – one of the following conditions or their duals:

(i) $u \equiv x_1 x_2 x_1$;

(ii) $u \equiv x_1 x_2 x_3 x_1, \quad v \equiv x_j^2$;

(iii) $u \equiv x_1 \ldots x_{k-1} x_k x_{k+1} x_k x_{k+2} \ldots x_n,$
$v \equiv x_{j(1)}^{\epsilon(1)} \ldots x_{j(m)}^{\epsilon(m)}, \quad \epsilon_1, \ldots, \epsilon_m \in \{1, 2\}$;

(iv) u is simple.

As, under the assumptions of our theorem, u and v are almost simple, case (iv) cannot occur, and both (ii) and (iii) reduce to (b) and (c) in the theorem, respectively. As for (i), Theorem 1 in [10] implies $v \not\equiv$ $\not\equiv x_2 x_1 x_2$ and, according to Theorem 4 in the same paper, $v \not\equiv x_1 x_2 x_1 v'$, $v \not\equiv v'' x_1 x_2 x_1$, either.

Sufficiency. We have to prove that an almost simple identity, which satisfies one of the conditions (i)–(iii), is h.f.b. In case (iii) this follows, as said above, from Propositions A–D, in case (ii) – from [9]. Now consider case (i). If $u = v$ is normal then $v \in \{x_1^2 x_2, x_1 x_2^2, x_2^2 x_1, x_2 x_1^2\}$ and, in virtue of Theorem 1 of [10], the identity is h.f.b. If, on the other hand, $u = v$ is not normal, then it follows easily from almost simplicity that either $u = v$ is not fulfilled by groups with more then 1 element, or (if $v \in \{x_j x_2 x_j, x_j^2 x_2, x_2 x_j^2\}$) it is not fulfilled by rectangular bands with more than 1 element. In both cases we have that every completely simple semigroup satisfying $u = v$ is a rectangular group over a group of exponent $\leqslant 2$ whence, in consequence of Theorem 5 and Corollary 1 of [10], $u = v$ is h.f.b.

2. A COMMON SCHEME FOR PROOFS OF THE H.F.B. PROPERTY

Every non-trivial syntactic proof of the hereditariness of the finite basis property known to the present rests on two circumstances: first, the existence of normal form for the elements of the free semigroup of the variety in question and, second, the possibility to suitably well-quasi-order these normal forms. This paper is not an exception in this respect. Therefore we want to give an exact form to this plan of a proof, reshaping it to a scheme which will prove convenient further on.

Let F be the (absolutely) free semigroup on a countable set of generators, $\mathfrak{V}$ a variety of semigroups, θ the fully invariant congruence on F corresponding to $\mathfrak{V}$. The variety is said to *admit a good standard form* if there is a subset $V \subseteq F$ which satisfies the following conditions:

(C) Completeness condition: for some $r \in \mathbf{N}$, every θ-class which has a non-empty intersection with F^r, has a non-empty intersection with V, too;

(O) Order condition: there is a fixed order relation $<$ defined on V such that $(V, <)$ is a well-ordered set;

(Q) Quasi-order condition: there is a fixed quasi-order relation $\preceq$ defined on $V^* = \{(u, v) \in V \times V \mid v < u\}$ such that $(V, \preceq)$ is a well-quasi-ordered set;*

(A) Ascension condition: if (u, v), $(u', v') \in V^*$ and $(u, v) \preceq$ $\preceq (u', v')$, then there is an element $w \in V$ such that $w < u'$ and $(u', w) \in \theta \vee \theta_{u,v}$.

Proposition 2.1. *If the variety* $\mathfrak{V}$ *admits a good standard form then every subvariety of* $\mathfrak{V}$ *is finitely based in* $\mathfrak{V}$. *In particular, if* $\mathfrak{V}$ *is finitely based then it is h.f.b.*

Proof. We have to prove that every fully invariant congruence $\kappa \supseteq \theta$ can be generated by θ and a finite number of pairs adjoined to it.

*A quasi-ordered Q set is said to be well-quasi-ordered (w.q.o.) if every strictly descending chain and every set of pairwise incomparable elements in Q is finite.

Denote by ρ the Rees congruence with kernel F^r and let first $\kappa \subseteq \theta \vee \rho$. According to (Q), a finite subset μ of $\kappa \cap V^*$ exists such that for every $(u', v') \in \kappa \cap V^*$ there is a $(u, v) \in \mu$ such that $(u, v) \preceq (u', v')$. Denote by κ' he fully invariant congruence generated by $\theta \cup \mu$; we are going to prove $\kappa' = \kappa$. In the opposite case $\kappa' \subset \kappa$. Consider a pair $(a, b) \in \in \kappa \setminus \kappa'$. It is easy to see that $\theta \vee \rho = \theta\rho\theta$ whence (as $\kappa \subseteq \theta \vee \rho$) the existence of $c, d \in F$ such that $(a, c) \in \theta$, $(c, d) \in \rho$, $(d, b) \in \theta$ follows. By definition, $(c, d) \in \rho$ means that either $c = d$ or $c, d \in F^r$. The first case is impossible since then $(a, b) \in \theta \subseteq \kappa'$. Thus, $(c, d) \in F^r$ and, by (C), one can find $\bar{c}, \bar{d} \in V$ such that $(c, \bar{c}), (d, \bar{d}) \in \theta$. Now $(\bar{c}, \bar{d}) \in \in \kappa \setminus \kappa'$, else we had $(c, d) \in \kappa'$ and, consequently, $(a, b) \in \kappa'$. Hence $(\kappa \setminus \kappa') \cap V^* \neq \phi$. Choose a pair $(u', v') \in (\kappa \setminus \kappa') \cap V^*$ with $<$-minimal u'. As noted above, there is a $(u, v) \in \mu$ such that $(u, v) \preceq (u', v')$. Making use of (A), we can find $(u', w) \in \kappa' \cap V^*$. Obviously $v' \not\equiv w$ whence either (v', w) or (w, v') lies in $\kappa \cap V^*$ and then, by the minimality of u', also in κ'. Hence $(u', v') \in \kappa'$, contrary to the assumption. Thus, $\kappa = \kappa'$.

In the general case $\kappa \cap (\theta \vee \rho)$ is finitely generated, as we have just shown. Denote by F_{2r-2} a free subsemigroup of F on $2r - 2$ letters and set $G = F_{2r-2} \cap (F \setminus F^r)$. Obviously, G is finite. Put, furthermore, $A = \kappa \cap (G \times G)$ and construct a set B as follows: for every $g \in G$ such that $\kappa \cap (\{g\} \times F^r) \neq \phi$ choose a pair $(g, f_g) \in \bar{\kappa}$, $f_g \in F^r$, and let $B = \{(g, f_g)\}$. It is clear that both A and B are finite. Let us show that $A \cup B \cup (\kappa \cap (\theta \vee \rho))$ generates κ. Indeed, suppose it generates $\kappa'' \subseteq \kappa$ and let $(u, v) \in \kappa$. If $u, v \in F^r$ then $(u, v) \in \kappa''$ is obvious. If $u, v \in \in F \setminus F^r$ then, for some automorphism $\alpha \in \operatorname{Aut} F$ we have $u\alpha, v\alpha \in G$ whence $(u\alpha, v\alpha) \in A$, and $(u, v) = (u\alpha, v\alpha)\alpha^{-1} \in \kappa''$. If $u \in F \setminus F^r$, $v \in F^r$ then, similarly, we have $u\beta \in G$ for some $\beta \in \operatorname{Aut} F$ whence $(u\beta, v\beta) \in \kappa \cap (G \times F^r)$, so that $f_{u\beta}$ exists. Hence $(f_{u\beta}, v\beta) \in \kappa \cap (\theta \vee \rho)$ and $(u\beta, v\beta) \in B \cdot (\kappa \cap (\theta \vee \rho)) \subseteq \kappa''$ which implies $(u, v) = (u\beta, v\beta)\beta^{-1} \in \in \kappa''$. This completes the proof.

Let us make a remark which will prove useful later. Suppose that $\mathfrak{B}$ is normal, and some subset $V \subseteq F$ fulfils (C), (Q), (A), and the following weaker version of (O):

(O′) there is a fixed order relation $<$ defined on V such that $(V \cap F_X, <)$ is a well-ordered set for every free subsemigroup F_X of F generated by a finite set of letters X.

Then the above proof, under evident modifications, remains valid for normal subvarieties of a normal variety. It is, however, well known [5], that every variety is finitely based in its normal closure. Therefore $\mathfrak{B}$ is h.f.b. if its normal subvarieties are all finitely based. Thus, in order to prove the h.f.b. property for a normal variety, it suffices to construct a set $V \subseteq F$ having properties (C), (O′), (Q) and (A); besides, (A) has to be verified only for normal pairs.

3. WELL-QUASI-ORDERED SETS

Let $\mathbf{N}$ be the set of positive integers, $\mathbf{N}_0 = \mathbf{N} \cup \{0\}$, and let Φ denote the set of order preserving injections $\mathbf{N} \to \mathbf{N}$. Choose a symbol $\omega \notin \mathbf{N}_0$ and put $\overline{\mathbf{N}_0} = \mathbf{N}_0 \cup \{\omega\}$. Extend the natural order relation on $\mathbf{N}_0$ to $\overline{\mathbf{N}_0}$ by putting $\nu < \omega$ for every $\nu \in \mathbf{N}_0$. Furthermore, extend every $\varphi \in \Phi$ to $\overline{\mathbf{N}_0}$ by setting $\varphi(0) = 0$, $\varphi(\omega) = \omega$ (as the extension is unique, no confusion will arise from denoting it by the same letter).

Denote by $\overline{W}_c^t$ the quo-set of vectors

$$(\lambda_1, \ldots, \lambda_c; \alpha_1, \ldots, \alpha_k)$$
$$(\lambda_1, \ldots, \lambda_c \in \overline{\mathbf{N}_0}; k, \alpha_1, \ldots, \alpha_k \in \mathbf{N}_0)$$

under the quasi-order

$$(\lambda_1, \ldots, \lambda_c; \alpha_1, \ldots, \alpha_k) \preceq (\mu_1, \ldots, \mu_c; \beta_1, \ldots, \beta_l) \stackrel{\text{def}}{=\!=}$$
$$\stackrel{\text{def}}{=\!=} \exists \varphi \in \Phi \quad (\varphi(\lambda_i) = \mu_i \wedge \alpha_j \leqslant \beta_{\varphi(j)} \wedge t \mid \beta_{\varphi(j)} - \alpha_j$$
$$\text{for} \quad i = 1, \ldots, c,\ j = 1, \ldots, k).$$

The relation $\alpha \leqslant \beta$, $t \mid \beta - \alpha$ will be written as $\alpha \underset{t}{\leqslant} \beta$.

Denote by W_c^t the subset of $\overline{W}_c^t$ that consists of those vectors having for the first c components positive integers. B r y a n t and V a u g h a n - L e e [2] have shown that W_c^t is w.q.o. This immediately yields

Lemma 3.1. $\bar{W}_c^t$ *is w.q.o.*

Indeed, it is easy to see that $\bar{W}_c^t$ is the disjoint union (the direct sum in the category of quo-sets) of a finite number of subsets isomorphic to some W_α^t, namely, of the

$$W(I,J) = \{(\lambda_1, \ldots, \lambda_c; \alpha_1, \ldots, \alpha_k) \mid \lambda_i = 0 \Leftrightarrow i \in I,$$

$$\lambda_j = \omega \Leftrightarrow j \in J\}.$$

where $I, J \subseteq \{1, \ldots, c\}$, $I \cap J = \phi$.

Another quo-set we shall make use of is the set $\mathbf{C}$ of all finite dimensional vectors over $\mathbf{N}$ under the quo-relation

$(\alpha_1, \ldots, \alpha_k) \lhd (\beta_1, \ldots, \beta_l)$ iff there is a monotone surjection $\{1, \ldots, l\} \to \{1, \ldots, k\}$ such that $\alpha_{\psi(i)} \leqslant \beta_i$ for $i = 1, \ldots, l$.

In other words, this means that the vector $(\beta_1, \ldots, \beta_n)$ can be divided in k sections $(\beta_1, \ldots, \beta_{r(1)}), \ldots, (\beta_{r(k-1)+1}, \ldots, \beta_n)$ such that α_i does not exceed any component of the i's section. From VIII, case $\lhd_h$ in [10] follows

Lemma 3.2. $\mathbf{C}$ *is w.q.o.*

4. SOME PROPERTIES OF THE IDENTITIES (1)–(5)

We consider a normal identity of the form

$$\begin{aligned} x_1 \ldots x_{l-1} x_l^2 x_{l+1} \ldots x_n &= \\ &= x_{1\tau} \ldots x_{(k-1)\tau} x_{k\tau} x_{(k+1)\tau} x_{k\tau} x_{(k+2)\tau} \ldots x_{n\tau}, \end{aligned} \tag{4.1}$$

where τ is a permutation of the indices $1, \ldots, n$. This form comprises all normal identities (1), (3)–(5). By ι we denote the identical permutation, and by $[a; b]$ the cycle $(a\; a+1 \ldots b-1\; b)$ or $(b\; b+1 \ldots a-1\; a)$ according to whether $a < b$ or $b < a$.

Lemma 4.1. *If* $\tau \notin \{\iota, [k; k+1], [k; l], [k+1; l]\}$ *then* (4.1) *implies some identity*

$$x_1 \ldots x_{l-1}x_l^2x_{l+1} \ldots x_s z_1 z_2 x_{s+1} \ldots x_n = \tag{4.2}$$
$$= x_1 \ldots x_{l-1}x_l^2x_{l+1} \ldots x_s z_2 z_1 x_{s+1} \ldots x_n$$

or its dual (where $s < l$*).*

Remark 1. The exceptional identities circumscribed by the condition are

$$x_1 \ldots x_{l-1}x_l^2x_{l+1} \ldots x_n = \tag{4.3}$$
$$= x_1 \ldots x_{k-1}x_k x_{k+1}x_k x_{k+2} \ldots x_n,$$

$$x_1 \ldots x_{l-1}x_l^2x_{l+1} \ldots x_n = \tag{4.4}$$
$$= x_1 \ldots x_{k-1}x_{k+1}x_k x_{k+1}x_{k+2} \ldots x_n,$$

$$x_1 \ldots x_{l-1}x_l^2x_{l+1} \ldots x_n = \tag{4.5}$$
$$= x_1 \ldots x_{k-1}x_l x_k x_l x_{k+1} \ldots x_{l-1}x_{l+1} \ldots x_n,$$

$$x_1 \ldots x_{l-1}x_l^2x_{l+1} \ldots x_n = \tag{4.6}$$
$$= x_1 \ldots x_{k-1}x_k x_l x_k x_{k+1} \ldots x_{l-1}x_{l+1} \ldots x_n,$$

and the duals of the two latter ones.

Proof. Set $I = \{i \mid i \in \{1, \ldots, n\},\ \ i\tau + 1 \neq (i+1)\tau\}$ where $(n+1)\tau = n+1$. Suppose first $I \not\subseteq \{k-1, k, k+1, (l-1)\tau^{-1}, l\tau^{-1}\} = K$ and let $i \in I \setminus K$. Then, applying (4.1) four times under the substitutions $x_{i\tau} \to x_{i\tau}z_1$, $x_{i\tau+1} \to z_2x_{i\tau+1}$; $x_{i\tau+1} \to z_2x_{i\tau+1}$, $x_{(i+1)\tau} \to z_1x_{(i+1)\tau}$; $x_{i\tau} \to x_{i\tau}z_2$, $x_{(i+1)\tau} \to z_1x_{(i+1)\tau}$; and $x_{i\tau} \to x_{i\tau}z_2z_1$, respectively, we obtain

$$x_1 \ldots (x_{i\tau}z_1)(z_2x_{i\tau+1}) \ldots x_{l-1}x_l^2x_{l+1} \ldots x_n =$$
$$= x_{1\tau} \ldots x_{k\tau}x_{(k+1)\tau}x_{k\tau} \ldots$$
$$\ldots x_{i\tau}(z_1x_{(i+1)\tau}) \ldots (z_2x_{i\tau+1}) \ldots x_{n\tau} =$$
$$= x_1 \ldots (x_{i\tau}z_2)x_{i\tau+1} \ldots (z_1x_{(i+1)\tau}) \ldots x_{l-1}x_l^2x_{l+1} \ldots x_n =$$
$$= x_{1\tau} \ldots x_{k\tau}x_{(k+1)\tau}x_{k\tau} \ldots (x_{i\tau}z_2z_1)x_{(i+1)\tau} \ldots x_{n\tau} =$$

$$= x_1 \dots x_{i\tau} z_2 z_1 x_{i\tau+1} \dots x_{l-1} x_l^2 x_{l+1} \dots x_n .$$

(Of course, $i\tau + 1 \in \{k\tau, (k+1)\tau\}$ and/or $(i+1)\tau = l$ are not excluded.)

Now let $I \subseteq K$. For the sake of unambiguousity, let $k \leqslant l$. It is easy to see that our condition implies that the subwords $u \equiv x_1 \dots x_{k-1}$, $v \equiv x_{k+2} \dots x_{l-1}$, $w \equiv x_{l+1} \dots x_n$ must unchangeably appear on the right side of (4.1), and, if $n \notin K$, $n\tau = n$. Dually, if $1 \notin K$, then $1\tau = 1$. Thus, (4.1) is of the form

$$x_1 \dots x_k \dots x_{l-1} x_l^2 x_{l+1} \dots x_n =$$

$$= x_1 \dots x_{k-1} x_{k\tau} x_{(k+1)\tau} x_{k\tau} x_{(k+2)\tau} \dots x_{l\tau} x_{l+1} \dots x_n$$

and v is a subword of $x_{(k+2)\tau} \dots x_{l\tau}$, i.e. either $x_{j\tau} \equiv x_j$ or $x_{(j+1)\tau} \equiv$ $\equiv x_j$ for $k + 2 \leqslant j \leqslant l - 1$. In the first case only $k, k+1$ and l are permuted, whence either $\tau = \iota$ or $\tau = [k; k+1]$ or $l \notin \{k, k+1\}$, $l\tau \in \{k, k+1\}$. This latter possibility yields

$$x_1 \dots (x_{l-1} z_1 z_2) x_l^2 x_{l+1} \dots x_n =$$

$$= x_1 \dots x_{k-1} x_{k\tau} x_{(k+1)\tau} x_{k\tau} x_{k+2} \dots$$

$$\dots (x_{l-1} z_1)(z_2 x_{l\tau}) x_{l+1} \dots x_n =$$

$$= x_1 \dots (x_{l\tau-1} z_2) x_{l\tau} \dots (x_{l-1} z_1) x_l^2 x_{l+1} \dots x_n =$$

$$= x_1 \dots (x_{l\tau-1} z_2) \dots x_{l-1} (z_1 x_{l\tau}) x_{l+1} \dots x_n =$$

$$= x_1 \dots x_{l\tau-1} (z_2 z_1 x_{l\tau}) \dots x_{l-1} x_l^2 x_{l+1} \dots x_n =$$

$$= x_1 \dots x_{k\tau} x_{(k+1)\tau} x_{k\tau} \dots (x_{l-1} z_2 z_1) x_{l\tau} x_{l+1} \dots x_n =$$

$$= x_1 \dots x_{l-1} z_2 z_1 x_l^2 x_{l+1} \dots x_n ,$$

where the parentheses indicate the substitutions. Finally, if $x_{(j+1)\tau} = x_j$ for $k + 2 \leqslant j \leqslant l - 1$ then $\{k\tau, (k+1)\tau, (k+2)\tau\} = \{k, k+1, l\}$ and either $\tau = [k; l]$ or $\tau = [k+1; l]$ or $(k+2)\tau \in \{k, l\}$; but for $(k+2)\tau = l$ we have

$$x_1 \dots x_{k+1} (z_1 z_2 x_{k+2}) \dots x_{l-1} x_l^2 x_{l+1} \dots x_n =$$

$$= x_1 \dots x_{k\tau} x_{(k+1)\tau} x_{k\tau} (x_l z_1)(z_2 x_{k+2}) \dots x_{l-1} x_{l+1} \dots x_n =$$

$$= x_1 \dots (x_{k+1} z_2) x_{k+2} \dots x_{l-1} (x_l z_1)^2 x_{l+1} \dots x_n =$$

$$= x_1 \dots (x_{k+1} z_2) \dots x_l (z_1 x_{k+2}) \dots x_{l-1} x_{l+1} \dots x_n =$$

$$= x_1 \dots x_{k+1} z_2 z_1 x_{k+2} \dots x_{l-1} x_l^2 x_{l+1} \dots x_n,$$

and for $(k+2)\tau = k$ even

$$x_1 \dots x_{k+1} (z x_{k+2}) \dots x_{l-1} x_l^2 x_{l+1} \dots x_n =$$

$$= x_1 \dots x_{k\tau} x_{(k+1)\tau} x_{k\tau} (x_k z) x_{k+2} \dots x_{l-1} x_{l+1} \dots x_n =$$

$$= x_1 \dots x_k z x_{k+1} x_{k+2} \dots x_{l-1} x_l^2 x_{l+1} \dots x_n,$$

which completes the proof.

Remark 2. Obviously, (4.2) implies also

$$\begin{aligned} &x_1 \dots x_r y^2 x_{r+1} \dots x_{2r} z_1 z_2 x_{2r+1} \dots x_{3r} = \\ &= x_1 \dots x_r y^2 x_{r+1} \dots x_{2r} z_2 z_1 x_{2r+1} \dots x_{3r} \end{aligned} \tag{4.7}$$

for arbitrary $r \geqslant \max(l-1, s-l+1, n-s)$. This form will be more convenient for our purposes.

Lemma 4.2. *Every (not necessarily non-balanced) almost simple identity* (5) *implies some identity*

$$x_1 \dots x_r x_{r+1} x_r \dots x_c = x_{i(1)} \dots x_{i(s)} x_{i(s+1)}^2 x_{i(s+2)} \dots x_{i(d)}, \tag{4.8}$$

where either $s \leqslant r$ *or* $\{i(1), \dots, i(s)\} = \{1, \dots, r+1\}$. *The second case occurs for non-balanced identities only.*

Proof. If $l \leqslant k+1$ the assertion is immediate. If $j(h) > k+1$ for some $h \leqslant l-1$ (in particular, if $l \geqslant k+3$), then, applying (5) and substituting $x_t \to a_t$ for the second time, where

$$a_t \equiv \begin{cases} y_t, \text{ if } t = j(g),\ g < h, \\ z_{j(h)} z_{j(h)+1} \dots z_n u x_{j(1)} \dots x_{j(h)}, \text{ if } t = j(h), \\ x_t, \text{ if } t = j(g),\ h < g \leqslant m \text{ or } t \neq j(g) \\ \quad \text{for } g = 1, \dots, m, \end{cases}$$

we obtain

$$w \equiv y_{j(1)} \dots y_{j(h-1)} z_{j(h)} z_{j(h)+1} \dots$$
$$\dots z_n u x_1 \dots x_k x_{k+1} x_k \dots x_n =$$
$$= y_{j(1)} \dots y_{j(h-1)} a_{j(h)} x_{j(h+1)} \dots$$
$$\dots x_{j(l-1)} x_{j(l)}^2 x_{j(l+1)} \dots x_{j(m)} =$$
$$= a_1 \dots a_k a_{k+1} a_k \dots a_n .$$

Now substitute

$$x_t \to b_t \equiv \begin{cases} a_t & \text{if } t < j(h), \\ z_t & \text{if } t \geqslant j(h), \end{cases}$$

and apply (5) once more. Then

$$w = a_1 \dots a_k a_{k+1} a_k \dots a_n =$$
$$= b_{j(1)} \dots b_{j(l-1)} b_{j(l)}^2 b_{j(l+1)} \dots b_{j(m)} v$$

where v is the empty word if $j(h) > n$, and

$$v \equiv u x_{j(1)} \dots x_{j(h)} a_{j(h)+1} \dots a_n$$

else. It is easy to check that, w being almost simple, so is the resulting word. Besides, if u is sufficiently long, then the prefix of w ending by x_k is longer than l, which proves the assertion.

Finally, if $l = k + 2$ and $\{j(1), \dots, j(l-1)\} = \{1, \dots, k+1\}$, then the second instance prevails, with $s = l - 1$, $r = k$. In this case (5) is non-balanced.

Corollary 4.1. *An arbitrary identity* (1) *(not excluding the case where* σ *is trivial) implies both*

$$x_1 \dots x_{r-1} x_r x_{r+1} x_r x_{r+2} \dots x_{2r} = \tag{4.9}$$
$$= x_{1\pi} \dots x_{s\pi} x_{(s+1)\pi}^2 x_{(s+2)\pi} \dots x_{(2r)\pi}$$

for some $r \geqslant \frac{n}{2}$, $s \leqslant r - 1$ *or* $s = r$, $i\pi = r + 1$ *with* $i \leqslant r$, *and the dual identity.*

Proof. As in this case $(s+1)\pi = r$, we have $r \notin \{1\pi, \ldots, s\pi\}$, and so $s > r-1$, $r+c \notin \{1\pi, \ldots, s\pi\}$ for $c > 1$ can hold only if $\{1\pi, \ldots, s\pi\} = \{1, \ldots, r-1, r+1\}$. Taking into account the proof of Lemma 4.2, this yields the assertion.

Lemma 4.3. *Identity* (3), *as well as its dual, imply*

$$x_1 \ldots x_{l-1} y^2 z x_l \ldots x_n = x_1 \ldots x_{l-1} z y^2 x_l \ldots x_n. \tag{4.10}$$

Proof. Indeed, we have

$$x_1 \ldots x_{l-1} y^2 (z x_l) \ldots x_n =$$
$$= x_1 \ldots (x_{l-1} z) x_l \ldots x_{k-1} y x_k y x_{k+1} \ldots x_n =$$
$$= x_1 \ldots x_{l-1} z y^2 x_l \ldots x_n,$$

where the parentheses indicate the substitutions.

Remark 3. As in case of Lemma 4.1, we shall mainly make use of the consequence of (4.10) of the form

$$x_1 \ldots x_r y^2 z x_{r+1} \ldots x_{2r} = x_1 \ldots x_r z y^2 x_{r+1} \ldots x_{2r}. \tag{4.11}$$

5. PROOF OF PROPOSITION A

Proposition A follows from Lemma 4.1 and the following

Theorem 5.1. *A variety* $\mathfrak{V}$, *defined by the identities* (1), (2) *and* (4.7) *or its dual, is h.f.b.*

Proof. To start with, we point out some properties of the identities fulfilled in $\mathfrak{V}$. According to Lemma 4.2, (1) implies an identity of the form (4.9) with some indices r', s' which exceed r in (4.7). We shall, however, use the original notation (the same r in (4.7) and in (4.9)), and shall suppose that (4.7) holds also for s instead of r. Furthermore, one can assume $p > 2r+2$, $q > r$. Then we obtain from (2) and (4.7)

$$y^t x_1 x_2 = y^p x_1 x_2 y^q = y^p x_2 x_1 y^q = y^t x_2 x_1, \tag{5.1}$$

where $t = p + q$.

In our constructions we shall use the following notations. The free generators of F will be denoted by x_i $(i = 1, 2, \ldots)$, the length of the word v by $|v|$, the set of its letters by $X(v)$, and the number of occurrences of x_i in v by $|v|_i$. As above, $\equiv$ will mean graphical identity, and θ the fully invariant congruence that corresponds to $\mathfrak{B}$.

We are going to prove that $\mathfrak{B}$ admits a good standard form. For this sake, we have to construct the set of standard words V, the order relation $<$ on V and the quasi-order $\preceq$ on V^*, and to check the fulfilment of the conditions (C), (O′), (Q) and (A).

Call u a *standard word of type* i $(i = 1, 2, 3)$, if the respective one of the following conditions is fulfilled:

(1) $u \equiv ax_{\epsilon(1)} \ldots x_{\epsilon(c)} x_\lambda^2 b x_1^{\alpha_1} \ldots x_\kappa^{\alpha_\kappa} \ldots x_\zeta^{\alpha_\zeta} a'$, where $s \leqslant |a| \leqslant r$, $|b| = |a'| = r$, $c = 0$ if $|a| < r$, $|x_{\epsilon(1)} \ldots x_{\epsilon(c)} x_\lambda^2 b x_1^{\alpha_1} \ldots x_\zeta^{\alpha_\zeta}|_{\epsilon(j)} = 1$ for $j = 1, \ldots, c$; $\alpha_1, \ldots, \alpha_\zeta \geqslant 0$, $\alpha_\kappa \geqslant t$;

(2) $u \equiv ax_{\epsilon(1)} \ldots x_{\epsilon(c)} x_\lambda^2 x_1^{\alpha_1} \ldots x_\zeta^{\alpha_\zeta} a'$, where $t \geqslant \alpha_1, \ldots, \alpha_\zeta \geqslant 0$, $a, b, c, \epsilon(j)$ as in (1);

(3) $u \equiv aeg$, where $|a| = r$, $|g| = 2r$, e is simple.

Let V be the set of all standard words, and denote by V_i the set of standard words of type i. Condition (C) is guaranteed by

Lemma 5.1. *If the intersection of some θ-class T with F^{3r+1} is non-empty, then $T \cap V \neq \phi$ holds, too.*

Proof. Let $v \in F^{3r+1}$, and put $v \equiv a'e'g'$, $|a'| = r$, $|g'| = 2r$. If e' is simple, we are done. Suppose $|e'|_\lambda > 1$ for some λ. Using (4.9) if necessary, we get $v \,\theta\, fe_1 x_\lambda^2 e_2 g$ with $|f| = s$, $|g| = 2r$. Write $e_2 g$ in the form $e_2 g \equiv bda'$, $|b| = |a'| = r$ and, taking into account (4.7), permute the letters in d so as to obtain a word $u \equiv fex_\lambda^2 bx_1^{\alpha_1} \ldots$ $\ldots x_m^{\alpha_m} a' \,\theta\, v$, and suppose it is one with minimal length $|e|$ of e. We claim that either $|fe| < r$ or $e \equiv qx_{\epsilon(1)} \ldots x_{\epsilon(c)}$, $|fq| = r$, $|x_{\epsilon(1)} \ldots x_{\epsilon(c)} x_\lambda^2 bx_1^{\alpha_1} \ldots x_m^{\alpha_m}|_{\epsilon(j)} = 1$ for $j = 1, \ldots, c$, i.e. u is a standard word of type 1 or 2. Indeed, in the opposite case we could apply (4.9) to u and obtain a word which contains the square of a variable more to the left then u does, but to the right from the first s letters. Using (4.7) if necessary to transform this word, we arrive to some $u' = = f'e'x^2b'x_1^{\gamma_1} \ldots x_\xi^{\gamma_\xi}a'' \theta v$ with $|e'| < |e|$, contrary to the assumption. This proves the lemma.

Now we introduce a relation $<$ on the set of finite dimensional vectors over $\mathbf{N}$ that we are going to call *lexicographical:* $(\lambda_1, \ldots, \lambda_l) < < (\mu_1, \ldots, \mu_m)$ iff there is an index i, $1 \leqslant i \leqslant m$ such that $\lambda_j = \mu_j$ for $j < i$, and either $i = l + 1$ or $i \leqslant l$, $\lambda_i < \mu_i$. We shall say that $(\lambda_1, \ldots, \lambda_l) < (\mu_1, \ldots, \mu_m)$ *at the place* ω in the first case and *at the place* i in the second.

Associate with every $u \in V$ three vectors as follows. Let

$$u \equiv \begin{cases} x_{\lambda(1)} \ldots x_{\lambda(q)} x_{\epsilon(1)} \ldots x_{\epsilon(c)} x_\lambda^2 x_{\lambda(q+1)} \ldots x_{\lambda(q+r)} x_1^{\alpha_1} \ldots \\ \qquad \ldots x_\xi^{\alpha_\xi} x_{\lambda(q+r+1)} \ldots x_{\lambda(q+2r)} & \text{if } u \in V_1 \cup V_2, \\ x_{\lambda(1)} \ldots x_{\lambda(r)} x_{\epsilon(1)} \ldots x_{\epsilon(c)} x_{\lambda(r+1)} \ldots x_{\lambda(3r)} & \text{if } u \in V_3 \end{cases}$$

$(s \leqslant q \leqslant r,\ c = 0$ if $q < r)$. Put

$$\mathbf{sk}\, u = \begin{cases} (\lambda(1), \ldots, \lambda(q+2r), \lambda) & \text{if } u \in V_1 \cup V_2, \\ (\lambda(1), \ldots, \lambda(3r)) & \text{if } u \in V_3; \end{cases}$$

$$\mathbf{low}\, u = \begin{cases} \phi & \text{if } c > 0, \\ (\epsilon(1), \ldots, \epsilon(c)) & \text{if } c > 0; \end{cases}$$

$$\mathbf{exp}\, u = \begin{cases} (\alpha_1, \ldots, \alpha_\xi) & \text{if } u \in V_1 \cup V_2, \\ \phi & \text{if } u \in V_3. \end{cases}$$

Call the first vector the *skeleton*, the second one the *lower vector*, the third one the *exponential vector* of u. Define the order relation on V by $u < v$ iff

$$\text{qu}\, u = \langle i(u); \mathbf{sk}\, u; \mathbf{low}\, u; \mathbf{exp}\, u \rangle <$$

$$< \langle i(v); \mathbf{sk}\,(v); \mathbf{low}\,(v); \mathbf{exp}\,(v) \rangle = \text{qu}\, v$$

in the lexicographical order of the quadruples, where $i(u)$, $i(v)$ are the types of u and v, resp., and the vectorial components of the quadruples are to be compared lexicographically. Clearly, we have

Lemma 5.2. *$(F_X \cap V, <)$ is a wo-set for every finite X.*

Thus, (O′) is fulfilled. Our next task is to construct the quasi-order $\preceq$. As generally, this is the most sophisticated part of the proof. The basic idea is to define $(u, v) \preceq (u', v')$ in such a way that there existed an endomorphism $\varphi \in \operatorname{End} F$ which maps u onto u' and the skeletons, lower vectors and exponential vectors of u and v onto the skeletons, lower vectors and exponential vectors of u' and v', resp. (more exactly, variables with indices in the skeletons etc. into similar variables), and preserves the lexicographical order of the quadruples $\operatorname{qu} u$ and $\operatorname{qu} v$. For this purpose, let $e \equiv x_{\epsilon(1)} \ldots x_{\epsilon(c)}$ be a simple word, and put $\epsilon(0) = 0$, $\epsilon(c+1) = \omega$, where x_0 and x_ω are some new variables. For a vector $\mathbf{a}$, denote by $\bar{a}$ the set of its components. A pair $(x_{\epsilon(i)}, x_{\epsilon(j)})$, $0 \leqslant i < j \leqslant$ $\leqslant c + 1$, is said to *mark off a block of length* $j - (i + 1)$ *of the word* e *with respect to the vector* $\mathbf{a} = (\lambda_1, \ldots, \lambda_p) \in \bar{\mathbf{N}}^p$ if $\epsilon(i), \epsilon(j) \in \bar{a} \cup \{0, \omega\}$ but $\epsilon(i+1), \ldots, \epsilon(j-1) \notin \bar{a}$. In particular, adjacent indices mark off a block of length 0. The *block vector of* e *with respect to* $\mathbf{a}$ will be the term for the vector $\mathbf{bl}_\mathbf{a}\, e = (\lambda_1, \ldots, \lambda_p, i_1, \ldots, i_p, l_1, \ldots, l_{\beta(e)})$ where $i_j = \omega$ if $\lambda_j \notin \{\epsilon(1), \ldots, \epsilon(c)\}$ and $\epsilon(i_j) = \lambda_j$ $(1 \leqslant i_j \leqslant c)$ else, and $l_1, \ldots, l_{\beta(e)}$ are the lengths of the blocks from left to right. If e is the empty word, define $\mathbf{bl}_\mathbf{a}\, e = (a, \omega, \ldots, \omega)$ a vector of length $2p$.

Example 1. If $e = x_1 x_3 x_5 x_7 x_9 x_8 x_6 x_4 x_2$, $\mathbf{a} = (9, 4, 7, \omega, 4, 11, 1)$, then $\mathbf{bl}_\mathbf{a}\, e = (9, 4, 7, \omega, 4, 11, 1; 5, 8, 4, \omega, 8, \omega, 1; 0, 2, 0, 2, 1)$.

The number $\beta(e)$ will be referred to as the *block length* of e. Note that $\beta(e) \leqslant p + 1$. As a matter of fact, all l_i's but $l_{\beta(e)}$ can be determined from $i_1, \ldots, i_p$, and they are added only because else the construction would become hard to be handled.

For $\mathbf{b} = (\lambda_1, \ldots, \lambda_p)$ and $q = x_{\lambda_1} \ldots x_{\lambda_p}$ put $q = \mathbf{b}^\#$, $\mathbf{b} = q^\natural$.

Associate with every pair $(u, v) \in V^*$ two vectors $\mathbf{a} = \mathbf{a}(u, v) =$ $= (\mathbf{sk}\, u, \mathbf{sk}\, v, \gamma, \delta)$ and $\mathbf{w} = \mathbf{w}(u, v)$ as follows.

(1) If $u \in V_3$, put $\mathbf{w} = \mathbf{bl_a}\,(\mathbf{low}\,u)^{\#}$, where

(a) $\gamma = \delta = \omega$ if either (α) $v \notin V_3$, or $v \in V_3$ and (β) $\mathbf{sk}\,v < \mathbf{sk}\,u$, or ($\gamma$) $\mathbf{sk}\,v = \mathbf{sk}\,u$, $\mathbf{low}\,v < \mathbf{low}\,u$ at the place ω;

(b) $\gamma = \epsilon(i)$, $\delta = \xi(i)$ if $v \in V_3$, $\mathbf{sk}\,v = \mathbf{sk}\,u$, and $\mathbf{low}\,v = (\xi(1), \ldots, \xi(d)) < \mathbf{low}\,u = (\epsilon(1), \ldots, \epsilon(c))$ at the place $i < \omega$;

(2) If $u \in V_1 \cup V_2$ and $\mathbf{exp}\,u = (\alpha_1, \ldots, \alpha_\zeta)$, put

$$\mathbf{w} = (\mathbf{bl_a}\,(\mathbf{low}\,u)^{\#}; d(1), \ldots, d(t-1); \mathbf{exp}\,u),$$

where $d(j) = \min\{\kappa \mid \kappa \notin \bar{a},\ \alpha_\kappa \equiv j \bmod t\}$ (here $\min \phi = \omega$) and

(a) $\gamma = \delta = \omega$ if either (α) u is of type 2, v is of type 1, or they are of the same type and (β) $\mathbf{sk}\,v < \mathbf{sk}\,u$ or (γ) $\mathbf{low}\,v < \mathbf{low}\,u$ at the place ω or (δ) $\mathbf{sk}\,v = \mathbf{sk}\,u$, $\mathbf{low}\,v = \mathbf{low}\,u$, $\mathbf{exp}\,v < \mathbf{exp}\,u$;

(b) $\gamma = \epsilon(i)$, $\delta = \xi(i)$ if $\mathbf{sk}\,v = \mathbf{sk}\,u$, $\mathbf{low}\,v = (\xi(1), \ldots, \xi(d)) < \mathbf{low}\,u = (\epsilon(1), \ldots, \epsilon(c))$ at the place $i < \omega$.

This definition ensures that, in case 2, $\mathbf{w} \in \bar{W}_l^t$ for some $l \leqslant 3(6r+4) + t$ (the length of $(\mathbf{bl_a}\,(\mathbf{low}\,u)^{\#}; d(1), \ldots, d(t-1))$). One can treat $\mathbf{w}$ as belonging to some $\bar{W}_l^t$ with $l \leqslant 3(6r+2)+1$ also in case 1 if one identifies a vector $(\delta_1, \ldots, \delta_l)$ with $(\delta_1, \ldots, \delta_l; \phi)$ where ϕ is the "empty vector".

Example 2 (case $(1, b)$). If $r = 3$,

$$u \equiv x_5 x_1 \cdot x_5 x_2 x_6 x_1 x_4 x_3 \cdot x_6^2 x_5^2,$$

$$v \equiv x_5 x_1 \cdot x_5 x_2 x_3 x_4 x_5 \cdot x_6^2 x_5^2$$

then

$$\mathbf{a} = (5, 1, 6, 6, 5, 5; 5, 1, 6, 6, 5, 5; 6, 3);$$

$$\mathbf{w} = (\mathbf{a}; 1, 4, 3, 3, 1, 1, 1, 4, 3, 3, 1, 1, 3, 6; 0, 1, 0, 1, 0).$$

Example 3 (case $(2, a, \alpha)$). If $r = 3$, $t = 4$,

$$u \equiv x_5 x_1 x_5 \cdot x_2 x_{10} x_1 x_4 x_7 x_3 \cdot x_6^3 x_5^4 x_8^3 x_9^2 \cdot x_8 x_1 x_{10},$$

$$v \equiv x_5^3 \cdot x_{12}x_1x_{11}x_4x_3x_7 \cdot x_5^5 \cdot x_2x_5^2x_8^4x_9^2x_{10} \cdot x_5^2x_1,$$

then

$$\mathbf{a} = (5, 1, 5, 6, 5, 5, 8, 1, 10, 6; 5, 5, 5, 5, 5, 5, 5, 5, 5, 1; \omega, \omega);$$

$$\mathbf{w} = (\mathbf{a}; \omega, 3, \omega, \omega, \omega, \omega, \omega, 3, 2, \omega, \omega, \omega, \omega, \omega, \omega, \omega, \omega, \omega,$$
$$3, \omega, \omega, \omega; 1, 0, 3; \omega, 9, \omega; 0, 0, 0, 0, 2, 0, 0, 3, 2).$$

Now let $(u, v), (u', v') \in V^*$, $\mathbf{a}' = \mathbf{a}(u', v') = (\mathbf{sk}\, u', \mathbf{sk}\, v', \gamma', \delta')$, $\mathbf{w}' = \mathbf{w}(u', v')$, and put $(u, v) \preceq (u', v')$ iff

(i) $\mathbf{w}$ and $\mathbf{w}'$ are constructed according to the same rule $(1, a, \alpha)$ or . . . or $(2, b)$;

(ii) $\mathbf{sk}\, u$ and $\mathbf{sk}\, u'$, $\mathbf{sk}\, v$ and $\mathbf{sk}\, v'$ as well as $\mathbf{bl}_{\mathbf{a}}\, (\mathbf{low}\, u)^{\#}$ and $\mathbf{bl}_{\mathbf{a}'}\, (\mathbf{low}\, u')^{\#}$ are of the same lengths, respectively (whence, in particular, $\mathbf{w}$ and $\mathbf{w}'$ belong to the same set $\bar{W}_l^t$);

(iii) $\mathbf{w} \preceq \mathbf{w}'$ in $\bar{W}_l^t$.

Now it is easy to see that there is a finite number of possibilities to satisfy (i) and (ii). Hence, Lemma 3.1 implies

Lemma 5.3. *$(V^*, \preceq)$ is a w.q.o. set.*

Thus, (Q) is fulfilled. The last step of the proof of Theorem 5.1 (to show that (A) is fulfilled) is accomplished by

Lemma 5.4. *If $(u, v), (u', v') \in V^*$, (u, v) is normal, and $(u, v) \preceq (u', v')$, then there is a $v'' \in V$ such that $v'' < v'$ and $(v', v'') \in \theta \vee \theta_{u,v}$.*

Proof.

1. Let $u \equiv x_{\lambda(1)} \ldots x_{\lambda(r)} x_{\epsilon(1)} \ldots x_{\epsilon(c)} x_{\lambda(r+1)} \ldots x_{\lambda(3r)}$, $u' \equiv x_{\lambda'(1)} \ldots x_{\lambda'(r)} x_{\epsilon'(1)} \ldots x_{\epsilon'(c')} x_{\lambda'(r+1)} \ldots x_{\lambda'(3r)}$ be words of type 3. Then $\mathbf{w} = \mathbf{bl}_{\mathbf{a}}\, x_{\epsilon(1)} \ldots x_{\epsilon(c)}$. From (ii) it follows that the block lengths $\beta\,(\mathbf{low}\, u)$ and $\beta\,(\mathbf{low}\, u')$ are equal and there is a mapping $\varphi \in \Phi$ such that $\varphi\,(\mathbf{sk}\, u) = \mathbf{sk}\, u'$ (i.e. $\varphi(\lambda(j)) = \lambda'(j)$ for $i = 1, \ldots, 3r$), $\varphi\,(\mathbf{sk}\, v) = \mathbf{sk}\, v'$, $\varphi(\gamma) = \gamma'$, $\varphi(\delta) = \delta'$, $\varphi(i_j) = i'_j$, $\varphi(l_j) = l'_j$, where i'_j and

l_j' have the obvious meaning. Denote by $\hat{\varphi}$ the endomorphism $\hat{\varphi}: x_k \mapsto x_{\varphi(k)}$ induced by φ. Then

$$u\hat{\varphi} \equiv \varphi(u^{\natural})^{\#} =$$

$$= x_{\lambda'(1)} \ldots x_{\lambda'(r)} x_{\varphi(\epsilon(1))} \ldots x_{\varphi(\epsilon(c))} x_{\lambda'(r+1)} \ldots x_{\lambda'(3r)}.$$

Besides, if $\epsilon(k) \in \bar{a}$, i.e. if $k = i_j$, $\epsilon(k) = \lambda_j$ for some j, then $\varphi(\epsilon(k)) = \varphi(\lambda(j)) = \lambda'(j)$, $\varphi(k) = \varphi(i_j) = i_j'$, whence $\epsilon'(\varphi(k)) = \lambda'(j) = \varphi(\epsilon(k))$. Furthermore, the monotonity of φ implies also $l_j \leqslant \varphi(l_j) = l_j'$, and, by definition, $l_j' = 0$ if and only if $l_j = 0$. Hence, as the image of

$$u_j \equiv x_{\varphi(\epsilon(k+1))} \ldots x_{\varphi(\epsilon(k+l_j))},$$

$(k = i_{j-1})$, of the j'th block of u, is a simple word of length $l_j < l_j'$, it is easy to find an endomorphism ψ_j which maps u_j onto the j'th block $x_{\epsilon'(\varphi(k)+1)} \ldots x_{\epsilon'(\varphi(k)+\varphi(l_j))}$ of u', and does not move variables $x_\tau \notin X(u_j)$: we can choose

$$x_\tau \psi_j \equiv \begin{cases} x_{\epsilon'(\varphi(k)+m)}, & \text{if } \tau = \varphi(\epsilon(k)+m),\ 1 \leqslant m < l_j, \\ x_{\epsilon'(\varphi(k)+l_j)} \ldots x_{\epsilon'(\varphi(k)+l_j')}, & \text{if } \tau = \varphi(\epsilon(k+l_j)), \\ x_\tau \ \text{ if } x_\tau \notin X(u_j), \end{cases}$$

if u_j is not the empty word; if u_j is empty, the identical automorphism will do. Putting

$$x_\tau \psi \equiv \begin{cases} x_\tau \psi_j, & \text{if } x_\tau \in X(u_j), \\ x_\tau, & \text{if } \tau \notin \overline{\mathbf{low}\, u\hat{\varphi}} \setminus \overline{\varphi(\mathbf{a})}, \end{cases}$$

where φ is applied to a vector componentwise, we obviously have $u\hat{\varphi}\psi \equiv u'$, and so $u' \, (\theta \vee \theta_{u,v}) \, v\hat{\varphi}\psi$.

We have to show that $v\hat{\varphi}\psi < u' = u\hat{\varphi}\psi$. From $v < u$ and the monotonity of φ follows $v\hat{\varphi} < u\hat{\varphi}$, therefore if suffices to deal with the effect of ψ on $v\hat{\varphi}$ and $u\hat{\varphi}$. Note first that the set $J_i = \{v \mid \exists v'(v' \, \theta \, v \wedge v' \in V_j \wedge j \leqslant i)\}$ is a fully invariant ideal for $i = 1, 2, 3$. Hence $v\hat{\varphi}\psi < u\hat{\varphi}\psi$ if $v \in V_1 \cup V_2$.

By the definitions of φ and ψ, $\mathbf{sk}\, v\hat{\varphi}\psi = \mathbf{sk}\, v\hat{\varphi} = \varphi(\mathbf{sk}\, v)$, $\mathbf{sk}\, u\hat{\varphi}\psi = \mathbf{sk}\, u\hat{\varphi} = \varphi(\mathbf{sk}\, u)$. Hence, if $v \in V_3$ and $\mathbf{sk}\, v < \mathbf{sk}\, u$, then $\mathbf{sk}\, v\hat{\varphi}\psi < \mathbf{sk}\, u\hat{\varphi}\psi$ and $v\hat{\varphi}\psi < u\hat{\varphi}\psi$.

Now suppose $v \in V_3$, $\mathbf{sk}\, v = \mathbf{sk}\, u$. Then the same holds for $v\hat{\varphi}\psi$ and $u\hat{\varphi}\psi$. If $\mathbf{low}\, v < \mathbf{low}\, u$ at the place ω, i.e. if $(\mathbf{low}\, v)^{\natural}$ is a prefix of $(\mathbf{low}\, u)^{\natural}$ then, obviously, the same holds for $\mathbf{low}\, v\hat{\varphi}\psi$ and $\mathbf{low}\, u\hat{\varphi}\psi$. Finally, if $\mathbf{low}\, v < \mathbf{low}\, u$ at the place $i < \omega$ then, putting $\mathbf{low}\, v = (\xi(1), \ldots, \xi(l))$, we have $\xi(j) = \epsilon(j)$ for $j < i$, $\delta = \xi(i) < \epsilon(i) = \gamma$. But $\varphi(\gamma) = \gamma'$, $\varphi(\delta) = \delta'$ and, since $\gamma, \delta \in \bar{a}$, we have $x_{\epsilon(i)}\hat{\varphi}\psi \equiv x_{\varphi(\epsilon(i))}\psi \equiv x_{\gamma'}\psi \equiv x_{\gamma'}$, $x_{\xi(i)}\hat{\varphi}\psi \equiv x_{\delta'}$. Hence $\mathbf{sk}\, u' = \mathbf{sk}\, u\hat{\varphi}\psi = \mathbf{sk}\, v\hat{\varphi}\psi$, and $(\mathbf{low}\, u)^{\#} \equiv u_1 x_\gamma u_2$, $(\mathbf{low}\, v)^{\#} \equiv u_1 x_\delta v_2$, $(\mathbf{low}\, u')^{\#} \equiv (u_1\hat{\varphi}\psi) x_{\gamma'} (u_2\hat{\varphi}\psi)$, $(\mathbf{low}\, v\hat{\varphi}\psi)^{\#} \equiv (u_1\hat{\varphi}\psi) x_{\gamma'} (v_2\hat{\varphi}\psi)$ for some $u_1, u_2, v_2 \in F$, which implies $v\hat{\varphi}\psi < u'$.

2. Let $u, u' \in V_2$, u have the same form as in (2), the same letters with primes will be used to denote the corresponding indices and exponents in u'. Now

$$\mathbf{w} = (\mathbf{bl}_{\mathbf{a}}\, (\mathbf{low}\, u)^{\#}, d(1), \ldots, d(t-1); \alpha_1, \ldots, \alpha_\zeta),$$

$$\mathbf{w}' = (\mathbf{bl}_{\mathbf{a}'}\, (\mathbf{low}\, u')^{\#}, d'(1), \ldots, d'(t-1); \alpha'_1, \ldots, \alpha'_{\zeta'}).$$

There is a $\varphi \in \Phi$ such that $\varphi(\mathbf{bl}_{\mathbf{a}}\, (\mathbf{low}\, u)^{\#}) = \mathbf{bl}_{\mathbf{a}'}\, (\mathbf{low}\, u')^{\#}$ (i.e. $\varphi(\mathbf{sk}\, u) = \mathbf{sk}\, u'$, $\varphi(\mathbf{sk}\, v) = \mathbf{sk}\, v'$, $\varphi(\gamma) = \gamma'$, $\varphi(\delta) = \delta'$, $\varphi(i_j) = i'_j$, $\varphi(l_j) = l'_j$), $\varphi(d(k)) = d'(k)$, and $\alpha'_{\varphi(l)} = \alpha_l$ for $l = 1, \ldots, \zeta$ (since $\alpha_l, \alpha'_{\varphi(l)} < t$). Define $\hat{\varphi}$, ψ_j, and ψ, as in case 1. Again, we obtain a pair of words $(u_0, v_0) \equiv (u\hat{\varphi}\psi, v\hat{\varphi}\psi)$, $\mathbf{sk}\, u_0 = \mathbf{sk}\, u'$, $\mathbf{low}\, u_0 = \mathbf{low}\, u'$, $(\mathbf{exp}\, u_0)^{\#} \equiv x_{\varphi(1)}^{\alpha_1} \ldots x_{\varphi(\zeta)}^{\alpha_\zeta}$ (note that here we make use of the fact that, if $\alpha_l \neq 0$ *then* $l \notin \overline{\mathbf{low}\, u}$ and therefore $x_{\varphi(l)}\psi \equiv x_{\varphi(l)}$, furthermore, $\varphi(l) \notin \overline{\mathbf{low}\, u'}$ as $\alpha'_{\varphi(l)} \neq 0$). We have $v_0 < u_0$: taking into account that $u = v$ is normal, and therefore $(\mathbf{low}\, v\hat{\varphi})^{\#}\psi = (\mathbf{low}\, v_0)^{\#}$, the same reasoning as above goes through, except for the case $v \in V_2$, $\mathbf{sk}\, v = \mathbf{sk}\, u$, $\mathbf{low}\, v = \mathbf{low}\, u$, $\mathbf{exp}\, v < \mathbf{exp}\, u$. However, if this latter inequality holds at the place ω, the assertion is obvious; if it holds at the place $i < \omega$, then $\mathbf{exp}\, v = (\alpha_1, \ldots, \alpha_{i-1}, \beta_i, \ldots, \beta_\eta)$, $\beta_i < \alpha_i$, whence $(\mathbf{exp}\, v\hat{\varphi})^{\#} \equiv x_{\varphi(1)}^{\alpha_1} \ldots x_{\varphi(i-1)}^{\alpha_{i-1}} x_{\varphi(i)}^{\beta_i} \ldots x_{\varphi(\eta)}^{\beta_\eta}$. Now for $j < i$ obviously holds

$x_{\varphi(j)}\psi \equiv x_{\varphi(j)}$ if $\alpha_j \neq 0$, but the same holds for $j \geqslant i$, $\beta_j \neq 0$, too, because then $j \notin \overline{\mathbf{low}\, v} = \overline{\mathbf{low}\, u}$. Thus, $\mathbf{exp}\, v\hat{\varphi}\psi = \mathbf{exp}\, v\hat{\varphi} < \mathbf{exp}\, u\hat{\varphi} = \mathbf{exp}\, u\hat{\varphi}\psi$ at the place $\varphi(i)$.

We obviously have $\mathbf{sk}\, u_0 = \mathbf{sk}\, u'$, $\mathbf{low}\, u_0 = \mathbf{low}\, u'$. Put $\mathbf{exp}\, u_0 = (\gamma_1, \ldots, \gamma_{\zeta'})$ where $\gamma_\kappa = \alpha'_\kappa = \alpha_l$ if $\kappa = \varphi(l)$ and $\gamma_\kappa = 0$ else. Suppose $\gamma_\kappa = 0$, $\alpha'_\kappa \neq 0$ for some κ; we are going to show that there is an i such that $\varphi(i) < \kappa$ (i.e. $i < l$) and $\alpha_i = \gamma_\kappa$. Indeed, $\kappa \in \{\varphi(k) \mid k \in \bar{a}\}$ because else we had $\beta_\kappa = \alpha_\pi = \alpha'_\kappa$. Hence, the set $\{\pi \mid \alpha'_\pi \equiv \alpha'_\kappa \bmod t\}$ is non-empty, and its minimal element $i' = d'(\alpha'_\kappa) \leqslant \kappa$ is a component of $\mathbf{w}(u', v')$. This yields $\varphi(i) = i'$ for $i = d'(\alpha'_\kappa) = d(\alpha_l)$ whence $\alpha_i = \alpha'_{i'} = \alpha'_\kappa$, and $i < l$ because $\gamma_{\varphi(i)} = \alpha_i$, $\gamma_{\varphi(l)} \neq \alpha_l$.

Consider the endomorphism χ_κ $(\kappa \leqslant \zeta', \alpha'_\kappa \neq 0 = \gamma_\kappa)$ defined by

$$(5.2) \qquad x_\tau \chi_\kappa \equiv \begin{cases} x_\tau x_\kappa, & \text{if } \tau = d'(\alpha'_\kappa), \\ x_\tau & \text{else,} \end{cases}$$

and set $\chi = \prod_\kappa \chi_\kappa$. Using (4.7), it is easy to see that $(u_0\chi)\, \theta\, u'$ and, since $(u_0\chi)\, \theta_{u,v}\, (v_0\chi)$, we have $u'\, (\theta \vee \theta_{u,v})\, v_1$ for the standard word $v_1\, \theta\, (v_0\chi)$. *

(a) If $v_0 \in V_1$ then $v_1 \in V_1$.

(b) If $\mathbf{sk}\, v_0 < \mathbf{sk}\, u_0$ then $v_1 < u'$ follows from the fact that $\mathbf{sk}\, v_1 = \mathbf{sk}\, v_0$, $\mathbf{sk}\, u' = \mathbf{sk}\, u_0$.

(c) Let $\mathbf{sk}\, u_0 = \mathbf{sk}\, v_0$, $\mathbf{low}\, u_0 > \mathbf{low}\, v_0$. By the definition of φ and ψ, $\mathbf{low}\, u_0 = \mathbf{low}\, u'$, whence $d'(k)$ does not enter into $\mathbf{low}\, u_0$. If, in addition, $d'(k) \notin \overline{\mathbf{low}\, v_0}$ for $k = 1, \ldots, \zeta'$, either, then $\mathbf{low}\, v_1 = \mathbf{low}\, v_0$ and we are done. Now suppose that $\xi_0(j)$ is the first component of $\mathbf{low}\, v_0$ that equals to some $d = d'(k)$. Then $\mathbf{low}\, v_0 < \mathbf{low}\, u_0$ at some place $i \leqslant j$, and so $\xi_0(i)$ coincides with the i'th component of $\mathbf{low}\, v_1$ (consult (5.2)!), so that $\mathbf{low}\, v_1 < \mathbf{low}\, u'$ at this very place.

*χ is defined in such a way that u' can be obtained from $u_0\chi$ by rearranging the variables in its "commutative part". The normality of the pair (u_0, v_0) implies the same for v_1 and $v_0\chi$, because if $x_\tau\chi = x_\tau x_{\kappa(1)} \ldots x_{\kappa(s)}$ then $x_{\kappa(j)} \notin X(u_0) = X(v_0)$.

(d) Now let $i(u_0) = i(v_0)$, $\mathbf{sk}\, u_0 = \mathbf{sk}\, v_0$, $\mathbf{low}\, u_0 = \mathbf{low}\, v_0$, $\mathbf{exp}\, u_0 >$ $> \mathbf{exp}\, v_0$. If the inequality holds at the place ω, there is nothing to be proved, so let $\mathbf{exp}\, v_0 = (\gamma_1, \ldots, \gamma_{k-1}, \delta_k, \ldots, \delta_\pi)$, $\delta_k < \gamma_k$. Denote $\mathbf{exp}\, v_1 = (\delta'_1, \ldots, \delta'_\rho)$. Sure enough, $\gamma_k \neq 0$, whence no new items of x_k are introduced by χ, i.e. $\alpha'_k = \gamma_k$, $\delta'_k = \delta_k$. But for $j < k$ either $\gamma_j = \alpha'_j$, and then $\delta'_j = \gamma_j$ since no χ_j is defined, or $\gamma_j = 0$, $\alpha'_j \neq 0$, and then $x_\tau \chi_j = x_\tau x_j$ for some $\tau < j$. However, this implies $\delta'_j = \gamma_\tau = \alpha'_j$. Thus, $\mathbf{exp}\, v_1 < \mathbf{exp}\, u'$ at the place k.

3. Finally, let u and u' be words of type 1. The pair $(u_0, v_0) \in$ $\in \theta \vee \theta_{u,v}$ can be constructed exactly as in the previous case. Let again $\mathbf{exp}\, u_0 = (\gamma_1, \ldots, \gamma_{\zeta'})$, and put $w \equiv x_1^{\alpha'_1 - \gamma_1} \ldots x_{\zeta'}^{\alpha'_{\zeta'} - \gamma_{\zeta'}}$. Since $u_0, v_0 \in V_1$, we can apply (5.1) to the pair $(u_0 w, v_0 w)$, and transform $u_0 w$ to u', and $v_0 w$ to some standard word v_1 of type 1, where $\mathbf{sk}\, v_1 = \mathbf{sk}\, v_0$, $\mathbf{low}\, v_1 = \mathbf{low}\, v_0$, $\mathbf{exp}\, v_1 = \mathbf{exp}\, v_0 + w^\natural$. Similarly, $\mathbf{sk}\, u' =$ $= \mathbf{sk}\, u_0$, $\mathbf{low}\, u' = \mathbf{low}\, u_0$, $\mathbf{exp}\, u' = \mathbf{exp}\, u_0 + w^\natural$. Hence $v_0 < u_0$ implies $v_1 < u'$. This completes the proof of the lemma.

Theorem 5.1 follows now easily from Proposition 2.1. It generalizes Theorem 2 of [11]. Note that, by the way, we have also proved

Theorem 5.2. *The variety defined by* (1), (4.7), *and*

(5.3) $\qquad x^p = x^{p+q}$

is h.f.b.

Indeed, substitute $t = p + q$ and reason as above. We needed (5.1) only for manipulating words of type 1; however, owing to (5.3), they are missing here, since no exponents t will occur. This result is analogous with Perkins' well-known result [6].

From the proof of Theorem 1 one can conclude also:

Theorem 5.3. *Every variety defined by a normal but non-balanced identity* (4.1) *and another one of the form* (4.7) *is h.f.b.*

Indeed, we made use of (4.9) only in proving Lemma 5, however, the same goal can be attained by the help of (4.8), which follows from (5) according to Lemma 4.2.

6. PROOF OF PROPOSITION B

Proposition B follows from

Theorem 6.1. *A variety* $\mathfrak{V}$, *defined by* (1), (2) *and* (4.11) *is h.f.b.*

Proof. It runs analogously to the proof of Theorem 5.1. Therefore we confine ourselves to the basic constructions and lemmata. Let us start with analysing the identities of $\mathfrak{V}$. In virtue of Lemma 4.2, we can suppose that also (4.9) is fulfilled in $\mathfrak{V}$ and, in view of Theorem 5.1, it suffices to consider the case when σ is identical. Then (if we choose r to be sufficiently large), one of the identities

(6.1) $$x_1 \dots x_{r-1} y x_r y x_{r+1} \dots x_{2r} = x_1 \dots x_{r-1} y^2 x_r \dots x_{2r},$$

(6.2) $$x_1 \dots x_{r-1} y x_r y x_{r+1} \dots x_{2r} = x_1 \dots x_r y^2 x_{r+1} \dots x_{2r}$$

is fulfilled. Indeed, if

$$x_1 \dots x_{s-1} y x_s y x_{s+1} \dots x_{2s} =$$
$$= x_1 \dots x_{s-l-1} y x_{s-l} y x_{s-l+1} \dots x_{2s}$$

then, with the obvious substitutions,

$$x_1 \dots x_{s+l-1} y x_{s+l} y x_{s+l+1} \dots x_{2(s+l)} =$$
$$= x_1 \dots x_{s-1} y^2 x_s \dots x_{2(s+l)} =$$
$$= x_1 \dots x_{s+2l-1} y_{s+2l} y x_{s+2l+1} \dots x_{2(s+l)} =$$
$$= x_1 \dots x_{s+l-1} y^2 x_{s+l} \dots x_{2(s+l)}.$$

Next we show that in $\mathfrak{V}$ the identity

(6.3) $$x_1 \dots x_r y^k z x_{r+1} \dots x_{2r} = x_1 \dots x_r z y^k x_{r+1} \dots x_{2r}$$

holds. For $k = 2$ this is (4.11). Suppose it is fulfilled for $k-1$, and let e.g. (6.2) be fulfilled. Substitute $x_{r-1} \to x_{r-1} x_r$, $x_r \to z y^{k-2}$ in (6.2). This yields (6.3) as follows:

$$x_1 \dots x_r y^k z x_{r+1} \dots x_{2r} = x_1 \dots x_r y z y^{k-1} x_{r+1} \dots x_{2r} =$$
$$= x_1 \dots x_r z y^k x_{r+1} \dots x_{2r}.$$

We can suppose $p \geqslant r$, $q \geqslant r$. Putting again $p + q = t$, (2) and (6.3) imply

$$(6.4) \qquad y^t z_1 z_2^k = y^p z_1 z_2^k y^q = y^p z_2^k z_1 y^q = y^t z_2^k z_1$$

for every $k \geqslant 2$.

Call $u \in F$ a *standard word of type* i if the respective one of the following three conditions is fulfilled:

(1) $u \equiv a x_{\epsilon(1)} \ldots x_{\epsilon(c)} x_1^{\alpha_1} \ldots x_\kappa^{\alpha_\kappa} \ldots x_\zeta^{\alpha_\zeta} a'$, where $|a| = |a'| = r$, $0 \leqslant \alpha_1, \ldots, \alpha_\zeta \neq 1$, $\alpha_\kappa \geqslant t$, $|x_{\epsilon(1)} \ldots x_{\epsilon(c)} x_1^{\alpha_1} \ldots x_\zeta^{\alpha_\zeta}|_{\epsilon(j)} = 1$ for $j = 1, \ldots, c$;

(2) $u \equiv a x_{\epsilon(1)} \ldots x_{\epsilon(c)} x_1^{\alpha_1} \ldots x_\zeta^{\alpha_\zeta} a'$, where $0 \leqslant \alpha_1, \ldots, \alpha_\zeta < t$, $\alpha_1, \ldots, \alpha_\zeta \neq 1$, $a, a', \epsilon(j)$ as in (1);

(3) $u = aea'$, $|a| = |a'| = r$, e simple.

Denote by V the set of standard words, by V_i the set of standard words of type i. Analogously to Lemma 5.1, one can prove

Lemma 6.1. *If the intersection of some θ-class T with F^{2r+1} is non-empty, then $T \cap V \neq \phi$, too.*

Note that here the case $|a| < r$ does not occur because of the absence of the "buffer" $x_\lambda^2 b$ (of course, $c = 0$ is admitted).

Define $\mathbf{sk}\, u = (aa')^\natural$, $\mathbf{low}\, u$, $\mathbf{exp}\, u$, $i(u)$, qu u, and the relation $<$ as in Section 5. As there, Lemma 5.2 holds, too.

We define the vectors $\mathbf{a}(u, v)$ and $\mathbf{w}(u, v)$ in the same way as in Section 5. The quasi-order relation $\preceq$ also can be defined analogously. We have to remark only, that part of (ii) (concerning the **sk** vectors) is here fulfilled automatically, so we might as well drop it. Now from Lemma 3.1 follows Lemma 5.3 for this case.

Finally, also Lemma 5.4 holds. The proof is the same as in Section 5 with the only modification, that now (4.11) and (6.3) are to be used instead of (4.7) and (5.1).

Theorem 6.1 follows now from the lemmas and Proposition 2.1. As in Section 5, we have the by-products

Theorem 6.2. *The variety defined by* (2), (4.11) *and* (5.3) *is h.f.b.*

Theorem 6.3. *Every variety defined by a normal but non-balanced identity* (5) *and another one of the form* (4.11) *is h.f.b.*

7. PROOF OF PROPOSITION C

Denote by $\mathfrak{B}$ the variety defined by (4). The word $u \in F$ will be said to be standard if $u \equiv x_{\lambda(1)} \ldots x_{\lambda(n-1)} x_{\xi(1)}^{\alpha_1} \ldots x_{\xi(c)}^{\alpha_c}$ $(\xi(i) \neq \xi(j)$ if $i \neq j)$. (Here n is the parameter of (4).) As before, the set of standard words will be denoted by V. We obviously have

Lemma 7.1. *If the intersection of some* θ*-class* T *with* F^n *is non-empty, then* $T \cap V \neq \phi$ *holds, too.*

Put $\mathbf{sk}\, u = (\lambda(1), \ldots, \lambda(n))$, $\mathbf{supp}\, u = (\xi(1), \ldots, \xi(c))$, $\mathbf{exp}\, u = (\alpha_1, \ldots, \alpha_c)$ and $\mathrm{tr}\, u = \langle \mathbf{sk}\, u, \mathbf{supp}\, u, \mathbf{exp}\, u \rangle$ which plays here the same role as $\mathrm{qu}\, u$ in both preceding sections, i.e. we put $v < u$ if $\mathrm{tr}\, v < \mathrm{tr}\, u$ in the lexicographical order. Obviously, Lemma 5.2 holds for this definition.

Associate with every pair $(u, v) \in V$, $q + 2$ vectors as follows. Put $\mathbf{a} = \mathbf{a}(u, v) = (\mathbf{sk}\, u, \mathbf{sk}\, v, \gamma, \delta)$ where $\gamma = \gamma(u, v) = \xi(i)$, $\delta = \delta(u, v) = \eta(i)$ if $\mathbf{sk}\, u = \mathbf{sk}\, v$, $\mathbf{supp}\, u > \mathbf{supp}\, v$ at the i'th place, and $\xi(i)$, $\eta(i)$ are the i'th components of $\mathbf{supp}\, u$ and $\mathbf{supp}\, v$, respectively; in all other cases put $\gamma = \delta = \omega$. Furthermore, define $\mathbf{w} = \mathbf{w}(u, v) = (\mathbf{bl}_{\mathbf{a}} (\mathbf{supp}\, u)^{\#}; \mathbf{exp}\, u)$, and $\mathbf{b}_j = \mathbf{b}_j(u, v) = (\alpha_{i(j-1)+1}, \ldots, \alpha_{i(j)-1})$, where $b_j = x_{\xi(i(j-1)+1)} \ldots \ldots x_{\xi(i(j)-1)}$ is the j'th block of $(\mathbf{supp}\, u)^{\#}$ with respect to $\mathbf{a}$. Obviously, $\mathbf{w} \in \overline{W}_l^1$ for some $l \leqslant 3(2n + 2)$.

Now let $(u, v) \preceq (u', v')$ iff $\mathbf{w}$ and $\mathbf{w}' = \mathbf{w}(u', v')$ belong to the same $\overline{W}_l^1$ (i.e. have the same number of blocks q), and there $\mathbf{w} \mathrel{\_} \mathbf{w}'$, furthermore, $\mathbf{b}_j \lhd \mathbf{b}_j' = \mathbf{b}_j(u', v')$ (see Section 3) for $j = 1, \ldots, q$. As the direct product of w.q.o. sets is w.q.o., Lemmas 3.1 and 3.2 imply Lemma 5.3 for our case.

The only thing that remains to be proved is Lemma 5.4. For this sake, put $(u,v) \preceq (u',v')$, and let $\varphi \in \Phi$ be the mapping which guarantees $w \preceq w'$, $\hat{\varphi}$ the induced endomorphism of F. Then $u\hat{\varphi} = x_{\lambda'(1)} \ldots x_{\lambda'(n)} x^{\alpha_1}_{\varphi(\xi(1))} \ldots x^{\alpha_c}_{\varphi(\xi(c))}$. Let ψ_j denote the surjection $\psi_j\colon \{1,\ldots,l'_j\} \to \{1,\ldots,l_j\}$ which figures in the definition of $\mathbf{b}_j \lhd \mathbf{b}'_j$ (here l_j and l'_j are the lengths of the j'th blocks of $(\mathbf{supp}\, u)^{\#}$ and $(\mathbf{supp}\, u')^{\#}$, resp.), and put for simplicity $\mathbf{b}_j = (\alpha_{j1},\ldots,\alpha_{jl_j})$, $\mathbf{b}'_j = (\alpha'_{j1},\ldots,\alpha'_{jl'_j})$, $b_j = x_{\xi(j,1)} \ldots x_{\xi(j,l_j)}$, $b'_j = x_{\xi'(j,1)} \ldots x_{\xi'(j,l'_j)}$. Define $\hat{\psi}_j \in \mathrm{End}\, F$ by

$$x_\tau \hat{\psi}_j = \begin{cases} x_{\xi'(j,k+1)} \ldots x_{\xi'(j,k+r)}, & \text{if } \tau = \varphi(\xi(j,i)), \\ & i\psi_j^{[-1]} = \{k+1,\ldots,k+r\}, \\ x_\tau & \text{else,} \end{cases}$$

where $i\psi_j^{[-1]}$ is the inverse image of i. Put $\hat{\psi} = \hat{\psi}_1 \ldots \hat{\psi}_q$. Then

$$u\hat{\varphi}\hat{\psi}\ \theta\ x_{\lambda'(1)} \ldots x_{\lambda'(n)} x^{\beta_1}_{\xi'(1)} \ldots x^{\beta_{c'}}_{\xi'(c')} \equiv u_1,$$

where $\beta_k = \alpha_k$ if $\xi(k) \in \bar{a}$ and $\beta_k = \alpha_{\bar{k}}$ if $x_{\xi(\bar{k})}$ belongs to the j'th block and $x_{\xi'(k)} \in X(x_{\xi(\bar{k})}\hat{\varphi}\hat{\psi}_j)$. By the definition of ψ_j, we have $\beta_k \leqslant \alpha'_k$ for all k. Furthermore, $v_1 < u_1$ for the standard word $v_1\ \theta\ v\hat{\varphi}\hat{\psi}$: this can be seen as in Section 5 if $\mathbf{sk}\, v < \mathbf{sk}\, u$ or $\mathbf{supp}\, v < \mathbf{supp}\, u$; if $\mathbf{sk}\, v = \mathbf{sk}\, u$, $\mathbf{supp}\, v = \mathbf{supp}\, u$, and $\mathbf{exp}\, v < \mathbf{exp}\, u$ at the i'th place, then $\mathbf{sk}\, v_1 = \mathbf{sk}\, u_1$, $\mathbf{supp}\, v_1 = \mathbf{supp}\, u_1$, and

$$u\hat{\varphi}\hat{\psi} \equiv (x_{\lambda(1)} \ldots x_{\lambda(n)} x^{\alpha_1}_{\xi(1)} \ldots x^{\alpha_{i-1}}_{\xi(i-1)})\hat{\varphi}\hat{\psi} \cdot (x_{\xi(i)}\hat{\varphi}\hat{\psi})^{\alpha_i} \cdot u_0,$$

$$v\hat{\varphi}\hat{\psi} \equiv (x_{\lambda(1)} \ldots x_{\lambda(n)} x^{\alpha_1}_{\xi(1)} \ldots x^{\alpha_{i-1}}_{\xi(i-1)})\hat{\varphi}\hat{\psi} \cdot (x_{\xi(i)}\hat{\varphi}\hat{\psi})^{\gamma_i} \cdot v_0$$

for some $u_0, v_0 \in F$ and $\gamma_i < \alpha_i$. Finally, put $\omega_j = \alpha'_j - \beta_j$, $w \equiv x^{\omega_1}_{\xi'(1)} \ldots x^{\omega_{c'}}_{\xi'(c')}$, and denote by v'' the "standard form" of $v_1 w$. In virtue of (4) we have $u_1 w\ \theta\ u'$ and $v'' < u'$, q.e.d.

8. PROOF OF PROPOSITION D

$\mathfrak{B}$ will denote the variety defined by (5). First we consider the case where (5) is normal; we shall write it in the form (4.1) or

$$(8.1)\qquad \begin{aligned} &x_1 \ldots x_{k-1}x_kx_{k+1}x_kx_{k+2} \ldots x_n = \\ &= x_{1\sigma} \ldots x_{(l-1)\sigma}x_{l\sigma}^2x_{(l+1)\sigma} \ldots x_{n\sigma}. \end{aligned}$$

We start with

Lemma 8.1. *If* (4.2) *and a non-balanced identity* (8.1) *hold in* $\mathfrak{B}$, *then either Lemma* 5.1 *or the duals of both* (4.2) *and Lemma* 5.1 *hold, too.*

Proof. In the proof of Lemma 5.1 we needed only (4.7) (which is equivalent to (4.2)) and (4.9). Therefore Lemma 5.1 certainly holds for those identities (8.1) which imply (4.9). From the proof of Lemma 4.2 one can see that (4.9) holds if either $l \notin \{k+1, k+2\}$ or $l = h\sigma < l\sigma$ for some $h < l$. These conditions are fulfilled for all identities of the form (8.1) but

$$(8.2)\qquad \begin{aligned} &x_1 \ldots x_{k-1}x_kx_{k+1}x_kx_{k+2} \ldots x_n = \\ &= x_{1\sigma} \ldots x_{k\sigma}x_{(k+1)\sigma}^2x_{(k+2)\sigma} \ldots x_{n\sigma}, \\ &\{1\sigma, \ldots, k\sigma\} = \{1, \ldots, k\}; \end{aligned}$$

$$(8.3)\qquad \begin{aligned} &x_1 \ldots x_{k-1}x_kx_{k+1}x_kx_{k+2} \ldots x_n = \\ &= x_{1\sigma} \ldots x_{(k+1)\sigma}x_{(k+2)\sigma}^2 \ldots x_{n\sigma}, \\ &\{1\sigma, \ldots, (k+1)\sigma\} = \{1, \ldots, k+1\}. \end{aligned}$$

The first part of the proof of Lemma 4.1 shows that (8.2) implies the dual of (4.2) whenever ($1\sigma \neq 1$ or) $(i+1)\sigma \neq i\sigma + 1$ for some $i < k - 1$; however, this is equivalent to $j\sigma \neq j$ for some $j \leqslant k$. On the other hand, (8.2) implies the dual of (4.9). Hence, in this case the dual of Lemma 5.1 holds. The same argument goes through for (8.3) except if $j\sigma = j$ for $j \leqslant k - 1$. Thus, the problem is reduced to the identities

$$(8.4)\qquad \begin{aligned} &x_1 \ldots x_kx_{k+1}x_k \ldots x_n = \\ &= x_1 \ldots x_kx_{(k+1)\sigma}^2x_{(k+2)\sigma} \ldots x_{n\sigma}, \end{aligned}$$

$$(8.5)\qquad \begin{aligned} &x_1 \ldots x_kx_{k+1}x_k \ldots x_n = \\ &= x_1 \ldots x_{k+1}x_{(k+2)\sigma}^2x_{(k+3)\sigma} \ldots x_{n\sigma}, \end{aligned}$$

$$(8.6)\quad x_1 \ldots x_k x_{k+1} x_k \ldots x_n = \\ = x_1 \ldots x_{k-1} x_{k+1} x_k x^2_{(k+2)\sigma} x_{(k+3)\sigma} \ldots x_{n\sigma}.$$

Now (4.9) can be deduced from (8.5), too. Indeed, we have

$$\begin{aligned} & x_1 \ldots x_{k+1} y_1 y_2 (y_3 y_4)^2 x_{k+2} \ldots x_n = \\ & = x_1 \ldots x_{k+1} y_1 y_2 y_3 y_4 y_3 x^2_{(k+2)\sigma} x_{(k+3)\sigma} \ldots x_{n\sigma} = \\ & = x_1 \ldots x_{k+1} y_1 \ldots y_4 x^2_{(k+2)\sigma^2} x_{(k+3)\sigma^2} \ldots \\ & \qquad \ldots x^2_{(k+2)\sigma} \ldots x_{n\sigma} = \\ & = x_1 \ldots x_{k+1} y_1 \ldots y_4 y_1 x^2_{(k+2)\sigma} x_{(k+3)\sigma} \ldots x_{n\sigma} = \\ & = x_1 \ldots x_{k+1} (y_1 \ldots y_4)^2 x_{k+2} \ldots x_n = \\ & = x_1 \ldots x_k x_{k+1} x_k z_{k+2} \ldots z_{n+4} = \\ & = x_1 \ldots x_{k+1} z^2_{(k+2)\sigma} z_{(k+3)\sigma} \ldots z_{n\sigma} z_{n+1} \ldots z_{n+4}, \end{aligned}$$

where $\{z_{k+2}, \ldots, z_{n+4}\} = \{x_{k+2}, \ldots, x_n, y_1, \ldots, y_4\}$. Hence, if $r \geqslant \max(k+2, n-k+4)$,

$$\begin{aligned} & x_1 \ldots x_r x_{r+1} x_r \ldots x_{2r} = \\ & = x_1 \ldots x_{r+1} (x_{(r+2)\tau} x_{(r+2)\tau+1})^2 x_{(r+4)\tau} \ldots x_{(2r)\tau} = \\ & = x_1 \ldots x_{r-1} x^2_{r\pi} x_{(r+1)\pi} \ldots x_{(2r)\pi} \end{aligned}$$

for some permutations τ and π.

For (8.6) we prove Lemma 5.1 directly. By means of (8.6) and (4.7), every word of F^{3r+1} which is not a standard word of type 3 can be transformed into a word $u \equiv p_1 x_{\epsilon(1)} \ldots x_{\epsilon(c)} x^2_\mu p_2 x_1^{\alpha_1} \ldots x_d^{\alpha_d} p_3$, where $|p_1| = |p_2| = |p_3| = r$ and $\epsilon(j) \neq \epsilon(j+1)$. If u is not standard then $x_{\epsilon(i)} \in X(x_{\epsilon(i+1)} \ldots x_{\epsilon(c)} x^2_\mu p_2 x_1^{\alpha_1} \ldots x_d^{\alpha_d})$ for some $i \leqslant c$. Put $u \equiv \equiv p_1 x_{\epsilon(1)} \ldots x_{\epsilon(i)} v x_{\epsilon(i)} w p_3$ where $x_{\epsilon(i)} \notin X(v)$ if either $\epsilon(i) = \epsilon(j)$ for some $i < j \leqslant c$ or $\epsilon(i) \neq \mu$, and $v = v' x_{\epsilon(i)}$, $x_{\epsilon(i)} \notin X(v')$ in the opposite case (i.e. $v' \equiv x_{\epsilon(i+1)} \ldots x_{\epsilon(c)}$). Substitute $x_k \to x_{\epsilon(i)}$, $x_{k+1} \to v$, $x_j \to x_{j'}$ (a single variable) if $j \notin \{k, k+1\}$ in (8.6) so as to obtain a subword of u on the left side, and apply the identity to u. We obtain

$u = u' \equiv p_1 x_{\epsilon(1)} \ldots x_{\epsilon(i-1)} v x_{\epsilon(i)} x^2_{[(k+2)\sigma]'} q$, and u' contains the square of a variable more to the left then u does. Thus, the argument of the proof of Lemma 5.1 goes through.

Similarly, using in case (8.4) the same notation as above, choose i to be the minimal index with $x_{\epsilon(i)} \in X(x_{\epsilon(i+1)} \ldots x_{\epsilon(c)} x^2_\mu p_2 x_1^{\alpha_1} \ldots x_d^{\alpha_d})$, and put $x_{\epsilon(i)} \notin X(v)$. The same substitution as in the foregoing case gives either $u = u' \equiv p_1 x_{\epsilon(1)} \ldots x_{\epsilon(i)} x^2_{[(k+1)\sigma]'} q$ (if $(k+1)\sigma \neq k+1$) or $u' \equiv p_1 x_{\epsilon(1)} \ldots x_{\epsilon(i)} v^2 q$. If in the first case $[(k+1)\sigma]' = \epsilon(i)$, the application of (4.2) transforms u' in a standard word of type 1 or 2; in the second case as well as in the first one if $[(k+1)\sigma]' \neq \epsilon(i)$, using (4.2), we obtain a word u'' of the same form as u, which contains $x_{\epsilon(i)}$ a less number of times than u does. Besides, as $x_{\epsilon(j)} \notin X(v)$ for $j < i$ by the choice of i, the number of occurrences of $x_{\epsilon(j)}$ can increase only if $(k+1)\sigma \neq k+1$, $[(k+1)\sigma]' = \epsilon(j)$; however, then the next step yields a standard word. Therefore, repeating the above process, we arrive in a finite number of iterations (either to a standard word or) to a word

$$u^{(n)} = p_1 x_{\epsilon(1)} \ldots x_{\epsilon(i)} x_{\eta(i+1)} \ldots x_{\eta(c')} x_\nu^2 q_2 x_1^{\beta_1} \ldots x_e^{\beta_e} q_3$$

of the same form as u, such that

$$x_{\epsilon(j)} \notin X(x_{\eta(i+1)} \ldots x_{\eta(c')} x_\nu^2 q_2 x_1^{\beta_1} \ldots x_e^{\beta_e} q_3)$$

for $1 \leqslant j \leqslant i$. As (8.6), (4.2) are normal, the proof can be accomplished by induction.

For normal identities, except for those of the forms (4.3), (4.4) and (4.6), Proposition D immediately follows from Lemma 8.1 and Lemma 4.1. As for the exceptional identities, next we prove

Lemma 8.2. *Each one of the identities* (4.3), (4.4), (4.6) *implies an identity*

(8.7) $$x_1 \ldots x_r x_{r+1} x_r \ldots x_{2r} = x_1 \ldots x_r x_{r+1}^2 x_{r+2} \ldots x_{2r}.$$

Proof. As (4.4) is dual to (4.3), it is sufficient to prove the assertion for this latter one and for (4.6). If $l < k-1$ in (4.3), we have for $\kappa - \lambda = k - l = d - 1$, $\lambda \geqslant l + d$, $\nu - \kappa \geqslant n - k$,

$$x_1 \ldots x_\lambda^2 \ldots x_\nu =$$

$$= x_1 \ldots (x_{\kappa-1-d} \ldots x_{\kappa-1}) x_\kappa x_{\kappa+1} x_\kappa \ldots x_\nu =$$

$$= x_1 \ldots x_{\lambda-d}^2 \ldots x_\nu = x_1 \ldots x_{\lambda-1} x_\lambda x_{\lambda-1} \ldots x_\nu,$$

where the word between brackets is that to be substituted for x_{k-1} in (4.3). This clearly entails (8.7). The same argument prevails if $l > k+1$. If $l = k-1$ and $\kappa \geqslant k+1$, $\nu - \kappa \geqslant n-k$, we have

$$x_1 \ldots x_{\kappa-1}^2 \ldots x_\nu = x_1 \ldots x_\kappa x_{\kappa+1} x_\kappa \ldots x_\nu =$$

$$= x_1 \ldots (x_{\kappa-2} x_{\kappa-1})^2 \ldots x_\nu \equiv$$

(8.8)

$$\equiv x_1 \ldots x_{\kappa-2} x_{\kappa-1} x_{\kappa-2} x_{\kappa-1} \ldots x_\nu =$$

$$= x_1 \ldots x_{\kappa-2}^2 x_{\kappa-1} x_{\kappa-2} \ldots x_\nu = x_1 \ldots x_{\kappa-2}^3 \ldots x_\nu.$$

Substituting in (8.8) $x_{\kappa-1} \to x_{\kappa-2}$, $x_\kappa \to x_{\kappa-1} x_\kappa$, $x_i \to x_i$ $(i \neq \kappa-1, \kappa)$, we obtain

$$x_1 \ldots x_{\kappa-2}^3 \ldots x_\nu = x_1 \ldots x_{\kappa-2}^4 \ldots x_\nu$$

and, consequently,

$$x_1 \ldots x_{\kappa-2}^3 \ldots x_\nu = x_1 \ldots x_{\kappa-2}^{3+c} \ldots x_\nu$$

for every $c \geqslant 0$. On the other hand, applying (8.8) and both latter identities under the condition $\kappa \geqslant k+2$,

$$x_1 \ldots x_{\kappa-1}^2 \ldots x_\nu = x_1 \ldots x_{\kappa-2}^3 \ldots x_\nu =$$

$$= x_1 \ldots x_{\kappa-3}^3 x_{\kappa-2}^2 \ldots x_\nu = x_1 \ldots x_{\kappa-3}^5 \ldots x_\nu =$$

$$= x_1 \ldots x_{\kappa-3}^3 \ldots x_\nu = x_1 \ldots x_{\kappa-2}^2 \ldots x_\nu.$$

This reduces our case to the one where $k - l > 1$. A similar (but simpler) argument helps us to deduce (8.7) if $l = k+1$.

If (4.6) holds, and, say, $l \leqslant k-1$, we have for the same κ, λ, ν, d as above:

$$x_1 \ldots x_\lambda^2 \ldots x_\nu =$$

$$= x_1 \ldots (x_{\lambda-2} x_{\lambda-1} x_{\lambda+1} \ldots x_{\kappa-1}) x_\kappa x_\lambda x_\kappa \ldots x_\nu =$$

$$= x_1 \dots x_{\lambda-1-d} x_\lambda^2 x_{\lambda-d} \dots x_{\lambda-1} x_{\lambda+1} \dots x_\nu =$$

$$= x_1 \dots x_{\lambda-1} x_\lambda x_{\lambda-1} \dots x_\nu,$$

which completes the proof of the lemma.

Now let (8.7) hold in $\mathfrak{B}$. Then, for $s \geqslant r + 1$,

$$x_1 \dots x_s x_{s+1}^3 x_{s+2} \dots x_{2s+1} = x_1 \dots (x_s x_{s+1})^2 \dots x_{2s+1} =$$

$$= x_1 \dots x_s x_{s+1} x_s^2 \dots x_{2s+1} =$$

$$= x_1 \dots x_s x_{s+1}^2 x_s \dots x_{2s+1} = x_1 \dots x_{s+1}^4 \dots x_{2s+1} =$$

(8.9)
$$= x_1 \dots x_s x_{s+1} x_s x_{s+1}^2 \dots x_{2s+1} =$$

$$= x_1 \dots (x_s x_{s+1})^2 x_s \dots x_{2s+1} =$$

$$= x_1 \dots x_{s-1} x_s x_{s+1} x_{s-1} x_s \dots x_{2s+1} =$$

$$= x_1 \dots x_s x_{s+1}^2 x_{s+2} \dots x_{2s+1}.$$

Denote by V the set of all words $u \in F$ of the form

$$u \equiv p x_{\xi(1)}^{\alpha_1} \dots x_{\xi(c)}^{\alpha_c} q, \quad |p| = |q| = r + 1,$$

$$\xi(i) \neq \xi(j) \text{ for } i \neq j, \ \alpha_i \in \{1, 2\}.$$

(8.7) and (8.9) imply that, if θ is the fully invariant congruence that corresponds to $\mathfrak{B}$, then for every θ-class T such that $T \cap F^{2r+3} \neq \phi$ also $T \cap V \neq \phi$. Put $p \equiv x_{\lambda(1)} \dots x_{\lambda(r+1)}$, $q \equiv x_{\lambda(r+2)} \dots x_{\lambda(2r+2)}$, and define **sk** u, **supp** u, **exp** u, tr u and the relation $<$ as in Section 7 (with $n = 2r + 2$). Clearly, $(F_X \cap V, <)$ is well-ordered for every finite X. To every pair $(u, v) \in V^*$, where

$$v \equiv x_{\mu(1)} \dots x_{\mu(r+1)} x_{\eta(1)}^{\beta_1} \dots x_{\eta(d)}^{\beta_d} x_{\mu(r+2)} \dots x_{\mu(2r+2)},$$

attach the block vector $\mathbf{bl}_{\mathbf{a}} (\mathbf{supp}\, u)^\# = \mathbf{w}(u, v) = \mathbf{w}$ of $x_{\xi(1)} \dots x_{\xi(c)}$ with respect to

$$\mathbf{a} = \mathbf{a}(u, v) = (\lambda(1), \dots, \lambda(2r + 2), \lambda, \mu(1), \dots, \mu(2r + 2), \mu);$$

here $\lambda = \xi(i)$, $\mu = \eta(i)$ if **sk** u = **sk** v and **supp** $u >$ **supp** v at the place i, $\lambda = \xi(i) = \eta(i)$, $\mu = \omega$ if **sk** u = **sk** v, **supp** u = **supp** v and **exp** $u >$ **exp** v

at the place i, and $\lambda = \mu = \omega$ else. The length of $\mathbf{w}$ is $\leqslant 3(4r + 6)$. Let $(u', v') \in V^*$, $\mathbf{w}, \mathbf{w}'\ (= \mathbf{w}(u', v')) \in \bar{W}_s^1$, $\mathbf{w} \preceq \mathbf{w}'$, and denote by $\varphi\ (\in \Phi)$ the monotone mapping that guarantees this inequality, and by $\hat{\varphi}$ the endomorphism it induces. Denote, furthermore, the parameters connected with (u', v') by the respective primes, and suppose $\alpha'_{\varphi(i)} = \alpha_i$ for $i \in \overline{\mathbf{a}(u, v)}$. Finally, put

$$u \equiv p b_0 x_{j(1)}^{\gamma_1} \ldots x_{j(l)}^{\gamma_l} b_l q, \quad u' \equiv p' b'_0 x_{j'(1)}^{\gamma'_1} \ldots x_{j'(l)}^{\gamma'_l} b'_l q',$$

where $l + 1$ is the block length of $\mathbf{supp}\, u$ (and $\mathbf{supp}\, u'$), $b_0, \ldots, b_l$, and $b'_0, \ldots, b'_l$ their blocks, and $\gamma_i = \alpha_{j(i)}$ (note that $j'(i) = \varphi(j(i))$!), and suppose $\mathbf{exp}\, b_i \lhd \mathbf{exp}\, b'_i$ for $i = 0, \ldots, l$ in the sense of $\mathbf{C}$ (cf. Section 3). Under these conditions (i.e. if $\mathbf{w} \preceq \mathbf{w}'$, $\alpha'_{\varphi(i)} = \alpha_i$ and $\mathbf{exp}\, b_i \lhd$ $\lhd\, \mathbf{exp}\, b'_i$) we put $(u, v) \preceq (u', v')$. Using Lemmas 3.1 and 3.2, it is easy to see that $(V^*, \preceq)$ is w.q.o.

The only thing to be checked is (Q). Denote by ψ_i the surjection ψ_i: $\{1, \ldots, |b'_i|\} \to \{1, \ldots, |b_i|\}$ that guarantees $\mathbf{exp}\, b_i < \mathbf{exp}\, b'_i$ and define the endomorphism $\hat{\psi}_i$ as follows. Let $b_i \equiv y_1^{\delta_1} \ldots y_k^{\delta_k}$, $b'_i \equiv z_1^{\xi_1} \ldots z_m^{\xi_m}$ (the y's and z's stand for variables), and put $\bar{g} =$ $= \{g' \mid g'\psi_i = g\}$,

$$\mu(g) = \begin{cases} \max\,(h \mid h\psi_i = g - 1), & \text{if } 1 \leqslant g \leqslant |b_i|, \\ 0 & \text{if } g = 0. \end{cases}$$

Put $z_0 = x_{j'(i)}$ (if $i = 0$, z_0 be the last letter of p'). Now set

$$x_\nu \hat{\psi}_i \equiv \begin{cases} \prod_{h \in \bar{g}} z_h^{\alpha'_h}, & \text{if } x_\nu \equiv y_g \text{ and } \delta_g = 1. \\ \prod_{h \in \bar{g}} (z_h z_{\mu(g)}), & \text{if } x_\nu = y_g \text{ and } \delta_g = 2, \\ x_\nu, & \text{if } x_\nu \notin X(b_i), \end{cases}$$

$$x_\nu \hat{\psi} = \begin{cases} x_\nu \hat{\varphi}, & \text{if } \nu \in \overline{\mathbf{a}(u, v)}, \\ x_\nu \hat{\psi}_i, & \text{if } x_\nu \in X(b_i), \\ x_\nu, & \text{if } x_\nu \notin X(u). \end{cases}$$

We claim that $u\hat{\psi} \equiv u'$. As

$$u\hat{\psi} \equiv p\hat{\varphi} \cdot b_0\hat{\psi}_0 \cdot x_{\varphi(j(1))}^{\alpha_{j(1)}} \dots x_{\varphi(j(l))}^{\alpha_{j(l)}} \cdot b_l\hat{\psi}_l \cdot q\hat{\varphi} \equiv$$

$$\equiv p' \cdot b_0\hat{\psi}_0 \cdot x_{j'(1)}^{\alpha_{j'(1)}} \cdot \dots \cdot x_{j'(l)}^{\alpha_{j'(l)}} \cdot b_l\hat{\psi}_l \cdot q',$$

it suffices to show that $w_1 z_0 \cdot b_i\hat{\psi}_i \cdot w_2 = w_1 z_0 b_i' w_2$ if $|w_1|, |w_2| \geqslant r$. But

$$b_i\hat{\psi}_i \equiv \bar{z}_1 \dots \bar{z}_k, \quad k = |b_i'|,$$

$$\bar{z}_j \equiv \begin{cases} z_j^{\alpha'_j}, & \text{if } j \in \bar{g},\ \delta_g = 1, \\ z_j z_{\mu(g)}, & \text{if } j \in \bar{g},\ \delta_g = 2. \end{cases}$$

By the definition of ψ_i, $\delta_g = 2$ implies $\zeta_j = 2$ for $j \in \bar{g}$. Thus, everything will be o.k. if we can show

$$v_1 x z_1 x z_2 \dots x z_t x v_2 = v_1 x z_1^2 \dots z_t^2 v_2$$

for $|v_1|, |v_2| \geqslant r$. However, this follows immediately from (8.7).

Let us show that $v\hat{\psi} \preceq u'$. If $\mathbf{sk}\, v < \mathbf{sk}\, u$, this follows from the monotonity of φ. If $\mathbf{sk}\, v = \mathbf{sk}\, u$ and, consequently, $\mathbf{sk}\, v\hat{\psi} = \mathbf{sk}\, u'$, and $\mathbf{supp}\, v < \mathbf{supp}\, u$ at the place ω, the same holds for $v\hat{\psi}$ and u'; if $\mathbf{supp}\, v < \mathbf{supp}\, u$ at the place i, then $\mu = \eta(i) < \xi(i) = \lambda < \omega$, whence $\mu' = \varphi(\mu) < \varphi(\lambda) = \lambda' < \omega$, and $u' \equiv w x_{\varphi(\lambda)} w'$, $u\hat{\psi} = w x_{\varphi(\mu)} w''$ for some $w, w', w'' \in F$. Similarly, if $\mathbf{sk}\, v = \mathbf{sk}\, u$, $\mathbf{supp}\, v = \mathbf{supp}\, u$, $\mathbf{exp}\, v < \mathbf{exp}\, u$ at the place i, then $\mathbf{sk}\, u' = \mathbf{sk}\, v\hat{\psi}$, and $\mathbf{supp}\, u' = \mathbf{supp}\, v\hat{\psi}$, because the order of the first occurrences of the letters is preserved by (8.7). Furthermore, $\lambda = \xi(i) = \eta(i)$, $\alpha_\lambda = 2$, $\beta_\lambda = 1$, whence $\varphi(\lambda) = \lambda' = \xi'(i') < \omega$, $x_\lambda \hat{\psi} = x_{\lambda'}$, and $\alpha'_{\lambda'} = 2$, so that $\mathbf{exp}\, v\hat{\psi} < \mathbf{exp}\, u'$ at the place i'. This completes the proof of Proposition D for normal identities.

Now we turn to the case of a non-normal identity (5).

Lemma 8.3. *Every non-normal identity* (5) *implies a normal identity of the same form.*

Proof. It is well known that the free semigroup $\bar{F}(X)$ of a non-normal variety $\mathfrak{V}$ contains a completely simple fully invariant ideal I consisting of those elements $\bar{u}$ which are images of some word u occurring in an identity $u = v$, $\mathfrak{V} \models u = v$. In our case this ideal is generated by the image $\bar{u}_0$ of $u_0 \equiv x_1 \ldots x_k x_{k+1} x_k \ldots x_n$ and is either a rectangular band of 2-groups – namely, if (5) is of the form

$$(8.10) \qquad x_1 \ldots x_{k-1} y x_k y x_{k+1} \ldots x_n = x_{1\sigma} \ldots x_{(l-1)\sigma} z^2 x_{l\sigma} \ldots x_{n\sigma},$$

σ a permutation – or a rectangular band (in all other cases). For (8.10) the assertion is obvious, so we can restrict ourselves to the other alternative.

Next suppose $j(1) \neq 1$. Then I is a right group. Let $x_{n+1}, \ldots, x_{n+d}$ be all variables which occur in $v_0 \equiv x_{j(1)} \ldots x_{j(l)}^2 \ldots x_{j(m)}$ but do not occur in u_0. As $\bar{u}\bar{v} = \bar{v}$ holds for $\bar{u}, \bar{v} \in I$, we have

$$v_0 = x_{n+1} \ldots x_{n+d} v_0^2 = x_{n+1} \ldots x_{n+d} u_0^2 =$$

$$= x_{n+1} \ldots x_{n+d} u_0$$

in $\mathfrak{V}$. Performing the same operations for u_0 and v_0 with the roles changed, we obtain a normal identity.

This conclusion can be carried over to all identities (5) satisfying $j(i) \neq i$ for some $i \leqslant \min(k, l)$ in virtue of the following obvious

Lemma 8.4. *If the identity* $u = v$ *implies* $u' = v'$ *and* $|u|_\lambda = |v|_\lambda = 0$, *then* $x_\lambda u = x_\lambda v$ *implies* $x_\lambda u' = x_\lambda v'$.

What has been proven up to now, together with its dual, yields the assertion for all identities but

$$(8.11) \qquad x_1 \ldots x_k x_{k+1} x_k \ldots x_n = x_1 \ldots x_k^2 \ldots x_n,$$

$$(8.12) \qquad \begin{aligned} &x_1 \ldots x_k x_{k+1} x_k \ldots x_n = \\ &= x_1 \ldots x_k x_{i(1)} \ldots x_{i(d)} x_{k+t}^2 \ldots x_n \qquad (t \geqslant 2), \end{aligned}$$

and the dual of (8.12). If in (5) x_k and $x_{j(l)}$ occur on both sides (as in (8.11) and (8.12)), some of the other variables must occur only on one of them; suppose it is x_h, $k+1 < h < n$ (the case where it is one of the $x_{j(i)}$'s is analogous). Then, substituting $x_h \to y_h \ldots y_{n+1}$, we have

$$x_1 \dots x_k x_{k+1} x_k \dots x_n =$$

$$(8.13) \qquad = x_1 \dots x_k x_{k+1} x_k \dots x_{h-1} y_h \dots y_{n+1} x_{h+1} \dots x_n =$$

$$= z_{j(1)} \dots z_{j(l)}^2 \dots z_{j(n)} y_{n+1} x_{h+1} \dots x_n ,$$

where

$$z_i = \begin{cases} y_i, & \text{if } h < j \leqslant n, \\ x_i & \text{else,} \end{cases}$$

and both sides of (8.13) contain variables at the first power which do not enter in the other side. Replacing these variables with appropriate (simple) words, we obtain a normal consequence of (8.12).

If x_{k+1} is the only letter in (5) which does not enter in the other side, then

$$x_1 \dots x_k x_{k+1} x_k \dots x_{n+3} =$$

$$(8.14) \qquad = x_1 \dots x_k x_{k+3} (x_k x_{k+2}) x_{k+3} \dots x_{n+3} =$$

$$= x_1 x_{t(1)} \dots x_{t(l)}^2 \dots x_{t(u)} ,$$

where

$$t(i) = \begin{cases} j(i) + 1, & \text{if } j(i) \leqslant k, \\ j(i) + 2, & \text{if } j(i) > k, \end{cases}$$

and x_{k+2} does not occur on the right side, so we reobtain the former case, and our lemma is proved.

If the normal consequence of (5) is non-balanced, we are done. If it is balanced, then (5) is either of the form (8.10) or $|u_0|_h = 1$, $|v_0|_h = 0$ (or the other way round) for some h. In virtue of (8.13) and (8.14), we can assume that $h = j(i)$, $l < i \leqslant m$. Performing the substitutions $x_{j(i)} \to yz$, $x_{j(i)} \to zy$, one obtains an identity of the form (4.2). Furthermore, (5) also implies some identity (5.3) (in our case with $q \leqslant 2$). Thus, we can apply Theorem 5.2 to find that $\mathfrak{B}$ is h.f.b.

Finally, for (8.10), if σ is non-trivial, we can argue as above in virtue of Lemma 4.1. However, if

$$x_1 \dots x_{k-1} y x_k y x_{k+1} \dots x_n = x_1 \dots x_{l-1} z^2 x_l \dots x_n$$

then, substituting $y \to x_{k+2}$, we obtain

$$x_1 \dots x_{k-1} y x_k y x_{k+1} \dots x_{n+2} =$$

$$= x_1 \dots x_{k-1} x_{k+2} x_k x_{k+2} x_{k+1} x_{k+2} \dots x_{n+2} =$$

$$= = x_1 \dots x_{l+1} z^2 \dots x_{k-1} x_{k+2} x_k x_{k+1} x_{k+3} \dots x_n ,$$

an identity with non-trivial σ, which completes the proof of Proposition D.

REFERENCES

[1] A.Ja. Aĭzenštat, On permutative identities, *Sovremennaja Algebra,* vyp. 3 (1975), 3–12 (in Russian).

[2] R.M. Bryant – M.R. Vaughan-Lee, Soluble varieties of Lie algebras, *Quart. J. Math.,* 23 (1972), 107–112.

[3] G. Higman, Ordering by divisibility in abstract algebras, *Proc. London Math. Soc.,* (3) 2 (1952), 326–336.

[4] E.S. Ljapin, On the inclusion of semigroup identities in irreducible systems, *Mat. Zametki,* 12 (1972), 95–104 (in Russian).

[5] I.I. Mel'nik, On varieties and lattices of varieties of semigroups, *Issledovanija po algebre,* vyp. 2, Saratov, 1970 (in Russian).

[6] P. Perkins, Bases for equational theories of semigroups, *J. Algebra,* 11 (1968), 298–314.

[7] G. Pollák, On hereditarily finitely based varieties of semigroups, *Acta Sci. Math.,* 37 (1975), 339–348.

[8] G. Pollák, On identities which define hereditarily finitely based varieties of semigroups, *Algebraic Theory of Semigroups,* Proc. Conf. Szeged, 1976, Coll. Math. Soc. János Bolyai, North-Holland, Amsterdam, vol. 20, 1979, 447–452..

[9] G. Pollák, On a class of hereditarily finitely based varieties of semigroups, *Algebraic Theory of Semigroups,* Proc. Conf. Szeged, 1976, Coll. Math. Soc. János Bolyai, North-Holland, Amsterdam, vol. 20, 1979, 433–445.

[10] G. Pollák, On two classes of hereditarily finitely based varieties of semigroups, *Semigroup Forum,* 25 (1982), 9–33.

[11] B.M. Vernikov – M.V. Volkov, Finite basis property for some varieties satisfying a permutative identity, *Issledovanie Algebraičeskih Sistem po Svoistvam Ih Podsistem,* Mat. Zapiski Ural Gos. Univ., Sverdlovsk, 12 (2) 1980, 3–23 (in Russian).

G. Pollák

Mathematical Research Institute of the Hungarian Academy, Szeged, Hungary.

M.V. Volkov

Ural State University, Sverdlovsk, USSR.

COLLOQUIA MATHEMATICA SOCIETATIS JÁNOS BOLYAI
39. SEMIGROUPS, SZEGED (HUNGARY), 1981.

A STRUCTURE THEOREM FOR COMBINATORIAL PSEUDO-INVERSE SEMIGROUPS

A.R. RAJAN

This paper is devoted to an exposition of some relations between semigroups and categories. It is shown that some standard constructions in categories give rise to general structure theorems for certain classes of semigroups. The structure theorem for combinatorial pseudo-inverse semigroups is obtained from some comma categories with adjunctions, satisfying certain conditions (cf. Theorem 3.5). Further, transformations of the above adjunctions provide homomorphisms of the corresponding semigroups. In Section 2, we construct pseudo-semilattices starting from partially ordered sets. Nambooripad [9] and Meakin and Pastijn [7] have given constructions of pseudo-semilattices, in terms of partially ordered sets and their ideals. Here it is shown that a pseudo-semilattice is a reflective subcategory of a preorder $I \times \Lambda$ where I and Λ are partially ordered sets. Structure theorems for locally testable regular semigroups, normal bands and combinatorial inverse semigroups are obtained as particular cases of the general theorem. In the last section we give two examples to illustrate the construction given here.

1. PRELIMIARIES

The notations and terminology used in this paper are as far as possible as in [4] and [5].

If $\mathscr{C}$ is a category we write $A \in \mathscr{C}$ to mean that A is an object of $\mathscr{C}$ and f in $\mathscr{C}$ to mean that f is a morphism in $\mathscr{C}$. 1_A will denote the identity morphism from A to A. Sometimes we write 1 to indicate identity morphisms. The composition of arrows is written as they appear in a commutative diagram. That is, fg means f acts first. If $\mathscr{C}$ is a category $\mathscr{C}^*$ will denote the dual category.

We denote by **Set** the category whose class of objects is the class of all small sets and morphisms are mappings between these sets. For any category $\mathscr{C}$ there exists a functor $H^\alpha: \mathscr{C} \to \mathbf{Set}$ for each $\alpha \in \mathscr{C}$, called the hom-functor corresponding to α. H^α is defined by

$$\beta \longmapsto [\alpha, \beta]_{\mathscr{C}} \quad \text{and} \quad \theta\colon \beta \to \gamma \longmapsto H^\alpha(\theta)\colon [\alpha, \beta]_{\mathscr{C}} \to [\alpha, \gamma]_{\mathscr{C}}$$

such that $uH^\alpha(\theta) = u\theta$ where $[\alpha, \beta]_{\mathscr{C}}$ denotes the morphism set of all morphism from α to β in $\mathscr{C}$. Further if $f\colon \beta \to \alpha$ in $\mathscr{C}$, then $H^f\colon H^\alpha \to H^\beta$ is a natural transformation whose components are $H^f_\gamma\colon [\alpha, \gamma] \to [\beta, \gamma]$ defined by $u \longmapsto fu$.

The following proposition characterises natural transformations to any functor from a hom-functor.

Proposition 1.1 (cf. [5], Yoneda Lemma, p. 61). *If $K\colon \mathscr{D} \to \mathbf{Set}$ is a functor and R is an object of $\mathscr{D}$ then there exists a bijection $Y\colon [H^R, K] \to K(R)$ where $[H^R, K]$ is the set of all natural transformations from H^R to K. Further Y is defined by $Y(\alpha) = (1_R)\alpha_R \in K(R)$ for each natural transformation $\alpha\colon H^R \to K$. Further the assignment $R \longmapsto H^R$ and $f \longmapsto H^f$ defines a faithful functor $Y\colon \mathscr{D}^* \to \mathbf{Set}^{\mathscr{D}}$.*

Y is called the Yoneda functor.

For any category $\mathscr{C}$, we denote by $\delta\colon \mathscr{C} \to \mathscr{C} \times \mathscr{C}$ the diagonal functor defined by $\alpha \longmapsto (\alpha, \alpha)$ and $\theta \longmapsto (\theta, \theta)$ for objects α and morphism θ. If $S, T\colon \mathscr{D} \to \mathbf{Set}$ are functors, $S * T$ will denote the product of S and T in the functor category $\mathbf{Set}^{\mathscr{D}}$.

A category $\mathscr{C}$ is called a preorder if for each pair $\alpha, \beta \in \mathscr{C}$, $[\alpha, \beta]_{\mathscr{C}}$ contains at most one element. If $[\alpha, \beta] \neq \square$ we say $\beta \leqslant \alpha$ in $\mathscr{C}$. If $\leqslant$ is a partial order we call $\mathscr{C}$ to be a partial order.

If $\langle F, G, \varphi \rangle$: $\mathscr{C} \to \mathscr{D}$ is an adjunction, then we denote by η the unit and ϵ the co-unit of the adjunction. For any $X \in \mathscr{C}$, $\eta_X : X \to FG(X)$ and for any $A \in \mathscr{D}$, $\epsilon_A : GF(A) \to A$ are universal arrows from X to G and F to A, respectively.

A subcategory $\mathscr{A}$ of $\mathscr{C}$ is said to be a reflective subcategory if the inclusion functor has a left adjoint. When $\mathscr{C}$ is a partial order this left adjoint is unique. Recall that a biordered set is a partial algebra $(E, \wedge)$ satisfying certain axioms (cf. [8]). $\mathscr{S}(e, f)$ denotes the sandwich set of e and f in a biordered set E.

Now we state some results on pseudo-semilattices and pseudo-inverse semigroups which are needed here.

Schein [11] defined a pseudo-semilattice to be a structure (E, ω^l, ω^r) where E is a set, and ω^l, ω^r are quasiorders (that is reflexive, transitive relations) on E such that for all $e, f \in E$, there exists a unique element $e \wedge f \in E$, with

(1.1) $\quad \omega^l(e) \cap \omega^r(f) = \omega(e \wedge f)$,

where $\omega = \omega^l \cap \omega^r$ is a partial order.

The following is an alternate characterization of pseudo-semilattices. Psudo-semilattices which can be embedded as the biordered sets of regular semigroups were called partially associative pseudo-semilattices by Nambooripad [9]. Partially associative pseudo-semilattices can be characterized as a variety of algebras as follows.

Proposition 1.2 (cf. [9]). *Let* $(E, \wedge)$ *be an algebra. Then* $(E, \wedge)$ *is both a pseudo-semilattice and a biordered set if and only if it satisfies the following identities and their duals.*

(a) $x \wedge x = x$,

(b) $(x \wedge y) \wedge (x \wedge z) = (x \wedge y) \wedge z$,

(c) $(x \wedge y) \wedge ((x \wedge z) \wedge (x \wedge u)) = ((x \wedge y) \wedge (x \wedge z)) \wedge (x \wedge u)$.

If $(E, \wedge)$ is a pseudo-semilattice as above, then ω^l and ω^r defined as follows are quasiorders such that $(E, \wedge) = (E, \omega^l, \omega^r)$.

(1.2) $\qquad e\, \omega^l f \Leftrightarrow f \wedge e = e \quad$ and $\quad e\, \omega^r f \Leftrightarrow e \wedge f = e.$

A useful alternate characterization for partially associative pseudo-semilattices is given by the following proposition.

Proposition 1.3 ([8], Theorem 7.6). *Let* $(E, \wedge)$ *be a biordered set. Then the following statements are equivalent.*

(i) E *is a pseudo-semilattice.*

(ii) *For all* $e \in E$, $\mathscr{S}(e,f)$ *contains exactly one element.*

(iii) *For all* $e \in E$ *the biordered subset* $\omega^r(e)$ *is right regular and* $\omega^l(e)$ *is left regular.*

(iv) *For all* $e \in E$, $\omega(e)$ *is a semilattice.*

(Note that a biordered set E is right [left] regular if and only if $\omega^l \subseteq \omega^r$ $[\omega^r \subseteq \omega^l]$.)

2. CONSTRUCTION OF PSEUDO-SEMILATTICES

Meakin and Pastijn [7] have given a construction of pseudo-semilattices starting from partially ordered sets. Another construction, using residuated subsets of partially ordered sets, is given by Nambooripad [9]. The present construction is similar to the one given by Nambooripad, in the sense that residuated subsets of partially ordered sets are just reflective subcategories.

Let I and Λ be partially ordered sets and Δ be a reflective subcategory of $I \times \Lambda$. Define ω^l and ω^r on Δ as follows.

(2.1) $\qquad (i, \lambda)\, \omega^r\, (j, \mu) \;\; [(i, \lambda)\, \omega^l\, (j, \mu)] \Leftrightarrow i \leqslant j \;\; [\lambda \leqslant \mu].$

Clearly ω^l and ω^r are quasiorders on Δ such that $\omega^l \cap \omega^r$ is a partial order. The next theorem gives a construction for pseudo-semilattices.

Theorem 2.1. *Let I and Λ be partially ordered sets and Δ be a reflective subcategory of $I \times \Lambda$. Then $(\Delta, \omega^l, \omega^r)$ where ω^l and ω^r are defined by* (2.1), *is a pseudo-semilattice.*

Proof. Since Δ is a reflective subcategory of the preorder $I \times \Lambda$ there exists a unique left adjoint $F\colon I \times \Lambda \to \Delta$ to the inclusion functor. For $e = (i, \lambda)$, $f = (j, \mu) \in \Delta$, define

$$(2.2) \qquad e \wedge f = F(j, \lambda).$$

Clearly $F(j, \lambda) \leqslant (j, \lambda)$ in $I \times \Lambda$ and if $(k, \nu) \in \Delta$ with $k \leqslant j$ and $\nu \leqslant \lambda$ in I and Λ respectively, then $(k, \nu) \leqslant F(j, \lambda)$. Thus $\omega^l(e) \cap \omega^r(f) = \omega(e \wedge f)$. Therefore $(\Delta, \omega^l, \omega^r)$ is a pseudo-semilattice.

Now we proceed to construct partially associative pseudo-semilattices. Let A be a graph with vertex set I. For each vertex $i \in A$, $A_{i*} = \bigcup_{j \in I} [i, j]_A$ of all edges with source i is called the star of i (cf. [3], p. 9). If $\theta\colon A \to B$ is a graph map, then $\theta_{i*}\colon A_{i*} \to B_{(i\theta)*}$ is also a graph map which we call the star of θ. θ is said to be star-bijective if θ_{i*} is a bijection for each vertex i.

If $P_I\colon I \times \Lambda \to I$ and $P_\Lambda\colon I \times \Lambda \to \Lambda$ are projections, then for any $E \subseteq I \times \Lambda$ we denote by $P_I\colon E \to I$ and $P_\Lambda\colon E \to \Lambda$ the composites $\subseteq \circ P_I$ and $\subseteq \circ P_\Lambda$ respectively.

Theorem 2.2. *Let I and Λ be partially ordered sets and Δ be a reflective subcategory of the preorder $I \times \Lambda$ such that*

(P) $P_I\colon \Delta \to I$ *and* $P_\Lambda\colon \Delta \to \Lambda$ *are star-bijections of the corresponding preorders.*

For each $e, f \in \Delta$, define

$$(2.3) \qquad e \wedge f = F(P_I f, P_\Lambda e)$$

where $F\colon I \times \Lambda \to \Delta$ is the left adjoint to the inclusion functor $\subseteq\colon \Delta \to I \times \Lambda$. Then (2.3) *defines a binary operation on Δ such that $(\Delta, \wedge)$ is a partially associative pseudo-semilattice. Conversely, every partially associative pseudo-semilattice can be constructed in this way.*

The following two lemmas will be useful in proving the theorem. In the following I, Λ, Δ etc. are as in Theorem 2.2.

Lemma 2.3. *If* $(i, \lambda), (i, \mu) \in \Delta$ *with* $\lambda \leqslant \mu$, *then* $\lambda = \mu$.

Proof. Since $P_I: \Delta \to I$ is a star-bijection, the restriction of P_I (denoted by P_I itself) to the principal ideal $[\leftarrow, (i, \mu)]_\Delta$ is an order isomorphism $P_I: [\leftarrow, (i, \mu)]_\Delta \to [\leftarrow, i]_I$. Therefore $(i, \lambda) \leqslant (i, \mu)$ implies $(i, \lambda) = (i, \mu)$. Hence $\lambda = \mu$.

Lemma 2.4. *If* $(i, \lambda), (j, \mu) \in \Delta$, $i \leqslant j$ *and* $\lambda, \mu \leqslant \nu$ *for some* $\nu \in \Lambda$, *then* $(i, \lambda) \leqslant (j, \mu)$.

Proof. Let $F(j, \nu) = (l, m)$. Then $l \leqslant j$ and $m \leqslant \nu$. Now $(j, \mu) \in \Delta$ and $\mu \leqslant \nu$, so that $(j, \mu) \leqslant F(j, \nu) = (l, m)$. Thus $j \leqslant l$ and so $j = l$. Further $\mu \leqslant m$. Now $F(j, \nu) = (j, m)$ and $(j, \mu), (j, m) \in \Delta$ with $\mu \leqslant m$. Therefore by Lemma 2.3 $\mu = m$. Thus $F(j, \nu) = (j, \mu)$. Also $(i, \lambda) \in \Delta$, with $i \leqslant j$, $\lambda \leqslant \nu$. Therefore $(i, \lambda) \leqslant F(j, \nu) = (j, \mu)$. Hence the lemma.

Proof of Theorem 2.2. We prove that the identities (a), (b) and (c) of Proposition 1.2 hold. Let $x = (i, \lambda)$, $y = (j, \mu)$, $z = (k, \nu)$ and $u = (l, m)$ be elements of Δ. (a) follows since for every $x \in \Delta$, $F(x) = x$.

(b) Let

$$x \wedge y = F(j, \lambda) = (j', \lambda')$$

$$x \wedge z = F(k, \lambda) = (k', \lambda'')$$

$$(x \wedge y) \wedge (x \wedge z) = F(k', \lambda') = (p, q)$$

and

$$x \wedge (y \wedge z) = F(k, \lambda') = (r, s).$$

Since $k' \leqslant k$ we have $F(k', \lambda') \leqslant F(k, \lambda')$ and so $(p, q) \leqslant (r, s)$. Also since $(r, s) = F(k, \lambda')$ we get $r \leqslant k$ and $s \leqslant \lambda' \leqslant \lambda$ and $(r, s) \in \Delta$. Therefore $(r, s) \leqslant (k, \lambda)$ in $I \times \Lambda$ so that $(r, s) \leqslant F(k, \lambda) = (k', \lambda'')$. Thus $r \leqslant k'$ and $s \leqslant \lambda'$. Therefore $(r, s) \leqslant F(k', \lambda') = (p, q)$ and so $(p, q) = (r, s)$. Hence (b).

(c) Let

$$x \wedge y = F(j, \lambda) = (j', \lambda')$$

$$x \wedge z = F(k, \lambda) = (k', \lambda'')$$

$$x \wedge u = F(l, \lambda) = (l', \lambda''')$$

$$(x \wedge z) \wedge (x \wedge u) = F(l', \lambda'') = (l'', \lambda_2)$$

$$(x \wedge y)((x \wedge z) \wedge (x \wedge u)) = F(l'', \lambda') = (p, q)$$

$$(x \wedge y) \wedge (x \wedge z) = F(k', \lambda') = (k'', \lambda_1)$$

and

$$((x \wedge y) \wedge (x \wedge z)) \wedge (x \wedge u) = F(l', \lambda_1) = (r, s).$$

We will prove that $(r, s) = (p, q)$.

Since $k'' \leqslant k' \leqslant k$ and $\lambda_1 \leqslant \lambda' \leqslant \lambda$ we get $(k'', \lambda_1) \leqslant (k, \lambda)$ so that $(k'', \lambda_1) \leqslant F(k, \lambda) = (k', \lambda'')$. Thus $\lambda_1 \leqslant \lambda''$. Now $r \leqslant l'$, $s \leqslant \lambda_1 \leqslant \lambda''$ and so $(r, s) \leqslant F(l'', \lambda'') = (l'', \lambda_2)$. Thus $r \leqslant l''$ and $s \leqslant \lambda'$ so that $(r, s) \leqslant F(l'', \lambda') = (p, q)$.

Now we proceed to prove the reverse inequality. Since $(p, q), (l'', \lambda_2) \in \Delta$, $p \leqslant l''$ and $q, \lambda_2 \leqslant \lambda$, it follows from Lemma 2.4 that $(p, q) \leqslant (l'', \lambda_2)$. Therefore $q \leqslant \lambda_2 \leqslant \lambda''$. Also $(k', \lambda'') \in \Delta$. Hence by condition (P) of the theorem there exists $p' \in I$ such that $(p', q) \leqslant$ $\leqslant (k', \lambda'')$ in Δ. Then $p' \leqslant k'$. Already $q \leqslant \lambda'$. Therefore $(p', q) \leqslant$ $\leqslant F(k', \lambda') = (k'', \lambda_1)$. Thus $q \leqslant \lambda_1$. Also $p \leqslant l'' \leqslant l'$. Hence $(p, q) \leqslant$ $\leqslant F(l', \lambda_1) = (r, s)$. Thus (c) holds.

Conversely let $(E, \wedge)$ be a partially associative pseudo-semilattice. Let ω^l, ω^r be quasiorders on E induced by the binary operation $\wedge$ as in (1.2). Let $\omega = \omega^r \cap \omega^l$, $R = \omega^r \cap (\omega^r)^{-1}$ and $L = \omega^l \cap (\omega^l)^{-1}$. Let $I = E/R$, $\Lambda = E/L$. Define $\leqslant$ on I and Λ as follows.

(2.4) $\quad R_e \leqslant R_f \Leftrightarrow e\, \omega^r f$ and $L_e \leqslant L_f \Leftrightarrow e\, \omega^l f$.

Clearly I and Λ are partially ordered by these relations. Let $\Delta =$ $= \{(R_e, L_e) \colon e \in E\}$. Then Δ is a partially ordered subset of $I \times \Lambda$ such that the mapping $\theta: E \to \Delta$ defined by $e \mapsto (R_e, L_e)$ is an order-isomorphism between the partially ordered sets (E, ω) and Δ. Because of this isomorphism we can treat E itself to be a partially ordered subset of $I \times \Lambda$ and θ may be treated as an inclusion functor. To prove that E is a reflective subcategory of the preorder $I \times \Lambda$ it is sufficient to produce a left adjoint to the functor θ. Define $F: I \times \Lambda \to E$ by $F(R_e, L_f) = f \wedge e$. We show that F is a left adjoint to θ. Let $g \in E$ with $g\theta = (R_g, L_g) \leqslant (R_e, L_f)$ in $I \times \Lambda$. Then $g \in \omega^l(e) \cap \omega^r(f) = \omega(f \wedge e)$. Therefore $g\, \omega\, f \wedge e$ in E. And if $g\, \omega\, F(R_e, L_f)$ in E, then $g\theta = (R_g, L_g) \leqslant (R_e, L_f)$ in $I \times \Lambda$. Hence F is a left adjoint to θ. To prove condition (P) of the theorem, observe that by Proposition 1.3(iii) if $e\,R\,f$ $[e\,L\,f]$ and $e, f \in \omega(g)$, then $e = f$. Further since E is a bi-ordered set if $R_e \leqslant R_f$ then there exists e' such that $R_e = R_{e'}$ and $e'\,\omega\, f$. From these it follows that (P) holds. This completes the proof of the theorem.

3. STRUCTURE OF COMBINATORIAL PSEUDO-INVERSE SEMIGROUPS

A regular semigroup S is said to be a pseudo-inverse semigroup if $\mathscr{S}(e,f)$ contains exactly one element for every $e, f \in E(S)$. A semigroup is said to be combinatorial if the subgroups are trivial. The following characterization of pseudo-inverse semigroups is due to Nambooripad [9].

Proposition 3.1. *Let S be a regular semigroup. Then the following are equivalent.*

(1) *S is pseudo-inverse.*

(2) *$E(S)$ is a pseudo-semilattice.*

(3) *For every $e \in E(S)$, eSe is an inverse semigroup.*

Now we introduce some notations and terminology regarding categories to be used in what follows. A category $\mathscr{D}$ is called strictly skeletal if the only isomorphism in $\mathscr{D}$ is the identity. A functor $P: \mathscr{D} \to \mathbf{Set}$ is said to be disjoint if $P(\alpha) \cap P(\beta) = \square$ for all $\alpha, \beta \in \mathscr{D}$, with $\alpha \neq \beta$. P is said to be non-empty if $P(\alpha) \neq \square$ for all $\alpha \in \mathscr{D}$. A functor $\Delta: \mathscr{D} \to \mathbf{Set}$ is a subdirect product of P_1 and $P_2: \mathscr{D} \to \mathbf{Set}$ if $\Delta(\alpha)$ is a subdirect product of $P_1(\alpha)$ and $P_2(\alpha)$ for all $\alpha \in \mathscr{D}$ and $\Delta(\theta) = P_1(\theta) \times P_2(\theta)|_{\Delta(\alpha)}$ for $\theta\colon \alpha \to \beta$ in $\mathscr{D}$.

Let $\mathscr{D}$ be a category and $K\colon \mathscr{D} \to$ **Set** be a functor. Let $Y\colon \mathscr{D}^* \to$ $\to \mathbf{Set}^{\mathscr{D}}$ be the Yoneda functor (cf. Proposition 1.1). Consider the comma category $(Y \downarrow K)$ (cf. [5], p. 46). From the definition of comma category objects of $(Y \downarrow K)$ are natural transformations $m\colon H^\alpha \to K$ for $\alpha \in \mathscr{D}$ and a morphism $\sigma\colon m \to m'$ for $m'\colon H^\beta \to K$ is a natural transformation $H^\theta\colon H^\alpha \to H^\beta$ for some $\theta\colon \beta \to \alpha$ in $\mathscr{D}$ ($\theta\colon \alpha \to \beta$ in $\mathscr{D}^*$) such that $m = H^\theta m'$. That is the following diagram is commutative.

$$\begin{array}{ccc} H^\alpha & \xrightarrow{\quad m \quad} & K \\ {\scriptstyle H^\theta}\big\downarrow & & {\scriptstyle m'} \\ H^\beta & & \end{array}$$

We use the notation $(\mathscr{D} \downarrow K)$ to denote the comma category $(Y \downarrow K)$ so as to specify the domain category explicitly.

Let $\mathscr{D}$ be a small category, $P_1, P_2\colon \mathscr{D} \to$ **Set** be functors and Δ be a subfunctor of $P_1 * P_2$ such that for each $\alpha \in \mathscr{D}$, $\Delta(\alpha)$ is a subdirect product of $P_1(\alpha)$ and $P_2(\alpha)$. Let $p_1\colon \Delta \to P_1$ and $p_2\colon \Delta \to P_2$ be natural transformations whose components are the corresponding projections. Now $(\mathscr{D} \downarrow \Delta)$, $(\mathscr{D} \downarrow P_1)$ and $(\mathscr{D} \downarrow P_2)$ are comma categories as described above. Let $\delta\colon (\mathscr{D} \downarrow \Delta) \to (\mathscr{D} \downarrow P_1) \times (\mathscr{D} \downarrow P_2)$ be defined by

$$m \longmapsto (mp_1, mp_2) \quad \text{and} \quad H^\theta \longmapsto (H^\theta, H^\theta).$$

Clearly δ is a functor.

Now we define a P-category, which provides the structure of combinatorial pseudo-inverse semigroups.

Definition 3.2. Let $\mathscr{D}$ be a strictly skeletal small category $P_1, P_2\colon$ $\mathscr{D} \to$ **Set** be functors which are non-empty and disjoint and Δ be a subdirect product of P_1 and P_2. We call $\mathbf{P} = \langle \mathscr{D}, P_1, P_2, \Delta \rangle$ a P-category if the following axioms hold.

I. $(\mathscr{D} \downarrow P_1)$ and $(\mathscr{D} \downarrow P_2)$ are preorders.

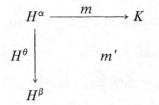

II. The functor $\delta\colon (\mathscr{D}\downarrow\Delta)\to(\mathscr{D}\downarrow P_1)\times(\mathscr{D}\downarrow P_2)$ has a right adjoint $\bar{\delta}$.

Remark. Let ϵ be the co-unit of the above adjunction from $(\mathscr{D}\downarrow\Delta)$ to $(\mathscr{D}\downarrow P_1)\times(\mathscr{D}\downarrow P_2)$. Therefore for $(x,y)\in(\mathscr{D}\downarrow P_1)\times(\mathscr{D}\downarrow P_2)$, $\epsilon_{(x,y)}\colon \delta(\bar{\delta}(x,y))\to(x,y)$ is a universal arrow from δ to (x,y). Let $\epsilon_{(x,y)}=(H^\theta,H^\varphi)$. We say (H^θ,H^φ) is universal for (x,y), to mean that $\epsilon_{(x,y)}=(H^\theta,H^\varphi)$.

If $K\colon\mathscr{D}\to\mathbf{Set}$ is a functor, define $\leqslant$ on $[\mathscr{D},K]=\mathrm{Obj}\,(\mathscr{D}\downarrow K)$ as follows.

$$m\leqslant n \Leftrightarrow \text{there exists a natural transformation } H^\theta : H^\alpha\to H^\beta \tag{3.1}$$
$$\text{such that } m=H^\theta n, \text{ where } m\colon H^\alpha\to K \text{ and } n\colon H^\beta\to K.$$

In what follows we assume that $\mathbf{P}=\langle\mathscr{D},P_1,P_2,\Delta\rangle$ is a P-category.

Lemma 3.3. *The sets* $[\mathscr{D},P_1]$, $[\mathscr{D},P_2]$ *and* $[\mathscr{D},\Delta]$ *are partially ordered by* (3.1).

Proof. Let $x,y\in[\mathscr{D},P_1]$ with $x\leqslant y$ and $y\leqslant x$. Then there exists θ,φ such that $x=H^\theta y$ and $y=H^\varphi x$. Therefore $x=H^\theta H^\varphi x=$ $=H^{\varphi\theta}x$. Now by axiom I, $(\mathscr{D}\downarrow P_1)$ is a preorder and so $H^{\varphi\theta}=1$. Hence $\varphi\theta=1$. Similarly $\theta\varphi=1$. Since $\mathscr{D}$ is strictly skeletal it follows that $\theta=\varphi=1$. Hence $x=y$. Therefore $[\mathscr{D},P_1]$ is a partially ordered set. Similarly the other cases may be proved.

Lemma 3.4. *The partial order* $[\mathscr{D},\Delta]$ *is a reflective subcategory of the preorder* $[\mathscr{D},P_1]\times[\mathscr{D},P_2]$ *and so* $[\mathscr{D},\Delta]$ *is a pseudo-semilattice.*

Proof. In view of Lemma 3.4, the preorders $(\mathscr{D}\downarrow P_1)$, $(\mathscr{D}\downarrow P_2)$ and $(\mathscr{D}\downarrow\Delta)$ are partial orders. Now axiom II implies that $(\mathscr{D}\downarrow\Delta)$ is a reflective subcategory of $(\mathscr{D}\downarrow P_1)\times(\mathscr{D}\downarrow P_2)$. Hence the lemma follows from Theorem 2.1.

Theorem 3.5. *Let* $\mathbf{P}=\langle\mathscr{D},P_1,P_2,\Delta\rangle$ *be a P-category. Let*

$$S=\mathbf{S}(\mathbf{P})=\bigcup_{\alpha\in\mathscr{D}}[H^\alpha,P_1]\times[H^\alpha,P_2]$$

where $[H^\alpha,P_i]$ *denotes the set of all natural transformations of* H^α

into P_i *for* $i = 1, 2$. *Define a binary operation on* S *by*

$$(3.2) \qquad (x, y)(u, v) = (H^{\theta} x, H^{\varphi} v)$$

where $(H^{\varphi}, H^{\theta})$ *is universal for* (u, y). *Then* S *with this product is a combinatorial pseudo-inverse semigroup whose set of idempotents is isomorphic to the pseudo-semilattice* $[\mathscr{D}, \Delta]$. *Conversely every combinatorial pseudo-inverse semigroup is isomorphic to one constructed in this way.*

Proof. We prove that eq. (3.2) defines an associative binary operation on S. Let $(x, y), (u, v), (w, z) \in S$ and let the domains of x, u, w be H^{α}, H^{β} and H^{τ} respectively. Let $\bar{\delta}(u, y) = r\colon H^{\gamma} \to \Delta$ and $(H^{\varphi}, H^{\theta})$ be universal for (u, y). Then $H^{\varphi} u = r_1$, $H^{\theta} y = r_2$ and

$$(x, y)(u, v) = (H^{\theta} x, H^{\varphi} v).$$

Now let $(H^{\varphi'}, H^{\theta'})$ be universal for $(w, H^{\varphi} v)$ with $\bar{\delta}(w, H^{\varphi} v) = s\colon H^{\eta} \to \Delta$. Then $H^{\varphi'} w = s_1$ and $H^{\theta'} H^{\varphi} v = s_2$, and

$$((x, y)(u, v))(w, z) = (H^{\theta} x, H^{\varphi} v)(w, z) = (H^{\theta'} H^{\theta} x, H^{\varphi'} z).$$

Also

$$(x, y)((u, v)(w, z)) = (x, y)(H^{\theta''} u, H^{\varphi''} z)$$

where $(H^{\varphi''}, H^{\theta''})$ is universal for (w, z). Let $\bar{\delta}(w, v) = r' = H^{\gamma'} \to \Delta$. Then $H^{\varphi''} w = r_1'$ and $H^{\theta''} v = r_2'$. Since $(H^{\varphi'}, H^{\theta'} H^{\varphi})\colon \delta(s) \to (w, v)$ is a morphism in $(\mathscr{D} \downarrow P_1) \times (\mathscr{D} \downarrow P_2)$, by the universality of $(H^{\varphi''}, H^{\theta''})$ there exists $\psi\colon \gamma' \to \eta$ in $\mathscr{D}$ such that $H^{\psi} H^{\varphi''} = H^{\varphi'}$ and $H^{\psi} H^{\theta''} = H^{\theta'} H^{\varphi}$. Now we prove that $(H^{\psi}, H^{\theta\theta'})\colon \delta(H^{\theta'} r) \to (H^{\theta''} u, y)$ is universal for $(H^{\theta''} u, y)$. First note that

$$H^{\theta'} r_1 = H^{\theta'} H^{\varphi} u = H^{\psi} H^{\theta''} u$$

and

$$H^{\theta'} r_2 = H^{\theta'} H^{\theta} y = H^{\theta\theta'} y.$$

Therefore $(H^{\psi}, H^{\theta\theta'})\colon \delta(H^{\theta'} r) \to (H^{\theta''} u, y)$ is a morphism in $(\mathscr{D} \downarrow P_1) \times (\mathscr{D} \downarrow P_2)$. Now let $(H^{\varphi'''}, H^{\theta'''})\colon \delta(s') \to (H^{\theta''} u, y)$ in $(\mathscr{D} \downarrow P_1) \times (\mathscr{D} \downarrow P_2)$ for some $s'\colon H^{\eta'} \to \Delta$ in $(\mathscr{D} \downarrow \Delta)$. Then $H^{\varphi'''} H^{\theta''} u = s_1'$ and

$H^{\theta'''}y = s_2'$. Since $(H^{\varphi}, H^{\theta})$ is universal for (u, y) and $(H^{\varphi'''}H^{\theta''}, H^{\theta'''})$: $\delta(s') \to (u, y)$ is a morphism, there exists ψ': $\gamma \to \eta'$ in $\mathscr{D}$ such that

$$H^{\theta'''} = H^{\psi'}H^{\theta} \quad \text{and} \quad H^{\varphi'''}H^{\theta''} = H^{\psi'}H^{\varphi}.$$

Further $H^{\varphi'''}r_1' = H^{\varphi'''}H^{\varphi''}w$ and $H^{\varphi'''}r_2' = H^{\varphi'''}H^{\theta''}v = H^{\psi'}H^{\varphi}v$. Thus $(H^{\varphi'''}H^{\varphi''}, H^{\psi'})$: $\delta(H^{\varphi'''}r') \to (w, H^{\varphi}v)$ is a morphism. Then by the universality of $(H^{\varphi'}, H^{\theta'})$ there exists ψ'': $\eta \to \eta'$ in $\mathscr{D}$ such that

$$H^{\varphi'''}H^{\varphi''} = H^{\psi''}H^{\varphi'} \quad \text{and} \quad H^{\psi'} = H^{\psi''}H^{\theta'}.$$

Then clearly $H^{\psi''}H^{\theta\theta'} = H^{\psi''}H^{\theta'}H^{\theta} = H^{\psi'}H^{\theta} = H^{\theta'''}$. Also

$$H^{\psi''}H^{\psi}r_1' = H^{\psi''}H^{\psi}H^{\varphi''}w = H^{\psi''}H^{\varphi'}w = H^{\varphi'''}H^{\varphi''}w = H^{\varphi'''}r_1'.$$

Therefore $H^{\psi''}H^{\psi} = H^{\psi\psi''}$, $H^{\varphi'''}$: $H^{\varphi'''}r_1' \to r_1'$, are morphisms in $(\mathscr{D} \downarrow P_1)$. Since $(\mathscr{D} \downarrow P_1)$ is a preorder, we get

$$H^{\psi''}H^{\psi} = H^{\varphi'''}.$$

Hence $(H^{\psi}, H^{\theta\theta'})$ is universal for the pair $(H^{\theta''}u, y)$. Therefore

$$(x, y)((u, v)(w, z)) = (x, y)(H^{\theta''}u, H^{\varphi''}z) = (H^{\theta\theta'}x, H^{\psi}H^{\varphi''}z) =$$

$$= (H^{\theta'}H^{\theta}x, H^{\varphi'}z) = ((x, y)(u, v))(w, z).$$

This proves that S is a semigroup.

Now we prove that $E(S) = \{\delta(m)\colon m \in (\mathscr{D} \downarrow \Delta)\}$. Let $(x, y) \in E(S)$. Then

$$(x, y)(x, y) = (H^{\theta}x, H^{\varphi}y) = (x, y)$$

where $(H^{\varphi}, H^{\theta})$ is universal for the pair (x, y). Since $(\mathscr{D} \downarrow P_i)$ for $i = 1, 2$ is a preorder $\theta = 1 = \varphi$. Let $\bar{\delta}(x, y) = r$. Then $r_1 = H^{\varphi}x = x$ and $r_2 = H^{\theta}y = y$. Therefore $(x, y) = \delta(r)$. It is clear that if $m \in (\mathscr{D} \downarrow \Delta)$ then $\delta(m)$ is an idempotent in S. Thus $E(S) = \{\delta(m)$: $m \in (\mathscr{D} \downarrow \Delta)\}$. Now it is easy to see that the quasiorders on $[\mathscr{D}, \Delta]$ are the same as the quasiorders ω^l and ω^r on the biordered set $E(S)$. Therefore by Lemma 3.4 we get that $E(S)$ is a pseudo-semilattice. Hence S is a pseudo-inverse semigroup. The fact that S is combinatorial will follow from the observation that

$$(x,y)\ \mathscr{R}\ (u,v) \Leftrightarrow x = u$$

and

$$(x,y)\ \mathscr{L}\ (u,v) \Leftrightarrow y = v.$$

Conversely let S be a combinatorial pseudo-inverse semigroup. Let $\varphi = \Phi(S)$ be the combinatorial functor corresponding to S (cf. [10]). Let $\mathscr{D} = \operatorname{im}\varphi$. Define $P_1, P_2\colon \mathscr{D} \to \mathbf{Set}$ by

$$P_1(\varphi(e)) = I_e, \quad P_2(\varphi(e)) = \Lambda_e$$

and

$$P_1(\varphi(e,f)) = \varphi_1(e,f), \quad P_2(\varphi(e,f)) = \varphi_2(e,f)$$

where $\varphi(e) = (I_e, \Lambda_e; \Delta_e)$ is the representation of the combinatorial Rees groupoid $\varphi(e)$. Define $\Delta(\varphi(e)) = \Delta_e$. Then $\mathbf{P} = \langle \mathscr{D}, P_1, P_2, \Delta\rangle$ satisfies the requirements of a P-category and $\mathbf{S}(\mathbf{P}) = S$. This completes the proof.

Now we proceed to describe the homomorphisms of combinatorial pseudo-inverse semigroups in terms of transformations of the corresponding P-categories. First of all we derive a result on the morphisms between the comma categories of the form $(\mathscr{D} \downarrow K)$.

The class of all pairs $(\mathscr{D}, K)$ where $\mathscr{D}$ is a small category and $K\colon \mathscr{D} \to \mathbf{Set}$ is a functor, forms a category with morphisms defined as follows. If $(\mathscr{D}, K)$ and $(\mathscr{D}', K')$ are two objects, a morphism from $(\mathscr{D}, K)$ to $(\mathscr{D}', K')$ is a pair (T, σ) where $T\colon \mathscr{D} \to \mathscr{D}'$ is a functor and $\sigma\colon K \to TK'$ is a natural transformation.

Lemma 3.6. *If* $(T,\sigma)\colon (\mathscr{D}, K) \to (\mathscr{D}', K')$ *is a morphism as described above, then define* $(T \downarrow \sigma)\colon (\mathscr{D} \downarrow K) \to (\mathscr{D}' \downarrow K')$ *as follows. For objects* $m\colon H^\alpha \to K$ *and morphisms* H^θ *in* $(\mathscr{D} \downarrow K)$,

$$(T \downarrow \sigma)(m) = m^* \quad \text{and} \quad (T \downarrow \sigma)(H^\theta) = H^{T(\theta)}$$

where $m^*\colon H^{T\alpha} \to K'$ *is defined by* $(1_{T\alpha})m^*_{T\alpha} = (1_\alpha)m_\alpha\sigma_\alpha$. *Then* $(T \downarrow \sigma)$ *is a functor. Further the assignments* $(\mathscr{D}, K) \mapsto (\mathscr{D} \downarrow K)$ *and* $(T,\sigma) \mapsto (T \downarrow \sigma)$ *are functorial.*

Proof. Let $m\colon H^\alpha \to K$ and $n\colon H^\beta \to K$ be objects in $(\mathscr{D} \downarrow K)$. Let $\theta\colon \beta \to \alpha$ in $\mathscr{D}$ be such that $m = H^\theta n$. Now we show that $H^{T\theta}\colon m^* \to n^*$ is a morphism in $(\mathscr{D}' \downarrow K')$. That is, $m^* = H^{T(\theta)} n^*$ where $m^* = (T \downarrow \sigma)(m)$ and $n^* = (T \downarrow \sigma)(n)$. By the definition of m^* and n^* we have

$$(1_{T\alpha})(H^{T(\theta)}n^*)_{T\alpha} = (T(\theta))n^*_{T\alpha} = (1_{T\beta})n_\beta\sigma_\beta K'(T(\theta)) =$$
$$= (1_\beta)n_\beta K(\theta)\sigma_\alpha = (\theta)n_\alpha\sigma_\alpha = (1_\alpha)(H^\theta n)_\alpha\sigma_\alpha =$$
$$= (1_\alpha)m_\alpha\sigma_\alpha = (1_{T\alpha})m^*_{T\alpha}.$$

Hence $H^{T(\theta)}\colon m^* \to n^*$ is a morphism in $(\mathscr{D}' \downarrow K')$. Since T is a functor it follows that $(T \downarrow \sigma)$ is a functor from $(\mathscr{D} \downarrow K)$ to $(\mathscr{D}' \downarrow K')$. Clearly the assignments $(\mathscr{D}, K) \mapsto (\mathscr{D} \downarrow K)$, $(T, \sigma) \mapsto (T \downarrow \sigma)$ are functorial.

Definition 3.7. A morphism of P-categories from $\mathbf{P} = \langle \mathscr{D}, P_1, P_2, \Delta \rangle$ to $\mathbf{P}' = \langle \mathscr{D}', P_1', P_2', \Delta' \rangle$ is a functor $(T \downarrow \sigma)\colon (\mathscr{D} \downarrow \Delta) \to (\mathscr{D}' \downarrow \Delta')$ satisfying the following condition:

(i) There exist natural transformations $\sigma^1\colon P_1 \to TP_1'$ and $\sigma^2\colon P_2 \to TP_2'$ such that the pair $((T \downarrow \sigma^1) \times (T \downarrow \sigma^2), (T \downarrow \sigma))$ is a transformation of adjunctions from $(\delta, \bar{\delta}, \varphi)$ to $(\delta', \bar{\delta}', \varphi')$ where $(\delta, \bar{\delta}, \varphi)$ and $(\delta', \bar{\delta}', \varphi')$ are the adjunctions corresponding to $\mathbf{P}$ and $\mathbf{P}'$ given by axiom II of Definition 3.2.

Theorem 3.8. *Let* $(T \downarrow \sigma)\colon (\mathscr{D} \downarrow \Delta) \to (\mathscr{D}' \downarrow \Delta')$ *be a morphism of P-categories. Define* $\mathbf{S}(T \downarrow \sigma)\colon S = \mathbf{S}(\mathbf{P}) \to S' = \mathbf{S}(\mathbf{P}')$ *by*

(3.3) $\qquad (x, y)\mathbf{S}(T, \sigma) = (x^*, y^*)$

where x^* *and* y^* *are as in Lemma* 3.6. *Then* $\mathbf{S}(T \downarrow \sigma)$ *is a homomorphism of combinatorial pseudo-inverse semigroups. Conversely if* $h\colon S \to S'$ *is a homomorphism of combinatorial pseudo-inverse semigroups, then there exists a unique morphism* $(T \downarrow \sigma)\colon \mathbf{P}(S) \to \mathbf{P}(S')$ *of P-categories such that* $\mathbf{S}(T \downarrow \sigma) = h$.

Proof. The fact that $\mathbf{S}(T \downarrow \sigma)$ is a homomorphism, follows from the observation that a transformation of adjunctions preserves the unit and co-unit. Conversely if $h\colon S \to S'$ is given, then by the equivalence of categories **GF** and **GRS** (cf. [10], Theorem 3.7) there exists (θ, τ): $\Phi(S) \to \Phi(S')$ such that $\Sigma(\theta, \tau) = h$. Define T: $\mathscr{D}_S \to \mathscr{D}_{S'}$ by $\varphi(e) \mapsto$ $\mapsto \varphi'(e\theta)$ and $\varphi(e, f) \mapsto \varphi'(e\theta, f\theta)$, where $\varphi = \Phi(S)$ and $\varphi' = \Phi(S')$. Further σ: $\Delta \to T\Delta'$ is defined by $\sigma_{\varphi(e)} = \tau e \mid \Delta(\varphi(e))$. Then $(T \downarrow \sigma)$: $\mathbf{P}(S) \to \mathbf{P}(S')$ is a morphism of P-categories with $\mathbf{S}(T \downarrow \sigma) = h$. Hence the theorem.

4. PARTICULAR CASES

Locally testable regular semigroups (cf. [12]) form an important subclass of the class of combinatorial pseudo-inverse semigroups. The following characterization of locally testable regular semigroups is useful.

Proposition 4.1 (cf. [9]). *Let S be a regular semigroup. Then the following are equivalent.*

(1) *S is locally testable.*

(2) *S is combinatorial and satisfies the following condition. If D and D' are two $\mathscr{D}$-classes of S, then for $x \in D$ there exists at most one $y \in D'$ such that $y \leqslant x$ in the natural partial order on S.*

(3) *For every $e \in E(S)$, $e\,S\,e$ is a semilattice.*

In view of this, Theorem 3.5 yields the following structure theorem for locally testable regular semigroups.

Theorem 4.2. *Let $\mathscr{D}$ be a small category which is a partial order and P_1, P_2, Δ: $\mathscr{D} \to$ **Set** be functors as in Definition 3.2, such that*

(L) *the functor δ: $(\mathscr{D} \downarrow \Delta) \to (\mathscr{D} \downarrow P_1) \times (\mathscr{D} \downarrow P_2)$ has a right adjoint $\bar{\delta}$.*

Then

$$S = \bigcup_{\alpha \in \mathscr{D}} [H^\alpha, P_1] \times [H^\alpha, P_2]$$

with product defined by eq. (3.2) is a locally testable regular semigroup. Conversely every locally testable regular semigroup is isomorphic to one constructed in this way.

Now we give a structure theorem for normal bands. A band B satisfying the permutation identity $efgh = egfh$ for every $e, f, g, h \in B$, is called a normal band.

Zalcstein has shown that normal bands are precisely those bands which are locally testable (cf. [12], Theorem 5). Therefore from Theorem 4.2, we get the following structure theorem for normal bands.

Theorem 4.3. *Let* $\mathscr{D}$ *be a semilattice and* $P_1, P_2 \colon \mathscr{D} \to$ **Set** *be functors which are disjoint and non-empty. Then*

$$S = \bigcup_{\alpha \in \mathscr{D}} [H^\alpha, P_1] \times [H^\alpha, P_2]$$

with product defined by eq. (3.2) *is a normal band. Conversely every normal band is isomorphic to one constructed in this way.*

Lastly we provide a structure theorem for combinatorial inverse semigroups. A preorder $\mathscr{C}$ is said to be a semilattice if $\mathscr{C}$ is a partial order and every pair of objects has product. Since in an inverse semigroup the set of $\mathscr{R}$-classes and the set of $\mathscr{L}$-classes are in one-one correspondence, and the set of idempotents is a semilattice, we obtain from Theorem 3.5 the following structure theorem for combinatorial inverse semigroups.

Theorem 4.4. *Let* $\mathscr{D}$ *be a strictly skeletal small category and* $P \colon \mathscr{D} \to$ **Set** *be a functor which is disjoint and non-empty such that*

(I) $(\mathscr{D} \downarrow P)$ *is a semilattice.*

Then on

$$S = \bigcup_{\alpha \in \mathscr{D}} [H^\alpha, P] \times [H^\alpha, P]$$

define a binary operation by

$$(4.1) \qquad (x, y)(u, v) = (H^\theta x, H^\varphi v)$$

where $H^\theta \colon u \times y \to y$ *and* $H^\varphi \colon u \times y \to u$ *are the projections. Then* S *with this binary operation, is a combinatorial inverse semigroup whose semilattice of idempotents is isomorphic to* $[\mathscr{D}, P]$. *Conversely every combinatorial inverse semigroup may be constructed in this way.*

5. EXAMPLES

Example 1. Here we construct the fundamental four-spiral semigroup SP_4. SP_4 is the fundamental regular idempotent generated (r.i.g.) semigroup on the four-spiral biordered set E_4 (cf. [2]). Let $\mathscr{D}$ be a category with one object α and $[\alpha, \alpha] = \langle \theta \rangle$ the infinite cyclic semigroup with identity. Let $P: \mathscr{D} \to \mathbf{Set}$ be defined by $P(\alpha) = \mathbf{N}$ and $P(\theta): n \mapsto n + 2$ and $\Delta: \mathscr{D} \to \mathbf{Set}$ is defined by $\Delta(\alpha) = \{(n, n), (n, n+1): n \in \mathbf{N}\}$, $\Delta(\theta) = P(\theta) \times P(\theta)|_{\Delta(\alpha)}$. Then clearly $(\mathscr{D} \downarrow P)$ is a preorder. Let $\delta: (\mathscr{D} \downarrow \Delta) \to (\mathscr{D} \downarrow P) \times (\mathscr{D} \downarrow P)$ be defined by $m \mapsto (mp_1, mp_2)$ where p_1 and p_2 are the projections. Define $\bar{\delta}: (\mathscr{D} \downarrow P) \times (\mathscr{D} \downarrow P) \to (\mathscr{D} \downarrow \Delta)$ as follows. Let $x, y \in (\mathscr{D} \downarrow P)$. $(1_\alpha)x_\alpha = m$ and $(1_\alpha)y_\alpha = n$. Define $u \in (\mathscr{D} \downarrow \Delta)$ by setting

$$(1_\alpha)u_\alpha = \begin{cases} (n, n) & \text{if } n - m \geqslant 0 \text{ and even} \\ (n, n+1) & \text{if } n - m \geqslant 0 \text{ and odd} \\ (m-1, m) & \text{if } m - n \geqslant 0 \text{ and odd} \\ (m, m) & \text{if } m - n \geqslant 0 \text{ and even.} \end{cases}$$

Then clearly $\bar{\delta}$ extends to a functor which is right adjoint to δ. Therefore

$$S = [H^\alpha, P] \times [H^\alpha, P]$$

with product defined by eq. (3.2) is a combinatorial pseudo-inverse semigroup with biordered set $\bar{\Delta} = [H^\alpha, \Delta]$. Clearly the mapping

$$e_i \mapsto u(2i+1, 2i+1)$$

$$f_i \mapsto u(2i+1, 2i+2)$$

$$g_i \mapsto u(2i+2, 2i+2)$$

$$h_i \mapsto u(2i+2, 2i+3)$$

extends to an isomorphism of the biordered set E_4 onto $\bar{\Delta}$, where $u(i,j)$ is the element in $(\mathscr{D} \downarrow \Delta)$ such that $(1_\alpha)u_\alpha = (i, j)$. Hence S is isomorphic to SP_4.

Example 2. Here we consider the construction of naturally partially ordered r.i.g. semigroups with a greatest idempotent. An ordered semigroup (cf. [1]) is said to be naturally partially ordered if the order on it is an extension of the natural partial order on idempotents (cf. [6]). If S is naturally partially ordered having a greatest idempotent u, then for $e, f \in E(S)$

$$e \leqslant f \Rightarrow efe = e.$$

Further in this case each $x \in S$ has a greatest inverse x° and $x^\circ = ux'u$ for any $x' \in V(x)$, where $V(x)$ is the set of inverse of x.

Let $\bar{E}$ be a naturally partially ordered r.i.g. semigroup, with a greatest idempotent u. The following results may be proved easily.

Proposition 5.1. *For* $\bar{e}, \bar{f} \in \bar{E}$

(i) $\bar{e} \mathscr{R} \bar{f} \Leftrightarrow \bar{e}u = \bar{f}u$.

(ii) $\bar{e} \mathscr{L} \bar{f} \Leftrightarrow u\bar{e} = u\bar{f}$.

(iii) $\bar{e} \mathscr{H} \bar{f} \Leftrightarrow \bar{e} = \bar{f}$.

(iv) $\bar{e} \mathscr{D} \bar{f} \Leftrightarrow u\bar{e}u = u\bar{f}u$.

(v) $\bar{e}^\circ = u\bar{e}u$ is the greatest idempotent of $D_{\bar{e}}$.

(vi) $\bar{e} = \bar{e}u\bar{e}$.

Now it is easy to see that $\bar{E}$ is a locally testable regular semigroup. The following proposition gives an insight into the position of idempotents in $\bar{E}$.

Proposition 5.2. *Let* D *be a* $\mathscr{D}$*-class of* $\bar{E}$ *and* $\bar{e} \in D$. *If* $\bar{f} \mathscr{R} u\bar{e}u$ $[\bar{f} \mathscr{L} u\bar{e}u]$ *then* $\bar{f} \in E(D)$.

It follows from Proposition 5.1(vi) that the subsemigroups $\bar{E}u$ and $u\bar{E}$ of $\bar{E}$ are bands. Both $\bar{E}u$ and $u\bar{E}$ contain u and the greatest element in each $\mathscr{D}$-class of $\bar{E}$. Further $\bar{E}u$ intersects each $\mathscr{D}$-class D of $\bar{E}$ in the $\mathscr{L}$-class containing the greatest element of D and $u\bar{E}$ meets each $\mathscr{D}$-class D of $\bar{E}$ in the $\mathscr{R}$-class containing the greatest element of D. Therefore the structure semilattices of $\bar{E}u$ and $u\bar{E}$ are the same. If

Γ is this structure semilattice, it is also seen that the partially ordered set Γ is isomorphic to the partially ordered set of $\mathscr{D}$-classes of $\bar{E}$. Clearly $\bar{E}u$ is a left normal band and $u\bar{E}$ is a right normal band.

Then we have the following theorem.

Theorem 5.3. *Let S be an r.i.g. semigroup which is naturally partially ordered and having a greatest idempotent u, which is a right identity. Then S is a left normal band. Dually if u is a left identity, then S is a right normal band.*

Let **O Set*** denote the small category whose objects are partially ordered sets, each with a greatest element, and whose morphisms are order preserving mappings which preserve the greatest elements. Let Γ be a semilattice with identity. The following is the structure theorem for naturally partially ordered r.i.g. semigroups having a greatest idempotent which is right identity.

Theorem 5.4. *Let $P\colon \Gamma \to$ **O Set*** be a functor which is disjoint and non-empty. On*

$$E = \cup \{P(\alpha)\colon\ \alpha \in \Gamma\}$$

define a product by

$$ab = aP(\alpha, \alpha\beta)$$

where $a \in P(\alpha)$ and $b \in P(\beta)$. Then E is an r.i.g. semigroup. Further define $\leqslant$ on E by

$$a \leqslant b \Leftrightarrow \alpha \leqslant \beta \quad \text{and} \quad a \leqslant bP(\beta, \alpha) \quad \text{in} \quad P(\alpha)$$

where $a \in P(\alpha)$ and $b \in P(\beta)$. Then $\leqslant$ is a partial order and E with this order, is a naturally partially ordered semigroup having a greatest idempotent which is a right identity. Conversely every naturally partially ordered r.i.g. semigroup with a greatest idempotent which is a right identity, may be constructed in this way.

Now we consider the general situation. First we set up some notations. Let Γ be a semilattice with identity and $P_1, P_2\colon \Gamma \to$ **O Set*** be functors each of which are disjoint and non-empty. Let $\leqslant'$ be defined on

$$I = \bigcup_{\alpha \in \Gamma} P_1(\alpha) \quad \text{and} \quad \Lambda = \bigcup_{\alpha \in \Gamma} P_2(\alpha)$$

as follows. For $i, j \in I$ $[\lambda, \mu \in \Lambda]$ $i \leqslant' j$ $[\lambda \leqslant' \mu] \Leftrightarrow \alpha \leqslant \beta$ and $i = jP_1(\beta, \alpha)$ $[\lambda = \mu P_2(\beta, \alpha)]$. Define $\leqslant$ on I and Λ by

$$i \leqslant j \ [\lambda \leqslant \mu] \Leftrightarrow \alpha \leqslant \beta \quad \text{and} \quad i \leqslant jP_1(\beta, \alpha) \ [\lambda \leqslant \mu P_2(\beta, \alpha)].$$

Clearly $\leqslant'$ and $\leqslant$ are partial orders and $\leqslant$ is an extension of $\leqslant'$. We denote by 1 the identity of Γ. The greatest element in $P_1(1)$ and $P_2(1)$ are also denoted by 1 itself. Let $\Delta\colon \Gamma \to$ **O Set*** be a functor such that $\Delta(\alpha)$ is a subdirect product of $P_1(\alpha)$ and $P_2(\alpha)$ for each $\alpha \in \Gamma$ and let $\bar{\Delta} = \bigcup_{\alpha \in \Gamma} \Delta(\alpha)$.

The following is our main result.

Theorem 5.5. *Let* Γ *be a semilattice with identity, and* P_1, P_2: $\Gamma \to$ **O Set*** *be functors which are disjoint and non-empty. Let* Δ *be a subdirect product of* P_1 *and* P_2 *satisfying the following conditions.*

(i) $(1, 1) \in \Delta(1)$.

(ii) $(\bar{\Delta}, \leqslant')$ *is a reflective subcategory of* $(I \times \Lambda, \leqslant')$ *such that the corresponding left adjoint* $F\colon I \times \Lambda \to \bar{\Delta}$ *preserves the extended order* $\leqslant$ *on* $I \times \Lambda$. *On*

$$S = \bigcup_{\alpha \in \Gamma} P_1 * P_2(\alpha)$$

define a binary operation as follows. For $(i, \lambda) \in P_1 * P_2(\alpha)$ *and* $(j, \mu) \in P_1 * P_2(\beta)$

$$(i, \lambda)(j, \mu) = (iP_1(\alpha, \nu), \mu P_2(\beta, \nu))$$

where $F(j, \lambda) \in \Delta(\nu)$. *Then* S *with the cartesian order of* $\leqslant$ *is a naturally partially ordered r.i.g. semigroup with a greatest idempotent. Conversely every naturally partially ordered r.i.g. semigroup with a greatest idempotent is isomorphic to one constructed in this way.*

Acknowledgement. The material in this paper forms part of the authors doctoral dissertation entitled 'Structure of Combinatorial Regular Semigroups', University of Kerala, Kariavattom Trivandrum, 1981. The author is grateful to his supervisor Dr. K.S.S. Nambooripad, for his guidance in the preparation of this paper.

REFERENCES

[1] T.S. Blyth – M.F. Janowitz, *Residuation Theory,* Pergamon Press, Oxford, 1972.

[2] K. Byleen – J. Meakin – F. Pastijn, The fundamental four-spiral semigroup, *J. Algebra,* 54 (1978), 6–24.

[3] P. Higgins, *Categories and Groupoids,* Van Nostrand Reinhold Company, London, 1971.

[4] J.M. Howie, *An Introduction to Semigroup Theory,* Academic Press, London, 1976.

[5] S. MacLane, *Categories for the Working Mathematician,* Springer-Verlag, New York, 1971.

[6] D.B. McAlister, On certain Dubreil–Jacotin semigroups, preprint.

[7] J. Meakin – F. Pastijn, Structure of pseudo-semilattices, preprint.

[8] K.S.S. Nambooripad, Structure of regular semigroups, I, *Mem. Amer. Math. Soc.,* 224 (1979).

[9] K.S.S. Nambooripad, Pseudo-semilattices and biordered sets, I, *Simon Stevin,* 55 (1981), 103–110; II, *Simon Stevin,* 56 (1982), 143–159; III, *Simon Stevin,* to appear.

[10] K.S.S. Nambooripad – A.R. Rajan, Structure of combinatorial regular semigroups, *Quart. J. Math. Oxford,* 29 (2) (1978), 489–504.

[11] B.M. Schein, Pseudo-semilattices and pseudo-lattices, *Izv. Vysš. Učebn. Zaved. Mat.,* 117 (1972), 81–94 (in Russian).

[12] Y. Zalcstein, Locally testable semigroups, *Semigroup Forum*, 5 (1973), 216–227.

A.R. Rajan

Department of Mathematics, University of Kerala, Kariavattom 695 581, Trivandrum, India.

COLLOQUIA MATHEMATICA SOCIETATIS JÁNOS BOLYAI
39. SEMIGROUPS, SZEGED (HUNGARY), 1981.

KERNEL AND TRACE IN INVERSE SEMIGROUPS

N.R. REILLY

1. THE BASIC CONCEPTS

Inverse semigroups have various descriptions: for example, as regular semigroups (for each $a \in S$, there exists an $x \in S$ such that $axa = a$) in which idempotents commute or as algebras endowed with a binary operation and a unary operation $x \to x^{-1}$ such that the following identities hold:

$$x(yz) = (xy)z$$

$$(x^{-1})^{-1} = x$$

$$(xy)^{-1} = y^{-1}x^{-1}$$

$$xx^{-1}x = x$$

$$xx^{-1}yy^{-1} = yy^{-1}xx^{-1}.$$

It is well known that the early pioneer work on the class of inverse semigroups as an abstract class was initiated by Vagner [46] and Preston [37] in the early fifties.

As with the investigation of any abstract class of algebras, attention quickly focused on the characterization of congruences. A good understanding of congruence is important to any study of algebraic systems.

Here, I wish to show how recent developments have led to an improved understanding of congruences on inverse semigroups and how this new approach gives at one and the same time a coherence to some existing theories and pointers to important lines of future investigations.

Vagner [46] established that any congruence on an inverse semigroup is completely determined by those classes containing idempotents while Preston [37] gave the following characterization in terms of kernel normal systems:

Notation 1.1. If ρ is a congruence on an inverse semigroup S, let

$$\mathcal{K}(\rho) = \{e\rho \mid e \in E_S\}.$$

Such collections of subsets of S can be characterized abstractly, and we thus have the following concept.

Definition 1.2. Let $\mathcal{K}$ be a family of pairwise disjoint inverse subsemigroups of an inverse semigroup S satisfying

(i) $E_S \subseteq \bigcup_{K \in \mathcal{K}} K$,

(ii) for each $a \in S$ and $K \in \mathcal{K}$, there exists $L \in \mathcal{K}$ such that $a^{-1}Ka \subseteq L$,

(iii) if $a, ab, bb^{-1} \in K$, $K \in \mathcal{K}$, then $b \in K$.

Then $\mathcal{K}$ is a *kernel normal system* for S. For such a family $\mathcal{K}$, define a relation $\rho_{\mathcal{K}}$ on S by

$$a \,\rho_{\mathcal{K}}\, b \Leftrightarrow aa^{-1}, bb^{-1}, ab^{-1} \in K \text{ for some } K \in \mathcal{K}.$$

Theorem 1.3. *Let* S *be an inverse semigroup. If* $\mathcal{K}$ *is a kernel normal system for* S, *then* $\rho_{\mathcal{K}}$ *is the unique congruence* ρ *on* S *for which* $\mathcal{K}(\rho) = \mathcal{K}$. *Conversely, if* ρ *is a congruence on* S, *then* $\mathcal{K}(\rho)$ *is a kernel normal system for* S *and* $\rho_{\mathcal{K}(\rho)} = \rho$.

This characterization of congruences was tremendously important as it gave early researchers into the theory of inverse semigroups the confidence that the theory was indeed tractable.

In the case where ρ is idempotent separating, so that each member of $\mathscr{K}(\rho)$ is a subgroup, the conditions simplify considerably and this influenced the study of idempotent separating extensions or $\mathscr{H}$-coextensions of inverse semigroups as considered by D'Alarcao [5], Houghton [10], Meakin [17], Munn [20], Širjaev [42]–[44] and others.

However, in its full generality, the characterization seems to have been too complicated and the relationship between the different conditions too intractable for it to be the starting point of deeper investigations.

Nor did it prove to be a useful tool in the study of the lattice of congruences on an inverse semigroup either in general or in specific instances.

The beginnings of a new approach appeared in Reilly and Scheiblich [40] with the introduction of the concept of the trace, as it is now called, of a congruence. The key step, due to Scheiblich [41], was to combine the concept of the trace of a congruence with that of the kernel to obtain an alternative description of congruences on an inverse semigroup. This approach has been refined by Green [8] and Petrich [29] to yield the following elegant description from the book by Petrich [33].

Notation 1.4. For any inverse semigroup S, E_S denotes the semilattice of idempotents of S.

We associate with each congruence ρ on an inverse semigroup S two parameters, which uniquely determine ρ.

Definition 1.5. For a congruence ρ on an inverse semigroup S, we define the *kernel* and the *trace* of ρ by

$$\ker \rho = \{a \in S \mid a\,\rho\,e \text{ for some } e \in E_S\}, \quad \operatorname{tr} \rho = \rho|_{E_S},$$

respectively.

This associates to each congruence ρ on S the ordered pair $(\ker \rho, \operatorname{tr} \rho)$.

Definition 1.6. Let S be an inverse semigroup. A subset K of S is *full* if $E_S \subseteq K$; it is *self-conjugate* if $s^{-1}Ks \subseteq K$ for all $s \in S$. A full, self-conjugate inverse subsemigroup of S is a *normal subsemigroup* of S.

A congruence τ on E_S is *normal* if for any $e, f \in E_S$ and $s \in S$, $e \, \tau \, f$ implies $s^{-1}es \, \tau \, s^{-1}fs$.

The pair (K, τ) is a *congruence pair* for S if K is a normal subsemigroup of S, τ is a normal congruence on E_S and these two satisfy:

(i) $ae \in K, \; e \, \tau \, a^{-1}a \Rightarrow a \in K \quad (a \in S, \; e \in E_S)$,

(ii) $k \in K \Rightarrow kk^{-1} \, \tau \, k^{-1}k$.

In such a case, define a relation $\rho_{(K,\tau)}$ on S by

$$a \, \rho_{(K,\tau)} \, b \Leftrightarrow a^{-1}a \, \tau \, b^{-1}b, \quad ab^{-1} \in K.$$

Theorem 1.7. *Let* S *be an inverse semigroup. If* (K, τ) *is a congruence pair for* S, *then* $\rho_{(K,\tau)}$ *is the unique congruence* ρ *on* S *for which* $\ker \rho = K$ *and* $\operatorname{tr} \rho = \tau$. *Conversely, if* ρ *is a congruence on* S, *then* $(\ker \rho, \operatorname{tr} \rho)$ *is a congruence pair for* S *and* $\rho_{(\ker \rho, \operatorname{tr} \rho)} = \rho$.

In contrast to the kernel normal system approach, this approach associates with each congruence other natural algebraic systems: an inverse subsemigroup and a congruence on the semilattice of S. In certain respects this new approach is not so far removed from the kernel normal system approach. After all, $\ker \rho$ is simply the union of the elements of $\mathscr{K}(\rho)$ etc. However, this rearrangement of the factors or change of perspective has already opened up many possibilities for the study of congruences themselves, for extension theories and for classifying varieties. It is likely that there will be many further such applications.

My limited objective here is to illustrate, in contrast to the limited impact of the kernel normal system approach to congruence on inverse semigroups, the variety of disparate ways in which the kernel/trace approach has influenced the development of new areas, served as a useful tool and enhanced our understanding of existing results.

This is in no way intended to be a comprehensive review of the theory of congruences on inverse semigroups or of the extensive literature relating to various extension theories for inverse semigroups. For a comprehensive treatise of such topics the interested reader is referred to the book on the subject by M. Petrich which should appear shortly.

2. A FRAMEWORK FOR THE LATTICE OF CONGRUENCES

If the kernel/trace approach to the study of congruences is to be useful then one would expect that the first area in which it would be of service would be in the study of the lattice of congruences on an inverse semigroup.

Notation 2.1. For any inverse semigroup S, let $\mathscr{C}(S)$ be the lattice of all congruences on S and let $\mathscr{N}(E_S)$ be the lattice of all normal congruences on E_S. Let $\mathscr{K}(S)$ denote the lattice of all full inverse subsemigroups K of S satisfying

$$ab \in K \Rightarrow aKb \subseteq K \qquad (a, b \in S).$$

With this notation we can give two basic homomorphisms of $\mathscr{C}(S)$ in relation to the kernel and trace concepts. The first is due to Reilly and Scheiblich [40] and the second to Green [8].

Theorem 2.2. *Let* S *be an inverse semigroup. Define a mapping* tr *by*

$$\text{tr}: \rho \longmapsto \text{tr}\,\rho \qquad (\rho \in \mathscr{C}(S)).$$

Then tr *is a complete homomorphism of* $\mathscr{C}(S)$ *onto* $\mathscr{N}(E_S)$. *Let* θ *denote the congruence on* $\mathscr{C}(S)$ *induced by* tr. *Then each* θ*-class is a complete modular lattice* (with commuting elements).

Theorem 2.3. *Let* S *be an inverse semigroup. Define a mapping* ker *by*

$$\ker: \rho \longmapsto \ker\rho \qquad (\rho \in \mathscr{C}(S)).$$

Then ker *is a complete* $\cap$*-homomorphism of* $\mathscr{C}(S)$ *onto* $\mathscr{K}(S)$. *Let* κ *be the equivalence relation on* $\mathscr{C}(S)$ *induced by* ker. *Then each* κ*-class is a complete sublattice of* $\mathscr{C}(S)$.

The congruences induced by these homomorphisms provide a natural framework for the study of $\mathscr{C}(S)$ for any inverse semigroup S.

In earlier developments, particularly where the study of $\mathscr{C}(S)$ in its entirety has been involved, the concept of trace seems to have been more useful than that of kernel. Examples of this can be found in the study of $\mathscr{C}(S)$ for one-parameter inverse semigroups by Eberhart and Selden [6] and in the study of congruence free inverse semigroups by Munn [21], Baird [3] and Trotter [45]. In the latter investigation, much of the effort is devoted to characterizing the class of inverse semigroups which have only the two trivial traces (ι_{E_S} and $E_S \times E_S$). Two instances where the classifications by both trace and kernel are used in the study of $\mathscr{C}(S)$ are the characterization of congruences on simple ω-semigroups by Petrich [30] and the study of congruences on bisimple inverse semigroups by Petrich and Reilly [36].

3. THE min AND max NETWORKS

Using the concepts of trace and kernel some interesting patterns can be obtained in $\mathscr{C}(S)$.

Notation 3.1. For $\rho \in \mathscr{C}(S)$:

$\rho_{\min}$ = the minimum congruence λ on S with $\operatorname{tr}\lambda = \operatorname{tr}\rho$;

$\rho_{\max}$ = the maximum congruence λ on S with $\operatorname{tr}\lambda = \operatorname{tr}\rho$;

$\rho^{\min}$ = the minimum congruence λ on S with $\ker\lambda = \ker\rho$;

$\rho^{\max}$ = the maximum congruence λ on S with $\ker\lambda = \ker\rho$.

For an arbitrary congruence ρ we have (Green [8]) a result which strengthens the relationship between ρ and the congruences introduced in 3.1.

Lemma 3.2. $\rho = \rho_{\min} \vee \rho^{\min} = \rho_{\max} \cap \rho^{\max}$.

On the other hand, if we take $\rho = \omega = S \times S$, we may obtain two sequences of congruences as follows

$$\omega, \omega_{\min}, (\omega_{\min})^{\min}, ((\omega_{\min})^{\min})_{\min}, \ldots$$

and

$$\omega, \omega^{\min}, (\omega^{\min})_{\min}, ((\omega^{\min})_{\min})^{\min}, \ldots .$$

The set of congruences in these two sequences has been called the min network of $\mathscr{C}(S)$, Petrich and Reilly [35]. Denoting the elements of the first sequence by $\omega, \omega_1, \omega_2, \omega_3, \ldots$ and those of the second by $\omega, \omega^1, \omega^2, \omega^3, \ldots$, we have

Lemma 3.3. $\omega_{i-1} \cap \omega^{i-1} = \omega_i \vee \omega^i$.

Thus the min network, together with the intersections of corresponding pairs is a sublattice of $\mathscr{C}(S)$. We obtain the following diagram:

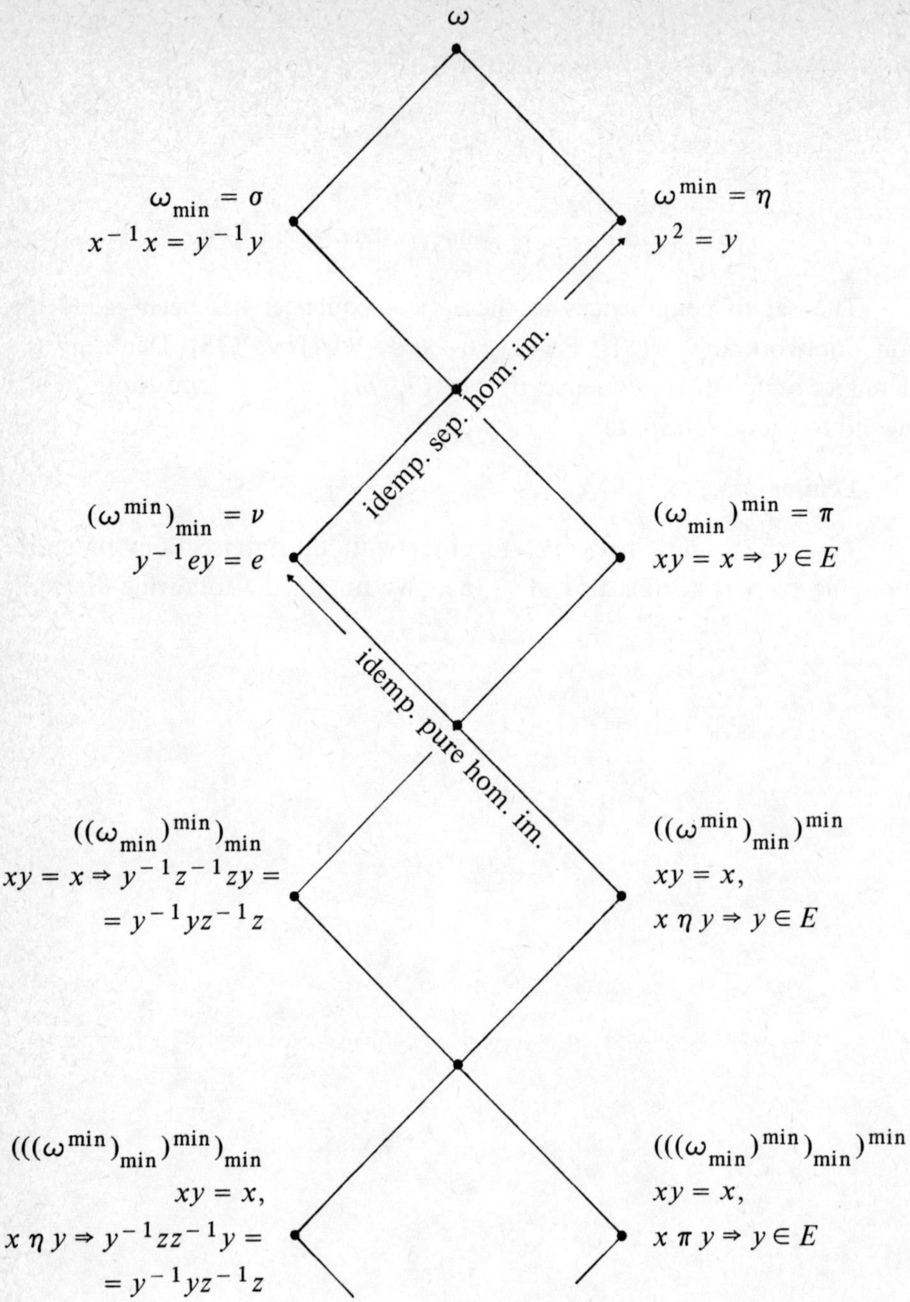

Diagram 1: The min network

Such a natural network of congruences leads to many questions. For instance one might ask what one can say about the class of inverse semigroups S such that some particular member of the min network is the identity congruence on S.

Notation 3.4. For any member ρ of the min network, or for any intersection of members of the min network, let $M(\rho)$ denote the class of inverse semigroups for which ρ is the identity congruence. Then

(1) $M(\omega_{\min})$ = all groups

(2) $M((\omega_{\min})^{\min})$ = all E-unitary inverse semigroups

(3) $M(\omega^{\min})$ = all semilattices

(4) $M((\omega^{\min})_{\min})$ = all semilattices of groups

(5) $M(((\omega^{\min})_{\min})^{\min})$ = all semilattices of E-unitary inverse semigroups

Equivalent to these observations are the statements. For any inverse semigroup S:

(1′) $\omega_{\min}$ is the minimum group congruence;

(2′) $(\omega_{\min})^{\min}$ is the minimum E-unitary congruence;

(3′) $\omega^{\min}$ is the minimum semilattice congruence;

(4′) $(\omega^{\min})_{\min}$ is the minimum semilattice of groups congruence;

(5′) $((\omega^{\min})_{\min})^{\min}$ is the minimum semilattice of E-unitary inverse semigroups congruence.

The classes (1), (3) and (4) are, of course, varieties. In general one has:

Theorem 3.5. *Each class $M(\rho)$ is a quasivariety.*

Quite specific equational implications can be derived for the first few classes. See Diagram 1.

The first four classes are well known and have been extensively studied. The fifth class has been investigated by O'Carroll [24], [25], [26] who called its members *strongly* *E-reflexive.*

Regarding the class $M(\rho)$, $\rho = \bigcap \omega_i$ we have the following observation.

Proposition 3.6. *Let* $\rho = \bigcap \omega_i$. *Then* $M(\rho)$ *satisfies the implication:* $x^3 = x^2 \Rightarrow x^2 = x$. *In particular, this implies that* $M(\rho)$ *contains no Brandt semigroup (which is not a group with zero).*

Of course, dual to the min network, there is a max network commencing with the identity congruence ι:

$$\iota_{max}, (\iota_{max})^{max}, ((\iota_{max})^{max})_{max}, \ldots$$

$$\iota^{max}, (\iota^{max})_{max}, ((\iota^{max})_{max})^{max}, \ldots .$$

Clearly ι_{max} is the maximum idempotent separating congruence while ι^{max} is the maximum congruence whose kernel is E_S, both of which have been extensively studied in the literature (for the former see, in particular, Preston [37], Howie [11] and Munn [18], [19] and for the latter see Green [7] and [8] and O'Carroll [23]).

The max network can be investigated in a manner dual to the min network, but is somewhat less well behaved.

For further information along the lines of this section see Petrich and Reilly [35] and Petrich [33].

Several natural possible lines of development suggest themselves here.

1. Investigate the structure of elements of the class $M((\omega_{min})^{min})_{min})$ and of the classes $M(\rho)$ for lower members ρ of the min network and dually for the max network.

2. What can one say about the class of semigroups for which the intersection of the min network is the identity congruence and dually for the max network?

4. EXTENSION THEORY

Just as the characterization of congruences on a group in terms of normal subgroups led to the development of a theory of group extensions, it is to be expected that the characterization of congruences on an inverse semigroup in terms of subsystems should lead to similar theories.

However, prior to the kernel-trace approach to congruences on an inverse semigroup, only two extreme cases had really proved amenable which might be said to have been offshoots of, or to have drawn significantly on, the theory of kernel normal systems. First there was the case of idempotent separating extensions, also called $\mathcal{H}$-coextensions by some authors (where the corresponding kernel normal system is a union of groups) considered, in particular, by D'Alarcao [5], Coudron [4], Grillet [9], Meakin [17] and Širjaev [42]–[44] and then there was the case of E-unitary inverse semigroups (where the corresponding kernel normal system has just one member and that equals E_S) which has been investigated by many authors including McFadden and O'Carroll [16], O'Carroll [22], McAlister [14], [15], Petrich and Reilly [34] and Žitomirskiĭ [47].

Of course other significant extension theories for inverse semigroups were developed. For example, there is an extensive and fruitful theory of ideal extensions, translational hulls etc. There are also well developed theories for certain special classes. For a comprehensive treatment, see [33]. But I think that one could reasonably claim that these developments were not influenced significantly in their genesis, development or interpretation by the theory of kernel normal systems.

Since it is not my purpose to review all these theories but rather to spotlight the strength and importance of the concepts of kernel and trace, I shall now elaborate in some detail the very comprehensive, but apparently unappreciated, theory due to D. Allouch.

The kernel-trace approach to congruences introduces a new subobject, the kernel, and it is natural to seek an extension theory in relation to this subobject. This search has been successfully conducted, with many interesting features by Allouch [1], [2].

We first describe the general extension problem.

Note. In this section, mappings are written as left operators, since they were so written in the main sources for this section (Allouch [1], [2]).

Let ρ be a congruence on an inverse semigroup S, $K = \ker \rho$, π: $S \rightarrow S/\rho$ be the natural homomorphism, $\alpha = \pi|_K$ and ι denote the embedding of K into S. Let us also write $\ker \pi$ for $\ker \pi^{-1} \circ \pi$. Then

$$\iota K = \ker \pi, \qquad \pi\iota = \alpha.$$

This leads us to the definition:

Definition 4.1. Let K and Q be inverse semigroups and let α be an epimorphism of the inverse semigroup K onto E_Q. Then (K, α, Q) is an *extension triple.* An *extension of* K *by* Q *via* α is a triple (ι, S, π) where S is an inverse semigroup, ι is a monomorphism of K into S and π is a homomorphism of S onto Q such that

$$\iota K = \ker \pi, \qquad \pi\iota = \alpha:$$

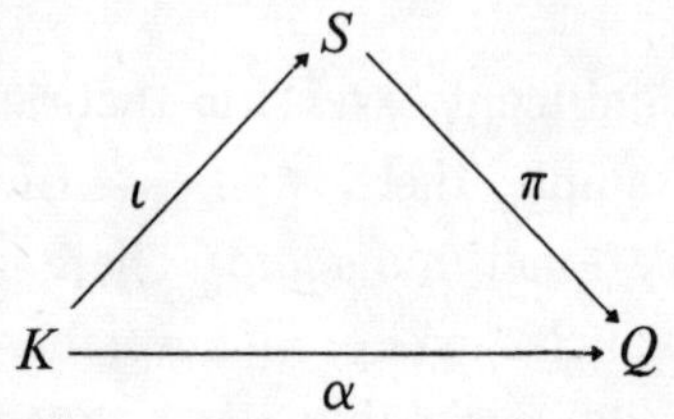

Two extensions of K by Q via α, say (ι, S, π) and (ι', S', π') are said to be *equivalent* if there exists an isomorphism β: $S \rightarrow S'$ such that the following diagram commutes:

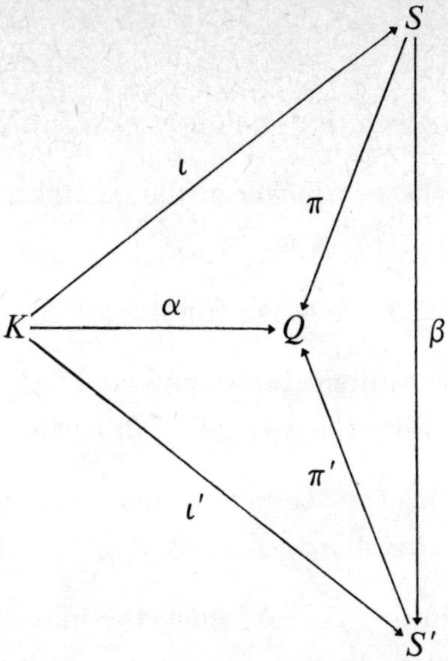

The *extension problem* is to construct for every extension triple (K, α, Q), that is for every pair of inverse semigroups K and Q and epimorphism $\alpha: K \to E_Q$, a representative of each equivalence class of extensions of K by Q via α.

We shall need the following preliminaries.

Definition 4.2 (Petrich [32]). The *normal hull* $\Phi(S)$ of an inverse semigroup S is the inverse subsemigroup of the symmetric inverse semigroup on S consisting of those mappings α such that there exist $e, f \in E_S$ with

$$\text{Domain of } \alpha = eSe, \quad \text{Range of } \alpha = fSf$$

and α an isomorphism of eSe onto fSf.

In the particular case where S is a semilattice E, say, $\Phi(E)$ coincides with the Munn semigroup T_E of E (see [18]).

Lemma 4.3 ([18]). *Let S be an inverse semigroup and $E = E_S$. For each $z \in S$, let θ_z denote the mapping of $Ez^{-1}z \to Ezz^{-1}$ defined by*

$$\theta_z e = zez^{-1} \qquad (e \in Ez^{-1}z).$$

Then the mapping $\theta\colon z \to \theta_z$ *is a homomorphism of* S *onto a full (that is,* $\theta(S) \supseteq E_{T_E}$*) inverse subsemigroup of* T_E *such that* $\theta^{-1} \circ \theta$ *is the maximum idempotent separating congruence* μ_S *on* S.

Definition 4.4. The *centralizer* of the idempotents of an inverse semigroup S is

$$C_S(E_S) = \{z \in S\colon\ ze = ez \quad \text{for all} \quad e \in E\}.$$

The maximum idempotent separating congruence on an inverse semigroup was characterized by Howie [11]. In particular,

Lemma 4.5 ([11]). *The kernel of the maximum idempotent separating congruence* μ_S *on an inverse semigroup* S *is* $C_S(E_S)$.

An inverse semigroup S for which the maximum idempotent separating congruence is the identity congruence or, equivalently, for which $C_S(E_S) = E_S$ is said to be *fundamental.* Always S/μ_S is fundamental.

Notation 4.6. For any congruence ρ on an inverse semigroup S we denote by $\rho^\natural$ the natural homomorphism

$$\rho^\natural\colon a \to a\rho \qquad (a \in S)$$

of S onto S/ρ.

Let (ι, S, π) be an extension of K by Q via α:

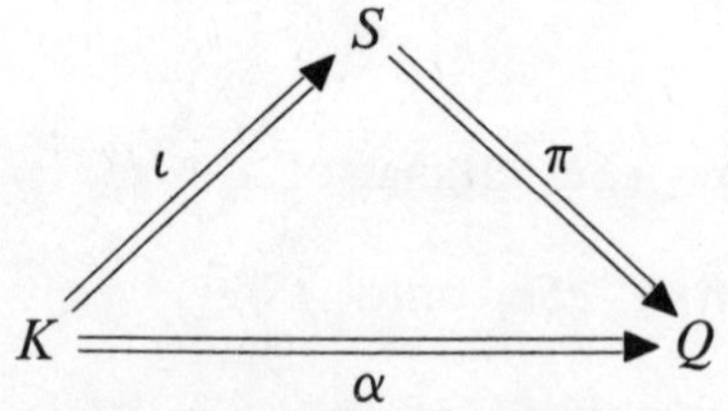

We will essentially decompose this extension into two extensions with more special features; the first is an extension of $C_K(E_K)$, a semilattice of groups, and the second is an extension of K/μ_K, a fundamental inverse semigroup.

Let $\kappa\colon C_K(E_K) \to K$ be the inclusion mapping. Then it is easily verified that $L = \iota\kappa(C_K(E_K))$ is the kernel of a unique idempotent separating congruence on S, namely, $\rho = \rho_{(L,\epsilon)}$ (in the notation of Theorem 1.7). This leads to the diagram:

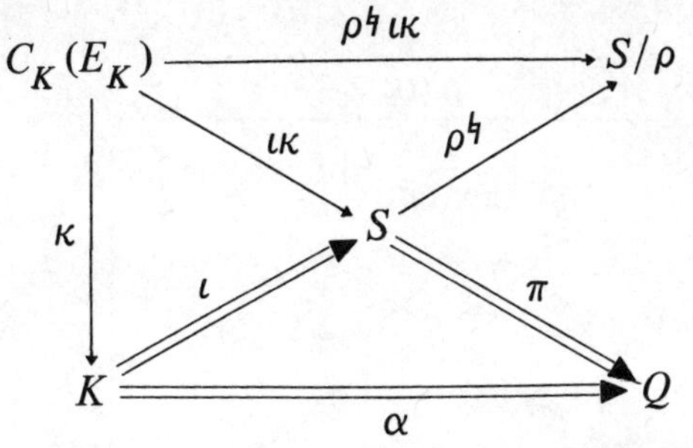

Here $(\iota\kappa, S, \rho^\natural)$ is an extension of $C_K(E_K)$ by S/ρ via $\rho^\natural\iota\kappa$.

Since they have the same kernel and trace, $(\rho^\natural\iota)^{-1} \circ (\rho^\natural\iota) = \mu_K$ so that there exists a monomorphism η: $K/\mu_K \to S/\rho$ such that $\eta\mu_K^\natural = \rho^\natural\iota$; writing $\mu = \mu_K$, we have the diagram:

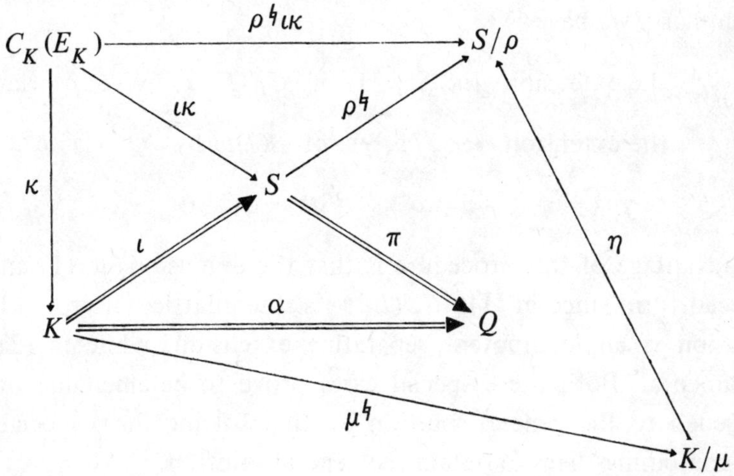

On the other hand, $\ker\alpha = K$ and so $\mu \subseteq \alpha^{-1} \circ \alpha$. Consequently, there exists a homomorphism α': $K/\mu \to Q$ with $\alpha'\mu^\natural = \alpha$ so that, in particular, $\alpha'(K/\mu) = E_Q$. Since $\ker\pi = \iota K$, we have $\rho \subseteq \pi^{-1} \circ \pi$. Hence there exists γ: $S/\rho \to Q$ with $\gamma\rho^\natural = \pi$ and $\gamma(S/\rho) = Q$. Now,

$$\gamma\eta\mu^\natural = \gamma\rho^\natural\iota = \pi\iota = \alpha = \alpha'\mu^\natural$$

so that $\gamma\eta = \alpha'$. This can be interpreted as saying that $(\eta, S/\rho, \gamma)$ is an extension of K/μ by Q via α'. We thus arrive at the following diagram:

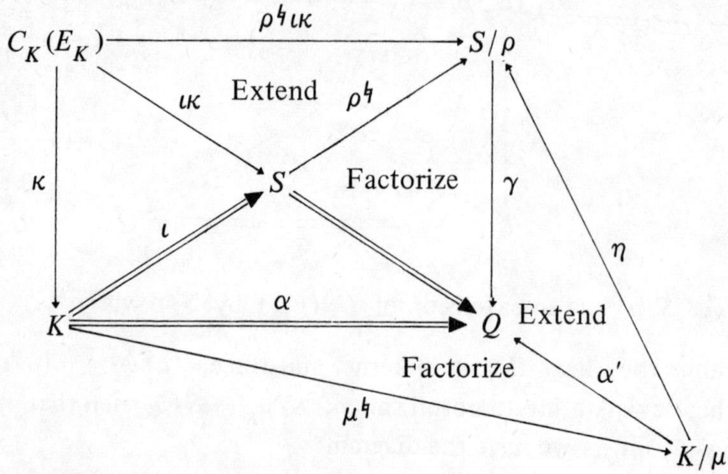

In summary, we have

(1) the extension $(\iota\kappa, S, \rho^\natural)$ of $C_K(E_K)$ by S/ρ via $\rho^\natural\iota\kappa$,

(2) the extension $(\eta, S/\rho, \gamma)$ of K/μ by Q via α',

$$\gamma\rho^\natural = \pi, \quad \gamma\eta = \alpha', \quad \alpha'\mu^\natural = \alpha.$$

The advantage of this procedure is that the extensions in (1) and (2) have a special form since in (1), $C_K(E_K)$ is a semilattice of groups so that the extension is an idempotent separating extension, while in (2) K/μ is fundamental. Both these special cases prove to be amenable and Allouch proceeded to the general solution by first solving these special cases and then combining them to obtain the general solution.

The building process may be described as follows. To find all extensions (ι, S, π) of K by Q via α, first factor α as $\alpha = \alpha'\mu^\natural$ where $\mu^\natural\colon K \to K/\mu$. Then form all extensions (ι_2, M, π_2) of K/μ by Q via α':

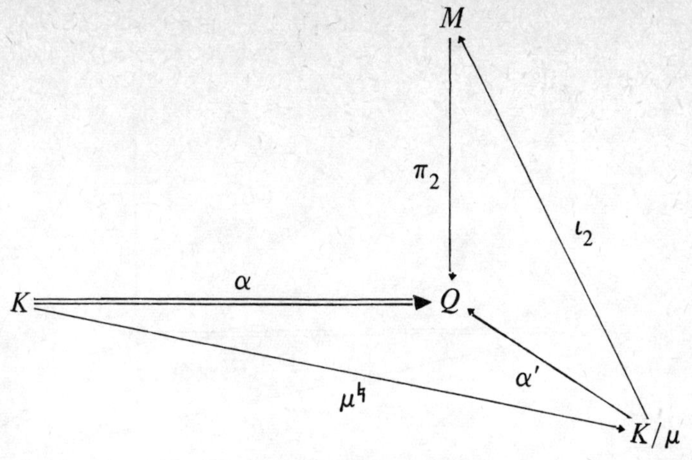

Next embed the kernel $C_K(E_K)$ of $\mu^\natural$ into K via κ:

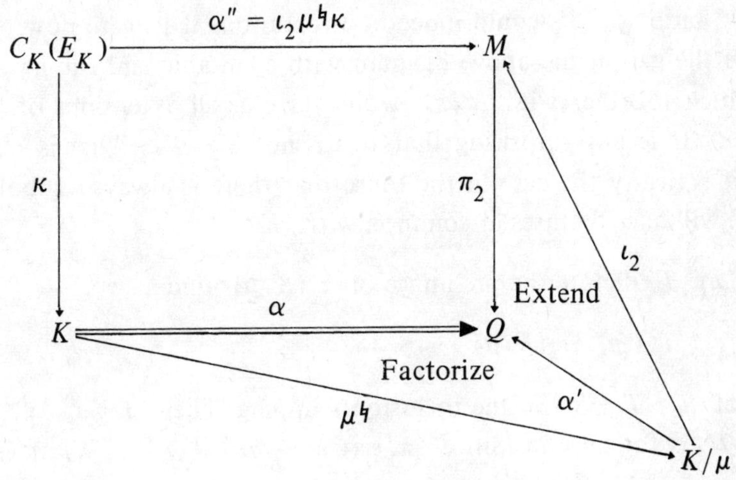

Since ι_2 maps K/μ onto a full inverse subsemigroup of M, the composite mapping $\alpha'' = \iota_2\mu^\natural\eta$ carries $C_K(E_K)$ onto the semilattice of idempotents of M. We can therefore consider all extensions (ι_1, S, π_1) of $C_K(E_K)$ by M via α''.

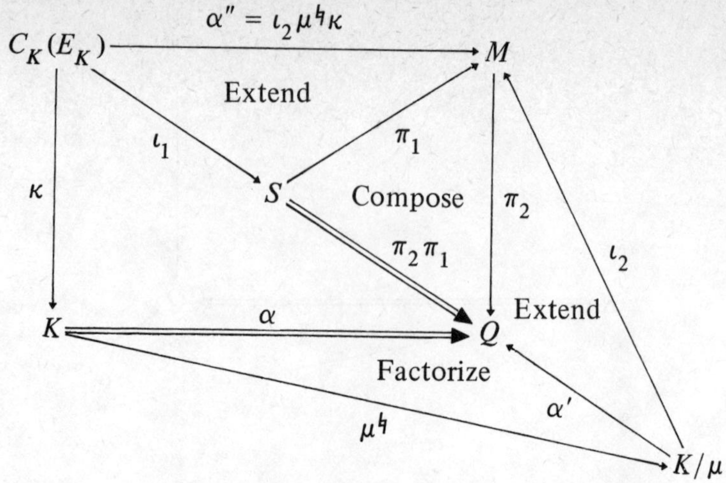

We note that by combining π_1 and π_2 we obtain a mapping $\pi_2\pi_1$ of S onto Q. It would indeed be fortuitous if it were now possible to bridge the gap in the above diagram with a suitable embedding of K into S which together with $\pi_2\pi_1$ would give us all extensions of K by Q via α. It is not surprising that this is not the case. What is surprising is that it is nearly the case in the sense that there is always a subobject of S which will have features in common with K.

Let T denote the pre-image of $\iota_2(K/\mu)$ under π_1,

$$T = \pi_1^{-1}\iota_2(K/\mu)$$

and let $i_2\colon T \to S$ be the inclusion mapping. Then $T = i_2^{-1}\pi_1^{-1}\iota_2(K/\mu)$ and T is full in S. Since $\pi_1\iota_1(C_K(E_K)) = E_M \subseteq \iota_2(K/\mu)$ it follows that

$$\iota_1(C_K(E_K)) \subseteq T.$$

Let i_1 be the mapping of $C_K(E_K)$ into T (that is, the mapping obtained from ι_1 by replacing the codomain S by T), so that

$$i_2 i_1 = \iota_1 .$$

This leads us to the following diagram (excluding β).

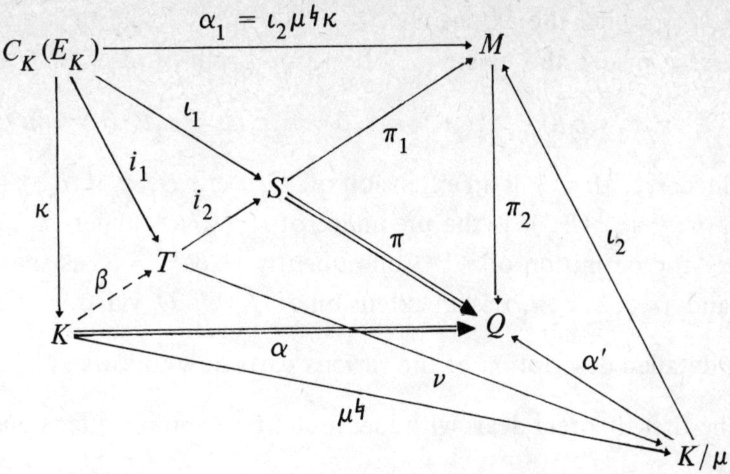

Since

$$\pi_1 i_2(T) = \pi_1 i_2 i_2^{-1}\pi_1^{-1}\iota_2(K/\mu) = \iota_2(K/\mu)$$

we can define a homomorphism $\nu\colon T \to K/\mu$ by

$$\nu = \iota_2^{-1}\pi_1 i_2.$$

Then

$$\nu i_1 = \iota_2^{-1}\pi_1 i_2 i_1 = \iota_2^{-1}\pi_1\iota_1 = \iota_2^{-1}\iota_2\mu^\natural\kappa = \mu^\natural\kappa$$

where $\mu^\natural\kappa$ is a homomorphism of $C_K(E_K)$ onto the semilattice of idempotents of K/μ. Thus we have the two extensions (i_1, T, ν) and $(\kappa, K, \mu^\natural)$ of $C_K(E_K)$ by K/μ via $\mu^\natural\kappa$.

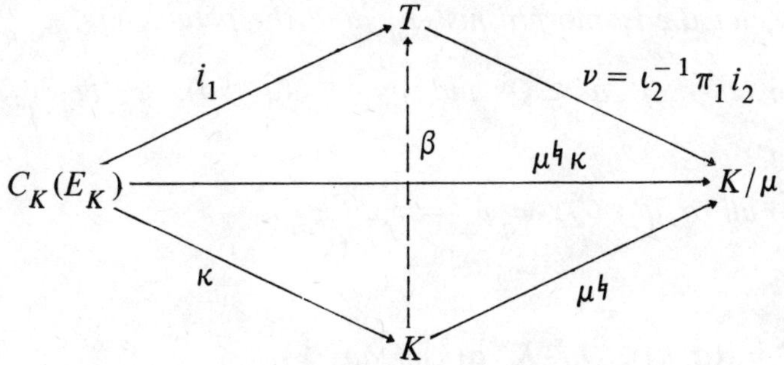

By restricting the extensions (ι_1, S, π_1) of $C_K(E_K)$ by M via α_1

to those for which the extensions (i_1, T, ν) and $(\kappa, K, \mu^\natural)$ are *equivalent* we thereby ensure the existence of an isomorphism $\beta\colon K \to T$ such that

$$(\pi_2\pi_1)(i_2\beta) = \pi_2\iota_2\iota_2^{-1}\pi_1 i_2\beta = \pi_2\iota_2\nu\beta = \pi_2\iota_2\mu^\natural = \alpha'\mu^\natural = \alpha.$$

Since (ι_2, M, π_2) is an extension of K/μ by Q, $\pi_2^{-1}(E_Q) = \iota_2(K/\mu)$ so that $\pi_1^{-1}\pi_2^{-1}(E_Q)$ is the pre-image of $\iota_2(K/\mu)$ under π_2, which is precisely the definition of T. Consequently $i_2\beta(K) = T$ is the kernel of $\pi_2\pi_1$ and $(i_2\beta, S, \pi_2\pi_1)$ is an extension of K by Q via α.

A detailed description of the various parts now follows.

The first theorem deals with idempotent separating extensions.

Theorem 4.7. *Let (K, α, Q) be an extension triple such that K is a semilattice of groups and such that $\alpha|_{E_K}: E_K \to E_Q$ is an isomorphism. Let $\xi = \alpha|_{E_K}$. For each $q \in Q$, let $\varphi_q \in \Phi(K)$ be an isomorphism of $\xi^{-1}(q^{-1}q)K$ onto $\xi^{-1}(qq^{-1})K$ such that*

(1) *for all $e \in E_Q$, φ_e is the identity on eK,*

(2) *for all $x \in E_K \cap$ Domain φ_q, $\varphi_q(x) = \xi^{-1}(q\alpha(x)q^{-1})$.*

For each pair $(q, q') \in Q \times Q$ let $g_{q,q'}$ be an element of the subgroup G_i of K where $i = \xi^{-1}[(qq')(qq')^{-1}]$ such that:

(3) *for all $e, f \in E_Q \times E_Q$, $g_{e,f} = ef$,*

and let $\tau_{q,q'} \in \Phi(K)$ be defined by $x \mapsto g_{q,q'}xg_{q,q'}^{-1}$ $(x \in iK)$.

Finally, let the isomorphisms φ_q and the parameters $g_{q,q'}$ satisfy:

(4) *for all $q, q', q'' \in Q$ and $f = \xi^{-1}(q^{-1}q)$, $\varphi_q(fg_{q',q''})g_{q,q'q''} = g_{q,q'}g_{qq',q''}$,*

(5) *for all $q, q' \in Q$, $\varphi_q\varphi_{q'} = \tau_{g_{q,q'}}\varphi_{qq'}$.*

Let

$$\Sigma = \{(q, k) \in Q \times K\colon \alpha(k) = qq^{-1}\}.$$

Then we have

(I) *The set* Σ *endowed with the multiplication*

$$(q, k)(q', k') = (qq', k\varphi_q(fk')g_{q,q'})$$

where $f = \xi^{-1}(q^{-1}q)$ *is an inverse semigroup.*

(II) *The mapping* $i\colon k \mapsto (\alpha(k), k)$ *of* K *into* Σ *is a monomorphism while the mapping* $\pi\colon (q, k) \mapsto q$ *of* Σ *into* Q *is an epimorphism such that* $\pi i = \alpha$. *Thus* (i, Σ, π) *is an extension of* K *by* Q *via* α.

(III) *Every extension of* K *by* Q *via* α *is equivalent to an extension of the form* (i, Σ, π) *as constructed above.*

The second theorem deals with extensions of fundamental inverse semigroups.

Theorem 4.8. *Let* (K, α, Q) *be an extension triple where* K *is fundamental. Let* ν *denote the projection of* $Q \times \Phi(E_K)$ *onto* Q *and* $\theta\colon K \to \Phi(E_K)$, *the Munn representation of* K.

(I) *The homomorphism* $\iota\colon K \to E_Q \times \Phi(E_K)$ *defined by* $\iota(k) = (\alpha(k), \theta_k)$ *is one-to-one.*

(II) *For each inverse subsemigroup* Σ *of* $Q \times \Phi(E_K)$ *such that*

(i) $\Sigma \cap (E_Q \times \Phi(E_K)) = \iota(K)$,

(ii) $\nu|_\Sigma$ *is surjective,*

$(\iota, \Sigma, \nu|_\Sigma)$ *is an extension of* K *by* Q *via* α.

(III) *Every extension of* K *by* Q *via* α *is equivalent to an extension of the form described in* (II).

The third theorem combines the previous two, to yield the general solution.

Theorem 4.9. *Let* (K, α, Q) *be an extension triple.*

(I) *The mapping* $\beta\colon K \to E_Q \times \Phi(E_K)$ *defined by* $\beta(k) = (\alpha(k), \theta_k)$ *is an idempotent separating homomorphism of* K *into* $E_Q \times \Phi(E_K)$.

Furthermore, there exists a monomorphism $\iota\colon k \mapsto (\alpha(k), \theta_k, g_k)$ *of* K *into* $E_Q \times \Phi(E_K) \times C_K(E_K)$, *where* $C_K(E_K)$ *is the centralizer of* E_k *in* K.

(II) *Let* Σ *be an inverse subsemigroup of* $Q \times \Phi(E_K)$ *such that* $\Sigma \cap (E_Q \times \Phi(E_K)) = \beta(K)$ *and such that the projection* $\tau\colon \Sigma \to Q$ *is a surjection. Let* Λ *be an extension of* $C_K(E_K)$ *by* Σ *via* $(\beta\tau)|_{C_K(E_K)}$ *such that* $\iota(K) \subseteq \Lambda \subseteq \Sigma \times C_K(E_K)$.

Then Λ *is an extension of* K *by* Q *via* α *and every extension of* K *by* Q *via* α *is equivalent to one obtained in this way.*

For the detailed proofs of these and other related results see Allouch [2].

If the results of this section represent the right generalization of the Schreier theory of group extensions to inverse semigroups, one might ask whether other aspects of the theory generalize.

5. VARIOUS HULLS

In any theory of extensions it is natural to consider special types of extensions and, in certain cases to consider the existence of maximal extensions of particular types. This has led to the introduction into the theory of inverse semigroups of the concept of essential extensions, a concept familiar in the study of groups, rings and modules. In this investigation some interesting new objects that can be associated with an inverse semigroup play important roles.

Definition 5.1 (Petrich [31], [32]). Let K be an inverse subsemigroup of an inverse semigroup S. Then S is a *conjugate extension* of K if K is self-conjugate in S and S is a *normal extension* of K if K is normal in S (recall: normal = conjugate and full).

Definition 5.2. For any inverse semigroup S, the set consisting of isomorphisms among subsemigroups of S of the form $\lambda S \rho$ with $(\lambda, \rho) \in$ $\in E_{\Omega(S)}$, where $\Omega(S)$ is the *translational hull* of S, together with composition of these isomorphisms as right operators is the *conjugate hull*

$\Psi(S)$ of S. The inverse subsemigroup $\Phi(S)$ of $\Psi(S)$ consisting of those isomorphisms between subsemigroups of the form eSe $(e \in E_S)$ is the normal hull of S (Definition 4.2).

The next result gives an important representation of an inverse semigroup on any self-conjugate inverse subsemigroup generalizing both the Munn representation and the representation of a group by conjugation on a normal subgroup.

Theorem 5.3 (Petrich [31]). *Let S be a conjugate extension of K. Then the mapping $\theta = \theta(S\colon K)\colon a \to \theta^a$ $(a \in S)$ where θ^a is defined by*

$$\theta^a\colon k \to a^{-1}ka \qquad (k \in aKa^{-1})$$

is a homomorphism of S into $\Psi(K)$.

For the special case $K = E_S$, $\theta(S\colon E_S)$ is essentially the familiar Munn representation of S (the only difference being that the codomain is $\Psi(E_S)$ and not $\Phi(E_S)$).

Definition 5.4 (Petrich [31]). The kernel of the congruence induced by $\theta(S\colon S)$ is the *metacentre* of S and is denoted by $M(S)$. Thus

$$M(S) = \{a \in S\colon\ a^{-1}za = z \ \text{ for } \ z \in aSa^{-1}\}.$$

If $M(S) = E_S$, we say that S has *idempotent metacentre.*

Definition 5.5. Let S be a conjugate (respectively, normal) extension of K. Then S is an *essential* conjugate (respectively, normal) extension if the equality relation on S in the only congruence on S whose restriction to K is the equality relation on K.

We can now state the result from which the representation $\theta(S\colon K)$ derives much of its importancc.

Proposition 5.6 (Petrich [31]). *Let S be a conjugate extension of an inverse semigroup K. Then $\theta(S\colon K)$ is one-to-one if and only if the metacentre of K is E_K and S is an essential conjugate extension of K.*

Notation 5.7. Let $\Theta(S)$ denote the image of S under $\theta(S\colon S)$.

The objects $\Theta(S)$, $\Phi(S)$ and $\Psi(S)$ are related as follows (the first observation being due to Pastijn [28] and the second to Petrich [31]).

Theorem 5.8. *For any inverse semigroup* S,

(i) $\Phi(S)$ *is an ideal of* $\Psi(S)$,

(ii) $\Phi(S)$ *is the maximum inverse subsemigroup of* $\Psi(S)$ *with semilattice of idempotents equal to* $E_{\Theta(S)}$.

In certain circumstances, the relationship between $\Theta(S)$, $\Phi(S)$ and $\Psi(S)$ can be made even stronger.

Theorem 5.9 (Petrich [32]). *Let* S *be an inverse semigroup with idempotent metacentre. Then* $\Phi(S)$ *is a maximal essential normal extension of* $\Theta(S)$. *Every inverse subsemigroup of* $\Phi(S)$ *containing* $\Theta(S)$ *is an essential normal extension of* $\Theta(S)$. *Every essential normal extension of* S *is isomorphic to a subsemigroup of* $\Phi(S)$ *containing* $\Theta(S)$.

The above theorem remains true if the word "normal" is replaced throughout by the word "conjugate" and $\Phi(S)$ is replaced throughout by $\Psi(S)$ (Petrich [31]). These results establish the existence of a largest essential normal (respectively, conjugate) extension of S when $M(S) = E_S$, since then $S \cong \Theta(S)$. By a powerful "tour de force", Pastijn [27] established the converse:

Theorem 5.10. *Let* S *be an inverse semigroup. If* S *has either a largest essential conjugate extension or a largest essential normal extension then* $M(S) = E_S$.

For an inverse semigroup S, the conjugate hull $\Psi(S)$ is somewhat reminiscent of the translational hull $\Omega(S)$ of S. Results relating $\Omega(S)$ to $\Theta(S)$, $\Phi(S)$ and $\Omega(S)$ will be found in Petrich [31], [33] and Pastijn [28].

6. VARIETIES

The kernel-trace approach to congruences also has a role to play in the theory of varieties of inverse semigroups. The following notation will be useful here:

$\mathcal{G}$	– the variety of all groups,
$\mathcal{I}$	– the variety of all inverse semigroup,
$\mathcal{L}(\mathcal{V})$	– the lattice of subvarieties of the variety $\mathcal{V}$,
$\mathcal{A}\mathcal{G}$	– the class of all fundamental inverse semigroups,
$[\mathcal{U}, \mathcal{V}]$	– the interval of all varieties $\mathcal{W}$ with $\mathcal{U} \leqslant \mathcal{W} \leqslant \mathcal{V}$,
F_X	– the free inverse semigroup on a countably infinite set of generators X,
$\rho_{\mathcal{V}}$	– the fully invariant congruence on F_X defining the variety $\mathcal{V}$.

Lemma 6.1 (Kleiman [12]). *The mappings*

$$\varphi_1 : \mathcal{V} \to \mathcal{V} \vee \mathcal{G}$$

$$\varphi_2 : \mathcal{V} \to \mathcal{V} \cap \mathcal{G}$$

are retractions of $\mathcal{L}(\mathcal{I})$ *onto the sublattices* $[\mathcal{G}, \mathcal{I}]$ *and* $\mathcal{L}(\mathcal{G})$ *of* $\mathcal{L}(\mathcal{I})$, *respectively.*

The homomorphisms φ_1 and φ_2 induce two very important congruences ν_1 and ν_2, respectively, on $\mathcal{L}(\mathcal{I})$:

$$(\mathcal{U}, \mathcal{V}) \in \nu_1 \Leftrightarrow \mathcal{U} \vee \mathcal{G} = \mathcal{V} \vee \mathcal{G},$$

$$(\mathcal{U}, \mathcal{V}) \in \nu_2 \Leftrightarrow \mathcal{U} \cap \mathcal{G} = \mathcal{V} \cap \mathcal{G}.$$

A useful alternative characterization of ν_2 is (Kleiman [12])

$$(\mathcal{U}, \mathcal{V}) \in \nu_2 \Leftrightarrow \mathcal{U} \cap \mathcal{A}\mathcal{G} = \mathcal{V} \cap \mathcal{A}\mathcal{G}.$$

These congruences have played a key role to date in the study of $\mathcal{L}(\mathcal{I})$ (see, for instance Kleiman [12], [13], Reilly [37], [38]). To describe the connection between ν_1, ν_2 and the kernel-trace approach to congruences we need the followong concept:

Definition 6.2. For any inverse subsemigroup S of an inverse semigroup T, the *closure* of S (in T) is $\bar{S} = \{t \in T\colon et \in S$ for some $e \in E_S\}$.

Proposition 6.3.

(1) $(\mathcal{U}, \mathcal{V}) \in \nu_1 \Leftrightarrow \operatorname{tr} \rho_{\mathcal{U}} = \operatorname{tr} \rho_{\mathcal{V}}$.

(2) $(\mathcal{U}, \mathcal{V}) \in \nu_2 \Leftrightarrow \overline{\ker \rho_{\mathcal{U}}} = \overline{\ker \rho_{\mathcal{V}}}$.

Thus these most useful decompositions of $\mathcal{L}(\mathcal{I})$ can be traced back to the "influence" of kernel and trace.

An interesting question here is what can be said of the relation

$$(\mathcal{U}, \mathcal{V}) \in \eta \Leftrightarrow \ker \rho_{\mathcal{U}} = \ker \rho_{\mathcal{V}} ?$$

REFERENCES

[1] D. Allouch, Extensions de demi-groupes inverses, *Semigroup Forum*, 16 (1978), 111–116.

[2] D. Allouch, *Sur les extensions de demi-groupes strictement réguliers,* Doctoral Dissertation, Univ. of Montpellier, France, 1979.

[3] G.R. Baird, Congruence-free inverse semigroups with zero, *J. Austral. Math. Soc.*, 28 (1979), 107–119.

[4] A. Coudron, Sur les extensions de demi-groupes réciproques, *Bull. Soc. Roy. Sci. Liège,* 37 (1968), 409–419.

[5] H. D'Alarcao, Idempotent-separating extensions of inverse semigroups, *J. Austral. Math. Soc.*, 9 (1969), 211–217.

[6] C. Eberhart – J. Selden, One parameter inverse semigroups, *Trans. Amer. Math. Soc.*, 168 (1972), 53–66.

[7] D.G. Green, Extensions of a semilattice by an inverse semigroup, *Bull. Austral. Math. Soc.*, 9 (1973), 21–31.

[8] D.G. Green, The lattice of congruences on an inverse semigroup, *Pacific J. Math.*, 57 (1975), 141–152.

[9] P.A. Grillet, Left coset extensions, *Semigroup Forum,* 7 (1974), 200–263.

[10] C.H. Houghton, Embedding inverse semigroups in wreath products, *Glasgow Math. J.*, 17 (1976), 77–82.

[11] J.M. Howie, The maximum idempotent-separating congruence on an inverse semigroup, *Proc. Edinburgh Math. Soc.*, (2) 14 (1964), 71–79.

[12] E.I. Kleiman, On the lattice of varieties of inverse semigroups, *Izv. Vysš. Učebn. Zaved. Matematika,* 7 (1976), 106–109 (in Russian).

[13] E.I. Kleiman, On bases of identities of Brandt semigroups, *Semigroup Forum,* 13 (1977), 209–218.

[14] D.B. McAlister, Groups, semilattices and inverse semigroups, *Trans. Amer. Math. Soc.*, 192 (1974), 227–244.

[15] D.B. McAlister, Groups, semilattices and inverse semigroups II, *Trans. Amer. Math. Soc.*, 196 (1974), 351–370.

[16] R. McFadden – L. O'Carroll, *F*-inverse semigroups, *Proc. London Math. Soc.*, 22 (1971), 652–666.

[17] J. Meakin, Coextensions of inverse semigroups, *J. Algebra,* 46 (1977), 315–333.

[18] W.D. Munn, Uniform semilattices and bisimple inverse semigroups, *Quart. J. Math. Oxford Ser.* (2) 17 (1966), 151–159.

[19] W.D. Munn, Fundamental inverse semigroups, *Quart. J. Math. Oxford Ser.* (2) 21 (1970), 157–170.

[20] W.D. Munn, 0-bisimple inverse semigroups, *J. Algebra,* 15 (1970), 570–588.

[21] W.D. Munn, Congruence-free inverse semigroups, *Quart. J. Math. Oxford Ser.* (2) 26 (1975), 385–387.

[22] L. O'Carroll, Reduced inverse semigroups, *Semigroup Forum,* 8 (1974), 270–276.

[23] L. O'Carroll, Idempotent determined congruences on inverse semigroups, *Semigroup Forum,* 12 (1976), 233–243.

[24] L. O'Carroll, Strongly *E*-reflexive inverse semigroups, *Proc. Edinburgh Math. Soc.,* (2) 20 (1976–77), 339–354.

[25] L. O'Carroll, Strongly *E*-reflexive inverse semigroups II, *Proc. Edinburgh Math. Soc.,* (2) 21 (1978), 1–10.

[26] L. O'Carroll, A note on strongly *E*-reflexive inverse semigroups, *Proc. Amer. Math. Soc.,* 79 (1980), 352–354.

[27] F. Pastijn, Essential normal and conjugate extensions of inverse semigroups, *Canad. J. Math.,* 34 (1982), 900–909.

[28] F. Pastijn, L'enveloppe conjugée d'un demi-groupe inverse à métacentre idempotent, *Glasgow Math. J.,* 23 (1982), 123–130.

[29] M. Petrich, Congruences on inverse semigroups, *J. Algebra,* 55 (1978), 231–256.

[30] M. Petrich, Congruences on simple ω-semigroups, *Glasgow Math. J.,* 20 (1979) 87–101.

[31] M. Petrich, The conjugate hull of an inverse semigroup, *Glasgow Math. J.,* 21 (1980), 103–124.

[32] M. Petrich, Extensions normales de demi-groupes inverses, *Fund. Math.,* CXII (1981), 188–203.

[33] M. Petrich, *Inverse semigroups,* Wiley, New York, 1984.

[34] M. Petrich – N.R. Reilly, A representation of *E*-unitary inverse semigroups, *Quart. J. Math. Oxford Ser.* (2) 30 (1979), 339–350.

[35] M. Petrich – N.R. Reilly, A network of congruences on an inverse semigroup, *Trans. Amer. Math. Soc.,* 270 (1982), 309–325.

[36] M. Petrich – N.R. Reilly, Congruences on bisimple inverse semigroups, *J. London Math. Soc.,* (2) 22 (1980), 251–262.

[37] G.B. Preston, Inverse semi-groups, *J. London Math. Soc.*, 29 (1954), 396–403.

[38] N.R. Reilly, Varieties of completely semisimple inverse semigroups, *J. Algebra,* 65 (1980), 427–444.

[39] N.R. Reilly, Modular sublattices of the lattice of varieties of inverse semigroups, *Pacific J. Math.*, 89 (1980), 405–417.

[40] N.R. Reilly – H.E. Scheiblich, Congruences on regular semigroups, *Pacific J. Math.*, 23 (1967), 349–360.

[41] H.E. Scheiblich, Kernels of inverse semigroup homomorphisms, *J. Austral. Math. Soc.*, 18 (1974), 289–292.

[42] V.M. Širjaev, A certain class of inverse semigroups, *Vestnik Beloruss. Gos. Univ. Ser. I,* 1970, no. 2, 10–13, 74 (in Russian).

[43] V.M. Širjaev, Inverse semigroups with a given *G*-radical, *Dokl. Akad. Nauk BSSR,* 14 (1970), 782–785 (in Russian).

[44] V.M. Širjaev, A certain extension in the class of inverse semigroups, *Vestnik Beloruss. Gos. Univ. Ser. I,* 1970, no. 1, 15–23, 65 (in Russian).

[45] P.G. Trotter, Congruence-free inverse semigroups, *Semigroup Forum,* 9 (1974), 109–116.

[46] V.V. Vagner, The theory of generalized heaps and generalized groups, *Mat. Sbornik,* 32 (1953), 545–632 (in Russian).

[47] G.I. Žitomirskiĭ, Extensions of universal algebras, *Semigroup Theory and its Applications,* Saratov University, No. 4 (1978), 19–40 (in Russian).

N.R. Reilly

Department of Mathematics, Simon Fraser University, Burnaby, B. C., Canada.

COLLOQUIA MATHEMATICA SOCIETATIS JÁNOS BOLYAI
39. SEMIGROUPS, SZEGED (HUNGARY), 1981.

SEMIGROUP VARIETIES POSSESSING THE AMALGAMATION PROPERTY

M.V. SAPIR

Let $\{S_i \mid i \in J\}$ be a family of semigroups. Suppose that for any $i, j, k, n \in J$, $S_i \cap S_j = S_k \cap S_n = S$ whenever $i \neq j$ and $k \neq n$ and S is a subsemigroup in each of the S_i $(i \in J)$. Then the union $\bigcup_{i \in J} S_i$ of the semigroups S_i is called their amalgam (with core S). A semigroup variety $\mathfrak{N}$ is said to possess the amalgamation property if every amalgam of semigroups from $\mathfrak{N}$ can be embedded into a semigroup from $\mathfrak{N}$. The problem of describing semigroup varieties with this property was raised in [7] (Problem 2.13.b); the analogous question for groups was asked in [5] (Problem 6). The latter problem was answered by B.H. Neumann for locally finite varieties of groups: such a variety has the amalgamation property if and only if it consists of abelian groups. Recently M.V. Volkov solved the analogous problem for varieties of associative rings. Inverse semigroup varieties with the amalgamation property are described modulo groups in T.E. Hall [3].

In the present paper we describe the locally finite semigroup varieties which have the amalgamation property. Furthermore, the proof of the main theorem gives essential information about those varieties, too, which possess the amalgamation property but are not locally finite. In particular, it follows from this proof that a semigroup variety with the amalgamation property is locally finite if and only if all groups in it are locally finite. However, no example is known of a non-abelian group variety which has the amalgamation property but is not locally finite, except the variety of all groups.

Denote by L, R, N, I the two-element left zero, right zero, zero semigroups, and the two-element semilattice, respectively. Let C_n be the cyclic group of order n, and $M(n, k)$ be the completely simple semigroup over C_n with sandwich matrix $\begin{pmatrix} 1 & 1 \\ 1 & c \end{pmatrix}$, where $c \in C_n$ and k is the order of c.

Theorem. *The following conditions are equivalent for a variety* $\mathfrak{M}$ *of locally finite semigroups:*

(1) $\mathfrak{M}$ *has the amalgamation property;*

(2) $\mathfrak{M}$ *is generated by a subset of one of the following two sets of semigroups:* $\{L, R, N, I, C_n\}$ *or* $\{N, M(n, k)\}$ *for some* n *and* k, $k \mid n$;

(3) $\mathfrak{M}$ *consists either of inflations of orthodox normal bands of abelian groups with finite exponent, or of inflations of completely simple semigroups over abelian groups with finite exponent;*

(4) $\mathfrak{M}$ *satisfies, for some* n, *one of the following two systems of identities:* $\{x^{n+1}y = xy = xy^{n+1},\ xyzt = xzyt\}$ *or* $\{x^2yx = xyx^2,\ xz = x(yx)^n z = xz^{n+1}\}$.

The theorem will be proved according to the scheme (1) → (3) → → (2) → (4) → (3) → (1). Notice that essential use will be made of the main and auxiliary results from [2]. Notations and basic definitions are taken from [1].

Proof of (1) → (3). In what follows $\mathfrak{M}$ will stand for a fixed semigroup variety possessing the amalgamation property. Square brackets will be used to define varieties by identities. $\mathcal{M}_1, \mathcal{M}_2, \mathcal{M}_3$ and $\mathcal{M}(p)$ (p is a prime number) denote the varieties $[xyz = t^2,\ xy = yx]$, $[x^2 = x,\ xyx = xy]$, $[xyz = xy]$, $[xyx^2 = x^2yx,\ x = x(yx)^p]$, respectively. Notice that, as was shown in [6], $\mathcal{M}(p)$ is the variety of all completely simple semigroups over abelian groups of exponent p. For a semigroup variety $\mathfrak{M}$, we denote by $\mathfrak{M}^*$ the variety of semigroups dual to members of $\mathfrak{M}$. By P we denote the semigroup $\{\epsilon, \omega, \pi\}$ with multiplication table

	ϵ	ω	π
ϵ	ϵ	ω	π
ω	π	π	π
π	π	π	π

$\nu(S)$ stands for the variety generated by the semigroup S. The symmetric semigroup over a set X will be denoted by T_X.

Lemma 1. $\mathcal{M}_1 \nsubseteq \mathfrak{M}$.

Proof. In [1], vol. I, p. 139 one can find two semigroups whose amalgam cannot be embedded into any semigroup. It is easy to see that these semigroups belong to $\mathcal{M}_1$. Lemma 1 is proven.

Verification of the following lemma is a simple matter of routine.

Lemma 2. *Let X be an n-element set, $a \in X$. Put* $J_1(X) = \{f \in T_X \mid f^2 = f,\ f(a) = a\}$, $J_2(X) = \{f \in T_X \mid f^2(X) = \{a\}\}$. *Then the subsemigroups of T_X generated by $J_1(X)$ and $J_2(X)$, respectively, contain isomorphic copies of the symmetric semigroup on $[(n-1)/2]$ elements.*

Lemma 3. $\nu(P) = [x^2y = xy,\ x^2y^2 = y^2x^2]$.

Proof. The inclusion of the left-hand side in the other one is verified in a straightforward way. We prove the converse. By [2], Theorem 2, the variety $[x^2y = xy,\ x^2y^2 = y^2x^2]$ consists of finitely approximable semigroups. By [2], Theorem 1, it can be generated by a semigroup of the form $P \times C_n$ for some n. Since all the groups in this variety are trivial, it is generated even by P. Lemma 3 is proven.

Lemma 4. $\mathfrak{M}$ *does not contain the varieties* $\mathcal{M}_2$, $\mathcal{M}_3$, *and* $\nu(P)$.

Proof. Take an arbitrary set X, $a \in X$, and any function f from

$J_1(X)$ or $J_2(X)$. On the set $X \cup \{f^*\}$, where $f^* \notin X$, we define three groupoids $S_i(X, f)$, $i = 1, 2, 3$, as follows.

If $i = 1$ and $f \in J_1(X)$, then X is a set of left zeros in $S_1(X, f)$, $(f^*)^2 = f^*$, and $f^*x = f(x)$ for all $x \in X$.

If $i = 2$ and $f \in J_2(X)$, then X is a set of left zeros in $S_2(X, f)$, $(f^*)^2 = a$, and $f^*x = f(x)$ for all $x \in X$.

If $i = 3$ and $f \in J_1(X)$, then X is a zero subsemigroup in $S_3(X, f)$ with a as zero element for the whole $S_3(X, f)$, $(f^*)^2 = f^*$, $f^*x = f(x)$ and $xf^* = a$ for all $x \in X$.

It is easy to verify that the groupoids $S_1(X, f)$, $S_2(X, f)$, $S_3(X, f)$ are semigroups and satisfy the identities of the varieties $\mathcal{M}_2$, $\mathcal{M}_3$, $[x^2y = xy,\ x^2y^2 = y^2x^2]$ ($= v(P)$ by Lemma 3), respectively.

In what follows $S(X, f)$ will denote one of the semigroups $S_i(X, f)$, $i = 1, 2, 3$, and $J(X)$ one of the subsets $J_1(X)$ and $J_2(X)$ of T_X.

It is easy to see that for any f_1 and f_2 from $J(X)$ the operations in $S(X, f_1)$ and $S(X, f_2)$ coincide on X, hence X is an ideal in $S(X, f)$. Consider the amalgam of the semigroups $S(X, f)$, $f \in J(X)$, with core X. Suppose that this amalgam can be embedded into a semigroup V from $\mathfrak{M}$. It can be assumed that V is generated by its subsemigroups $S(X, f)$, $f \in J(X)$. Then X is an ideal of V. Consider the canonical representation φ of the semigroup V by left translations of the ideal X. By the definition of $S(X, f)$, the left translation of X defined by the element f^* acts exactly as the function f. Consequently, $\varphi(V)$ contains the semigroup of transformations of X generated by the set $J(X)$. Hence, by Lemma 2, $\varphi(V)$ contains the symmetric semigroup on $[(n-1)/2]$ elements, where $n = |X|$. Since $V \in \mathfrak{M}$, this symmetric semigroup belongs to $\mathfrak{M}$, and, as the finite set X was chosen in an arbitrary way, $\mathfrak{M}$ contains all finite symmetric semigroups, hence all finite semigroups. Therefore $\mathfrak{M}$ is the variety of all semigroups, which contradicts Lemma 1. Lemma 4 is proven.

For an arbitrary semigroup S, we denote by GS the set of all group elements in S (the union of the subgroups of S).

Lemma 5. *If a variety* $\mathfrak{N}$ *does not contain* $\mathscr{M}_1$ *then* GS *is a subsemigroup for every* $S \in \mathfrak{N}$.

Proof. By Lemma 1 in [2], if $\mathscr{M}_1 \nsubseteq \mathfrak{N}$ then $\mathfrak{N}$ satisfies one of the identities $x^{n+1}y = xy$, $xy^{n+1} = xy$, $xy = xyf(x, y)xy$, where $n \geqslant 1$ and f is a word of the variables x and y. In the last case $f(x, y)xy$ is a right unit of xy for any $x, y \in S$. Consequently, $f(xy, f(x, y)xy) = (xy)^k$ for some $k \geqslant 1$ independent of x and y. This implies the validity of the following identities in $\mathfrak{N}$ (for some $k \geqslant 1$):

$$xy = xy[f(x, y)xy] =$$

$$= xy[f(x, y)xy]f(xy, f(x, y)xy)xy \cdot f(x, y)xy =$$

$$= xy(xy)^k xy = (xy)^{k+2}.$$

Hence $GS = S^2$ is a subsemigroup of S. Let $\mathfrak{N}$ satisfy the identity $xy = x^{n+1}y$, and take $g, h \in GS$. Then $h^{n+1} = h$, since the groups in $\mathfrak{N}$ satisfy $x^{n+1} = x$. So we have

$$gh = gh^{n+1} = (gh)^{n+1}h^n = (gh)^n gh^{n+1} = (gh)^n gh = (gh)^{n+1},$$

whence $gh \in GS$, so GS is a subsemigroup of S. Dually we settle the case when $\mathfrak{N}$ satisfies $xy^{n+1} = xy$. Lemma 5 is proven.

Lemma 6. $v(I) \vee \mathscr{M}(p) \nsubseteq \mathfrak{M}$ *for some prime* p.

Proof. Consider the semigroups $S_1 = L^0 \times C_n$ and $S_2 = M(p, p)$. The subset $\{(g; i, 1) \mid g \in C_n, i = 1, 2\}$ is a subsemigroup of $M(p, p)$ which is isomorphic to $L \times C_n$. Suppose that the amalgam of S_1 and S_2 with core $L \times C_n$ can be embedded into a semigroup V from $\mathfrak{M}$. By Lemma 5, GV is a subsemigroup of V. Since S_1 and S_2 are Cliffordian semigroups (they are unions of groups), their amalgam can be embedded into GV. Hence V can be assumed to be Cliffordian. Then $M(p, p)$ is contained in one $\mathscr{D}$-class of V, and the subsemigroup $\{0\} \times C_n$ in another one. Denote the idempotents $(1; 1, 1)$, $(1; 2, 1)$, $(1; 1. 2)$, $(c^{-1}; 2, 2)$ from $M(p, p)$ by e, i, f, j, respectively, and the idempotent $(0, 1)$ from $\{0\} \times C_n$ by h; the element $(c; 1, 1)$ from $M(p, p)$ will be denoted by u. Then we have $u = fi$, $he = hi$, whence $(0, c) = hu = hfi = hei = he = h = (0, 1)$, a contradiction, which proves Lemma 6.

Lemma 7. *All groups in* $\mathfrak{M}$ *are abelian.*

Proof. If the amalgam of a family of groups G_i $(i \in J)$, can be embedded in a semigroup S then it is contained in an $\mathcal{H}$-class of S. This $\mathcal{H}$-class contains an idempotent and is therefore a subgroup of S. Thus the variety of groups in $\mathfrak{M}$ possesses the amalgamation property. Now we apply B.H. Neumann's theorem which was formulated in the introduction, and this completes the proof of Lemma 7.

Lemma 8. *If a variety* $\mathfrak{N}$ *does not contain the varieties* $\mathcal{M}_1 - \mathcal{M}_3$, $\mathcal{M}_1^* - \mathcal{M}_3^*$, $v(P)$, $v(P)^*$, $\mathcal{M}(p) \vee v(I)$ *(for all primes* p*) and if all groups in* $\mathfrak{N}$ *are abelian, then* $\mathfrak{N}$ *consists either of inflations of completely simple semigroups or of inflations of orthodox normal bands of groups.*

Proof. We shall consider two cases: $I \in \mathfrak{N}$ and $I \notin \mathfrak{N}$. Let first $I \in \mathfrak{N}$. By our assumption, $\mathfrak{N}$ contains none of the varieties $\mathcal{M}_1 - \mathcal{M}_3$, $\mathcal{M}_1^* - \mathcal{M}_3^*$, and $\mathcal{M}(p)$ (p is any prime number). Notice that P belongs to both of the varieties $\mathcal{M}_4 = [x^2y = xy, xyz^2 = xzy^2]$ and $\mathcal{M}_5 = [x^2y = xy, xyx = yx^2, xyz^2 = yxz^2]$ hence neither they nor their duals can be contained in $\mathfrak{N}$. Therefore Lemma 6 from [2] or its dual applies, and we obtain that either $\mathfrak{N}$ consists of inflations of orthodox normal bands of groups or $\mathfrak{N}$ satisfies the identities $x^{n+1}y = xy$ and $x^n y^n z = y^n x^n z$ or the duals of these identities. Since all groups in $\mathfrak{N}$ are abelian, Propositions 3, 4 and 5 from [2] allow us to conclude that $\mathfrak{N}$ consists of finitely approximable semigroups. By Theorem 1 from [2], $\mathfrak{N}$ can be generated by a finite semigroup of one of the 20 kinds listed in that theorem. The semigroups of the last four kinds from there contain P or $P^* = Q$, while the rest are inflations of orthodox normal bands of groups.

Let now $I \notin \mathfrak{N}$. Then $\mathfrak{N}$ satisfies a non-normal identity and therefore it consists of ideal extensions of completely simple semigroups by nil semigroups. Since $\mathfrak{N}$ does not contain $\mathcal{M}_1$, the nil semigroups in it must be zero semigroups (this is an easy consequence of Lemma 1 from [2]). Next we prove that $\mathfrak{N}$ satisfies an identity of the form

(1) $$w(x, y)xy = w_1(x, y)xy^k \qquad (k > 1).$$

In fact, since $\mathfrak{N}$ does not contain $\mathcal{M}_2$, it must satisfy either an identity of the form (1) or an identity of the form $w_1(x, y)x = w_2(x, y)y$, where it can be assumed without loss of generality that the words w_1 and w_2 contain the letters x and y. Multiplying both sides of the last identity by y on the right, we obtain an identity of the form (1).

Since $\mathfrak{N}$ consists of extensions of completely simple semigroups by zero semigroups, an identity of the form $xy = (xy)^{n+1}$, $n \geqslant 1$, holds in it. This implies that $\mathfrak{N}$ satisfies the identities $x^{kn} = x^{2kn}$ and $x^{kn+1} = x^{n+1}$. Multiplying both sides of (1) by x on the left and by xy^{kn} on the right, we obtain $xyxy^{n+1} = xw_1(x, y)xy^k y^{kn} = = xw_1(x, y)xy^k = xw(x, y)xy$. Thus an identity

(2) $$xw(x, y)xy^{n+1} = xw(x, y)xy \qquad (n \geqslant 1)$$

holds in $\mathfrak{N}$.

Let $S \in \mathfrak{N}$, e be an idempotent and a be an arbitrary element of S. Substituting e for x and a for y in (2), we obtain $ua = ua^{n+1}$ for some u such that $u^n = e$. Multiplying this equality by u^{n-1} on the left, we arrive at $ea = ea^{n+1}$. Let now b be an arbitrary element of S, and put $e = (ab)^n$, then we have $ba^{n+1} = baa^n = (ba)^{n+1}a^n$, whence $ba^{n+1} = (ba)^{n+1}a^n = b(ab)^n a^{n+1} = b(ab)^n a = (ba)^{n+1} = ba$ for arbitrary $a, b \in S$. Thus we have shown that an identity of the form $xy^{n+1} = xy$ holds in $\mathfrak{N}$. Dually one proves that $\mathfrak{N}$ satisfies $xy = x^{n+1}y$. Therefore $\mathfrak{N}$ consists of inflations of completely simple semigroups, as was to be proven.

Lemma 8 together with Lemmas 1, 4, 5 and 7 shows the validity of the implication (1) → (3).

Proof of (3) → (2). If the variety $\mathfrak{N}$ consists of inflations of orthodox normal bands of abelian groups of finite exponent, then by [2], Proposition 3, it is generated by a subset of $\{L, R, N, I, C_n\}$. If $\mathfrak{N}$ consists of inflations of completely simple semigroups then $\mathfrak{N}$ is the join of the variety of zero semigroups and a variety of completely simple semigroups. By [6], the latter variety can be generated either by a semigroup $M(n, k)$ for some n and k or by a subset of $\{L, R, C_n\}$, $n \geqslant 1$.

Proof of (2) → (4). It is easy to verify that each semigroup from the set $\{L, R, N, I, C_n\}$ satisfies the identities $\{x^{n+1}y = xy = xy^{n+1}, xyzt = xzyt\}$, while both N and $M(n, k)$ satisfy $\{x^2yx = xyx^2, xz = x(yx)^n z = xz^{n+1}\}$. This implies what has to be proven.

Proof of (4) → (3). It follows from [2], Theorem 3, that the variety $[x^{n+1}y = xy = xy^{n+1}, xyzt = xzyt]$ consists of inflations of orthodox normal bands of abelian groups of finite exponent. The variety $\mathfrak{N} = [x^2yx = xyx^2, x(yx)^n z = xz = xz^{n+1}]$ satisfies a non-normal identity, hence it consists of extensions of completely simple semigroups by nil semigroups. Since $xy = xy^{n+1}$ holds in $\mathfrak{N}$, every nil semigroup in $\mathfrak{N}$ must be a zero semigroup. Substituting x for y in $xz = x(yx)^n z$ and applying $xz = xz^{n+1}$, we obtain that $x^{n+1}z = xz$ is valid in $\mathfrak{N}$. Hence $\mathfrak{N}$ consists of inflations of completely simple semigroups. Now it suffices to notice that the groups satisfying $xyx^2 = x^2yx$ are abelian.

Proof of (3) → (1). The proof of (3) → (1) will be done in a series of lemmas. First of all we note the following obvious fact.

Lemma 9. *A variety $\mathfrak{N}$ of semigroups has the amalgamation property if and only if the amalgam of any two semigroups from $\mathfrak{N}$ can be embedded in a semigroup from $\mathfrak{N}$.*

Lemma 10. *Each variety of completely simple semigroups over abelian groups of finite exponent possesses the amalgamation property.*

Proof. Take an arbitrary variety $\mathfrak{N}$ of completely simple semigroups over abelian groups of exponent n. Then, by [6], there exists a natural number k, $k \mid n$, such that $\mathfrak{N}$ consists of all completely simple semigroups of the form $M(G; I, J, \Phi)$, where G is a group of exponent n and Φ is a normalized matrix all of whose entries have orders dividing k.

Let $S_1 = M(G_1; I_1, J_1, \Phi_1)$ and $S_2 = M(G_2; I_2, J_2, \Phi_2)$ be such semigroups and S be the amalgam of S_1 and S_2 with core $U = M(G; I, J, \Phi)$. We can assume without loss of generality that $G = G_1 \cap G_2$, $I = I_1 \cap I_2$, $J = J_1 \cap J_2$ and that for every $i \in I$, $j \in J$ we have $\Phi_1(i, j) = \Phi_2(i, j) = \Phi(i, j)$. As is well known, the amalgam of any two abelian groups of exponent n can be embedded into an abelian group of the same exponent. Let G_0 be an abelian group of exponent n in which the amalgam of G_1 and G_2 with core G is embedded. Then S can be embedded into the completely simple semigroup $M(G_0; I^*, J^*, \Phi^*) = T$, where $I^* = I_1 \cup I_2$, $J^* = J_1 \cup J_1$, and

$$\Phi^*(i,j) = \begin{cases} \Phi_1(i,j) & \text{if } i \in I_1,\ j \in J_1 \\ \Phi_2(i,j) & \text{if } i \in I_2,\ j \in J_2 \\ 1 & \text{otherwise.} \end{cases}$$

Since the order $|\Phi^*(i,j)|$ divides k for all $i \in I^*$, $j \in J^*$, we have $T \in \mathfrak{N}$. Hence the amalgam of any two semigroups from $\mathfrak{N}$ can be embedded in a semigroup from $\mathfrak{N}$. By Lemma 9 this implies that $\mathfrak{N}$ has the amalgamation property, which proves Lemma 10.

Lemma 11. *Every variety of orthodox normal bands of abelian groups of finite exponent possesses the amalgamation property.*

Proof. Let $\mathfrak{N}$ be such a variety, $\mathfrak{B}$ be the subvariety of bands in $\mathfrak{N}$, and $\mathfrak{L}$ be the subvariety of semilattices of groups in $\mathfrak{N}$. It follows from [2], Proposition 3, that every semigroup in $\mathfrak{N}$ is a subdirect product of semigroups from $\mathfrak{B}$ and $\mathfrak{L}$, respectively.

Let S be a semigroup in $\mathfrak{N}$. Green's equivalence $\mathscr{H}_S$ is a congruence on S, and $S/\mathscr{H}_S$ is the greatest idempotent homomorphic image of S. The greatest homomorphic image of S in $\mathfrak{L}$ is defined by the smallest congruence σ_S on S which glues together the idempotents in each $\mathscr{D}$-class. As is easy to check,

$$(a,b) \in \sigma_S \Leftrightarrow \exists e, f \in S\ (e^2 = e,\ f^2 = f,\ a = ebf,\ b = fae).$$

By the foregoing, σ_S and $\mathscr{H}_S$ have trivial intersection. Notice that the restrictions of σ_S and $\mathscr{H}_S$ to any subsemigroup U of S coincide with σ_U and $\mathscr{H}_U$, respectively.

Consider now an amalgam A of semigroups S and T from $\mathfrak{N}$ with core U. In view of what has been said before, the relations $\mathscr{H}_A$ and σ_A on A are correctly defined if we set them equal to $\mathscr{H}_S$ and σ_S on S and to $\mathscr{H}_T$ and σ_T on T, respectively. They are congruences of the partial groupoid A. The factor groupoids of A under these congruences are amalgams of semigroups from $\mathfrak{B}$ and $\mathfrak{L}$, respectively. As was shown in [4] and [3], both of these varieties have the amalgamation property, hence the corresponding factor groupoids can be embedded into semigroups S_1 and T_1 from $\mathfrak{B}$ and $\mathfrak{L}$, respectively, and then A can be embedded into $S_1 \times T_1$. This proves Lemma 11.

Let S_1 be an inflation of a semigroup S by means of the family of sets $\{X_a \mid a \in S\}$, and T_1 be an inflation of a semigroup T by means of the family of sets $\{Y_a \mid a \in T\}$. Let, further, $S \cap T = U$ be a subsemigroup in S and in T, and $S_1 \cap T_1 = U_1$ be an inflation of U. Suppose, finally, that the amalgam $S \cup T$ can be embedded into a semigroup R. For every $a \in R$ put

$$Z_a = \begin{cases} X_a \cup Y_a & \text{if } a \in U \\ X_a & \text{if } a \in S \setminus U \\ Y_a & \text{if } a \in T \setminus U \\ \{a\} & \text{if } a \in R \setminus (S \cup T). \end{cases}$$

Clearly, the amalgam $S_1 \cup T_1$ can be embedded in the inflation of R by means of the family of sets $\{Z_a \mid a \in R\}$. Therefore we have

Lemma 12. *If a variety* $\mathfrak{N}$ *has the amalgamation property then so does the variety consisting of inflations of semigroups from* $\mathfrak{N}$.

The validity of (3) → (1) follows from Lemmas 10, 11, 12.

The Theorem is proven.

The author thanks L.N. Ševrin for his guidance during this work and M.V. Volkov for valuable suggestions.

Added in proof. Some months after this paper had been submitted, we got to know of G.T. Clarke's preprint Semigroup Varieties with the Amalgamation Property (Algebra Paper, Nr. 53, Dept. of Math., Monash Univ., 1981) in which an analogous theorem is proven. However, our proof is different, and maybe this proof will be useful in other investigations, too, therefore we decided to publish our paper unchanged.

REFERENCES

[1] A.H. Clifford – G.B. Preston, *The Algebraic Theory of Semigroups,* I, II, Amer. Math. Soc. Providence, R. I., 1961, 1967.

[2] E.A. Golubov – M.V. Sapir, Varieties of finitely approximable semigroups, *Dokl. Akad. Nauk SSSR,* 247 (1979), 1037–1041 (in Russian).

[3] T.E. Hall, Inverse semigroup varieties with the amalgamation property, *Semigroup Forum,* 16 (1978), 37–51.

[4] T. Imaoka, Free product with amalgamation of bands, *Mem. Fac. Lit. Sci. Shimane Univ.,* 10 (1976), 7–17.

[5] H. Neumann, *Varieties of Groups,* Springer, Berlin, 1965.

[6] V.V. Rasin, On the lattice of varieties of completely simple semigroups, *Semigroup Forum,* 17 (1979), 113–122.

[7] *The Sverdlovsk Tetrad,* 2nd ed., Sverdlovsk Univ., Sverdlovsk, 1979 (in Russian).

M.V. Sapir

Uralskiĭ Gosudarstvennyĭ Universitet, Matfak, 620 083 Sverdlovsk, ul. Lenina 51, USSR.

COLLOQUIA MATHEMATICA SOCIETATIS JÁNOS BOLYAI
39. SEMIGROUPS, SZEGED (HUNGARY), 1981.

ON MAXIMAL A-IDEALS IN SEMIGROUPS

L. SATKO – O. GROŠEK

The notion of an A-ideal in semigroups has been introduced in [1]. In [2] and [3] minimal A-ideals and the smallest A-ideal of a semigroup have been studied.

In this paper we are dealing with two questions concerning maximal A-ideals of S. For convenience of the reader we recall the basic definitions.

An LA-ideal G of a semigroup S is a nonempty subset G of S such that for any $s \in S$ we have $sG \cap G \neq \phi$. In other words, for any $s \in S$ there exists an element $g \in G$ such that $sg \in G$. An RA-ideal G of S is defined analogously, i.e. $Gs \cap G \neq \phi$ for any $s \in S$. An A-ideal of a semigroup S is a nonempty subset of S which is both an LA-ideal and RA-ideal of S.

If G is an LA-ideal of S then any subset H of S such that $G \subset H \subset S$ is also an LA-ideal of S. A maximal LA-ideal of S, say G_m, is an LA-ideal of S such that $G_m \neq S$ and for any LA-ideal H satisfying $G_m \subset H$, $G_m \neq H$ we have $H = S$. This implies that any maximal LA-ideal of S can be written in the form $S \setminus x$ where x is a suitably chosen element of S. Note that there exist semigroups without (proper) LA-ideals (see [1]).

1. THE INTERSECTION OF ALL MAXIMAL LA-IDEALS

The intersection M_L of all LA-ideals of a given semigroup S is described in [3]. The following theorem holds:

Theorem 1 ([3]). *The intersection $M_L = M_L(S)$ of all LA-ideals of a semigroup S is*

(i) *either the empty set;*

(ii) *or the subsemigroup of all left zero elements of S (if such exist);*

(iii) *or one of the three nonempty subsets of a two-element group ideal of S (if such a group ideal exists).*

Remark 1. If the last possibility occurs, the whole (two-element) group ideal G is equal to M_L only in the trivial case $G = S$.

Notation. In accordance with [1] the class of all left zero semigroups together with the class of all groups of order 2 will be denoted by $\mathfrak{G}_{\mathscr{L}}$.

Denote by $Q_L = Q_L(S)$ the intersection of all maximal LA-ideals of a given semigroup S. If $S \in \mathfrak{G}_{\mathscr{L}}$ then $M_L = S$ and $Q_L = \phi$. If $S \notin \mathfrak{G}_{\mathscr{L}}$ then clearly $M_L \subset Q_L$. In this case we prove $M_L = Q_L$.

Let $M_L = \phi$ and $S \notin \mathfrak{G}_{\mathscr{L}}$. This is possible only if to any $x \in S$ there exists an LA-ideal G_x of S such that $x \notin G_x$. But then the set $S \setminus x \supset G_x$ is a maximal LA-ideal of S and this implies that Q_L is empty.

If $M_L \neq \phi$ and $S \notin \mathfrak{G}_{\mathscr{L}}$ suppose that $M_L \neq Q_L$. Let $x_0 \in Q_L \setminus M_L$. The set $S \setminus x_0$ cannot be a maximal LA-ideal of S since the intersection of all maximal LA-ideals of S (i.e. Q_L) contains x_0 (and $x_0 \notin S \setminus x_0$). But then the set $S \setminus x_0$ contains no LA-ideal of S. In other words if a subset $G \subset S$ does not contain the element x_0 it cannot be an LA-ideal of S. Thus any LA-ideal of S must contain the element x_0. This is a contradiction to $x_0 \notin M_L$.

Hence if $S \notin \mathfrak{G}_{\mathscr{L}}$, then $M_L = Q_L$ and we have proved the following lemma:

Lemma. *Let $S \notin \mathfrak{G}_{\mathscr{L}}$. The intersection of all maximal LA-ideals of S is equal to the intersection of all LA-ideals of S.*

This lemma together with Theorem 1 and Remark 1 imply the next theorem:

Theorem 2. *Let $S \notin \mathfrak{G}_{\mathscr{L}}$. Then the intersection Q_L of all maximal LA-ideals of a semigroup S is*

(i) *either the empty set;*

(ii) *or the (proper) subsemigroup of all left zero elements of S (if such exist);*

(iii) *or one of the elements of the (two-element) group ideal of S (if such a group ideal exists).*

Remark 2. If the last possibility occurs, then the one-element subsets of G are not LA-ideals of S. Thus Q_L is a proper LA-ideal of S only if Q_L is the set of all left zero elements of S and $Q_L \neq S$. It is known that in this case Q_L is the kernel of S.

An analogous theorem holds for (two-sided) A-ideals of S. Denote by $\mathfrak{G}_{\mathscr{R}}$ the class of all right-zero semigroups together with all groups of order 2. We have:

Theorem 3. *Let $S \notin \mathfrak{G}_{\mathscr{L}} \cup \mathfrak{G}_{\mathscr{R}}$. Then the intersection of all maximal A-ideals of S is*

(i) *either the empty set;*

(ii) *or the (proper) subsemigroup of all left zero elements of S (if such exist);*

(iii) *or the (proper) subsemigroup of all right zero elements of S (if such exist);*

(iv) *or one of the elements of the two-element group ideal of* S *(if such a group ideal exists).*

Remark 3. The intersection Q of all maximal A-ideals of a semigroup S is an A-ideal of S only if Q is either the subsemigroup of all left zero elements of S or the subsemigroup of all right zero elements of S.

2. THE EXISTENCE OF A GREATEST LA-IDEAL OF S

If a semigroup S contains a greatest LA-ideal G, then G is the unique maximal LA-ideal of S. Hence there exists an element $a \in S$ such that $G = S \setminus a$ and any proper LA-ideal of S is contained in $S \setminus a$. Of course in this case the intersection Q_L of all maximal LA-ideals of S is equal to $S \setminus a$ and this set is at the same time a proper LA-ideal of S. By Remark 2 this is possible only if $S \setminus a$ is the (proper) subsemigroup of all left zero elements of S. This result can be formulated in the following way:

Theorem 4. *If a semigroup* S *contains a greatest* LA*-ideal, then* $S = L \cup a$, *where* L *is the set of all left zero elements of* S *and* $a \notin L$.

The converse to Theorem 4 is given by the following theorem.

Theorem 5. *Let* L *be a left zero semigroup and* $a \notin L$. *Then we can define a multiplication on the set* $S = L \cup a$ *such that* S *is a semigroup and* L *is a greatest* LA*-ideal of* S.

Proof. Consider first necessary conditions for this multiplication. By Theorem 4, L is necessarily the set of all left zero elements of S and thus it is the kernel of S. Hence the left zero multiplication on L must be in any case preserved. We have only to define a^2, ta and at for any $t \in L$. Since any $t \in L$ is a left zero element of S, $ta = t$ for any $t \in L$.

Now we will consider two possibilities: $a^2 = a$ and $a^2 \neq a$.

(1) Let $a^2 = a$. Since L is an ideal (the kernel) of S, $at \in L$ for any $t \in L$. Let $L_1 = aL$ and $L_2 = L \setminus L_1$. Then $L = L_1 \cup L_2$, $L_1 \cap L_2 = \phi$ and $L_1 \neq \phi$. Now for any $s \in L_1$ there exists a $t \in L$

such that $s = at$ and thus $as = a(at) = (aa)t = at = s$ for any $s \in L_1$. For any $s \in L_2$ we have only one condition: $as \in L_1$. Thus in this case the necessary conditions for the multiplication to be defined on S are:

(i) $a^2 = a$;

(ii) $st = s$ for any $s, t \in L$;

(iii) $ta = t$ for any $t \in L$;

(iv) $L = L_1 \cup L_2$, $L_1 \cap L_2 = \phi$, $L_1 \neq \phi$ and $at = t$ for any $t \in L_1$ and $at \in L_1$ for any $t \in L_2$.

To complete the proof of Theorem 5 in this case we prove conversely that necessary conditions (i)–(iv) are also sufficient. I.e. if a multiplication on $S = L \cup a$ satisfies (i)–(iv), then $S = L \cup a$ is a semigroup with the greatest LA-ideal L. It is clear that we have only to check the associativity for this multiplication.

Let $s \in L$ and $r, t \in S$, then $(sr)t = st = s = s(rt)$. If $r \in L$ and $t \in S$, then $(ar)t = ar = a(rt)$. If $t \in S$, then $(aa)t = at = a(at)$ (since $at \in L_1$). This completes the proof in the case $a^2 = a$.

The semigroups considered can be described in the following way. Let $S = a \cup L$, $L = L_1 \cup L_2$, $L_1 \cap L_2 = \phi$, $L_1 \neq \phi$, $L_1 = \{x, y, z, \ldots\}$, $L_2 = \{u, v, w, \ldots\}$ then the multiplication table of the semigroup S is the following:

	a	x	y	z	$\ldots$	u	v	w	$\ldots$
a	a	x	y	z	$\ldots$	y	z	x	$\ldots$
x	x	x	x	x	$\ldots$	x	x	x	$\ldots$
y	y	y	y	y	$\ldots$	y	y	y	$\ldots$
.	.	.	.	.	$\ldots$	.	.	.	$\ldots$
u	u	u	u	u	$\ldots$	u	u	u	$\ldots$

(2) Suppose now, as the second possibility, that $a \neq a^2$. Then necessarily $a^2 = x_0 \in L$. Since L is an ideal of S, $aS \subset L$. Denote $aS = L_1$. Then $L = L_1 \cup L_2$, $L_1 \cap L_2 = \phi$ and $L_1 \neq \phi$. This implies that for any $t \in L_1$ there exists an $s \in S$ such that $t = as$ and

$at = a(as) = (aa)s = x_0 s = x_0$. For any $t \in L_2$ we have only one condition namely $at \in L_1$. Thus in this case necessary conditions for the multiplication to be defined on S are:

(i) $S = a \cup L$, $a \notin L$, $L = L_1 \cup L_2$, $L_1 \cap L_2 = \phi$ and $L_1 \neq \phi$;

(ii) $a^2 = x_0 \in L_1$;

(iii) $st = s$ for any $s, t \in L$;

(iv) $ta = t$ for any $t \in L$;

(v) $at = x_0$ for any $t \in L_1$ and

(vi) for any $t \in L_2$, at is an arbitrarily chosen element from L_1.

These necessary conditions are also in this case sufficient. I.e. if a multiplication on $S = L \cup a$ satisfies (i)–(vi) then S is a semigroup having L as its greatest LA-ideal. We have again only to prove the associativity of the given multiplication.

For any $s \in L$ and $r, t \in S$ we have $(sr)t = st = s = s(rt)$. If $r \in L$ and $t \in S$, then $(ar)t = ar = a(rt)$. If $t \in S$, then $(aa)t = x_0 t = x_0 = a(at)$ (since $at \in L_1$). This completes the proof of Theorem 5 in the case $a^2 \neq a$.

Let $L_1 = \{x_0, y, z, \ldots\}$, $L_2 = \{u, v, w, \ldots\}$. The semigroups just discussed have the following multiplication table:

	a	x_0	y	z	...	u	v	w	...
a	x_0	x_0	x_0	x_0	...	y	z	x_0	...
x_0	x_0	x_0	x_0	x_0	...	x_0	x_0	x_0	...
y	y	y	y	y	...	y	y	y	...
.	.	.	.	.	...	.	.	.	...
u	u	u	u	u	...	u	u	u	...

Remark 4. Note that in the preceding proof we have described all semigroups having a greatest LA-ideal.

Since the intersection Q of all (two-sided) A-ideals of a semigroup

S is an A-ideal only if Q is either the set of all left zero elements of S or the set of all right zero elements of S, an analogous theorem concerning the greatest A-ideal can be obtained in a similar way.

Theorem 6. *A semigroup S contains a greatest A-ideal H if and only if $S = H \cup a$ where H is either the subsemigroup of all left zero elements of S or the subsemigroup of all right zero elements of S and $a \notin H$.*

Examples of such semigroups are given in the proof of Theorem 5. Further examples can be obtained from the preceding ones by interchanging the rows and columns in the given multiplication tables.

REFERENCES

[1] O. Grošek – L. Satko, A new notion in the theory of semigroups, *Semigroup Forum,* 20 (1980), 233–240.

[2] L. Satko – O. Grošek, On minimal A-ideals of semigroups, *Semigroup Forum,* 23 (1981), 283–295.

[3] O. Grošek – L. Satko, Smallest A-ideals in semigroups, *Semigroup Forum,* 23 (1981), 297–309.

O. Grošek – L. Satko

Elektrotechnická fakulta SVŠT, Gottwaldovo nám. 19, 812 19 Bratislava, Czechoslovakia.

COLLOQUIA MATHEMATICA SOCIETATIS JÁNOS BOLYAI
39. SEMIGROUPS, SZEGED (HUNGARY), 1981.

ATTAINABILITY AND SOLVABILITY FOR CLASSES OF ALGEBRAS

L.N. ŠEVRIN – L.M. MARTYNOV

INTRODUCTION

Let $\mathfrak{X}$ be an arbitrary variety of algebras (of fixed type). For any algebra A of the given type, we call its verbal $\mathfrak{X}$-congruence the smallest congruence on A which yields a factor algebra in $\mathfrak{X}$ – this is known to exist. Distinguish those classes of the verbal $\mathfrak{X}$-congruence on A which are subalgebras; if there is none of them, then we are finished. On each of these subalgebras consider again the verbal $\mathfrak{X}$-congruence. It may happen that on all the distinguished subalgebras the verbal $\mathfrak{X}$-congruences are the universal relations (i.e. they coincide with the Cartesian squares of the corresponding subalgebras); in this case again we are done. Otherwise we continue to construct the "verbal chain". For the first case the famous Tamura – Petrich theorem [47], [59] yields an important example. It considers semigroups, and the role of $\mathfrak{X}$ is played by the variety of semilattices; the theorem states that the components of the finest semilattice decomposition of an arbitrary semigroup are semilattice indecomposable. In the second case it may occur that the process of construction of the verbal chain can be arbitrarily long. An important example of this kind is presented by

M a l c e v ' s famous theorem [30] which concerns groups and, as $\mathfrak{X}$, the variety of abelian groups; the theorem states that for an arbitrary ordinal number γ there is a group whose commutator series stabilizes exactly at this γ.

So the variety of semilattices in the semigroup case and that of abelian groups in the group case constitute examples of situations in diametrical opposition. How can one take in all the richness of existing situations in this picture? The aim of the present paper is just to make an attempt to show some ways in the direction of answering this general question.

That case is of special interest when the verbal chain descends to one-element subalgebras of the given algebra (and thus turns into a verbal series, as one could say). A classical example for such a situation is given by solvable groups and *RK*-groups: here $\mathfrak{X}$ is the variety of abelian groups, and the commutator series of the given group attains the one-element subgroup in a finite number of steps in the first case, and at some (possibly transfinite) step in the second case. How long can be such verbal series? Here, too, the varieties of semilattices and abelian groups, respectively, yield examples for opposed situations: in the first case the length is always at most 1, in the second case it can be an arbitrary ordinal number.

These problem settings are of certain interest from the point of view of classes of universal algebras: stress is laid in them upon properties of the "coarse structure" of algebras and upon the possibilities of "putting them together" by extensions with factors from the fixed variety. To speak in a descriptive language: how deep can be the "shafts" or how high can be the "sky-scrapers" if only blocks of the same fixed type of material are used for constructing them? It is visible that these questions are connected with problems about the multiplication of classes of algebras [31]: here we consider only powers (possibly infinite ones, too) but not arbitrary products. In particular, the case when our chain stabilizes at the first step (that was considered above) is related to idempotency of the corresponding class (and is often simply equivalent to idempotency); the same stabilization has also connections with radical theory (for more details about these connections see Section 2).

The term "attainability" (for varieties and also for quasi-varieties) was first used by Tamura [60], [61] to denote the situation when stabilization at the first step, as just considered, takes place in every algebra of the given type. In particular, it was proved in [61] that the variety of semilattices is the only non-trivial attainable variety in the class of all semigroups. A.I. Mal'cev [31], [32] considered a lot of properties of the notion of attainability for prevarieties in connection of his investigations of the operation of multiplication of classes of algebras. In order to make the above questions more precise, the notions of γ-attainability and γ-solvability (γ is an ordinal number) and some closely related ones like finite attainability were introduced in [52]; according to this terminology, attainability in the sense of [31], [32], [60], [61] is called 1-attainability. The above theorem of Tamura was sharpened in [52]: it turned out that the variety of semilattices is the only non-trivial finitely attainable variety of semigroups; the finitely attainable varieties of associative rings are also described there. The same paper outlined the basic problems of this field, on which a number of mathematicians started to work rather intensively during the years after. The first results in this direction were exposed by the first of the present authors in his survey talks at the Conference on Universal Algebra in Oberwolfach, 1971, and at the 12th All-Union Conference on Algebra in Sverdlovsk, 1973. Since that time many papers appeared which dealt with these questions. The present paper aims at presenting the basic results obtained in this field and at formulating several open problems.

The paper consists of 16 sections. In Section 1 we introduce the basic definitions and notations to be used throughout. In Section 2 we present the basic problem settings and formulate a number of general remarks and open problems. The rest is devoted to a survey of results and open problems for concrete varieties of algebras: semigroups, semigroups with zero, monoids, inverse semigroups, groups, modules, rings and linear algebras, linear Ω-algebras, lattices, groupoids, quasi-groups, loops, unars; the last section is a table of results.

The authors express their sincere gratitude to László Márki for discussions on the topic of the present paper and for his translating the Russian text into English.

1. BASIC DEFINITIONS AND NOTATIONS

All the classes of algebras that will be considered, are supposed to be abstract (i.e. closed under isomorphic images) and to be subclasses of the class of all algebras of a fixed type. Recall that a class of algebras is called a prevariety (in other terms, a replete class) if it is closed with respect to subalgebras, Cartesian products, and it contains the one-element algebra.

By a partial equivalence on a set M we mean a symmetric and transitive binary relation in M. If ϵ is a partial equivalence on M then for each element $x \in M$ we put $x^{\epsilon} = \{y \in M \mid (x, y) \in \epsilon\}$ and call x^{ϵ} the ϵ-class corresponding to x; if ϵ is not an equivalence (i.e. it is not reflexive) then there are empty ϵ-classes. The classes of a partial equivalence on a set M form a family of pairwise disjoint subsets of M; conversely, every family of this kind corresponds to a partial equivalence on M.

When considering partial equivalence, it is worth visualizing them as families of subsets like above. In what follows we shall deal mainly with such cases when the members of these families are subalgebras of an algebra A; for short, we shall call an arbitrary disjoint family of subalgebras a scatter, and the subalgebras it consists of will be named the components of the scatter. On the set of all scatters of a given algebra it is natural to define the partial order induced by the relation of inclusion of binary relations:

$$\{A_i \mid i \in I\} \leqslant \{B_j \mid j \in J\} \Leftrightarrow \forall i\, \exists j\, (A_i \subseteq B_j).$$

For the sake of generality we shall sometimes consider the empty subalgebra, too; in these cases it makes sense to speak of the empty scatter. This convention makes it always possible to speak of the complete lattice of scatters of a given algebra. A scatter will be called trivial if either it is empty or all its components are one-element algebras. Clearly, there can be many trivial scatters.

For a congruence ρ on an algebra A it will be important for us to consider the scatter whose components are exactly those ρ-classes which are subalgebras of A; this scatter will be called the kernel of the congruence ρ. Notice that the kernel may be empty.

Let $\mathfrak{X}$ be a subclass of a class $\mathfrak{K}$ of algebras, $A \in \mathfrak{K}$, and ρ be a congruence on A such that $A/\rho \in \mathfrak{X}$, then ρ is called an $\mathfrak{X}$-congruence.

Let us fix a prevariety $\mathfrak{K}$ of algebras and in it a subprevariety $\mathfrak{X}$. The set of all $\mathfrak{X}$-congruences of an arbitrary algebra A in $\mathfrak{K}$ has a smallest element, which will be called the preverbal $\mathfrak{X}$-congruence of A and will be denoted by $\rho(\mathfrak{X}, A)$. If $\mathfrak{X}$ is a variety (quasi-variety) then $\rho(\mathfrak{X}, A)$ will be called verbal (quasi-verbal).

We shall call the kernel of the congruence $\rho(\mathfrak{X}, A)$ the $\mathfrak{X}$-preverbal of the algebra A, and denote it by $\mathfrak{X}(A)$. If $\mathscr{F} = \{A_i \mid i \in I\}$ is a scatter of an algebra A then the scatter $\mathfrak{X}(\mathscr{F}) = \bigcup_{i \in I} \mathfrak{X}(A_i)$ will be called the $\mathfrak{X}$-preverbal of $\mathscr{F}$. We define the $\mathfrak{X}$-preverbal of step α of an algebra A for any ordinal number α by transfinite induction, and denote it by $\mathfrak{X}^\alpha A$: we put $\mathfrak{X}^0 A = A$; if α is not a limit ordinal then let $\mathfrak{X}^\alpha A = \mathfrak{X}(\mathfrak{X}^{\alpha-1} A)$; if α is a limit ordinal then put $\mathfrak{X}^\alpha A = \bigwedge_{\beta < \alpha} \mathfrak{X}^\beta A$ (here $\wedge$ is the symbol of intersection in the lattice of scatters). Thus we obtain a descending chain of scatters

$$(1.1) \qquad A = \mathfrak{X}^0 A \geqslant \mathfrak{X}^1 A \geqslant \ldots \geqslant \mathfrak{X}^\alpha A \geqslant \ldots,$$

which will be called the $\mathfrak{X}$-preverbal chain of the algebra A. For some ordinal number γ this chain stabilizes, i.e.

$$(1.2) \qquad \mathfrak{X}^\gamma A = \mathfrak{X}^{\gamma+1} A.$$

In this case we shall say that the class $\mathfrak{X}$ is γ-attainable on the algebra A, and the smallest γ with property (1.2) will be called the step of attainability of $\mathfrak{X}$ on A. If there is an ordinal number γ such that the class $\mathfrak{X}$ is γ-attainable on all algebras of the class $\mathfrak{K}$, then $\mathfrak{X}$ will be called γ-attainable in the class $\mathfrak{K}$, and the smallest γ with this property is said to be the step of attainability of $\mathfrak{X}$ in $\mathfrak{K}$. A class $\mathfrak{X}$ is called attainable in $\mathfrak{K}$ if it is γ-attainable in $\mathfrak{K}$ for some γ.

As was noticed above, the term "attainability" was introduced first in [60], [61] and considered in [31], [32] in the sense of 1-attainability in our terminology.

A prevariety $\mathfrak{X}$ is said to be finitely attainable in $\mathfrak{K}$ if for every algebra A from $\mathfrak{K}$ there is a natural number n such that $\mathfrak{X}$ is n-attainable on A.

The equality (1.2) is clearly satisfied if $\mathfrak{X}^{\gamma}A$ is a trivial scatter. In this case the algebra A is said to be γ-$\mathfrak{X}$-solvable, and the smallest ordinal number with this property is called the step of $\mathfrak{X}$-solvability of the algebra A. If there is an ordinal number γ such that all algebras from $\mathfrak{K}$ are γ-$\mathfrak{X}$-solvable, then the class $\mathfrak{K}$ is called γ-$\mathfrak{X}$-solvable, and the smallest ordinal number with this property is the step of $\mathfrak{X}$-solvability of the class $\mathfrak{K}$. If γ is a natural number then the γ-$\mathfrak{X}$-solvable algebras will also be called finitely $\mathfrak{X}$-solvable. A prevariety $\mathfrak{K}$ is said to be finitely $\mathfrak{X}$-solvable if each of its algebras is finitely $\mathfrak{X}$-solvable.

If the kernel of each congruence on every algebra from $\mathfrak{K}$ consists of one subalgebra (this is the case in a number of classical algebras: groups, rings, etc.) then we shall not distinguish the one-element family of subalgebras and this subalgebra itself, i.e. by the kernel we shall mean the subalgebra. Now it is clear that if $\mathfrak{K}$ is the class of all groups, $\mathfrak{X}$ is the variety of abelian groups, then the notion of finitely $\mathfrak{X}$-solvable group coincides with the well-known concept of solvable group (in fact, we took the name from this classical case), while the notion of $\mathfrak{X}$-solvable group is the same as that of RK-group [23].

Notice that the above concepts related to attainability and solvability were defined in the paper [52] in terms of partial equivalences.

By the A-spectrum of a prevariety $\mathfrak{X}$ in $\mathfrak{K}$ we mean the class $\mathrm{Spec}_A(\mathfrak{X}, \mathfrak{K})$ of all ordinal numbers which occur as steps of attainability of $\mathfrak{X}$ on some algebra from $\mathfrak{K}$. The S-spectrum of $\mathfrak{X}$ in $\mathfrak{K}$ is the class $\mathrm{Spec}_S(\mathfrak{X}, \mathfrak{K})$ of all ordinal numbers which occur as steps of $\mathfrak{X}$-solvability of some algebra from $\mathfrak{K}$.

For any ordinal number α we denote by $W(\alpha)$ the totally ordered set of all ordinals less than α (the order is the natural one); ω is the first nfinite ordinal number; $N = W(\omega)$ stands for the set of natural numbers, and W for the class of all ordinal numbers with the natural order.

A prevariety $\mathfrak{X}$ is called A-free or S-free in $\mathfrak{K}$ if $\mathrm{Spec}_A(\mathfrak{X}, \mathfrak{K}) = W$ or $\mathrm{Spec}_S(\mathfrak{X}, \mathfrak{K}) = W$, respectively.

Let us introduce also some other definitions which will be needed in what follows. We shall make no distinction in the notation of an algebra and of its base set. If B is a subalgebra of A then this will be denoted by writing $B \leqslant A$. For $M \subseteq A$, the subalgebra generated by M will be denoted by $\langle M \rangle$. If $A = \langle a \rangle$ for some element $a \in A$ then the algebra A will be called monogenic. An element e of an algebra A is said to be idempotent if $\langle e \rangle = \{e\}$. An algebra A is called simple if it has no congruences other than the identical relation Δ_A and the universal relation ∇_A.

A subclass $\mathfrak{X}$ of a class $\mathfrak{K}$ of algebras is called proper if $\mathfrak{X} \neq \mathfrak{K}$, and non-trivial if, in addition, $\mathfrak{X} \neq \mathfrak{E}$, where $\mathfrak{E}$ is the class of one-element algebras. If $\mathfrak{X}$ is a prevariety and there is an identity which is non-trivial in $\mathfrak{K}$ and is satisfied in all algebras from $\mathfrak{X}$, then we shall say that $\mathfrak{X}$ is bounded in $\mathfrak{K}$. For a family Q of quasi-identities (in particular, of identities) $[Q]$ denotes the quasi-variety of algebras defined by Q. A quasi-identity is called variable-preserving if each variable occurring in the premise appears in the conclusion. A quasi-variety of algebras is said to be variable-preserving if it can be defined by a system of variable-preserving quasi-identities. An identity $u = v$ is called normal if u and v contain the same variables; otherwise it is called non-normal. A variety of algebras is called normal if it admits a basis of normal identities, and non-normal if it does not.

A prevariety $\mathfrak{K}$ is said to be regular if every congruence on each algebra from $\mathfrak{K}$ is determined by any of its classes; $\mathfrak{K}$ is called regular for trivial $\mathfrak{X}$-preverbals if for any $A \in \mathfrak{K}$, $\rho(\mathfrak{X}, A) = \Delta_A$ whenever the $\mathfrak{X}$-preverbal $\mathfrak{X}(A)$ is trivial.

A prevariety $\mathfrak{K}$ is a Schreier prevariety if each subalgebra with more than one element of any free algebra in $\mathfrak{K}$ is itself free (in $\mathfrak{K}$).

For two subclasses $\mathfrak{A}$ and $\mathfrak{B}$ of a class $\mathfrak{K}$ of algebras, we denote by $\mathfrak{A} \underset{\mathfrak{K}}{\circ} \mathfrak{B}$ their $\mathfrak{K}$-product, i.e. the class of all those algebras from $\mathfrak{K}$ which admit a $\mathfrak{B}$-congruence such that all non-empty components of its kernel belong to $\mathfrak{A}$. If $\mathfrak{X} \underset{\mathfrak{K}}{\circ} \mathfrak{X} = \mathfrak{X}$ for a subclass $\mathfrak{X}$ in $\mathfrak{K}$, then $\mathfrak{X}$ is said to be idempotent in $\mathfrak{K}$ or extension closed in $\mathfrak{K}$.

If $\mathfrak{X}$ and $\mathfrak{K}$ are prevarieties, $\mathfrak{X} \subseteq \mathfrak{K}$, $A \in \mathfrak{K}$ and $\rho(\mathfrak{X}, A) = \nabla_A$, then the algebra A is said to be $\mathfrak{X}$-radical or $\mathfrak{X}$-indecomposable. Denote by $R\mathfrak{X}$ the class of all $\mathfrak{X}$-radical algebras from $\mathfrak{K}$. If $\mathfrak{X}$ is 1-attainable in $\mathfrak{K}$ then we have $R\mathfrak{X} \underset{\mathfrak{K}}{\circ} \mathfrak{X} = \mathfrak{K}$.

A prevariety $\mathfrak{K}$ is said to be transpreverbal for $\mathfrak{X}$ if the following condition is satisfied for every algebra $A \in \mathfrak{K}$ and every congruence ρ on A: if $\mathscr{K}$ is the kernel of ρ then $\mathfrak{X}(\mathscr{K})$ is the kernel of some congruence σ on A the restriction of which coincides on every component of $\mathscr{K}$ with the preverbal $\mathfrak{X}$-congruence of the given component; in this case we shall say that σ is an $\mathfrak{X}$-extension of ρ. A class $\mathfrak{K}$ is called transpreverbal if it is transpreverbal for each of its subprevarieties.

For a fixed prevariety $\mathfrak{K}$ of algebras and a subclass $\mathfrak{X}$ of $\mathfrak{K}$, we shall use the following notations.

$\mathfrak{X}^{\gamma}$ – the class of all γ-$\mathfrak{X}$-solvable algebras from $\mathfrak{K}$;

$\mathfrak{X}^{W}$ – the class of all $\mathfrak{X}$-solvable algebras from $\mathfrak{K}$;

$\mathfrak{X}^{(n+1)} = \mathfrak{X}^{(n)} \underset{\mathfrak{K}}{\circ} \mathfrak{X}$, $n = 1, 2, \ldots$; $\mathfrak{X}^{(1)} = \mathfrak{X}$;

$G_P(\mathfrak{K})$ – the groupoid of prevarieties from $\mathfrak{K}$ (with respect to the operation of $\mathfrak{K}$-product);

$G_Q(\mathfrak{K})$ – the partial groupoid of sub-quasi-varieties of $\mathfrak{K}$;

$G_V(\mathfrak{K})$ – the partial groupoid of subvarieties of $\mathfrak{K}$.

2. PROBLEM SETTING AND GENERAL REMARKS

The first natural problem to raise is

Problem IA (attainability).

The problem can be formulated as follows; adding the letters V, Q, P means that the problem is related to varieties, quasi-varieties, and prevarieties, respectively.

IAV. Characterize all subvarieties of a given variety $\mathfrak{K}$ of algebras which are attainable in $\mathfrak{K}$.

IAQ. The same for quasi-varieties.

IAP. The same for prevarieties.

For many classes $\mathfrak{K}$ of algebras, the attainable varieties in $\mathfrak{K}$ turned out to be the same as the 1-attainable varieties. It would be nice to have useful general conditions (at least sufficient ones) to describe varieties $\mathfrak{K}$ having the property that each of their subvarieties is either 1-attainable or not attainable in $\mathfrak{K}$; in this case we shall say that $\mathfrak{K}$ has the alternation property for attainability of subvarieties. This problem is interesting for quasi-varieties and prevarieties, too.

In any case, the description of 1-attainable classes has to be considered not only as a special case but as a problem in itself; the special interest in it is due to its connection with radical theory (see below).

Problem I_1A (1-attainability).

I_1AV. Characterize all subvarieties of a given variety $\mathfrak{K}$ of algebras which are 1-attainable in $\mathfrak{K}$.

I_1AQ. The same for quasi-varieties.

I_1AP. The same for prevarieties.

Problem I_1AV was the starting point of the investigations concerning attainability; the problems listed in [61] (see also [32]) are all special cases of I_1A related to some concrete classes $\mathfrak{K}$. In some cases it might be of interest to describe all γ-attainable classes for a fixed γ which does not necessarily equal 1 (this is Problem I_γA). Let us remark that Problem IA was formulated first in [52] (see question 3, p. 1370).

It is possible that the solution of problem IA does not always yield complete information on the A-spectra. Therefore it is worth raising the following problem independently.

Problem IIA (A-spectra).

IIAV. Find the A-spectra of all subvarieties of a given variety of algebras.

IIAQ. The same for quasi-varieties.

IIAP. The same for prevarieties.

Of course, special attention has to be paid to finite attainability. This leads to the following two problems.

Problem IIIA (finite attainability).

IIIAV. Characterize those subvarieties of a given variety $\mathfrak{R}$ of algebras which are finitely attainable in $\mathfrak{R}$.

IIIAQ. The same for quasi-varieties.

IIIAP. The same for prevarieties.

This problem was formulated first in [52] (question 2, p. 1370).

Problem IVA (hereditary finite attainability).

IVAV. Characterize the varieties of algebras in which all subvarieties are finitely attainable.

IVAQ. The same for quasi-varieties.

IVAP. The same for prevarieties.

Notice that finite attainability is a much weaker condition than n-attainability for a fixed natural number n ([52], p. 1369). This raises the following

Problem VA (equivalence of finite attainability and n-attainability).

VAV. Characterize those varieties of algebras in which finite attainability of subvarieties is equivalent to n-attainability for a fixed natural number n.

VAQ. The same for quasi-varieties.

VAP. The same for prevarieties.

The notions of attainability and solvability are closely related. In particular, if $\mathfrak{X}$ and $\mathfrak{R}$ are prevarieties, $\mathfrak{X} \subseteq \mathfrak{R}$ and $\mathfrak{R}$ is γ-$\mathfrak{X}$-solvable, then $\mathfrak{X}$ is γ-attainable in $\mathfrak{K}$; if $\mathfrak{X}$ is S-free in $\mathfrak{K}$, then it is also A-free in $\mathfrak{K}$; the converse statements are in general false. Therefore it makes sense to set analogous problems for solvability as were formulated above for attainability.

Problem IS (solvability).

ISV. Characterize those subvarieties $\mathfrak{X}$ of a given variety $\mathfrak{K}$ of algebras for which $\mathfrak{K}$ is $\mathfrak{X}$-solvable.

ISQ. The same for quasi-varieties.

ISP. The same for prevarieties.

Problem $\mathbf{I_\gamma S}$ (γ-solvability).

Let γ be an arbitrary but fixed ordinal number.

I_γSV. Characterize all subvarieties $\mathfrak{X}$ of a given variety $\mathfrak{K}$ of algebras for which $\mathfrak{K}$ is γ-$\mathfrak{X}$-solvable.

I_γSQ. The same for quasi-varieties.

I_γSP. The same for prevarieties.

Problem IIS (S-spectra).

The formulation of this problem is obtained from that of Problem IIA by changing the term A-spectrum to S-spectrum.

Problem IIIS (finite solvability).

IIISV. Characterize those subvarieties $\mathfrak{X}$ of a given variety $\mathfrak{K}$ of algebras for which $\mathfrak{K}$ is finitely $\mathfrak{X}$-solvable.

IIISQ. The same for quasi-varieties.

IIISP. The same for prevarieties.

Problem IVS (hereditary finite solvability).

IVSV. Characterize those varieties of algebras which are finitely $\mathfrak{X}$-solvable for each of their subvarieties $\mathfrak{X}$.

IVSQ. The same for quasi-varieties.

IVSP. The same for prevarieties.

Problem VS (equivalence of finite solvability and n-solvability).

VSV. Characterize those varieties of algebras for which finite $\mathfrak{X}$-solvability is equivalent to n-$\mathfrak{X}$-solvability for every subvariety $\mathfrak{X}$, where n is a fixed natural number.

VSQ. The same for quasi-varieties.

VSP. The same for prevarieties.

The Problems IS, I_γS, IIIS were raised in [52] (questions 2, 3 on p. 1370).

The prevarieties which are attainable in $\mathfrak{K}$ are mostly idempotent in $\mathfrak{K}$, and conversely, idempotents in $\mathfrak{K}$ are often 1-attainable in $\mathfrak{K}$; the classes of idempotent and 1-attainable prevarieties, respectively, in $\mathfrak{K}$ sometimes coincide (this is so in the case of groups, modules, etc). All this suggests the following two problems.

Problem VI (equivalence of 1-attainability and idempotency).

VIV. Find necessary and sufficient conditions for a variety $\mathfrak{K}$ of algebras in order that the idempotent subvarieties in $\mathfrak{K}$ be the same as the 1-attainable subvarieties.

VIQ. The same for quasi-varieties.

VIP. The same for prevarieties.

Problem VII (idempotency).

VIIV. Given a variety $\mathfrak{K}$, characterize the idempotents in $G_V(\mathfrak{K})$.

VIIQ. The same for quasi-varieties.

VIIP. The same for prevarieties.

Notice that Problem VIIP is hard to grasp for "large" classes since the class of idempotents in $G_P(\mathfrak{K})$ is often not a set. On the other hand, it is generally easy to describe the idempotents in $G_V(\mathfrak{K})$ – in a number of cases there are no non-trivial idempotent subvarieties.

Now we give some nearer comments on the problems formulated above, and notice some simple general facts about them; at the same time we shall also formulate some open questions concerning the general case, which we think to be interesting.

Problems IA and IS are, of course, of a "global" character. Concerning Problem IA, there are no general results as yet. As for Problem IS, there is quite a lot of classes of algebras for which it is easy to exhibit the non-existence of the desired subclasses. Namely, it holds:

Remark 2.1 ([52], p. 1371). If in a prevariety $\mathfrak{K}$ every algebra can be embedded in a simple one, then $\mathfrak{K}$ cannot be $\mathfrak{X}$-solvable for any of its proper subprevarieties $\mathfrak{X}$.

In particular, groups, semigroups, lattices, groupoids, quasi-groups, loops, etc. possess the property in Remark 2.1. Hence in these cases non-trivial concretizations of this problem must concern "smaller classes".

About Problem $I_\gamma A$ for $\gamma = 1$, the following two sufficient conditions can be immediately verified.

Remark 2.2 ([3], Proposition 4). Let $\mathfrak{K}$ be a variety of algebras and $\mathfrak{X}$ be a subvariety of $\mathfrak{K}$ such that the algebras from $\mathfrak{X}$ with more than one element contain no idempotents. Then $\mathfrak{X}$ is 1-attainable in $\mathfrak{K}$.

Remark 2.3 ([3], Proposition 1). Let $\mathfrak{K}$ be a regular Schreier variety of algebras containing at least one idempotent element. Then $\mathfrak{K}$ has no non-trivial 1-attainable subvarieties.

Notice that Remarks 2.2 and 2.3 are valid (practically with the same proof) also if we replace the word "variety" by "prevariety" in their formulations.

The concept of 1-attainability is of special interest from the point of view of radical theory. Recall that a prevariety $\mathfrak{X}$ of algebras determines a radical, namely, the one which assigns to each algebra from $\mathfrak{K}$ its preverbal $\mathfrak{X}$-congruence. If $\mathfrak{X}$ is 1-attainable in $\mathfrak{K}$ then, as we have mentioned already,

$$(2.1)\qquad R\mathfrak{X} \underset{\mathfrak{K}}{\circ} \mathfrak{X} = \mathfrak{K}.$$

This gives the following possibility of investigating the structure of algebras from $\mathfrak{K}$: investigate the structure of algebras from $\mathfrak{X}$ and from $R\mathfrak{X}$, and then study extensions of algebras from $R\mathfrak{X}$ by algebras from $\mathfrak{X}$ as described in (2.1). This approach has long been followed e.g. in ring theory. One of the most important examples of this approach for algebras which do not have a uniquely distinguished idempotent element, is the radical determined by the variety $\mathfrak{S}$ of semilattices in the class of all semigroups: many structure theorems in semigroup theory are formulated in terms of decompositions into semilattices of $\mathfrak{S}$-radical semigroups (a classical example for such a theorem is the decomposition of semigroups which are unions of groups into semilattices of completely simple semigroups).

In connection with Problem IIA one can ask the following

Question 2.1. Is every A-spectrum an initial segment of ordinals? In other words: does $\alpha \in \mathrm{Spec}_A(\mathfrak{X}, \mathfrak{K})$ always imply $\beta \in \mathrm{Spec}_A(\mathfrak{X}, \mathfrak{K})$ for $\beta < \alpha$?

Notice that the similar question on S-spectra has a negative answer; counterexamples can be found for varieties of lattices (see Section 11).

That case is of special interest when the A-spectrum is the class of all ordinals, i.e., when the corresponding prevariety is A-free. Such a prevariety is clearly non-attainable. The converse implication is not so obvious. In this connection we ask the following

Question 2.2. If $\mathfrak{X}$ is a variety, quasi-variety or prevariety which is not attainable in a given variety $\mathfrak{K}$, is $\mathfrak{X}$ necessarily A-free in $\mathfrak{K}$?

Of course, an affirmative answer to Question 2.1 would entail the same to Question 2.2. Both questions are answered affirmatively in any transpreverbal variety, as shown by the following statement.

Remark 2.4. If $\mathfrak{K}$ is a variety of algebras, $\mathfrak{X}$ is a subvariety of $\mathfrak{K}$, $\mathfrak{K}$ is transpreverbal for $\mathfrak{X}$, and there is an algebra A in $\mathfrak{K}$ on which $\mathfrak{X}$

is attainable of order γ, then for any $\alpha < \gamma$, there is an α-$\mathfrak{X}$-solvable algebra in $\mathfrak{K}$.

For the proof consider a chain

$$A = \mathfrak{X}^0 A > \mathfrak{X}^1 A > \ldots > \mathfrak{X}^\alpha A > \ldots > \mathfrak{X}^\gamma A$$

and construct the following chain of congruences of A:

$$\nabla_A \supset \rho_1 \supset \rho_2 \supset \ldots \supset \rho_\alpha \supset \ldots \supset \rho_\gamma,$$

for which the kernel of ρ_α is $\mathfrak{X}^\alpha A$, $\rho_{\alpha+1}$ is an $\mathfrak{X}$-extension of ρ_α and $\rho_\alpha = \bigcap_{\beta<\alpha} \rho_\beta$ if α is a limit ordinal. It is easy to check that A/ρ_α is an α-$\mathfrak{X}$-solvable algebra in $\mathfrak{K}$.

On Problems IIIA and IIIS one can make some obvious remarks.

Remark 2.5. Let $\mathfrak{K}$ be a prevariety of algebras. If for every non-trivial, bounded subprevariety $\mathfrak{X}$ in $\mathfrak{K}$, the corresponding preverbal chain on the free $\mathfrak{K}$-algebra on a countable set of generators does not stabilize at any natural number, then $\mathfrak{K}$ has no non-trivial, finitely attainable, bounded subprevarieties (in particular, subvarieties).

For varieties this is the case – as was noticed in [52], p. 1371 – if $\mathfrak{K}$ is a Schreier variety in which every free algebra which generates $\mathfrak{K}$ has the property that any of its non-trivial congruences has a class which is a subalgebra again generating $\mathfrak{K}$. Examples of such varieties are groups, loops, Lie algebras, etc. This is why free algebras take a prominent part in investigations of finite attainability. In fact, looking at verbal chains of free algebras (which is also interesting in itself in view of its connection with the study of properties of products of varieties) often helps in solving Problem IIIA.

Remark 2.6. For any subprevariety $\mathfrak{X}$ of a prevariety $\mathfrak{K}$ of algebras and for any natural number n,

(2.2) $\quad \mathfrak{X}^{(n)} \subseteq \mathfrak{X}^n \subseteq \mathfrak{X}^{(n+1)}.$

In particular, if $\mathfrak{X}^{(n)} \neq \mathfrak{X}^{(n+1)}$ for all n, then $\mathfrak{X}$ is not finitely attainable in $\mathfrak{K}$; if $\mathfrak{X}^{(n)} = \mathfrak{K}$ for some n, then $\mathfrak{K}$ is n-$\mathfrak{X}$-solvable and consequently $\mathfrak{X}$ is n-attainable in $\mathfrak{K}$.

The inclusions (2.2) are easy to prove by induction on n, taking into consideration the obvious relations

$$(2.3) \qquad \mathfrak{X} = \mathfrak{X}^{(1)} \subseteq \mathfrak{X}^{1},$$

$$(2.4) \qquad \mathfrak{X}^{n+1} = \mathfrak{X}^{n} \circ \mathfrak{X}$$

for all natural numbers n (we put $\mathfrak{X}^{(0)} = \mathfrak{E} = \mathfrak{X}^{0}$).

From Remark 2.6 it follows that if $\mathfrak{K}$ is a variety and $G_V(\mathfrak{K})$ is a free groupoid or a free semigroup with zero and identity, then there are no non-trivial finitely attainable subvarieties in $\mathfrak{K}$.

Concerning the inclusions (2.2) the question arises when the classes $\mathfrak{X}^{(n)}$ and $\mathfrak{X}^{n}$ coincide. The answer is given by the following statement, which can be readily verified.

Remark 2.7. Let $\mathfrak{X}$ and $\mathfrak{K}$ be prevarieties of algebras, $\mathfrak{X} \subseteq \mathfrak{K}$. The equality $\mathfrak{X}^{(n)} = \mathfrak{X}^{n}$ holds for any natural number n if and only if $\mathfrak{K}$ is regular for trivial $\mathfrak{X}$-preverbals.

It is also easy to prove the following.

Remark 2.8. Let $\mathfrak{X}$ be a subvariety of a variety $\mathfrak{K}$ of algebras, and $\mathfrak{X}^{n}$ be a variety for any natural number n. Then n-$\mathfrak{X}$-solvability of the free algebra on a countable set of generators implies n-$\mathfrak{X}$-solvability of $\mathfrak{K}$, in particular, $\mathfrak{X}$ is n-attainable in $\mathfrak{K}$.

In this connection one can ask:

Question 2.3. In which varieties $\mathfrak{K}$ of algebras is it true that $\mathfrak{X}^{n}$ is a variety for any subvariety $\mathfrak{X}$ of $\mathfrak{K}$ and any natural number n?

Notice that for any subprevariety $\mathfrak{X}$ of a prevariety $\mathfrak{K}$ of algebras and for any ordinal number γ, $\mathfrak{X}^{\gamma}$ is a prevariety (see [52], Proposition 1); if $\mathfrak{K}$ and $\mathfrak{X}$ are varieties, then even $\mathfrak{X}^{1}$ need not be a variety (see [52], p. 1367). However, it holds

Remark 2.9 ([52], Proposition 2). If each algebra in a variety $\mathfrak{K}$ has permutable congruences and contains and idempotent element, then $\mathfrak{X}^{n}$ is a variety for any subvariety $\mathfrak{X}$ of $\mathfrak{K}$ and any natural number n.

Concerning Problem IVA, the following can be said in view of Remark 2.2.

Remark 2.10. If the algebras in a prevariety $\mathfrak{K}$ contain no idempotents then every subvariety is 1-attainable in $\mathfrak{K}$.

The situation with Problem IVS is similar:

Remark 2.11. If the algebras in a prevariety $\mathfrak{K}$ contain no idempotents then $\mathfrak{K}$ is 1-$\mathfrak{X}$-solvable for every subprevariety $\mathfrak{X}$.

The following is related to Remark 2.8.

Remark 2.12. Let $\mathfrak{K}$ be a variety of algebras such that $\mathfrak{X}^n$ is a variety for every non-trivial subvariety $\mathfrak{X}$ and every natural number n, and suppose that for every subvariety $\mathfrak{X}$, the free $\mathfrak{K}$-algebra on a countable set of generators is finitely $\mathfrak{X}$-solvable. Then $\mathfrak{K}$ is hereditarily finitely solvable and, consequently, hereditarily finitely attainable.

On Problems VA and VS. As we have mentioned before, in [52] (p. 1369) there is an example of a prevariety $\mathfrak{K}$ and a subprevariety $\mathfrak{X}$ of $\mathfrak{K}$ such that all algebras from $\mathfrak{K}$ are finitely $\mathfrak{X}$-solvable but $\mathfrak{K}$ is not n-$\mathfrak{X}$-solvable for any natural number n. It would be desirable to know the answer to the following questions on varieties.

Question 2.4. Does finite attainability of a variety $\mathfrak{X}$ in a variety $\mathfrak{K}$ always imply n-attainability of $\mathfrak{X}$ in $\mathfrak{K}$ for some natural number n?

Question 2.5. For varieties $\mathfrak{K}$ and $\mathfrak{X}$ such that $\mathfrak{X} \subseteq \mathfrak{K}$, does finite $\mathfrak{X}$-solvability of $\mathfrak{K}$ always imply its n-$\mathfrak{X}$-solvability for some natural number n?

Concerning Problem VI, we have the following two statements.

Remark 2.13 ([31], p. 363). If a prevariety $\mathfrak{X}$ is 1-attainable in a prevariety $\mathfrak{K}$ of algebras and $\mathfrak{K}$ is regular for trivial $\mathfrak{X}$-preverbals, then $\mathfrak{X}$ is idempotent in $\mathfrak{K}$.

Remark 2.14 ([31], Theorem 11). If a subvariety (sub-quasi-variety) $\mathfrak{X}$ of a transverbal (trans-quasi-verbal) hereditary class $\mathfrak{K}$ is idempotent in $\mathfrak{K}$, then $\mathfrak{X}$ is 1-attainable in $\mathfrak{K}$.

Notice that the condition on the class $\mathfrak{K}$ in Remark 2.14 can be weakened in that it suffices to require of $\mathfrak{K}$ to be transverbal (trans-quasi-verbal) for $\mathfrak{X}$ only. This remark is valid also if $\mathfrak{X}$ is a prevariety.

Concerning Problem VII we formulate

Remark 2.15 ([39], Proposition 1). If a prevariety $\mathfrak{K}$ of algebras has the joint embedding property and has arbitrarily large simple algebras, then the idempotent prevarieties in $\mathfrak{K}$ do not form a set.

The class of idempotent quasi-varieties is often a set of cardinality continuum. In this connection we ask

Question 2.6. For what quasi-varieties $\mathfrak{K}$ of algebras is the set of idempotent quasi-varieties in $\mathfrak{K}$ of cardinality continuum?

Question 2.7. For what prevarieties $\mathfrak{K}$ of algebras is the class of idempotents from $G_P(\mathfrak{K})$ a set?

Finally, let us remark that the problems formulated above and many of the questions are interesting also for bounded subprevarieties of a fixed variety of algebras. On the other hand, these problems and questions are interesting in some concrete cases for modified notions of attainability and solvability, which are obtained by considering not all congruences on the algebras but only those which have a given property and form a complete sublattice in the lattice of all congruences (e.g., Rees congruences in the case of semigroups). Notice that other modifications of the notions of attainability and solvability can be considered, as well as their duals, which are connected with considering ascending chains of scatters (see [27], [28], [29]); we shall not deal with them in this survey. Finally, let us note that many of the considerations on groups in the papers [48], [49] can be very fruitful for the theory of universal algebras. These considerations are related to the topic treated here, and concern lower and upper products of families (also infinite ones) of classes, operations on functorials, etc.

3. SEMIGROUPS

Throughout this section, $\mathfrak{K}$ will denote the class of all semigroups and $\mathfrak{S}$ the variety of semilattices.

Theorem 3.1 ([52], Theorem 2). *$\mathfrak{S}$ is the only non-trivial, finitely attainable variety in $\mathfrak{K}$, and it is 1-attainable.*

This theorem solves Problems IIIAV and I_γAV (for any natural number γ) for semigroups, in particular, it generalizes the main result of [61]. Problems IIIAQ and IIIAP are still open for semigroups. A partial solution to Problem IIIAP is given by

Theorem 3.2 ([52], Theorem 1). *Let $\mathfrak{X}$ be a non-trivial prevariety of periodic semigroups. In order that $\mathfrak{X}$ be finitely attainable in $\mathfrak{K}$, it is necessary and sufficient that $\mathfrak{X}$ coincides with the variety $\mathfrak{S}$, which is 1-attainable in $\mathfrak{K}$.*

The following statement yields quite a lot of information about the Problems IAV, IIAV and IISV (recall that Problem IS has a trivial solution for the class of all semigroups in view of Remark 2.1 and the well-known fact that every semigroup can be embedded into a simple (i.e., congruence-free) one [1], [55]). Before presenting this result, let us introduce some definitions and notations.

A prevariety $\mathfrak{X}$ of semigroups is said to be of infinite type if the $\mathfrak{X}$-free monogenic semigroup is infinite; otherwise $\mathfrak{X}$ is said to be of finite type, and its type is defined to be the type of the $\mathfrak{X}$-free monogenic semigroup. We put

$\mathfrak{A} = [xy = yx]$, the variety of commutative semigroups;

$\mathfrak{B} = [x^2 = x]$, the variety of bands;

$\mathfrak{R} = [xyx = x]$, the variety of rectangular bands;

$\mathfrak{Z}_l = [xy = x]$, the variety of left zero semigroups;

$\mathfrak{Z}_r = [xy = y]$, the variety of right zero semigroups;

$\mathfrak{R}_l = [x^2 = x,\ xyx = yx]$, the variety of left regular bands;

$\mathfrak{R}_r = [x^2 = x,\ xyx = yx]$, the variety of right regular bands;

$\mathfrak{G}$ = the class of all groups.

Theorem 3.3 ([35], Theorem 2). *Let* $\mathfrak{X}$ *be a non-trivial variety of semigroups which is not a normal variety of type* $(r, 1)$, *where* $r > 1$. *Then:*

if $\mathfrak{X} \not\subseteq \mathfrak{B}$, *then* $\mathrm{Spec}_S(\mathfrak{X}, \mathfrak{K}) = W$;

if $\mathfrak{X}$ *is one of the varieties* $\mathfrak{B}, \mathfrak{N}, \mathfrak{Z}_l, \mathfrak{Z}_r, \mathfrak{S}, \mathfrak{R}_l, \mathfrak{R}_r$, *then* $\mathrm{Spec}_S(\mathfrak{X}, \mathfrak{K}) = \{0, 1\}$; *in all other cases* $\mathrm{Spec}_S(\mathfrak{X}, \mathfrak{K}) = \{0, 1, 2\}$.

In the proof of the first statement essential use is made in one of the cases of an auxiliary statement which is a direct consequence of the main result in Section 7 (see Theorem 7.1); this auxiliary statement is very useful for considerations of semigroups enriched with further operations, therefore we formulate it explicitly here.

Lemma 3.1 ([35], Lemma 11). *Let* $\mathfrak{X}$ *be a prevariety of semigroups, and put* $\mathfrak{H} = \mathfrak{G} \cap \mathfrak{X}$. *If the class* $\mathfrak{H}$ *is not idempotent in* $\mathfrak{G}$, *then for every ordinal number* γ *there exists a* γ-$\mathfrak{X}$*-solvable group; in particular,* $\mathfrak{X}$ *is S-free in* $\mathfrak{K}$.

Further, the proof of the second assertion of Theorem 3.3 makes use of the solution of Problem VIIV for semigroups, which is given by

Proposition 3.1 ([35], Lemma 10). *The idempotents of the partial groupoid* $G_V(\mathfrak{K})$ *are exactly the varieties,* $\mathfrak{K}, \mathfrak{B}, \mathfrak{N}, \mathfrak{S}, \mathfrak{Z}_l, \mathfrak{Z}_r, \mathfrak{R}_l, \mathfrak{R}_r, \mathfrak{E}$.

As for Problem VIIQ, the following is known.

Proposition 3.2 ([34], Proposition 5.1). *The quasi-variety of semigroups* $\mathfrak{Q}_P = [x^{p+1} = x \rightarrow x^2 = x \mid p \in P]$, *where* P *is an arbitrary non-empty set of prime numbers, is idempotent in* $\mathfrak{K}$; *thus the set of idempotent quasi-varieties of semigroups is of cardinality continuum.*

The class of idempotent prevarieties of semigoups is not a set by Remark 2.15.

The situation with attainable prevarieties is interesting in the class $\mathfrak{B}$ of all bands.

Theorem 3.4 ([21]). *The variety* $\mathfrak{R}$ *is the only prevariety of bands which is not finitely attainable in* $\mathfrak{B}$. *All the other prevarieties of bands are* 2*-attainable in* $\mathfrak{B}$, *while the* 1*-attainable prevarieties in* $\mathfrak{B}$ *are exactly the varieties* $\mathfrak{B}, \mathfrak{C}, \mathfrak{S}, \mathfrak{Z}_l, \mathfrak{Z}_r$.

This theorem solves Problem IIIA for bands; the main result of [62], which is the description of 1-attainable varieties in $\mathfrak{B}$, is a special case of it.

As we remarked at the end of Section 2, the notions of attainability and solvability for semigroups can be modified by considering only Rees congruences; the corresponding notions will be named 'ideal'.

Theorem 3.5 ([37], Theorem 2). *A non-trivial variety* $\mathfrak{X}$ *of semigroups is either ideal* 1*-attainable or ideal S-free in* $\mathfrak{K}$. *The first case takes place if and only if the following equivalent conditions are satisfied:*

(a) $\mathfrak{X}$ *is closed under ideal extensions.*

(b) $\mathfrak{X} = \mathfrak{M}$ *or* $\mathfrak{X} = \mathfrak{M} \underset{\mathfrak{K}}{\circ} \mathfrak{S}$, *where* $\mathfrak{M}$ *is an arbitrary variety of completely simple semigroups.*

Notice that the equivalence of conditions (a) and (b) is proven in [57], Proposition 1.

Semigroups which are unions of groups, will be called Cliffordian. Similarly, varieties of Cliffordian semigroups will be named Cliffordian.

Proposition 3.3 ([37], Proposition 1). *A non-trivial variety* $\mathfrak{X}$ *of commutative semigroups is either ideal* 1*-attainable or ideal S-free in* $\mathfrak{A}$. *The first case takes place if and only if* $\mathfrak{X}$ *is Cliffordian.*

Proposition 3.4 ([37], Proposition 2). *A variety* $\mathfrak{X}$ *of bands is either ideal* 1*-attainable or ideal S-free in* $\mathfrak{B}$. *The first case takes place if and only if* $\mathfrak{X}$ *is one of the varieties* $\mathfrak{B}, \mathfrak{C}, \mathfrak{S}, \mathfrak{R}, \mathfrak{Z}_l, \mathfrak{Z}_r, \mathfrak{R}_l, \mathfrak{R}_r$.

Notice that Theorem 3.5 and Propositions 3.3, 3.4 solve Problems IAV, ISV, I_γAV, I_γSV, IIAV, IISV, IIIAV, IIISV, VIIV related to ideal attainability and ideal solvability for the classes $\mathfrak{K}, \mathfrak{A}, \mathfrak{B}$.

We close this section by some open questions.

Question 3.1 ([58], Problem 2.43). Is every non-trivial variety of bands is different from $\mathfrak{S}$, non-attainable in $\mathfrak{K}$?

No example is known of a variety $\mathfrak{X}$ of bands and a semigroup S such that $\mathfrak{X}^{\omega}S \neq \mathfrak{X}^{\omega+1}S$.

Question 3.2. Is every normal variety of semigroups of type $(r, 1)$, where $r > 1$, S-free in $\mathfrak{K}$?

Notice that a positive solution to Problem 2.42 in [58] would imply an affirmative answer to Question 3.2. Further, affirmative answers to Questions 3.1 and 3.2 would involve for $\mathfrak{K}$ to have the alternation property for attainability of subvarieties, thus solving Problem 2.65 from [58].

Question 3.3. What are the A-spectra of the varieties of bands?

Question 3.4 ([58], Problem 2.44).

(a) What are the finitely attainable, bounded prevarieties in $\mathfrak{K}$?

(b) The same for attainability.

(c) The same for ideal attainability.

In view of Theorem 3.2, for the solution of Question 3.4(a) it suffices to consider the case of bounded prevarieties of semigroups of infinite type.

Question 3.5 ([58], Problem 2.68).

(a) What are the finitely attainable quasi-varieties in $\mathfrak{K}$?

(b) The same for attainability.

(c) The same for ideal attainability.

Question 3.6 ([58], Problem 2.45). What are the idempotents of the groupoid $G_Q(\mathfrak{K})$ of quasi-varieties of semigroups?

Question 3.7 ([58], Problem 2.46). Which quasi-varieties of semigroups are closed under ideal extensions?

Question 3.8.

(a) Does $\mathfrak{K}$ have the alternation property for attainability of subprevarieties?

(b) The same question for ideal attainability.

Question 3.9. What are the semigroup varieties in which every subvariety is finitely attainable?

Question 3.10. Which semigroup varieties are $\mathfrak{X}$-solvable for each of their non-trivial subvarieties $\mathfrak{X}$?

Question 3.11. Is the variety $\mathfrak{N}$ non-attainable in $\mathfrak{B}$?

4. SEMIGROUP WITH ZERO

Let $\mathfrak{K}$ denote the variety of semigroups with zero as a nullary operation. We shall preserve the notations $\mathfrak{A}$, $\mathfrak{B}$, $\mathfrak{S}$, $\mathfrak{R}_l$, $\mathfrak{R}_r$ of the previous section for the corresponding varieties of semigroups with zero.

Theorem 4.1 ([40], Theorem 1). *A non-trivial variety $\mathfrak{X}$ of semigroups with zero is finitely attainable in $\mathfrak{K}$ if and only if $\mathfrak{X}$ is a variety of Cliffordian semigroups with zero. In this case $\mathfrak{X}$ is 1-attainable in $\mathfrak{K}$.*

This theorem solves Problem IIIAV for semigroups with zero; in particular, it solves a more general question than one of those raised in about 1-attainability. Notice the fact that there are much more 1-attainable varieties of semigroups with zero (in fact, they are continuum many) than 1-attainable varieties of semigroups (cf. Theorem 3.1).

Theorem 4.2 ([40], Proposition 2). *A non-trivial prevariety of periodic semigroups with zero is either 1-attainable in $\mathfrak{K}$ (equivalently, is Cliffordian) or is not finitely attainable in $\mathfrak{K}$.*

The following statements are also of interest.

Proposition 4.1 ([40], Lemma 11). *If a bounded prevariety of semigroups with zero contains the variety of semigroups with zero multiplication, then it is not finitely attainable in $\mathfrak{K}$.*

Proposition 4.2 ([40], Proposition 3). *Any non-trivial prevariety of nil semigroups is non-attainable in* $\mathfrak{K}$.

Analogues of Theorems 4.1, 4.2 and Propositions 4.1, 4.2 hold also for the class of all commutative semigroups with zero [40].

Problem IS is solved by Remark 2.1, since every semigroup with zero is a subsemigroup with zero of some simple (i.e., congruence-free) semigroup with zero (see [39]).

As for Problem VII, we have

Theorem 4.3 ([39]).

(1) *The idempotents of the groupoid* $G_P(\mathfrak{K})$ *do not form a set.*

(2) *No non-trivial bounded prevariety of semigroups with zero is an idempotent in* $G_P(\mathfrak{K})$.

(3) *The set of idempotents in the groupoid* $G_Q(\mathfrak{K})$ *is of cardinality continuum.*

(4) $\mathfrak{K}$ *cannot be decomposed into any* $\mathfrak{K}$*-product of proper subprevarieties.*

(5) *If a class of bounded prevarieties in* $\mathfrak{K}$ *is a subgroupoid of* $G_P(\mathfrak{K})$ *then it consists of prevarieties of nil semigroups.*

Notice that the non-trivial 1-attainable varieties in $\mathfrak{K}$ are not idempotents in $\mathfrak{K}$.

Before formulating what is known about ideal attainability and ideal solvability for semigroups with zero, let us introduce the following definition. A class $\mathfrak{X}$ of semigroups with zero will be said to be semilattice closed if whenever S is a semigroup with zero which admits a semilattice decomposition into a system S_i $(i \in I)$ of subsemigroups such that $S_i^0 \in \mathfrak{X}$ for every $i \in I$, S itself belongs to $\mathfrak{X}$ (for an arbitrary semigroup S, we denote by S^0 the semigroup obtained from S by adjoining a zero element).

Theorem 4.4 ([40], Theorem 2). *A non-trivial variety* $\mathfrak{X}$ *of semigroups with zero is either ideal* 1*-attainable or ideal S-free in* $\mathfrak{K}$. *The first case takes place if and only if the following equivalent conditions are satisfied:*

(a) $\mathfrak{X}$ *is closed under ideal extensions;*

(b) $\mathfrak{X}$ *is a semilattice closed variety of Cliffordian semigroups with zero.*

Notice that the class of ideal attainable varieties turned out to be smaller than that of 1-attainable varieties for semigroups with zero; in the case of semigroups we have just the opposite inclusion between the corresponding classes.

Proposition 4.3 ([40], Proposition 4). *Any non-trivial subvariety* $\mathfrak{X}$ *of* $\mathfrak{A}$ *is either ideal* 1*-attainable of S-free in* $\mathfrak{A}$. *The first case takes place if and only if* $\mathfrak{X}$ *is Cliffordian.*

Proposition 4.4 ([40], Proposition 5). *Any subvariety* $\mathfrak{X}$ *of* $\mathfrak{B}$ *is either ideal* 1*-attainable or ideal S-free in* $\mathfrak{B}$. *The first case takes place if and only if* $\mathfrak{X}$ *is one of the varieties* $\mathfrak{B}$, $\mathfrak{E}$, $\mathfrak{S}$, $\mathfrak{R}_l$, $\mathfrak{R}_r$.

Theorem 4.4 and Propositions 4.3, 4.4 solve, for the classes $\mathfrak{K}$, $\mathfrak{A}$, $\mathfrak{B}$, Problems IAV, I_γAV, I_γSV, IIAV, IISV, IIIAV, IIISV, VIIV related to ideal attainability and ideal solvability.

Question 4.1. Is it true that a non-trivial variety of semigroups with zero which is neither Cliffordian nor consisting of nil semigroups only, cannot be attainable in $\mathfrak{K}$?

In view of Theorem 4.1 and Proposition 4.2, an affirmative answer for this question would imply that the variety of all semigroups with zero has the alternation property for attainability of subvarieties.

Question 4.2.

(a) What are the finitely attainable, bounded prevarieties in $\mathfrak{K}$?

(b) The same for attainability.

(c) The same for ideal attainability.

Question 4.3.

(a) What are the finitely attainable quasi-varieties in $\mathfrak{M}$?

(b) The same for attainability.

(c) The same for ideal attainability.

Question 4.4. What are the idempotents in the groupoid of quasi-varieties of semigroups with zero?

Question 4.5. Which quasi-varieties of semigroups with zero are closed under ideal extensions?

Question 4.6.

(a) What are the varieties of semigroups with zero in which every subvariety is finitely attainable?

(b) The same for ideal attainability.

Question 4.7.

(a) Which varieties of semigroups with zero are finitely $\mathfrak{X}$-solvable for each of their subvarieties $\mathfrak{X}$?

(b) The same for ideal solvability.

5. MONOIDS

Let $\mathfrak{M}$ denote the variety of monoids (i.e., semigroups with identity as a nullary operation).

Theorem 1 ([18], Theorem 1). *Any proper variety* $\mathfrak{X}$ *of monoids is either* 1*-attainable or S-free in* $\mathfrak{M}$. *The first case takes place if and only if the identity* $x^{r+1} = x^r$ *is satisfied in* $\mathfrak{X}$ *for some natural number* r.

This theorem solves Problems IAV, IIAV, IIIAV, I_γAV for monoids, in particular, it answers the questions (a) and (b) from [58], Problem 2.67. Its proof is essentially based on Lemma 3.1.

Theorem 5.2 ([18], Theorem 2). *A prevariety* $\mathfrak{X}$ *of monoids with an outer adjoined identity is either* 1*-attainable or non-attainable in* $\mathfrak{M}$. *The first case is possible if and only if* $\mathfrak{X}$ *satisfies the following condition:*

(i) *for an arbitrary monoid* S *from* $\mathfrak{X}$ *with more than one element, there exists a homomorphism* f *into some monoid such that* $S^* + f(S^*) + 1 \in \mathfrak{X}$ *(here* S^* *denotes the semigroup* $S \setminus 1$, *while* $S^* + f(S^*) + 1$ *stands for the ordinal sum of the semigroups* S^*, $f(S^*)$ *and* $\{1\}$*).*

Theorem 5.3 ([18], Theorem 3). *A bounded prevariety* $\mathfrak{X}$ *of monoids is* 1*-attainable in* $\mathfrak{M}$ *if and only if* $\mathfrak{X}$ *is a prevariety of monoids with an outer adjoined identity and condition* (i) *from Theorem* 5.2 *is satisfied.*

Question 5.1.

(a) Which bounded prevarieties of monoids are finitely attainable in $\mathfrak{M}$?

(b) The same for quasi-varieties.

(c) The same for prevarieties.

Question 5.2.

(a) Does $\mathfrak{M}$ have the alternation property for attainability of bounded prevarieties?

(b) The same for quasi-varieties.

(c) The same for prevarieties.

Notice that, by Lemma 3.1 and Theorem 5.2, it suffices to consider in Questions 5.1 and 5.2 only those prevarieties of monoids which contain the bicyclic monoid but do not contain groups with more than one elements.

Question 5.3. What are the idempotents in the groupoid $G_Q(\mathfrak{M})$ of quasi-varieties of monoids?

Remark 5.1. Non-trivial bounded prevarieties of monoids cannot be idempotent. In fact, if a bounded prevariety $\mathfrak{X}$ of monoids contains a group with more than one elements then it is well known that $\mathfrak{X} \underset{\mathfrak{K}}{\circ} \mathfrak{X} = \mathfrak{K}$, while otherwise it is clear that the free monoid of countable rank belongs to $\mathfrak{X} \underset{\mathfrak{K}}{\circ} \mathfrak{X}$, hence $\mathfrak{X} \underset{\mathfrak{K}}{\circ} \mathfrak{X} \neq \mathfrak{X}$.

Question 5.4. What are the prevarieties of monoids in which every subvariety is finitely attainable?

Question 5.5. Which monoid varieties are $\mathfrak{X}$-solvable for each of their non-trivial subvarieties $\mathfrak{X}$?

6. INVERSE SEMIGROUPS

Let $\mathfrak{K}$ denote the variety of inverse semigroups considered as algebras with one binary associative operation (multiplication) and one unary operation (taking the inverse element).

Theorem 6.1 ([50]). *In order that a non-trivial variety of inverse semigroups be finitely attainable in* $\mathfrak{K}$, *it is necessary and sufficient that it coincides either with the variety* $\mathfrak{G}$ *of all groups or with the variety of semilattices; both of them are* 1*-attainable in* $\mathfrak{K}$.

This theorem solves Problem IIIAV for inverse semigroups.

Remark 6.1. Independently, 1-attainable varieties of inverse semigroups have been described also in the unpublished paper "Attainable varieties of inverse semigroups" by L.M. Martynov and T.A. Martynova, where it is also shown that non-trivial combinatorial varieties, different from $\mathfrak{S}$, are not finitely attainable in $\mathfrak{K}$, and that the idempotent varieties of inverse semigroups are exactly $\mathfrak{K}$, $\mathfrak{E}$, $\mathfrak{G}$, $\mathfrak{S}$.

Problems IS, I_γS, IIIS have trivial solutions by Remark 2.1, for every inverse semigroup can be embedded into a simple (i.e., congruence-free) one [55].

Further, Lemma 3.1 implies that every variety of inverse semigroups having a non-trivial group variety as intersection with the variety of all groups, is S-free in $\mathfrak{K}$. On the other hand, we have

Proposition 6.1 ([50]). *If a non-trivial variety of inverse semigroups strictly contains the variety of all groups then it is not attainable in* $\mathfrak{K}$.

Hence, in order to solve Problem IAV, only the case of combinatorial inverse semigroup varieties has to be investigated.

Theorem 6.2 ([42]). *A non-trivial variety* $\mathfrak{X}$ *of inverse semigroups is either ideal* 1*-attainable (equivalently, closed under ideal extensions) or ideal S-free in* $\mathfrak{K}$. *The first case takes place if and only if* $\mathfrak{X}$ *is a Cliffordian variety.*

This theorem solves for inverse semigroups Problems IAV, ISV, I_γAV, I_γSV, IIAV, IISV, IIIAV, IIISV, VIIV related to ideal attainability and ideal solvability.

In connection with the above results we ask the following open questions.

Question 6.1 ([58], Problem 2.66). Is it true that every non-trivial variety of combinatorial inverse semigroups, different from $\mathfrak{S}$, is non-attainable in $\mathfrak{K}$?

A positive answer to this question would mean that $\mathfrak{K}$ has the alternation property for attainability of subvarieties.

Question 6.2.

(a) What are the finitely attainable, bounded prevarieties in $\mathfrak{K}$?

(b) The same for attainability.

(c) The same for ideal attainability.

Question 6.3.

(a) Which quasi-varieties of inverse semigroups are finitely attainable in $\mathfrak{K}$?

(b) The same for attainability.

(c) The same for ideal attainability.

Question 6.4. What are the idempotents of the groupoid $G_Q(\mathfrak{K})$ of quasi-varieties of inverse semigroups?

Question 6.5. Which quasi-varieties of inverse semigroups are closed under ideal extensions?

Question 6.6. What are the inverse semigroup varieties in which every subvariety is finitely attainable?

Question 6.7. Which inverse semigroup varieties are $\mathfrak{X}$-solvable for each of their non-trivial subvarieties $\mathfrak{X}$?

7. GROUPS

Let $\mathfrak{K}$ be the variety of all groups.

Theorem 7.1 ([35], [66]). *Any prevariety of groups is either 1-attainable (equivalently, idempotent) or S-free in* $\mathfrak{K}$.

This theorem yields a complete solution, for groups, for Problems IS, I_γS, IIA, IIS, IIIS, and reduces Problems IA, I_γA, IIIA to Problem VII.

As a direct consequence of Theorem 7.1, we have

Corollary 7.1. *Every non-trivial bounded prevariety (in particular, variety)* $\mathfrak{X}$ *is S-free in* $\mathfrak{K}$.

This corollary solves Problems IAV, I_γAV, I_γSV, IIIAV, IIISV, VIIV for groups; in this special case when $\mathfrak{X}$ is the variety of abelian groups, it gives back Mal'cev's well-known old result [30] that was mentioned in the introduction: for every ordinal number γ there is a group whose descending chain of commutators stabilizes exactly from this γ.

As for Problem VIIP, by Remark 2.15 we know that idempotent prevarieties of groups do not form a set (see also [65]). For quasi-varieties of groups we have

Proposition 7.1 ([35], Proposition 1). *For any non-empty set* P *of prime numbers, the quasi-variety of groups*

$$\mathfrak{Q}_P = [x^p = 1 \rightarrow x = 1 \mid p \in P]$$

is idempotent in $\mathfrak{K}$, *hence the set of idempotent quasi-varieties of groups is of cardinality continuum.*

For sub-quasi-varieties of the variety $\mathfrak{A}$ of abelian groups, the following proposition solves all the problems except IVA and IVS.

Proposition 7.2 ([35], Proposition 2). *A non-trivial quasi-variety* $\mathfrak{X}$ *of abelian groups is either* 1*-attainable (i.e., idempotent) or S-free in* $\mathfrak{A}$. *The first case takes place if and only if* $\mathfrak{X} = \mathfrak{A} \cap \mathfrak{Q}_P$ *for some set* P *of primes.*

The proof of this proposition makes heavy use of the fact that for any prime number p there exist reduced abelian p-groups of an arbitrarily fixed type (see [10], [19] or [20]) and also of the description of the lattice of quasi-varieties of abelian groups [64].

Proposition 7.2 has the following immediate

Corollary 7.2. *Every non-trivial variety of abelian groups is S-free in* $\mathfrak{A}$.

In connection with the results presented above, we ask:

Question 7.1. What are the idempotents of the semigroup $G_Q(\mathfrak{K})$ of quasi-varieties of groups?

Question 7.2. What are the group varieties in which every subvariety is finitely attainable?

Question 7.3. Which group varieties are $\mathfrak{X}$-solvable for each of their non-trivial subvarieties $\mathfrak{X}$?

8. MODULES

Let R be an arbitrary associative ring with unit element, and $\mathfrak{K}$ be the variety of all unitary left R-modules. It is well known that the lattice of subvarieties of $\mathfrak{K}$ is antiisomorphic to the lattice of two-sided ideals of R. This antiisomorphism is realized by the mapping φ which assigns to each variety $\mathfrak{X}$ of R-modules the ideal $I = \varphi(\mathfrak{X})$ that consists of the coefficients in the elements of the $\mathfrak{X}$-verbal of the free R-module F over a countable alphabet X. Now we have for an arbitrary R-module M, $\mathfrak{X}(M) = IM = \left\{ \sum a_i m_i \mid a_i \in I,\ m_i \in M \right\}$ and for any natural number n, $\mathfrak{X}^n M = I^n M$. Taking into consideration that

$$\mathfrak{X}^m F = \mathfrak{X}^n F \Leftrightarrow I^m = I^n$$

for any natural numbers m and n, we see the validity of

Proposition 8.1 ([41]). *A variety $\mathfrak{X}$ of R-modules is finitely attainable in $\mathfrak{R}$ if and only if there is a natural number n such that $I^n = I^{n+1}$, where $I = \varphi(\mathfrak{X})$.*

Hence Problem IIIAV and Problem I_γAV for natural numbers γ have no importance in themselves in the module case, as their solution can be traced back to the investigation of natural powers of the ideals of the ring R. Clearly, for $I = \varphi(\mathfrak{X})$ we have $\bigcap_{n<\omega} I^n = 0$, however, this does not in general imply $\mathfrak{X}(M) = 0$ for every R-module M; abelian groups yield examples for this situation (see Corollary 7.2). Therefore Problem IAV is of interest for modules, too.

Before formulating the main result of this section, we introduce some definitions. Let I be a maximal left ideal of the ring R. A left ideal of R is said to be I-primary if every nonzero homomorphic image of the R-module R/L contains a submodule isomorphic to R/I. A ring R will be called special if each of its nonzero maximal left ideals I satisfies the following conditions:

(1I) every I-primary left ideal of R is a two-sided ideal in R;

(2I) I/I^2 is an irreducible left and right R-module;

(3I) $I^n \neq I^{n+1}$ for any natural number n;

(4I) $\bigcap_{n<\omega} I^n = 0$.

Theorem 8.1 ([41]). *If R is a special ring then every non-trivial subvariety $\mathfrak{X}$ of $\mathfrak{R}$ is S-free in $\mathfrak{R}$.*

This theorem solves Problems IAV, I_γAV, I_γSV, IIIAV, IIISV, IIAV, IISV and VIIV for modules over a special ring; in the special case of $R = Z$

we obtain Corollary 7.2.

Since the variety of all modules over an arbitrary ring is transpreverbal, by Remark 2.4 we see that Questions 2.1 and 2.2 have an affirmative answer for modules.

Concerning possible generalizations of Theorem 8.1 for varieties, we think that case to be the most interesting when every non-zero ideal I of R satisfies conditions (3I) and (4I); such a ring R will be called a ring with ω-nilpotent ideals.

Question 8.1. Is every non-trivial variety of R-modules S-free in $\mathfrak{K}$ if R is a ring with ω-nilpotent ideals?

Question 8.2.

(a) Does the variety of all modules over a ring with ω-nilpotent ideals have the alternation property for attainability of sub-quasivarieties?

(b) The same for prevarieties.

Question 8.3. What are the idempotents in the semigroup of quasivarieties of modules over a ring with ω-nilpotent ideals?

9. RINGS AND LINEAR ALGEBRAS

Throughout this section, unless stated differently, R is an arbitrary associative and commutative ring with unit element different from zero. The results we are after for rings and linear algebras over fields (mainly associative ones) will be obtained as special cases of the corresponding results on linear algebras over R (R-algebras). The following notations will be used:

Ass (R) – the variety of all associative R-algebras;

Alt (R) – the variety of all alternative R-algebras;

Asd (R) – the variety of all power associative R-algebras;

Jord (R) – the variety of all Jordan R-algebras.

Theorem 9.1 ([41], Theorem 3.1). *Let R be a special ring. Any non-trivial variety $\mathfrak{X}$ of associative R-algebras is either 1-attainable or S-free in* Ass (R). *The first case takes place if and only if $\mathfrak{X}$ is a variety of associative R-algebras without (nonzero) nilpotent elements.*

The proof of this theorem makes essential use of Theorem 8.1 and of the following assertion which is of independent interest.

Theorem 9.2 ([41]). *Let $\mathfrak{X}$ be an arbitrary variety of associative R-algebras. Then every associative $\mathfrak{X}$-solvable R-algebra A has a descending chain of ideals starting from A and ending in the zero ideal, with all factors in $\mathfrak{X}$.*

In the special case $R = Z$ this theorem yields Theorem 1 from [36].

Before formulating the corollaries to Theorem 9.1, let us quote some statements from [41].

Proposition 9.1 ([41]). *A subvariety $\mathfrak{X}$ of* Asd (R) *is a variety of R-algebras without nilpotent elements if and only if the identity $x^n = x$ holds in $\mathfrak{X}$ for some natural number $n > 1$.*

Proposition 9.2 ([41]). Asd (R) *has a non-trivial subvariety consisting of R-algebras without nilpotent elements if and only if among the maximal ideals of the ring R there is one which is of finite index.*

Proposition 9.3 ([41]). *Every non-trivial variety of alternative R-algebras without nilpotent elements consists of R-algebras which are subdirect products of finite R-fields with uniformly bounded orders.*

Proposition 9.2 and Theorem 9.1 enable one to formulate

Proposition 9.4. *The variety* Ass (R), *where R is a special ring, has no non-trivial attainable subvarieties if and only if all maximal ideals of R are of infinite index.*

Taking into consideration that every Dedekind domain and, in particular, every principal ideal domain is a special ring, we obtain from Proposition 9.4 as a special case:

Corollary 9.1. *The variety of all associative algebras over an infinite field has no non-trivial attainable subvarieties.*

The following two assertions are immediate consequences of Theorem 9.1 and Proposition 9.2.

Corollary 9.2. *Any non-trivial variety* $\mathfrak{X}$ *of associative rings is either* 1*-attainable or non-attainable in* Ass (Z). *The first case takes place if and only if* $\mathfrak{X}$ *is generated by a finite number of finite fields.*

Corollary 9.3. *Any non-trivial variety* $\mathfrak{X}$ *of associative algebras over a finite field* R *is either* 1*-attainable or non-attainable in* Ass (R). *The first case takes place if and only if* $\mathfrak{X}$ *is generated by a finite extension of* R.

Notice that Theorem 9.1 solves Problems IAV, I_γAV, I_γSV, IIAV, IISV, IIIA, IIIS for associative algebras over a special ring and, in particular, the corresponding problems for associative rings or algebras over a field. Let us also remark that Corollary 9.2 generalizes Theorem 4 from [52], which describes the finitely attainable varieties of associative rings and gives a negative answer to the question in [52] about the existence of γ-attainable but not 1-attainable subvarieties in the variety of all associative rings.

Let us formulate some further statements from [41] concerning the problems listed above. For an ideal I in R, we denote by $\mathfrak{Z}_I$ the variety of all R-algebras with zero multiplication which are annihilated by I.

Proposition 9.5 ([41], *Let* $\mathfrak{K}$ *be a prevariety of (not necessarily associative) algebras over a special ring* R, *and suppose that* $\mathfrak{K}$ *contains the variety of all algebras with zero multiplication. Let* $\mathfrak{X}$ *be a subprevariety of* $\mathfrak{K}$ *containing* $\mathfrak{Z}_{I^k}$ *but not* $\mathfrak{Z}_{I^{k+1}}$ *for some maximal ideal* I *of* R *and some natural number* k. *Then* $\mathfrak{X}$ *is S-free in* $\mathfrak{K}$.

In the special case $R = Z$ this proposition yields Theorem 4 in [36] and the Proposition in [38].

Proposition 9.6 ([41]). *If* $\mathfrak{X}$ *is a prevariety of associative algebras containing the variety* $\mathfrak{Z}_P$ *for some prime ideal* P *of* R *and* $\mathfrak{X}$ *satis-*

fies a multilinear identity one of whose coefficients does not belong to P, then the S-spectrum of $\mathfrak{X}$ in Ass (R) *is not bounded.*

As a direct consequence of this proposition we obtain

Corollary 9.4 ([41]). *Let R be a semiprime ring and $\mathfrak{X}$ be a variety of associative algebras defined by identities containing no terms of the first degree. Then the S-spectrum of $\mathfrak{X}$ in* Ass (R) *is not bounded.*

Next we turn to Problem VII.

In view of Remark 2.13, every 1-attainable prevariety in the variety $\mathfrak{A}$ of R-algebras is an idempotent in $\mathfrak{A}$. On the other hand, we have

Proposition 9.7 ([11], Corollary 1.12). *Let R be a principal ideal domain, $\mathfrak{K}$ be a variety of R-algebras, and F be the free R-algebra of countable rank in $\mathfrak{K}$. Suppose that*

(i) *$\mathfrak{K}$ contains all R-algebras with zero multiplication;*

(ii) $\bigcap_{n=1}^{\infty} F^n = 0.$

Then every idempotent from $G_V(\mathfrak{K})$ is 1-attainable in $\mathfrak{K}$ in each of the following cases:

(a) *$\mathfrak{K}$ is the variety of associative R-algebras;*

(b) *$\mathfrak{K}$ is the variety of alternative R-algebras;*

(c) *$\mathfrak{K}$ is the variety of Jordan R-algebras and $1/2 \in R$.*

Remark 9.1. This assertion holds also if R is an arbitrary special ring (see [41]).

In view of this remark, Proposition 9.5 and Theorem 9.1 yield the solution of Problem VIIV for associative algebras over a special ring.

Proposition 9.8 ([41]). *Let R be a special ring. A proper subvariety $\mathfrak{X}$ of* Ass (R) *is an idempotent in* Ass (R) *if and only if $\mathfrak{X}$ is a variety of R-algebras without nilpotent elements.*

Corollary 9.5 ([36], Theorem 2; [16], Theorem 5). *The non-trivial* *varieties of associative rings which are idempotent in* Ass (Z) *are exactly the varieties generated by finitely many finite fields.*

It was proved already in [52] (see Lemma 10) that the varieties occurring in Corollary 9.5 are idempotent; the same paper presents also finite bases for the identities of these varieties (see also [15], [68]). In [36] it is proved that $G_V(\text{Ass}(R))$ contains no other non-trivial idempotents – this is an easy consequence of an analogue of Theorem 9.2 which is proved there for associative rings and of the following statement, proved in fact in [52] (see the proof of the necessity in Theorem 3 and Lemma 11 on pp. 1378–1379).

Proposition 9.9 ([52]). *If a bounded prevariety* $\mathfrak{X}$ *in* Ass (Z) *contains the variety* $\mathfrak{Z}_P$ *for some prime ideal* P *of the ring* Z, *then the free ring of countable rank in* Ass (Z) *is* ω-$\mathfrak{X}$*-solvable.*

The paper [16] gives a different proof for Corollary 9.5.

Remark 9.2. Theorem 15 from [16], which claims the non-existence of non-trivial 1-attainable subvarieties in Ass (Z), is false, since it contradicts the main result, Theorem 5, of the same paper.

In fact, if $\mathfrak{X}$ is a variety generated by finitely many finite fields then there are no non-zero nilpotent rings in $\mathfrak{X}$. Therefore the subring $\mathfrak{X}^2A$ is an ideal in A for any ring A (Ass (Z) is transpreverbal for $\mathfrak{X}$). Supposing $\mathfrak{X}^1A \neq \mathfrak{X}^2A$, we arrive at $A/\mathfrak{X}^2A \in \mathfrak{X} \underset{\text{Ass}(Z)}{\circ} \mathfrak{X}$ and $A/\mathfrak{X}^2A \notin \mathfrak{X}$, whereas Theorem 5 claims $\mathfrak{X} \underset{\text{Ass}(Z)}{\circ} \mathfrak{X} = \mathfrak{X}$. Theorem 5 in [63] is also false: it claims that the variety of Boolean rings is not 1-attainable in Ass (Z) (this was noticed already in [52], p. 1380); the author of [16] made use of this false statement.

The varieties generated by finitely many finite fields turned out to be distinguished in many other respects: they are exactly the non-trivial varieties of arithmetical rings (i.e. rings with distributive lattice of ideals) [43], the radical-semisimple classes (in the sense of Kuroš and Amitsur) of associative rings [15]; they form a semilattice in $G_V(\text{Ass}(Z))$ [36] and, together with the variety $\mathfrak{E}$, a distributive sublattice in $L(\text{Ass}(Z))$ (for its description see [36] and [68]); each of them has a solvable elementary theory [70]; and, as was said above, a finite basis of identities. The following system of identities is such a finite basis [15]:

$$p_1 p_2 \ldots p_m x = 0, \quad \hat{p}_i \prod_{n \in N(p_i)} (x^{p_i^n} - x) = 0$$

$$(i = 1, 2, \ldots, m),$$

where $p_1, p_2, \ldots, p_m$ are suitable prime numbers, the $N(p_i)$ are finite sets of positive integers, and $\hat{p}_i = p_1 \ldots p_{i-1} p_{i+1} \ldots p_m$ (for $m = 1$ we put $p_i = 1$).

The notions of 1-attainability and radical-semisimplicity are closely connected; this is expressed in

Proposition 9.10 ([11], Theorem 1.5). *Let* $\mathfrak{R}$ *be the variety of all R-algebras,* $\mathfrak{X} \subseteq \mathfrak{R}$. *The following conditions are equivalent:*

(i) $\mathfrak{X}$ *is a* 1*-attainable variety in* $\mathfrak{R}$*;*

(ii) $\mathfrak{X}$ *is a homomorphically closed semisimple class;*

(iii) $\mathfrak{X}$ *is a radical-semisimple class.*

From this proposition and from Theorem 9.1 we obtain a description of the radical-semisimple subclasses of Ass (R), where R is a special ring; this in turn includes as special cases the corresponding results on associative algebras over a field [11] and on associative rings [15], [56]. The same question is solved also for alternative R-algebras over a special ring by combining Proposition 9.7, Proposition 9.2, Remark 9.1, and the following

Proposition 9.11 ([41]). *Let* R *be a special ring. A proper subvariety* $\mathfrak{X}$ *of* Alt (R) *is idempotent in* Alt (R) *if and only if* $\mathfrak{X}$ *is a variety of R-algebras without nilpotent elements.*

This proposition solves Problems I_1AV and VIIV for alternative algebras over a special ring; as a special case it includes the description of 1-attainable varieties of alternative rings, announced in [38], Theorem 2, and obtained independently in [12], Theorem 4.3. Notice that, by Proposition 9.3, the sets of non-trivial idempotent varieties in Ass (R) and Alt (R), respectively, coincide if R is a special ring.

Problem IIIAV is still open for alternative rings or algebras over a field. Notice that the finitely attainable subvarieties in Alt (R) need not be 1-attainable. In [46] a proper subvariety $\mathfrak{X}$ of Alt (R) is constructed, where R is a ring containing $1/6$, such that $\mathfrak{X}^2 = \text{Alt}(R)$. By Remarks 2.9 and 2.8, $\mathfrak{X}$ is 2-attainable in Alt (R).

The papers [11] and [12] give some information about attainability of varieties of Jordan rings.

Proposition 9.12 ([11], Theorem 3.3). *Let $\mathfrak{X}$ be a variety of Jordan rings generated by finitely many finite fields and simple Jordan rings of the kind described in Proposition* 15.4 *of* [45], *all of the same prime characteristic $p \neq 2$. Then $\mathfrak{X}$ is* 1*-attainable in the class of all Jordan rings.*

Proposition 9.13 ([12], Theorem 4.4). *Let $\mathfrak{X}$ be a variety of Jordan rings such that*

$$\mathfrak{E} \neq \mathfrak{X} \quad \textit{and} \quad \mathfrak{X} \subseteq \{A \in \text{Jord}(Z) \mid pA = 0\}$$

for some odd prime number p. Then $\mathfrak{X}$ is 1*-attainable in* Jord (Z) *if and only if there is a finite set $N(p)$ of integers > 1 such that*

$$X = \left[px = 0, \prod_{n \in N(p)} (x^{p^n} - x) = 0\right].$$

Corollary 9.6 ([12], Corollary 4.5). *Let p be an odd prime number, $N(p)$ be a finite set of integers > 1. Then the variety of Jordan rings*

$$\left[px = 0, \prod_{n \in N(p)} (x^{p^n} - x) = 0, (xy)z - x(yz) = 0\right]$$

is 1*-attainable in* Jord (Z).

For an arbitrary ring variety $\mathfrak{X}$, we denote by $\mathfrak{X}_K$ the variety of all those K-algebras, where K is a field, which as rings belong to $\mathfrak{X}$.

Proposition 9.14 ([11], Theorem 3.5). *Let $\mathfrak{M}$ be a variety of rings containing all zero rings. If, for each field K, $\mathfrak{M}_K$ has no non-trivial* 1*-attainable subvarieties, then the same is true of $\mathfrak{M}$.*

Corollary 9.7 ([11], Corollary 3.6). *Let $\mathfrak{M}$ be a ring variety containing all zero rings, and suppose that $\mathfrak{M}_K$ is a Schreier variety for* *every field K. Then* $\mathfrak{M}$ *has no non-trivial* 1*-attainable subvarieties.*

Examples of Schreier varieties of algebras are, over any field, the varieties of all algebras [22], Lie algebras [53], [69], commutative and anticommutative algebras [54]. Thus Corollary 9.7 states that the corresponding ring varieties have no 1-attainable subvarieties. Notice that Corollary 9.7 is the special case $R = Z$ of Theorem 10.2 below for linear Ω-algebras over R.

Notice that for rings which are further from being associative, idempotent varieties need not be 1-attainable [13]. Now we present an example of the same kind for algebras with involution. Let $\mathfrak{B}$ and $\mathfrak{C}$ the varieties of R-algebras with involution $*$ defined by the identities $x^* = x$ and $x^* = -x$, respectively. In [14] it is proved that the varieties $\mathfrak{B}$ and $\mathfrak{C}$ are closed under extensions in the class $\mathfrak{A}$ of all R-algebras with involution (Theorem 1).

Proposition 9.15 ([14], Theorem 3). *If an extension-closed variety* $\mathfrak{X}$ *of R-algebras with involution does not contain non-zero nilpotent algebras from* $\mathfrak{B}$ *and* $\mathfrak{C}$, *then* $\mathfrak{X}$ *is* 1*-attainable in* $\mathfrak{A}$.

Proposition 9.16 ([14], Theorem 4). $\mathfrak{C}$ *is not* 1*-attainable in* $\mathfrak{A}$.

Question 9.1. What are the finitely attainable varieties of alternative rings and algebras over a field?

Question 9.2. What are the finitely attainable varieties of Jordan rings and algebras over a field?

Question 9.3. What are the finitely attainable varieties of power-associative rings and algebras over a field?

Question 9.4. What are the finitely attainable varieties of associative algebras over an arbitrary ring?

Question 9.5. Is every variety of Lie rings or Lie algebras S-free?

Question 9.6. Let R be a ring with ω-nilpotent ideals. Is it true that every variety of associative R-algebras which contains non-zero nilpotent algebras, is S-free in Ass (R)?

Question 9.7. What are those varieties of associative rings in which every subvariety is finitely attainable? The same question for varieties of alternative, Jordan, and Lie rings, respectively, and for the corresponding varieties of algebras over a field.

Question 9.8. Which varieties of associative rings are $\mathfrak{X}$-solvable for each of their subvarieties $\mathfrak{X}$? The same question for varieties of alternative, Jordan, and Lie rings, respectively, and for the corresponding varieties of algebras over a field.

Question 9.9. What are the idempotent quasi-varieties of associative rings? The same question for associative algebras over a field.

10. LINEAR Ω-ALGEBRAS

Let R be an arbitrary associative and commutative ring with identity, $\Omega = \bigcup_{n=0}^{\infty} \Omega_n$ be a set of operation symbols. An R-module A is called a linear Ω-algebra over R (R-Ω-algebra) if for every $\omega \in \Omega_n$ $(n = 1, 2, \ldots)$ a multilinear mapping $\omega\colon A^n \to A$ is given, where A^n is the n-th Cartesian power of the module A. The classical linear algebras, rings, and linear (binary) algebras over R (R-algebras) are special cases of linear Ω-algebras.

The first results on 1-attainability for R-Ω-algebras were obtained in [5]; before formulating them we have to introduce some definitions. In what follows we shall assume that every operation from Ω is at least binary. If we apply an operation ω from Ω_n to an ordered n-tuple $a_1, a_2, \ldots, a_n$ form an R-Ω-algebra A, then the result will be denoted by $a_1 a_2 \ldots a_n \omega$. Let $\mathfrak{A}_P$ be the variety of R-Ω-algebras defined by a system P of permutational identities, i.e. identities of the form

$$x_1 x_2 \ldots x_n \omega - \alpha(x_{\sigma(1)} x_{\sigma(2)} \ldots x_{\sigma(n)} \omega) = 0,$$

where $\alpha \in R$, $\omega \in \Omega_n$, and σ is a permutation of degree n. For a non-trivial subvariety $\mathfrak{X}$ of $\mathfrak{A}_P$ we shall consider the following condition:

(V) $\mathfrak{X}$ can be defined by a system of identities either not containing identities of the form $x + f(x) = 0$, or not containing identities of the form $f(x + f(x)) = 0$, where $f(x)$ is a non-constant element of the free monogenic algebra in $\mathfrak{A}_P$ generated by x.

An ideal I of a ring is said to be idempotent if $I^2 = I$.

Theorem 10.1 ([5], Theorem 7). *Let $\mathfrak{X}$ be a non-trivial subvariety of $\mathfrak{A}_P$ satisfying condition* (V). *$\mathfrak{X}$ is not 1-attainable in $\mathfrak{A}_P$ if and only if R is an integral domain without non-trivial idempotent ideals.*

Corollary 10.1 ([5], Corollary 1). *Let R be a local integral domain in which the maximal ideal is of infinite index (in particular, an infinite field). Then $\mathfrak{A}_P$ has no non-trivial 1-attainable subvarieties if and only if R has no non-trivial idempotent ideals.*

Corollary 10.2 ([5], Corollary 2). *Let R be a Noetherian integral domain. If $xx \ldots x\omega = 0$ follows from P for every $\omega \in \Omega$, then $\mathfrak{A}_P$ has no non-trivial 1-attainable subvarieties.*

In the paper [7] the notions of solvability and attainability are considered for linear R-Ω-algebras in connection with the operation of product of varieties. Let $\mathfrak{A}_\Sigma$ be a variety of linear Ω-algebras over a field R defined by a system of identities of the form $\sum_{i=1}^{m} \alpha_i x_{j_1} x_{j_2} \ldots x_{j_{n_i}} \omega_i = 0$. In [7] it is proven that the groupoid $G_V(\mathfrak{A}_\Sigma)$ is cancellative. From this fact we obtain

Proposition 10.1 ([7], Corollary 2). *$\mathfrak{A}_\Sigma$ has no non-trivial 1-attainable subvarieties.*

Let $\mathfrak{A}$ be a variety of linear Ω-algebras over a ring R, and $\mathfrak{M}$ be the class of those subvarieties of $\mathfrak{A}$ which can be defined by systems of identities of the form $\alpha x + w(x_1, x_2, \ldots, x_n) = 0$. In [7] it is proven that every variety from $\mathfrak{M}$ is a product of finitely many varieties from $\mathfrak{M}$ which are indecomposable in the class $\mathfrak{M}$. From this fact we conclude:

Proposition 10.2 ([7], Corollary 3). *A variety $\mathfrak{X}$ from $\mathfrak{M}$ is finitely attainable in $\mathfrak{A}$ if and only if $\mathfrak{A}$ is finitely $\mathfrak{X}$-solvable.*

Remark 10.1. Every simple algebra not contained in a variety $\mathfrak{X}$, is

$\mathfrak{X}$-radical. Hence if a variety $\mathfrak{A}$ satisfies the condition

(P) every $\mathfrak{A}$-free algebra can be embedded into a simple algebra in $\mathfrak{A}$, then $\mathfrak{A}$ is not $\mathfrak{X}$-solvable for any proper bounded subprevariety $\mathfrak{X}$ (cf. Remark 2.1).

In view of this remark and Proposition 10.2 we have

Proposition 10.3 ([7], Corollary 4). *Varieties of* R-Ω-*algebras satisfying condition* (P) *contain no non-trivial finitely attainable subvarieties belonging to* $\mathfrak{M}$.

In [7] it is said that the following varieties fulfil condition (P):

(1) varieties of linear R-Ω-algebras over an integral domain defined by permutational identities with coefficients ± 1 (see [5]);

(2) the variety of all Lie algebras over an integral domain (see [2], [8]);

(3) the variety of all associative algebras over an integral domain (see [8]);

(4) the variety of all commutative associative algebras over an integral domain (see [25]);

(5) the variety of all idempotent algebras over an integral domain (see [5], [44]).

Proposition 10.4 ([7], p. 474). *Let* R *be an integral domain without idempotent ideals. If* R *is an infinite field or the variety* $\mathfrak{A}$ *of* R-Ω-*algebras satisfies* $xx\ldots x\omega = 0$ *for every operation* ω, *then* $\mathfrak{A}$ *has no non-trivial finitely attainable subvarieties.*

Proposition 10.5 ([7], Lemma 4). *Schreier varieties of linear* R-Ω-*algebras have no non-trivial finitely attainable bounded subprevarieties.*

A variety $\mathfrak{A}$ of linear Ω-algebras over a principal ideal domain R is called an SP-variety if the intersection of $\mathfrak{A}$ and the variety of all linear Ω-algebras over an arbitrary field K is a Schreier variety and if this intersection satisfies condition (P) in case K is the quotient field of R.

Examples for SP-varieties are the varieties of all non-associative rings, Ω-rings, commutative or anticommutative rings, Lie rings, and Lie algebras, respectively (see [2], [4], [5], [6], [8], [24], [53]).

Theorem 10.2 ([7], Theorem 3). SP-*varieties over infinite rings have no non-trivial finitely attainable bounded subprevarieties.*

Question 10.1 ([7]). Which varieties of linear R-Ω-algebras have no non-trivial 1-attainable subvarieties?

11. LATTICES

Let $\mathfrak{K}$ denote the variety of all lattices.

Theorem 11.1 ([51]). *Every non-trivial prevariety of lattices is A-free in* $\mathfrak{K}$.

This theorem gives complete answer to Problems IA, $I_\gamma A$, IIA, IIIA. It contains as a very special case a result from [63] which states that the variety of distributive lattices is not 1-attainable in $\mathfrak{K}$. In fact, it holds

Proposition 11.1 ([51]). *The variety of distributive lattices is not* 1-*attainable in any lattice prevariety properly containing it.*

Notice also the following. It is well known that every lattice can be embedded into a simple one, hence Remark 2.1 shows that $\mathfrak{K}$ cannot be $\mathfrak{X}$-solvable for any of its proper subprevarieties. This observation solves Problems IS, $I_\gamma S$, IIIS.

Problem IIS is solved in [26]. By the degree of idempotence of a prevariety $\mathfrak{X}$ with respect to $\mathfrak{K}$ we mean the smallest ordinal $\gamma > 0$ such that $\mathfrak{X} \neq \mathfrak{X}^\gamma$. If $\mathfrak{X} = \mathfrak{X}^\gamma$ for every γ then $\mathfrak{X}$ is said to have absolute degree of idempotence. Clearly, the degree of idempotence of a prevariety equals 2 if and only if this prevariety is not idempotent.

An ordinal γ is said to be singular if $\gamma > 1$ and if a decomposition $\gamma = \gamma_1 + \gamma_2$ where $\gamma_1 \geqslant 1$ is possible only if $\gamma_2 \leqslant 1$ or $\gamma_2 = \gamma$. There are singular ordinals, e.g. such are the smallest ordinals belonging to infinite cardinalities. The infinite singular ordinals are either limit ordinals or direct successors of limit ordinals. There is only one finite singular ordinal: the number 2.

Theorem 11.2 ([26], Theorem 1). *An ordinal is the degree of idempotence of some prevariety of lattice if and only if it is singular.*

Theorem 11.3 ([26], Theorem 3). *The steps of* $\mathfrak{X}$*-solvability of lattices for some lattice prevariety* $\mathfrak{X}$ *can be the ordinals of the form* $\gamma\alpha + \epsilon$ *and only they. Here* γ *is the degree of idempotence of* $\mathfrak{X}$, α *is an arbitrary ordinal; if* $\gamma \neq 2$ *or* $\gamma\alpha$ *is not a limit ordinal then* $\epsilon = 0$, *otherwise* ϵ *equals* 0 *or* 1.

Corollary 11.1 ([26]). *If a prevariety* $\mathfrak{X}$ *of lattices is not idempotent in* $\mathfrak{K}$ *then it is S-free in* $\mathfrak{K}$.

Notice that Theorems 11.2 and 11.3 yield a complete solution to Problem IIS. In particular, they show that the *S*-spectra of lattice prevarieties need not be intervals in the class of ordinals. The following result is also interesting.

Theorem 11.4 ([26], Theorem 2). *The lattice of all non-trivial lattice prevarieties is cofinal with the family of all non-trivial lattice prevarieties of an arbitrarily given (in particular, absolute) degree of idempotence.*

From this theorem it follows, in particular, that the class of idempotents in the groupoid $G_p(\mathfrak{K})$ is not a set (this follows also from Remark 2.15). On the other hand, the following assertions are valid.

Proposition 11.3 ([26], p. 460). *The set of idempotent quasi-varieties of lattices is of continuum cardinality.*

Theorem 11.5 ([9]). *In the variety of all lattices there are no non-trivial bounded idempotent subprevarieties.*

This theorem solves Problem VIIV for lattices; in particular, it answers a question raised by the first author.

Question 11.1. What are the idempotent lattice quasi-varieties?

Question 11.2. What are the lattice varieties in which all subvarieties are finitely attainable?

Question 11.3. Which lattice varieties are finitely $\mathfrak{X}$-solvable for each of their subvarieties $\mathfrak{X}$?

Question 11.4 ([26]). Do there exist idempotent lattice quasi-varieties of non-absolute degree of idempotence?

Question 11.5. Does the analogue of Theorem 11.4 hold for quasi-varieties?

12. GROUPOIDS

Let $\mathfrak{K}$ be the variety of all groupoids. The first results on 1-attainability in $\mathfrak{K}$ appeared in [3]. First we introduce some definitions.

A groupoid G is said to be rigid if it admits only one endomorphism: the identical mapping. A groupoid is called a TS-quasigroup if the identities $x \cdot xy = y = yx \cdot x$ are satisfied in it.

Theorem 12.1 ([3]). *There exists a variety* $\mathfrak{M}$ *of* TS-*quasigroups which possesses the following properties:*

(1) *all subvarieties of* $\mathfrak{M}$ *are 1-attainable in* $\mathfrak{K}$;

(2) $\mathfrak{M}$ *has continuum many minimal sub-quasi-varieties, each of which contains a unique (up to isomorphism) minimal quasigroup which is at the same time rigid and simple.*

Further results are given in [17]. An element b of a groupoid G is said to be a proper divisor of an element a from G if there is an $x \in G$ such that at least one of the equalities $bx = a$, $xb = a$ holds, where $x \neq a$ if $b = a$. Let $\mathfrak{N}$ denote the class of those groupoids which satisfy the following condition: no principal ideal generated by an idempotent contains proper divisors of idempotents. This class $\mathfrak{N}$ is a quasi-variety ([17], Lemma 3).

Theorem 12.2 ([17], Theorem 3). *The quasi-variety* $\mathfrak{N}$ *is the largest non-trivial prevariety of groupoids which is attainable in* $\mathfrak{K}$. *It is* 1-*attainable in* $\mathfrak{K}$.

This theorem yields quite good information about Problem IA. The following theorem gives its complete solution as well as those of Problems $I_\gamma A$, IIIA for groupoid varieties.

Theorem 12.3 ([17], Theorem 4). *A non-trivial variety is attainable in* $\mathfrak{K}$ *if and only if the one-element groupoid is its only member which contains an idempotent element. In this case the variety is* 1*-attainable in* $\mathfrak{K}$.

This theorem solves, in particular, a question from [61].

Explicit collections of attainable groupoids are constructed in

Proposition 12.1 ([17], Proposition 1). *Let* $u(x)$, $v(x)$, $w(x)$ *be arbitrary terms in one variable* x, *and put* $\mathfrak{X} = [u(x) = u(y),\ v(x) = v(y),\ v(x)x = v(x)y,\ u(x)w(x) = x]$. *Then* $\mathfrak{X}$ *is* 1*-attainable in* $\mathfrak{K}$.

In order to formulate further results on attainability of groupoid quasi-varieties, we need some definitions. A term $U(x, x_1, \ldots, x_n)$ is called elementary if it is of the form

$$U(x, x_1, \ldots, x_n) \equiv$$
$$\equiv u_n(u_{n-1}(\ldots(u_2(u_1(x, x_1), x_2), x_3), \ldots, x_{n-1}), x_n),$$

where each of the terms $u_i(y, z)$ is either yz or zy or $u \equiv x$. Let a quasi-identity

$$f \equiv (p(x_1, \ldots, x_k) \to u(x_1, \ldots, x_k) = v(x_1, \ldots, x_k))$$

be given, where p is a conjunction of quasi-atomic formulas, u and v are terms, $x_1, \ldots, x_k$ are different variables which need not all occur in every term from f.

We shall consider the following conditions on a non-trivial quasi-variety $\mathfrak{X} = [Q]$ of groupoids, where Q is a system of quasi-identities.

(A) $\mathfrak{X} \subseteq \mathfrak{N}$.

(B) If $f \in Q$ then $\{p(x_1, \ldots, x_k)$ and $a^2 = a$ and

$a = U(u(x_1, \ldots, x_k), y_1, \ldots, y_n) \to u(x_1, \ldots, x_k) = a$:

U is an elementary term$\}$ is valid in $\mathfrak{X}$.

The following construction from [17] will also be needed.

Suppose that two groupoids A and B are given, A contains an idempotent a. Take an arbitrary mapping f of the set $\{(x, y) \in A \times A \mid xy = a\}$ into B. Define a multiplication on the set $(A \setminus a) \cup B$ as follows:

$$
\begin{array}{ll}
xy = xy & \text{if } x, y \in B; \\
xy = xy & \text{if } x, y \in A \setminus a,\ xy \neq a; \\
xy = f(x, y) & \text{if } x, y \in A \setminus a,\ xy = a; \\
xy = xa & \text{if } x \in A \setminus a,\ y \in B,\ xa \neq a; \\
xy = f(x, a) & \text{if } x \in A \setminus a,\ y \in B,\ xa = a; \\
xy = ay & \text{if } x \in B,\ y \in A \setminus a,\ ay \neq a; \\
xy = f(a, y) & \text{if } x \in B,\ y \in A \setminus a,\ ay = a.
\end{array}
$$

Denote the groupoid thus obtained by $[A, a] + B$. The mapping f is called the defining mapping.

This construction can be generalized as follows. Let $A_1, A_2, \ldots, A_\gamma$ be a transfinite sequence of groupoids, $a_\alpha \in A_\alpha$ $(1 \leqslant \alpha < \gamma)$ an idempotent element. Put $\Gamma_\alpha = \bigcup_{\alpha \leqslant \beta < \gamma} (A_\beta \setminus a_\beta) \cup A_\gamma$. We define a groupoid on the set Γ_1 so that $\Gamma_\alpha = [A_\alpha, a_\alpha] + \Gamma_{\alpha+1}$ is satisfied for some defining mappings.

For a congruence σ on B, we denote by σ^* the equivalence defined on $[A, a] + B$ which coincides with σ on B and is identical on $A \setminus a$. If $A, B \in \mathfrak{X}$ and $\mathfrak{X} \subseteq \mathfrak{N}$ then, obviously, σ^* is a congruence.

For an arbitrary ordinal $\gamma > 1$ and an arbitrary quasi-variety $\mathfrak{X}$ of groupoids we consider the following condition.

(C_γ) There exists a sequence of groupoids $A_1, \ldots, A_\gamma$ in $\mathfrak{X}$, A_α containing an idempotent a_α $(\alpha < \gamma)$, such that the following holds for suitable defining mappings and all $\alpha < \gamma$: if σ_α is a congruence on $\Gamma_{\alpha+1}$ different from $\nabla_{\Gamma_{\alpha+1}}$, further σ_α^* is a congruence on Γ_α and $\Gamma_{\alpha+1}/\sigma_\alpha \in \mathfrak{X}$, then $\Gamma_\alpha/\sigma_\alpha^* \notin \mathfrak{X}$.

Let $\neg(C_\gamma)$ denote the negation of condition (C_γ). The following theorem reduces the question about γ-attainability (for a fixed γ) of a groupoid quasi-variety $\mathfrak{X}$ to investigations of the construction described above.

Theorem 12.4 ([17], Theorem 5). *In order that a non-trivial quasi-variety $\mathfrak{X}$ of groupoids be γ-attainable in $\mathfrak{K}$, it is necessary and sufficient that the condition* (A), (B), (C_γ), $\neg(C_{\gamma+1})$ *be simultaneously fulfilled.*

For variable-preserving quasi-varieties the condition of 1-attainability admits a much simpler formulation.

Proposition 12.2. *A non-trivial variable-preserving quasi-variety of groupoids is* 1*-attainable in* $\mathfrak{K}$ *if and only if* (A), (B) *and the following condition* (D) *are satisfied:*

(D) *Let* $f \equiv p \to u = v$ *be the quasi-identity defining* $\mathfrak{X}$,

$u \equiv u_1 u_2$, $v \equiv v_1 v_2$. *Then the quasi-identities*

$p \,\&\, u^2 = u \to u_1 = v_1 \,\&\, u_2 = v_2$ *are valid in* $\mathfrak{X}$.

Proposition 12.3 ([17], Proposition 3). *The step of attainability of a variable-preserving quasi-variety is at most two.*

In [17] examples are given for not variable-preserving quasi-varieties of groupoids whose step of attainability in $\mathfrak{K}$ is 2 (see Proposition 4). Such is e.g. the quasi-variety

$$[xz = yz \to xy = yx,\ a^2 = a \,\&\, b^2 = b \to a = b,$$

$$a^2 = a \,\&\, xy = a \to x = y = a].$$

Notice that Remark 2.1 solves Problems IS, $I_\gamma S$, IIIS for groupoids since every groupoid can be embedded into a simple one (see e.g. [17], Lemma).

Question 12.1 ([17]). Does there exist a variable-preserving quasi-variety of groupoids which is 2-attainable in $\mathfrak{K}$?

Question 12.2. What are the A-spectra of groupoid quasi-varieties?

Question 12.3.

(a) What are the S-spectra of groupoid varieties?

(b) The same for quasi-varieties.

Question 12.4.

(a) What are the idempotent varieties of groupoids?

(b) The same for quasi-varieties.

Question 12.5. In which groupoid varieties is every subvariety finitely attainable?

Question 12.6. Which groupoid varieties are finitely $\mathfrak{X}$-solvable for each of their non-trivial subvarieties $\mathfrak{X}$?

13. QUASIGROUPS

Let $\mathfrak{K}$ be the variety of all quasigroups.

Theorem 13.1 ([17], Theorem 1). *Any non-trivial prevariety* $\mathfrak{X}$ *of quasigroups is either* 1*-attainable or* A*-free in* $\mathfrak{K}$. *The first case takes place if and only if the one-element quasigroup is the only member of* $\mathfrak{X}$ *which contains an idempotent element.*

This theorem provides the complete solution of Problems IA, $I_\gamma A$, IIIA for quasigroups. Taking into consideration that every quasigroup can be embedded into a simple one, Remark 2.1 solves Problems IS, $I_\gamma S$, IIIS.

Question 13.1. What are the S-spectra of prevarieties of quasigroups?

Question 13.2.

(a) What are the idempotent varieties of quasigroups?

(b) The same for quasi-varieties.

Question 13.3. In which quasigroup varieties is every subvariety finitely attainable?

Question 13.4. Which quasigroup varieties are $\mathfrak{X}$-solvable for each of their subvarieties $\mathfrak{X}$?

14. LOOPS

Theorem 14.1 ([17], Theorem 2). *Every non-trivial prevariety of loops is A-free in the class* $\mathfrak{K}$ *of all loops.*

This theorem gives the complete solution of Problems IA, I_γA, IIIA for loops. Remark 2.1 solves Problems IS, I_γS, IIIS in this case, too, since every loop can be embedded into a simple one.

Notice that the analogy with groups is not complete: none of the non-trivial idempotent prevarieties of loops (and these form a class which is not a set by Remark 2.15) is attainable in $\mathfrak{K}$ (in the group case all the idempotents are 1-attainable). In particular, by Remark 2.14, the class of all loops is not transpreverbal for any of its subprevarieties.

Question 14.1. What are the S-spectra of prevarieties of loops?

Question 14.2.

(a) What are the idempotent varieties of loops?

(b) The same for quasi-varieties.

Question 14.3. In which loop varieties is every subvariety finitely attainable?

Question 14.4. Which loop varieties are $\mathfrak{X}$-solvable for each of their subvarieties $\mathfrak{X}$?

15. UNARS

By a unar we mean an algebra with one unary operation. Let x' denote the result of the operation on the element x. Put $x^0 = x$, $x^1 = x'$ and $x^{n+1} = (x^n)'$ for $n = 1, 2, \ldots$.

Theorem 15.1 ([34], p. 19). *In order that a proper subvariety* $\mathfrak{X}$ *of the variety* $\mathfrak{K}$ *of all unars be finitely attainable in* $\mathfrak{K}$, *it is necessary and sufficient that there exists a natural number* $n > 0$ *such that*

$\mathfrak{X} = [x^n = x]$. *In this case* $\mathfrak{X}$ *is* 1*-attainable in* $\mathfrak{K}$.

This theorem solves Problems IIIAV and I_γAV for natural numbers γ.

Remark 15.1 In [34] it is noticed that the variety $[x^n = x]$ is not idempotent in $\mathfrak{K}$ for $n > 0$.

16. SUMMARY OF RESULTS

In order to give the reader a better overview, a table is presented at the end of our survey in which we describe the present situation of the problems treated above in the classes of algebras that were considered in the text. This table is structured as follows. In the utmost left column we give the name of the algebra, in the next one the type of class: V – variety, Q – quasi-variety, P – prevariety. Further on the top we see the numeration of the problems. Notice that, according to the nature of things, we often have theorems which solve Problems I_1A and IIIA (or I_γS and IIIS) at the same time; therefore we made an inversion in the order of columns so that these two ones be neighbouring. In the areas enclosed by closed contours (in most cases these areas are simply rectangles) we list those assertions in the text which yield information on the intersection of the given "rows" (types of classes) and "columns" (problems). Here we use the following abbreviations: T. – Theorem, P. – Proposition, C. – Corollary, L. – Lemma, R. – Remark. The presence of "?" together with references to assertions means that these assertions yield some information on the corresponding problems but do not solve any of them completely. If only the mark "?" is to be found, this indicates the absence of any valuable information on the special situation. If there is no "?" in the area, this means that the assertions listed there provide complete solutions to the corresponding problems.

We also included in the table the general problems VA, VS, VI about characterizations of classes in which certain ones of the considered notions coincide, in order to show the state of things in the classes of algebras dealt with in the survey. The answers ("coincide", "do not coincide") are reproduced by "yes" and "no", respectively, in case they are known. References to Remarks 2.1, 2.15 and 10.1, which have the form of implications, are made only if it is well known that their premises are valid in the given class of algebras. Finally, the sign ■ means that the corresponding problems have trivial solutions in view of well-known results.

PROBLEMS / ALGEBRAS	Type of classes	ATTAINABILITY						SOLVABILITY							
		IA	I_1A	IIIA	IIA	IVA	VA	IS	$I_\gamma S$	IIIS	IIS	IVS	VS	VI	VII
SEMIGROUPS	V		T.3.1			?	yes T.3.1				T.3.3	?	yes	no	P.3.1
	Q		T.3.2, T.3.3, L.3.1			R.2.10	?		R.2.1		L.3.1	R.2.11	R.2.1	T.3.1	P.3.2 R.2.15 ?
	P		?			?					?	?		P.3.1	
		IDEAL ATTAINABILITY						IDEAL SOLVABILITY							
	V		T.3.5				yes T.3.5				T.3.5		yes	yes T.3.5	T.3.5
	Q				?				R.2.1			?	R.2.1	?	P.3.2 R.2.15 ?
	P														
SEMIGROUPS WITH ZERO	V		T.4.1				T.4.1				P.4.2		yes	no	T.4.3
	Q		T.4.2, P.4.1, P.4.2			?			R.2.1		?	?	R.2.1	T.4.1	T.4.3
	P		?											T.4.3	?
		IDEAL ATTAINABILITY						IDEAL SOLVABILITY							
	V		T.4.4				yes T.4.4				T.4.4		yes	yes T.4.4	T.4.4
	Q				?				R.2.1			?	R.2.1	?	T.4.3
	P														?
MONOIDS	V		T.5.1				yes T.5.1			T.5.1				no	R.5.1
	Q		T.5.2, T.5.3			?			T.5.2, T.5.3				?	R.5.1	R.5.1
	P		?						?					T.5.1	?

<table>
<tr><th rowspan="2">PROBLEMS / ALGEBRAS</th><th rowspan="2">Type of classes</th><th colspan="6">ATTAINABILITY</th><th colspan="6">SOLVABILITY</th><th colspan="2"></th></tr>
<tr><th>IA</th><th>I_1A</th><th>IIIA</th><th>IIA</th><th>IVA</th><th>VA</th><th>IS</th><th>$I_\gamma S$</th><th>IIIS</th><th>IIS</th><th>IVS</th><th>VS</th><th>VI</th><th>VII</th></tr>
<tr><td rowspan="7">INVERSE SEMIGROUPS</td><td>V</td><td></td><td colspan="2">T.6.1</td><td></td><td></td><td>yes T.6.1</td><td colspan="3" rowspan="3">R.2.1</td><td rowspan="3">L.3.1 ?</td><td rowspan="3">?</td><td rowspan="3">yes R.2.1</td><td>yes T.6.1,R.6.1</td><td>R.6.1</td></tr>
<tr><td>Q</td><td colspan="4" rowspan="2">P.6.1, L.3.1 ?</td><td colspan="2" rowspan="2">?</td><td rowspan="2">?</td><td rowspan="2">P.7.1 R.2.15 ?</td></tr>
<tr><td>P</td></tr>
<tr><td></td><td colspan="6">IDEAL ATTAINABILITY</td><td colspan="6">IDEAL SOLVABILITY</td><td colspan="2"></td></tr>
<tr><td>V</td><td colspan="4">T.6.2</td><td></td><td>yes T.6.2</td><td colspan="3" rowspan="3">R.2.1</td><td>T.6.2</td><td></td><td rowspan="3">yes R.2.1</td><td>yes T.6.2</td><td>T.6.2</td></tr>
<tr><td>Q</td><td colspan="6" rowspan="2">?</td><td colspan="2" rowspan="2">?</td><td colspan="2" rowspan="2">?</td></tr>
<tr><td>P</td></tr>
<tr><td rowspan="3">GROUPS</td><td>V</td><td colspan="4">C.7.1</td><td rowspan="3">?</td><td rowspan="3">yes T.7.1</td><td colspan="3" rowspan="3">R.2.1</td><td rowspan="3">T.7.1</td><td rowspan="3">?</td><td rowspan="3">yes R.2.1</td><td rowspan="3">yes T.7.1</td><td>C.7.1</td></tr>
<tr><td>Q</td><td colspan="3" rowspan="2">T.7.1 ?</td><td rowspan="2">T.7.1</td><td rowspan="2">P.7.1 R.2.15 ?</td></tr>
<tr><td>P</td></tr>
<tr><td rowspan="3">MODULES OVER SPECIAL RINGS</td><td>V</td><td colspan="4">T.8.1</td><td></td><td>yes T.8.1</td><td></td><td colspan="3">T.8.1</td><td></td><td>yes T.8.1</td><td rowspan="3">yes R.2.13 R.2.14</td><td>T.8.1</td></tr>
<tr><td>Q</td><td colspan="6" rowspan="2">?</td><td colspan="6" rowspan="2">?</td><td rowspan="2">?</td></tr>
<tr><td>P</td></tr>
<tr><td rowspan="3">ASSOCIATIVE RINGS AND ALGEBRAS</td><td>V</td><td colspan="4">T.9.1</td><td></td><td>yes T.9.1</td><td colspan="3">R.10.1</td><td>T.9.1</td><td></td><td>yes T.9.1</td><td>yes P.9.7</td><td>P.9.8</td></tr>
<tr><td>Q</td><td colspan="4" rowspan="2">P.9.5, P.9.6 ?</td><td colspan="2" rowspan="2">?</td><td colspan="4" rowspan="2">P.9.5, P.9.6 ?</td><td colspan="2" rowspan="2">?</td><td colspan="2" rowspan="2">?</td></tr>
<tr><td>P</td></tr>
</table>

PROBLEMS / ALGEBRAS	Type of classes	ATTAINABILITY						SOLVABILITY							
		IA	I_1A	IIIA	IIA	IVA	VA	IS	$I_\gamma S$	IIIS	IIS	IVS	VS	VI	VII
ALTER-NATIVE RINGS AND ALGEBRAS	V		P.9.7 P.9.11											yes P.9.7	P.9.11
	Q		P.9.5			?			P.9.5 ?			?		?	
	P		?												
JORDAN RINGS AND ALGEBRAS	V		P.9.13 C.9.6											yes P.9.7	P.9.12 ?
	Q		P.9.12, P.9.5			?			P.9.5 ?			?		?	
	P		?												
LIE RINGS AND LIE ALGEBRAS	V		P.10.5				yes P.10.5		R.10.1					yes ■	■
	Q		P.9.5, P.10.5			?			P.9.5			?		?	
	P		?						?						
POWER-ASSOCI-ATIVE RINGS AND ALGEBRAS	V														
	Q		P.9.5 ?			?			P.9.5 ?			?		?	
	P														
LATTICES	V													yes T.11.1,T.11.5	T.11.5
	Q		T.11.1			?	yes T.11.1		R.2.1		T.11.2 T.11.3	?	yes R.2.1	no P.11.3 T.11.1	P.11.3 R.2.15 ?
	P														

<table>
<tr><th rowspan="2">PROBLEMS / ALGEBRAS</th><th rowspan="2">Type of classes</th><th colspan="6">ATTAINABILITY</th><th colspan="6">SOLVABILITY</th><th rowspan="2">VI</th><th rowspan="2">VII</th></tr>
<tr><th>IA</th><th>I_1A</th><th>IIIA</th><th>IIA</th><th>IVA</th><th>VA</th><th>IS</th><th>$I_\gamma S$</th><th>IIIS</th><th>IIS</th><th>IVS</th><th>VS</th></tr>
<tr><td rowspan="3">GROUPOIDS</td><td>V</td><td colspan="3">T.12.3</td><td>T.12.3</td><td rowspan="3">R.2.10
?</td><td rowspan="3">?</td><td colspan="3" rowspan="3">R.2.1</td><td rowspan="3">?</td><td rowspan="3">R.2.11
?</td><td rowspan="3">yes
R.2.1</td><td rowspan="3">no
■</td><td rowspan="3">R.2.15
?</td></tr>
<tr><td>Q</td><td colspan="3">T.12.4, P.12.2, ?</td><td>P.12.3</td></tr>
<tr><td>P</td><td colspan="3">T.12.2, ?</td><td>?</td></tr>
<tr><td rowspan="3">QUASI-GROUPS</td><td>V</td><td colspan="4" rowspan="3">T.13.1</td><td rowspan="3">R.2.10
?</td><td rowspan="3">yes
T.13.1</td><td colspan="3" rowspan="3">R.2.1</td><td rowspan="3">?</td><td rowspan="3">R.2.11
?</td><td rowspan="3">yes
R.2.1</td><td rowspan="3">no
■</td><td rowspan="3">R.2.15
?</td></tr>
<tr><td>Q</td></tr>
<tr><td>P</td></tr>
<tr><td rowspan="3">LOOPS</td><td>V</td><td colspan="4" rowspan="3">T.14.1</td><td rowspan="3">?</td><td rowspan="3">yes
T.14.1</td><td colspan="3" rowspan="3">R.2.1</td><td colspan="2" rowspan="3">?</td><td rowspan="3">yes
R.2.1</td><td>?</td><td rowspan="3">R.2.15
?</td></tr>
<tr><td>Q</td><td rowspan="2">no
T.14.1
R.2.15</td></tr>
<tr><td>P</td></tr>
<tr><td rowspan="3">UNARS</td><td>V</td><td></td><td colspan="2">T.15.1</td><td></td><td></td><td>yes
T.15.1</td><td colspan="6" rowspan="3">?</td><td rowspan="3">no
R.15.1</td><td rowspan="3">?</td></tr>
<tr><td>Q</td><td colspan="6" rowspan="2">?</td></tr>
<tr><td>P</td></tr>
</table>

REFERENCES

[1] L.A. Bokut', Some embedding theorems for rings and semigroups, I, *Sibirsk. Mat. Ž.*, 4 (1963), 500–518 (in Russian).

[2] L.A. Bokut', Embedding theorems in the theory of algebras, *Colloq. Math.*, 14 (1966), 349–353.

[3] A.D. Bol'bot, Attainable varieties of quasigroups, *Algebra i Logika,* 11 (1972), no. 6, 22–30 (in Russian).

[4] M.S. Burgin, Permutation products of linear Ω-algebras, *Izv. Akad. Nauk SSSR, Ser. Mat.*, 34 (1970), 977–999 (in Russian).

[5] M.S. Burgin, Linear Ω-algebras over commutative rings and attainable varieties, *Vestnik Mosk. Gos. Univ., Ser. Mat. Meh.*, (1972), no. 2, 56–63 (in Russian).

[6] M.S. Burgin, Schreier varieties of linear Ω-algebras, *Mat. Sb.*, 93 (1974), 555–573 (in Russian).

[7] M.S. Burgin, The cancellation law and attainable classes of linear Ω-algebras, *Mat. Zametki,* 16 (1974), 467–478 (in Russian).

[8] P.M. Cohn, *Universal Algebra,* Harper & Row, New York – London, 1965.

[9] A. Day, Idempotents in the groupoid of all SP of lattices, *Canad. Math. Bull.*, 21 (1978), 499–501.

[10] L. Fuchs, On the structure of abelian *p*-groups, *Acta Math. Acad. Sci. Hungar.*, 4 (1953), 267–288.

[11] B.J. Gardner, Semi-simple radical classes of algebras and attainability of identities, *Pacific J. Math.*, 61 (1975), 401–416.

[12] B.J. Gardner, Radical properties defined locally by polynomial identities, II, *J. Austral. Math. Soc.*, A 27 (1979), 274–283.

[13] B.J. Gardner, Extension-closed varieties of rings need not have attainable identities, *Bull. Malaysian Math. Soc.*, (2) 2 (1979), 37–39.

[14] B.J. Gardner, Extension-closure and attainability for varieties of algebras with involution, *Comment. Math. Univ. Carolin.*, 21 (1980), 285–292.

[15] B.J. Gardner – P.N. Stewart, On semi-simple radical classes, *Bull. Austral. Math. Soc.*, 13 (1975), 349–353.

[16] A.A. Iskander, Product of ring varieties and attainability, *Trans. Amer. Math. Soc.*, 193 (1974), 231–238.

[17] V.M. Kaplan, On attainable classes of groupoids, quasigroups, loops, *Sibirsk. Mat. Ž.*, 19 (1978), 604–616 (in Russian).

[18] O.V. Knjazev, On attainable classes of monoids, *Algebraic Systems and their Varieties,* Ural'sk. Gos. Univ., Sverdlovsk, 1982, 44–49 (in Russian).

[19] L.Ja. Kulikov, Generalized primary groups, I, *Trudy Mosk. Mat. Obšč.*, 1 (1952), 247–326 (in Russian).

[20] L. Ja. Kulikov, Generalized primary groups, II, *Trudy Mosk. Mat. Obšč.*, 2 (1953), 85–167 (in Russian).

[21] G.E. Kupčik – L.M. Martynov, Finite attainability in the class of bands, *Algebraic Systems and their Varieties,* Ural'sk. Gos. Univ., Sverdlovsk, 1982, 65–69 (in Russian).

[22] A.G. Kuroš, Non-associative free algebras and free products of algebras, *Mat. Sb.*, 20 (1947), 239–262 (in Russian).

[23] A.G. Kuroš, *The theory of groups,* Nauka, Moscow, 1967 (in Russian).

[24] A.G. Kuroš, Multioperator rings and algebras, *Uspehi Mat. Nauk,* 24 (1969), no. 1, 3–15 (in Russian).

[25] S. Lang, *Algebra,* Addison–Wesley, Reading, 1965.

[26] V.B. Lender, On the orders of solvability of lattices and the degrees of idempotence of lattice prevarieties, *Mat. Sb.*, 95 (1974), 445–460. Correction. *ibid.* 99 (1976), 477 (in Russian).

[27] V.B. Lender, On the orders of RA-solvability of lattices and the degrees of A-idempotence of lattice prevarieties, *Investigations in Modern Algebra,* Ural'sk. Gos. Univ., Sverdlovsk, 1976, 37–53 (in Russian).

[28] V.B. Lender, On normally solvable lattices, *Izv. Vysš. Učebn. Zaved. Matematika,* 1977, no. 12, 44–53 (in Russian).

[29] V.B. Lender, On RNA-solvable lattices, *Investigations in Modern Algebra,* Ural'sk. Gos. Univ., Sverdlovsk, 1979, 71–95 (in Russian).

[30] A.I. Mal'cev, Generalized nilpotent algebras and the groups associated with them, *Mat. Sb.,* 25 (1949), 347–366 (in Russian).

[31] A.I. Mal'cev, On the multiplication of classes of algebraic systems, *Sibirsk. Mat. Ž.,* 8 (1967), 346–365 (in Russian).

[32] A.I. Mal'cev, *Algebraic systems,* Nauka, Moscow, 1970 (in Russian).

[33] L. Márki, Radical semisimple classes and varieties of semigroups, *Algebraic Theory of Semigroups* (Proc. Conf. Szeged, 1976), Colloq. Math. Soc. J. Bolyai, vol. 20, North-Holland, Amsterdam, 1979, 357–369.

[34] L.M. Martynov, *Verbal chains in universal algebras,* Candidate thesis, Sverdlovsk, 1972 (in Russian).

[35] L.M. Martynov, On attainable classes of groups and semigroups, *Mat. Sb.,* 90 (1973), 235–245 (in Russian).

[36] L.M. Martynov, On solvable rings, *Mat. Zapiski Ural'sk. Gos. Univ.,* 8 (1973), no. 3, 82–93 (in Russian).

[37] L.M. Martynov, On ideal solvability and attainability in semigroups, *Mat. Zapiski Ural'sk. Gos. Univ.,* 9 (1974), no. 1, 40–55 (in Russian).

[38] L.M. Martynov, On attainable classes of rings, Abstracts, 14th All-Union Conference on Algebra, vol. 2, Novosibirsk, 1977, 56–58 (in Russian).

[39] L.M. Martynov, On the groupoid of prevarieties of semigroups with zero, *Izv. Vysš. Učebn. Zaved. Matematika,* 1981, no. 3, 80–82 (in Russian).

[40] L.M. Martynov, On attainable classes of semigroups with zero, *Izv. Vysš. Učebn. Zaved. Matematika,* 1981, no. 9, 29–37 (in Russian).

[41] L.M. Martynov, On attainability and solvability for varieties of modules and associative algebras, to appear (in Russian).

[42] L.M. Martynov – T.A. Martynova, Ideal attainable varieties of inverse semigroups, *Izv. Vysš. Učebn. Zaved. Matematika,* 1978, no. 8, 67–73 (in Russian).

[43] G. Michler – R. Wille, Die primitiven Klassen arithmetischer Ringe, *Math. Z.,* 113 (1970), 369–372.

[44] B.H. Neumann, Embedding non-associative rings in division rings, *Proc. London Math. Soc.,* 1 (1951), 241–256.

[45] J.M. Osborn, Varieties of algebras, *Advances in Math.,* 8 (1972), 163–369.

[46] S.V. Pčelincev, On the metaideals of alternative algebras, to appear (in Russian).

[47] M. Petrich, The maximal semilattice decomposition of a semigroup, *Math. Z.,* 85 (1964), 68–82.

[48] B.I. Plotkin, On functorials, radicals, and coradicals in groups, *Mat. Zapiski Ural'sk. Gos. Univ.,* 7 (1970), no. 3, 150–182 (in Russian).

[49] B.I. Plotkin, Radicals in groups, operations on classes of groups, and radical classes, *Selected Questions in Algebra and Logic,* Novosibirsk, 1973, 205–244 (in Russian).

[50] V.V. Rasin, Finitely attainable varieties of inverse semigroups, *Izv. Vysš. Učebn. Zaved. Matematika,* 1978, no. 8, 80–87 (in Russian).

[51] L.N. Ševrin – V.B. Lender, On attainable classes of lattices, *Izv. Vysš. Učebn. Zaved. Matematika,* 1972, no. 12, 11–115 (in Russian).

[52] L.N. Ševrin – L.M. Martynov, On attainable classes of algebras, *Sibirsk. Mat. Ź.,* 12 (1971), 1363–1381 (in Russian).

[53] A.I. Širšov, Subalgebras of free Lie algebras, *Mat. Sb.,* 33 (1953), 441–452 (in Russian).

[54] A.I. Širšov, Subalgebras of free commutative and free anticommutative algebras, *Mat. Sb.,* 34 (1954), 81–88 (in Russian).

[55] E.G. Šutov, Embedding semigroups into simple and complete semigroups, *Mat. Sb.,* 62 (1963), 496–511 (in Russian).

[56] P.N. Stewart, Semi-simple radical classes, *Pacific J. Math.,* 32 (1970), 249–254.

[57] E.V. Suhanov, On the closedness of semigroup varieties under some constructions, *Investigations in Modern Algebra,* Sverdlovsk, 1978, 182–189 (in Russian).

[58] *The Sverdlovsk Tetrad* (Unsolved problems in semigroup theory), Sverdlovsk, 1979 (in Russian).

[59] T. Tamura, The theory of construction of finite semigroups, I, *Osaka Math. J.,* 8 (1956), 243–261.

[60] T. Tamura, Report on attainability of systems of identities, *Bull. Amer. Math. Soc.,* 71 (1965), 555–558.

[61] T. Tamura, Attainability of systems of identities on semigroups, *J. Algebra,* 3 (1966), 261–276.

[62] T. Tamura, Note on attainability of identities of bands, *J. Algebra,* 28 (1974), 1–9.

[63] T. Tamura – F.M. Yaqub, Example related to attainability of identities on lattices and rings, *Math. Japon.*, 10 (1965), 35–39.

[64] A.A. Vinogradov, Quasi-varieties of abelian groups, *Algebra i Logika,* 4 (1965), no. 6, 15–19 (in Russian).

[65] S.M. Vovsi, On prevarieties of groups, *Some problems in group theory,* Riga, 1971, 5–13 (in Russian).

[66] S.M. Vovsi, On infinite products of classes of groups, *Sibirsk. Mat. Ž.*, 13 (1972), 272–285 (in Russian).

[67] S.M. Vovsi, Triangular products, prevarieties and radical classes of linear representations of groups, *Doklady Akad. Nauk SSSR,* 224 (1975), 27–30 (in Russian).

[68] H. Werner – R. Wille, Charakterisierungen der primitiven Klassen arithmetischer Ringe, *Math. Z.*, 115 (1970), 197–200.

[69] E. Witt, Die Unterringe der freien Lieschen Ringe, *Math. Z.*, 64 (1956), 195–216.

[70] A.P. Zamjatin, Prevarieties of associative rings whose elementary theory is solvable, *Sibirsk. Mat. Ž.*, 19 (1978), 1266–1282 (in Russian).

L.N. Sevrin

Ural State University, 620 083 Sverdlovsk, Lenin ul. 51, USSR.

L.M. Martynov

Omsk Pedagogical Institute, Omsk, USSR.

COLLOQUIA MATHEMATICA SOCIETATIS JÁNOS BOLYAI
39. SEMIGROUPS, SZEGED (HUNGARY), 1981.

ON MATRIX CHARACTERIZATIONS OF PRIMITIVE REGULAR SEMIGROUPS

I. SZABÓ

INTRODUCTION

It is known that every primitive regular semigroup can be embedded in a completely 0-simple semigroup, namely, in the following way (see F. Pastijn [2]). If $S = \dot{\bigcup}_{\gamma \in \Gamma} S_\gamma$ is a primitive regular semigroup (the symbol $\dot{\bigcup}$ means 0-disjoint union) and the S_γ are completely 0-simple subsemigroups of S, $S_\gamma \cong M(G_\gamma^0; I_\gamma, \Lambda_\gamma; P_\gamma)$, then S can be embedded in $M\left(\prod_{\gamma \in \Gamma} G_\gamma^0; \bigcup_{\gamma \in \Gamma} I_\gamma, \bigcup_{\gamma \in \Gamma} \Lambda_\gamma; P\right)$. In this paper we give another matrix representation, namely, in the generalized Rees matrix semigroup $M\left(\dot{\bigcup}_{\gamma \in \Gamma} G_\gamma^0; \max_{\gamma \in \Gamma} I_\gamma, \max_{\gamma \in \Gamma} \Lambda_\gamma; P^4\right)$ with 4-dimensional sandwich matrix (Theorem 3). This representation uses matrices of smaller size than Pastijn's one, and in an interesting special case S is not just embedded but isomorphic to a generalized Rees matrix semigroup (Theorem 1). We also give a similar characterization for primitive inverse semigroups.

MATRIX CHARACTERIZATIONS

A semigroup S with zero is called (r*)-*decomposable* if

(1) $$S = \dot{\bigcup_{i\in I}} R_i = \dot{\bigcup_{\lambda\in\Lambda}} L_\lambda,$$

where the R_i $[L_\lambda]$ are right [left] similar right [left] ideals of S, and there exists an index $1 \in I \cap \Lambda$ such that R_1 is a semigroup with relative left identity elements and $H = R_1 \cap L_1$ is a semigroup with relative identity elements. (We say that a semigroup S is a *semigroup with relative [right, left] identity elements* if, for every element a in S, there exist elements d^a, e^a in S with $a = ad^a = e^a a$ $[a = ad^a, a = e^a a]$.)

An (r*)-decomposable semigroup S is called (r*q*)-*decomposable* if S is a semigroup with relative identity elements.

We say that a semigroup S is *similarly decomposable* (see O. Steinfeld [3]) if S satisfies condition (1) and $R_i = e_i R$ $(e_i^2 = e_i \neq 0,$ $i \in I)$, $L_\lambda = Se_\lambda$ $(e_\lambda^2 = e_\lambda \neq 0,$ $\lambda \in \Lambda)$ and $1 \in I \cap \Lambda$.

Let H be a semigroup with zero and relative identity elements, I, Λ be two index sets, and $P^4 = (p[a, \lambda, j, b])$, $a, b \in H$, $\lambda \in \Lambda$, $j \in I$ be a mapping of $H \times \Lambda \times I \times H$ into H such that for every element a, b, s, t of H and for any indices $j \in I$, $\lambda \in \Lambda$ the equality

(2) $$sap[a, \lambda, j, b]bt = sap[sa, \lambda, j, bt]bt$$

is satisfied.

Define a multiplication on the set of all Rees matrices over H by

(3) $$(a)_{i\lambda} \circ (b)_{j\mu} = (ap[a, \lambda, j, b]b)_{i\mu}.$$

From [4] Lemma 4.1 it follows that this multiplication is associative. So we have obtained in this way a semigroup which we term (*)-*generalized Rees matrix semigroup* and denote by $\mathcal{M} = M(H; I, \Lambda; P^4)$.

We say that a (*)-generalized Rees matrix semigroup $M(H; I, \Lambda; P^4)$ is (R*)-*regular* if it has the following property: there exists an index $1 \in$ $\in I \cap \Lambda$ such that for each element $s \in H$ the equality

$$q_1 p[q_1, 1, 1, s]s = sp[s, 1, 1, q_2]q_2 = s$$

holds for some $q_1, q_2 \in H$.

A (Q*)-*regular* Rees matrix semigroup is a (*)-generalized Rees matrix semigroup satisfying the following conditions:

(a) for each $a \in H$, $\lambda \in \Lambda$ there exist $j \in I$ and $q \in H$ such that $a = ap[a, \lambda, j, q]q$;

(b) for each $a \in H$, $j \in I$ there exist $\lambda \in \Lambda$ and $q \in H$ such that $a = qp[q, \lambda, j, a]a$.

These two notions are generalizations of the notion of local regularity (see O. Steinfeld [3]).

By Theorems 4.2 and 4.3 of [4], a semigroup S with zero is (r*)- [(r*q*)-] decomposable if and only if it is isomorphic to an (R*)- [(R*)- and (Q*)-] regular (*)-generalized Rees matrix semigroup over a semigroup with zero and relative identity elements.

As it is well known, a semigroup S with zero ($|S| \geqslant 2$ where $|S|$ denotes the cardinality of S) is primitive regular if and only if $S = \dot{\bigcup}_{\gamma \in \Gamma} S_\gamma$, where S_γ's are completely 0-simple subsemigroups of S. Then we obtain:

$$S = \dot{\bigcup}_{\gamma \in \Gamma} \dot{\bigcup}_{i \in I_\gamma} e_i S_\gamma = \dot{\bigcup}_{\gamma \in \Gamma} \dot{\bigcup}_{\lambda \in \Lambda_\gamma} S_\gamma e_\lambda .$$

We say that a primitive regular semigroup S is *primitive regular of identical index sets* if $I_\gamma = I_{\gamma'}$ and $\Lambda_\gamma = \Lambda_{\gamma'}$ for all $\gamma, \gamma' \in \Gamma$.

Theorem 1. *A semigroup S with zero is primitive regular of identical index sets if and only if it is isomorphic to an* (R*)- *and* (Q*)-*regular Rees matrix semigroup* $\mathcal{M} = M\left(H = \dot{\bigcup}_{\gamma \in \Gamma} H_\gamma ; I, \Lambda; P^4\right)$, *where the base semigroup H is a 0-direct union of its ideals which are subgroups with zero.*

This theorem will be deduced from a more general result (following Theorem 5).

We say that a $(*)$-generalized Rees matrix semigroup $\mathcal{M} = M(H; I, \Lambda; P^4)$ is *(Q^*_Γ)-regular* (where Γ is a set of indices) if $H = \dot{\bigcup}_{\gamma\in\Gamma} H_\gamma$ (H_γ are ideals of H), $I_\gamma \subseteq I$ and $\Lambda_\gamma \subseteq \Lambda$ for all $\gamma \in \Gamma$, and we have

(i) for all $a \in H_\gamma$, $\lambda \in \Lambda_\gamma$ there exist an index $j(\lambda, a) \in I_\gamma$ and an element $q \in H_\gamma$ such that $a = ap[a, \lambda, j(\lambda, a), q]q$;

(ii) for all $a \in H_\gamma$, $j \in I_\gamma$ there exist an index $\lambda(j, a) \in \Lambda_\gamma$ and an element $q \in H_\gamma$ such that $a = qp[q, \lambda(j, a), j, a]a$;

(iii) for all indices $\lambda \in \Lambda_\gamma$, $j \in I_{\gamma'}$ if $\lambda \notin \Lambda_\gamma \cap \Lambda_{\gamma'}$ or $j \notin I_\gamma \cap I_{\gamma'}$ then $p[a, \lambda, j, b] = 0$ for all elements a, b of H.

We say that an element a of a semigroup S with zero is a *bi-annihilator* if $aba = 0$ for all b in S.

Theorem 2. *Let S be a semigroup with zero ($|S| \geqslant 2$). The following two conditions are equivalent:*

(I) $S = \dot{\bigcup}_{\gamma\in\Gamma} S_\gamma$, *that is, S is a 0-direct union of its ideals S_γ which are similarly decomposable semigroups, and 0 is the only bi-annihilator of S;*

(II) *there exists an* (R^*)- *and* (Q^*_Γ)*-regular matrix semigroup $\mathcal{M} = M(H; I, \Lambda; P^4)$ $(1 \in I \cap \Lambda)$ such that H is a 0-direct union of its ideals H_γ $\left(H = \dot{\bigcup}_{\gamma\in\Gamma} H_\gamma\right)$ having identity elements, 0 is the only bi-annihilator of H, and if $\mathcal{M}_b$ is the subsemigroup of all bi-annihilators of $\mathcal{M}$, then S is isomorphic to $\bar{\mathcal{M}} = (\mathcal{M} \setminus \mathcal{M}_b) \cup (0)$.*

Proof. Suppose that S satisfies condition (I). Then from Theorem 4.1 of O. Steinfeld [3] we get that

$$S \cong \bar{\mathcal{M}} = \dot{\bigcup}_{\gamma\in\Gamma} \mathcal{M}_\gamma = \dot{\bigcup}_{\gamma\in\Gamma} M(H_\gamma; I_\gamma, \Lambda_\gamma; P_\gamma = (p_\gamma[\lambda, j])).$$

Suppose that $H = \dot{\bigcup}_{\gamma\in\Gamma} H_\gamma$ is a 0-direct union of its ideals H_γ. Denote by 1 the element $1_\gamma = I_\gamma \cap \Lambda_\gamma$ for all $\gamma \in \Gamma$, and choose some index sets I, Λ satisfying $I \supseteq I_\gamma$, $\Lambda \supseteq \Lambda_\gamma$ for all indices $\gamma \in \Gamma$ (and $1 \in I \cap \Lambda$).

Consider

$$\mathcal{M} = M(H; I, \Lambda; P^4 = (p[a, \lambda, j, b]))$$

where

$$p[a, \lambda, j, b] =$$

$$= \begin{cases} p_\gamma[\lambda, j] & \text{if } a, b \in H_\gamma \setminus 0,\ \lambda \in \Lambda_\gamma,\ j \in I_\gamma \text{ for some } \gamma \in \Gamma; \\ 0 & \text{otherwise.} \end{cases}$$

Condition (2) being fulfilled, $\mathcal{M}$ is a matrix semigroup with multiplication (3).

First we show that $\mathcal{M}$ is an (R*)-regular matrix semigroup. For any element $s \neq 0$ in $H = \dot{\bigcup}_{\gamma \in \Gamma} H_\gamma$ there exists an index $\gamma \in \Gamma$ such that $s \in H_\gamma$ and there exist elements $q_1, q_2 \in H_\gamma$ such that $s = q_1 p_\gamma[1, 1]s = sp_\gamma[1, 1]q_2$, because $\mathcal{M}_\gamma$ is a locally regular matrix semigroup. Similarly, we can easily prove that $\mathcal{M}$ is (Q^*_Γ)-regular, too. (If, for example, there exist an element a of H and indices $\gamma, \gamma' \in \Gamma$, $\gamma \neq \gamma'$, $\lambda \in \Lambda$ such that $a \in H_\gamma$ and $\lambda \in \Lambda_{\gamma'}$, $\lambda \notin \Lambda_\gamma$ then $\mathcal{M}$ is not (Q*)-regular, because for any index $j(\lambda, a) \in I$ and element $q \in H$, $0 = ap[a, \lambda, j(\lambda, a), q]q \neq a$.)

We show that there are no bi-annihilators of H. Let a be an arbitrary non-zero element of H. Because of $S = \dot{\bigcup}_{\gamma \in \Gamma} S_\gamma$ there are no bi-annihilators, so we get that there exist an index $\gamma \in \Gamma$ and matrix $(b)_{j\mu} \in \mathcal{M}_\gamma$ such that $(a)_{11} \circ (b)_{j\mu} \circ (a)_{11} \neq (0)$, that is, $ap_\gamma[1, j]bp_\gamma[\mu, 1]a = aca \neq \neq 0$, where $c = p_\gamma[1, j]bp_\gamma[\mu, 1] \in H$. So we obtain that a is not a bi-annihilator of H.

Consider any element $(a)_{i\lambda} \in (\mathcal{M} \setminus \bar{\mathcal{M}}) \cup (0)$. For any element $(b)_{j\mu} \in \mathcal{M}$, the condition

$$(a)_{i\lambda} \circ (b)_{j\mu} \circ (a)_{i\lambda} = (ap[a, \lambda, j, b]bp[b, \mu, i, a]a)_{i\lambda} \neq (0)$$

is satisfied if and only if $a, b \in H_\gamma$, $\lambda, \mu \in \Lambda_\gamma$, $i, j \in I_\gamma$, but then

$(a)_{i\lambda} \in \mathcal{M}_\gamma \subseteq \bar{\mathcal{M}}$. Hence it follows that $(\mathcal{M} \setminus \bar{\mathcal{M}}) \cup (0)$ is the subsemigroup of all bi-annihilators of $\mathcal{M}$.

Conversely, suppose that condition (II) of the theorem holds.

Then there are no non-zero bi-annihilators of S. Let $\mathcal{M}_\gamma$ $(\gamma \in \Gamma)$ be the following matrix semigroups: $\mathcal{M}_\gamma = M(H_\gamma; I_\gamma, \Lambda_\gamma; P^4_\gamma)$, where $H = \dot{\bigcup}_{\gamma \in \Gamma} H_\gamma$; $I_\gamma \subseteq I$, $\Lambda_\gamma \subseteq \Lambda$ (according to the definition of (Q^*_Γ)-regularity) and $P^4_\gamma = (p[a, \lambda, j, b]) = (p_\gamma[a, \lambda, j, b])$, where $a, b \in H_\gamma$, $\lambda \in \Lambda_\gamma$, $j \in I_\gamma$ for some $\gamma \in \Gamma$; and it can be supposed that $p_\gamma[a, \lambda, j, b] \in H_\gamma$ if $ap[a, \lambda, j, b]b \neq 0$, and if $ap[a, \lambda, j, b]b = 0$ then put $p_\gamma[a, \lambda, j, b] = 0 \ (\in H_\gamma)$.

Consider $\bar{\mathcal{M}} = \dot{\bigcup}_{\gamma \in \Gamma} \mathcal{M}_\gamma$, then it is easy to show that $\bar{\mathcal{M}}$ is a 0-direct union of the $\mathcal{M}_\gamma$.

We shall prove that $\mathcal{M}_b = (\mathcal{M} \setminus \bar{\mathcal{M}}) \cup (0)$ is the subsemigroup of all bi-annihilators of $\mathcal{M}$. If $(a)_{i\lambda} \in \mathcal{M}_b$ then $(a)_{i\lambda}$ is clearly a bi-annihilator of $\mathcal{M}$. If $(a)_{i\lambda} \in \mathcal{M}_\gamma$ $(\gamma \in \Gamma)$ is a bi-annihilator of $\mathcal{M}$ then for all $(b)_{j\mu} \in \mathcal{M}$ we have $(a)_{i\lambda} \circ (b)_{j\mu} \circ (a)_{i\lambda} = (0)$, and from the condition of (Q^*_Γ)-regularity we get the existence of indices $j(\lambda, a) \in I_\gamma$, $\lambda(i, a) \in \Lambda_\gamma$, and of elements $q_1, q_2 \in H_\gamma$ such that

$$a = ap_\gamma[a, \lambda, j(\lambda, a), q_1]q_1 = q_2 p_\gamma[q_2, \lambda(i, a), i, a]a.$$

For an arbitrary element x of H_γ we have now

$$(a)_{i\lambda} \circ (q_1 x q_2)_{j(\lambda, a)\lambda(i, a)} \circ (a)_{i\lambda} = (0),$$

therefore

$$ap_\gamma[a, \lambda, j(\lambda, a), q_1]q_1 x q_2 p_\gamma[q_2, \lambda(i, a), i, a]a = axa = 0,$$

that is, a is a bi-annihilator of $H \setminus 0$. This is a contradiction, therefore $\mathcal{M}_b$ is the subsemigroup of all bi-annihilators of $\mathcal{M}$.

So we get that S is isomorphic to $\bar{\mathcal{M}} = \dot{\bigcup}_{\gamma \in \Gamma} \mathcal{M}_\gamma$.

If $\mathcal{M}_\gamma$ is an (R^*)- and (Q^*_Γ)-regular matrix subsemigroup of $\mathcal{M}$, then it is a locally regular matrix semigroup, and from Theorem 4.1 of

O. Steinfeld [3] it follows that S is a 0-direct union of some of its ideals which are similarly decomposable subsemigroups.

The proof of Theorem 2 is finished.

This theorem implies the following.

Corollary 3. *Let* S *be a semigroup with zero* $(|S| \geq 2)$. *The following two conditions are equivalent:*

(i) S *is a primitive regular semigroup;*

(ii) *the semigroup* S *can be embedded in a semigroup* T *such that*

(a) $(T \setminus S) \cup \{0\}$ *is the subsemigroup of all bi-annihilators of* T;

(b) T *is isomorphic to an* (R^*)- *and* (Q_Γ^*)-*regular Rees matrix semigroup* $\mathcal{M} = M(H; I, \Lambda; P^4)$, *where* H *is a* 0-*direct union of groups with zero.*

It is well known that a semigroup S with zero $(|S| \geq 2)$ is primitive inverse if and only if it is a 0-direct union of completely 0-simple inverse semigroups.

From Corollary 3 and Theorem 3.9 of Clifford – Preston [1] we get:

Corollary 4. *Let* S *be a semigroup with zero* $(|S| \geq 2)$, *then the following two conditions are equivalent:*

(i) S *is a primitive inverse semigroup;*

(ii) *the semigroup* S *can be embedded in a semigroup* T *such that*

(a) $(T \setminus S) \cup \{0\}$ *is the subsemigroup of all bi-annihilators of* T;

(b) T *is isomorphic to an* (R^*)- *and* (Q_Γ^*)-*regular Rees matrix semigroup* $\mathcal{M} = M(H; I, I; \Delta^4)$, *where* H *is a* 0-*direct union of groups with zero* G_γ^0, *and the sandwich matrix is the following:*

$$\Delta^4 = (p[a,i,j,b]) = \begin{cases} e_\gamma & \text{if } a, b \in G_\gamma^0,\ i = j, \\ 0 & \text{otherwise.} \end{cases}$$

Theorem 2 has the following result as a special case.

Theorem 5. *For a semigroup* S *with zero,* $|S| \geqslant 2$, *the following two conditions are equivalent:*

(I) S *is a* 0-*direct union of similarly decomposable semigroups* S_γ *such that*

$$S_\gamma = \dot{\bigcup_{i \in I_\gamma}} e_i S_\gamma = \dot{\bigcup_{\lambda \in \Lambda_\gamma}} S_\gamma e_\lambda$$

$(1_\gamma \in I_\gamma \cap \Lambda_\gamma)$ *and* $I_\gamma = I_{\gamma'}$, $\Lambda_\gamma = \Lambda_{\gamma'}$ *for all indices* $\gamma, \gamma' \in \Gamma$, *and there are no non-zero bi-annihilators of* S;

(II) S *is isomorphic to an* (R*)- *and* (Q*)-*regular Rees matrix semigroup* $\mathcal{M} = M(H; I, \Lambda; P^4)$, *where* $H = \dot{\bigcup_{\gamma \in \Gamma}} H_\gamma$ *is a* 0-*direct union of its ideals which are subsemigroups with zero and identity elements, and there are no non-zero bi-annihilators of* H.

Proof. Consider the proof of Theorem 2. From condition (I) of this theorem it follows that $S \subseteq T \cong M(H; I, \Lambda; P^4)$ and $S \cong \bar{\mathcal{M}} = \dot{\bigcup_{\gamma \in \Gamma}} \mathcal{M}_\gamma$. If $(a)_{i\lambda}$ is a bi-annihilator of $\mathcal{M}$ from $\mathcal{M} \setminus \bar{\mathcal{M}} \cup (0)$ then $a \in H_\gamma$ for some $\gamma \in \Gamma$ and $i \in I_\gamma = I$, $\lambda \in \Lambda_\gamma = \Lambda$, so from local regularity we get $(a)_{i\lambda} = 0$, so $S \cong \bar{\mathcal{M}} = \dot{\bigcup_{\gamma \in \Gamma}} \mathcal{M}_\gamma$.

Conversely, (Q*)-regularity implies $I_\gamma = I_{\gamma'}$, $\Lambda_\gamma = \Lambda_{\gamma'}$, for all $\gamma, \gamma' \in \Gamma$.

This theorem implies Theorem 1.

We say that a primitive inverse semigroup S is *primitive inverse of identical index sets* if it is primitive regular of identical index sets.

Corollary 6. *A semigroup* S *with zero* $(|S| \geqslant 2)$ *is primitive inverse of identical index sets if and only if it is isomorphic to an* (R*)- *and* (Q*)-*regular Rees matrix semigroup* $\mathcal{M} = M\left(H = \dot{\bigcup_{\gamma \in \Gamma}} G_\gamma^0; I, I; \Delta^4\right)$, *where the base semigroup* H *is a* 0-*direct union of groups with zero.*

REFERENCES

[1] A.H. Clifford – G.B. Preston, *The algebraic theory of semigroups* I, Amer. Math. Soc., Providence, R. I., 1961.

[2] F. Pastijn, Idempotente elementen in reguliere semigroepen, *Communication and cognition monographies,* Dienst Hogere Meetkunde, Rijksuniversiteit te Gent, 1980.

[3] O. Steinfeld, On a generalization of completely 0-simple semigroups, *Acta Sci. Math. (Szeged),* 28 (1967), 135–145.

[4] I. Szabó, Rees matrix semigroups with 4-dimensional sandwich matrices, *Acta Math. Acad. Sci. Hungar.*, 35 (1980), 339–350.

I. Szabó

H-1125 Budapest, Nógrádi u. 10, Hungary.

COLLOQUIA MATHEMATICA SOCIETATIS JÁNOS BOLYAI
39. SEMIGROUPS, SZEGED (HUNGARY), 1981.

ALGEBRAIC SYSTEMS ASSOCIATED WITH INVOLUTED REGULAR SEMIGROUPS*

T. TAMURA

1. INTRODUCTION

By a groupoid we mean a set with a binary operation. A congruence ρ on a groupoid G is called a group-congruence if G/ρ is a group. A group-congruence ρ_0 on G is called smallest if $\rho_0 \subseteq \rho$ for all group-congruences ρ on G. Not all groupoids have a smallest group-congruence, for example, the semigroup of all positive integers under addition does not have a smallest group-congruence. This is due to the fact that the axiom of groups cannot be expressed by a set of implications. However, it is known that groups are characterized in terms of algebraic systems satisfying identities, the so-called varieties. First, the most obvious is:

$A(\cdot, ')$ with a binary operation and involution satisfying

(1.1) $(x')' = x, \quad (xy)' = y'x', \quad (xy)z = x(yz), \quad (xy)y' = x,$

$x'(xy) = y.$

*The abstract of this paper was sent to the organizers of Conference by the author under the title: Algebraic systems associated with inverse semigroups.

In the following, a groupoid $A(*)$ is related to left division and $A(\circ)$ is related to right division. See Lorenzen [8], Ward [12], Shoda [11], Hall [5], also see [1] at page 154.

$$(1.2) \quad A(*) \text{ satisfying } (z * x) * (z * y) = x * y, \ (x * x) * x = x, \ x * x = y * y.$$

$$(1.2') \quad A(\circ) \text{ satisfying } (x \circ z) \circ (y \circ z) = x \circ y, \ x \circ (x \circ x) = x, \ x \circ x = y \circ y.$$

Furthermore (1.2′) is equivalent to the following two laws (1.3) and also equivalent to the single law (1.4). These are due to Higman and Neumann [6].

$$(1.3) \quad (x \circ z) \circ (y \circ z) = x \circ y, \ (x \circ x) \circ ((y \circ y) \circ y) = y.$$

$$(1.4) \quad x \circ \{[((x \circ x) \circ y) \circ z] \circ [((x \circ x) \circ x) \circ z]\} = y.$$

The reader can find similar forms for $*$. Also see [2] with respect to (1.4).

Recently N. Kimura and M.K. Sen [7] have characterized groups in terms of systems with two binary operations, called *bigroupoids* $K(*, \circ)$ satisfying

$$(1.5) \quad (x \circ y) * z = y \circ (z * x), \ (x \circ y) * x = y.$$

In virtue of these examples it is natural to consider mappings of groupoids to groups through the medium of varieties closely related to groups.

Schein [9], [10] characterized inverse semigroups in terms of groupoids with involution satisfying the following:

$$(1.6) \quad (xy)z = x(yz), \ (x')' = x, \ (xy)' = y'x', \ x = xx'x, \ (xx')(x'x) = (x'x)(xx').$$

Moreover Schein proved in [10] that inverse semigroups cannot be characterized in terms of a single n-ary operation.

In this paper we consider the problem for regular semigroups with involution and inverse semigroups as a generalization of (1.5). We characterize regular semigroups with involution [inverse semigroups] in terms of bigroupoids satisfying four [five] identities. With a slight modification of the concept of homomorphism and congruence, we can say that any bigroupoid has a smallest inverse semigroup equivalence with respect to some variety.

2. EQUIVALENCES

A bigroupoid $G(\theta_1, \theta_2)$ is defined to be a set G with two binary operations θ_1, θ_2 and possibly with a set Γ of identities involving θ_1 or θ_2 or both where $x\,\theta_i\,y$ denotes the product of x and y under θ_i. Γ may be a set of identities, $W_\alpha = W_\alpha$, the equalities of same words; this is the case where $G(\theta_1, \theta_2)$ has just two operations without any condition. For each $i = 1, 2$, $G(\theta_i)$ denotes the groupoid with respect to θ_i in $G(\theta_1, \theta_2)$. Groupoids are regarded as special case of bigroupoids $G(\theta_1, \theta_2)$ with $\theta_1 = \theta_2$.

Let $G(\theta_1, \theta_2)$ and $G'(\theta_1', \theta_2')$ be bigroupoids. A homomorphism is defined to be a couple (π, h) of a permutation π of the set $\{1, 2\}$ and a mapping h of the set G to G' such that

$$h(x\,\theta_i\,y) = h(x)\,\theta'_{\pi i}\,h(y) \quad \text{for all } x, y \in G \qquad (i = 1, 2).$$

If there is a homomorphism of $G(\theta_1, \theta_2)$ to $G'(\theta_1', \theta_2')$, we say $G(\theta_1, \theta_2)$ is homomorphic to $G'(\theta_1', \theta_2')$. If h is a bijection, the homomorphism is called an isomorphism. If π is fixed and if there is no fear of confusion we identify $\theta'_{\pi i}$ with θ_i, so we say, "$G(\theta_1, \theta_2)$ is homomorphic to $G'(\theta_1, \theta_2)$" or "h is a homomorphism of $G(\theta_1, \theta_2)$ to $G'(\theta_1, \theta_2)$", and so on.

A congruence on $G(\theta_1, \theta_2)$ is an equivalence ρ on the set G such that

$$x\,\rho\,y, \; z\,\rho\,u \;\text{ imply }\; x\theta_i z\,\rho\,y\theta_i u \;\text{ for all } i = 1, 2,$$

equivalently, left and right compatible. It goes without saying that the fundamental theorem holds between homomorphisms and congruences. Let $\mathscr{T}$ be a finite set of identities involving θ_1 or θ_2 or both. If $G(\theta_1, \theta_2)$ is homomorphic to $G'(\theta_1, \theta_2)$ and if $G'(\theta_1, \theta_2)$ satisfies $\mathscr{T}$, then the homomorphism is called a $\mathscr{T}$-homomorphism. If ρ is a congruence on $G(\theta_1, \theta_2)$ and if $G(\theta_1, \theta_2)/\rho$ satisfies $\mathscr{T}$, then ρ is called a $\mathscr{T}$-congruence on $G(\theta_1, \theta_2)$. If ρ_0 is a $\mathscr{T}$-congruence on $G(\theta_1, \theta_2)$ and if $\rho_0 \subseteq \rho$ for all $\mathscr{T}$-congruences ρ on $G(\theta_1, \theta_2)$ then ρ_0 is called the smallest $\mathscr{T}$-congruence on $G(\theta_1, \theta_2)$. The existence of ρ_0 follows from the fact that if $\{\rho_\xi \colon \xi \in \Xi\}$ is a family of $\mathscr{T}$-congruences on $G(\theta_1, \theta_2)$ then $\bigcap_{\xi \in \Xi} \rho_\xi$ is a $\mathscr{T}$-congruence on $G(\theta_1, \theta_2)$. Then $G(\theta_1, \theta_2)/\rho_0$ is called the greatest $\mathscr{T}$-homomorphic image of $G(\theta_1, \theta_2)$. If $G(\theta_1, \theta_2)$ is a groupoid with $\theta_1 = \theta_2$, then $\mathscr{T}$ should be modified by letting $\theta_1 = \theta_2$ in the identities of $\mathscr{T}$.

Lemma 2.1. *Given $\mathscr{T}$, any bigroupoid has a smallest $\mathscr{T}$-congruence.*

Let $G(\cdot)$ and $G'(\cdot)$ be groupoids. If there are three bijections $f\colon G \to G'$, $g\colon G \to G'$, $h\colon G \to G'$ such that

$$f(x)g(y) = h(xy) \quad \text{for all} \quad x, y \in G,$$

then $G(\cdot)$ is said to be *isotopic* to $G'(\cdot)$ and the triplet (f, g, h) is called an *isotopy* of $G(\cdot)$ to $G'(\cdot)$ [2]. The concept of isotopy is an equivalence relation. Recall $G(\theta_i)$ is defined in $G(\theta_1, \theta_2)$. If each of $G(\theta_1)$ and $G(\theta_2)$ is isotopic to $S(\cdot)$ then $G(\theta_1, \theta_2)$ is said to be *isotopic* to $S(\cdot)$. If (f_i, g_i, h_i) is an isotopy of $G(\theta_i)$ to $S(\cdot)$ $(i = 1, 2)$, define $(\bar{f}, \bar{g}, \bar{h})$ by $\bar{f} = (f_1, f_2)$, $\bar{g} = (g_1, g_2)$, $\bar{h} = (h_1, h_2)$. Then $(\bar{f}, \bar{g}, \bar{h})$ is called the isotopy of $G(\theta_1, \theta_2)$ to $S(\cdot)$.

Assume h is a homomorphism of $H(\theta_1, \theta_2)$ to $G(\theta_1, \theta_2)$ and $(\bar{f}, \bar{g}, \bar{h})$ is an isotopy of $G(\theta_1, \theta_2)$ to $S(\cdot)$. Then consider a "composition" of h and $(\bar{f}, \bar{g}, \bar{h})$; it is called a *weak homomorphism* of $H(\theta_1, \theta_2)$ to $S(\cdot)$. More precisely, a weak homomorphism of $H(\theta_1, \theta_2)$ to $S(\cdot)$ is defined by the triple $((\xi, \eta, \lambda))$ of pairs where

$$\xi = (f_1 h, f_2 h), \quad \eta = (g_1 h, g_2 h), \quad \lambda = (h_1 h, h_2 h).$$

If $\Phi = (\xi, \eta, \lambda)$ is a weak homomorphism of $H(\theta_1, \theta_2)$ into S such that $\Phi\colon H(\theta_1, \theta_2) \xrightarrow{h} G(\theta_1, \theta_2) \xrightarrow{\varphi} S$, where h is a homomorphism and $\varphi = (\bar{f}, \bar{g}, \bar{h})$ $(\bar{f} = (f_1, f_2),\ \bar{g} = (g_1, g_2),\ \bar{h} = (h_1, h_2))$ is an isotopy, then Φ determines the equivalence ρ on the set H as follows: $x \,\rho\, x'$ iff

$\varphi h(x) = \varphi h(x')$, that is, for $i = 1, 2$,

$$f_i h(x) = f_i h(x'), \quad g_i h(x) = g_i h(x'), \quad h_i h(x) = h_i h(x').$$

However, this equivalence ρ is essentially the same as the equivalence on H determined by h.

Assume $G(\theta_1, \theta_2)$ satisfies $\mathcal{T}$, and $S(\cdot)$ belongs to a class $\mathcal{S}$ of groupoids. We denote $G(\theta_1, \theta_2) \in \mathcal{T}$, that is, the class of all bigroupoids $G(\theta_1, \theta_2)$ satisfying $\mathcal{T}$ is still denoted by $\mathcal{T}$. (The class is called a variety.) Further assume there is a bijection $\mu\colon \mathcal{T} \to \mathcal{S}$ such that $G(\theta_1, \theta_2)$ is isotopic to $\mu(G(\theta_1, \theta_2))$. Then the smallest $\mathcal{T}$-congruence ρ_0 on any bigroupoid $H(\theta_1, \theta_2)$ is called the *smallest* $\mathcal{S}$*-equivalence* on $H(\theta_1, \theta_2)$ *relative* to $\mathcal{T}$. A weak homomorphism of $H(\theta_1, \theta_2)$ to $S \in \mathcal{S}$ is called an $\mathcal{S}$-weak homomorphism. An $\mathcal{S}$-weak homomorphism of $H(\theta_1, \theta_2)$ corresponding to the smallest $\mathcal{S}$-equivalence ρ_0 is called the greatest $\mathcal{S}$-weak homomorphism of $H(\theta_1, \theta_2)$, relative to $\mathcal{T}$, and its image is called the greatest $\mathcal{S}$-weakly homomorphic image of $H(\theta_1, \theta_2)$ relative to $\mathcal{T}$. Note $\mathcal{T}$ and $\mathcal{S}$ contain the trivial groupoid, i.e. the one-element groupoid.

If there is $\mu\colon \mathcal{T} \to \mathcal{S}$ as stated above, we say $\mathcal{T}$ *is isotopic to* $\mathcal{S}$.

3. INVOLUTED REGULAR SEMIGROUPS AND INVERSE SEMIGROUPS

By an involuted regular semigroup we mean a semigroup S with a unary operation $x \to x^{-1}$ satisfying the following conditions:

$$(x^{-1})^{-1} = x, \quad (xy)^{-1} = y^{-1}x^{-1}, \quad x = xx^{-1}x$$

for all $x, y \in S$.

Theorem 3.1. *Let* $S(\cdot, -1)$ *be an involuted regular semigroup. Define a bigroupoid* $G(*, \circ)$ *on the set* $G = S$ *as follows:*

$$x * y = x^{-1}y, \quad x \circ y = xy^{-1}.$$

Then $G(*, \circ)$ *satisfies the following four identities:*

(3.1.1) $\quad (x \circ y) * z = y \circ (z * x)$,

(3.1.2) $\quad (x \circ x) * x = x,$

(3.1.3) $\quad x \circ ((y * y) \circ y) = (x * (x \circ x)) * y,$

(3.1.4) $\quad y \circ x = (x \circ y) * ((x \circ y) \circ (x \circ y)).$

Conversely assume a bigroupoid $G(*, \circ)$ *satisfies* (3.1.1) *through* (3.1.4). *If we define a unary operation* -1 *and a binary operation* $\cdot$ *on the set* $S = G$ *by*

$$x^{-1} = x * (x \circ x), \quad x \cdot y = x \circ y^{-1}$$

then $S(\cdot, -1)$ *is an involuted regular semigroup. In particular* $S(\cdot, -1)$ *is an inverse semigroup if and only if* $G(*, \circ)$ *satisfies* (3.1.1) *through* (3.1.4) *and*

(3.1.5) $\quad (x \circ x) \circ (y \circ y) = (y \circ y) \circ (x \circ x).$

Thus there is a bijection μ [*bijection* μ'] *from the class* $\mathscr{B}$ [*the class* $\mathscr{B}'$] *of all bigroupoids satisfying* (3.1.1) *through* (3.1.4) [(3.1.1) *through* (3.1.5)] *to the class* $\mathscr{S}$ [*the class* $\mathscr{S}'$] *of all involuted regular semigroups* [*inverse semigroups*] *such that* $G(*, \circ)$ *is isotopic to* $\mu(G(*, \circ)) = S(\cdot, -1)$ $[\mu'(G(*, \circ)) = S(\cdot, -1)]$.

Proof. Assume $S(\cdot, -1)$ is an involuted regular [inverse] semigroup. Then (3.1.1) through (3.1.4) can be derived by using $(ab)^{-1} = b^{-1}a^{-1}$ and $aa^{-1}a = a$; (3.1.5) follows from the fact that idempotents commute.

Conversely assume $G(*, \circ)$ satisfies (3.1.1) through (3.1.4). First define a unary operation $x \to x^{-1}$ by

(3.1.6) $\quad x^{-1} = x * (x \circ x).$

Then (3.1.4) implies

(3.1.7) $\quad (x \circ y)^{-1} = y \circ x$

for all $x, y \in G$. By (3.1.1) and (3.1.2),

(3.1.2′) $\quad x = x \circ (x * x)$

and hence every element x of G has the form $a \circ b$, that is, $G \circ G = G$. For any $x \in G$, let $x = a \circ b$; then

$$(x^{-1})^{-1} = ((a \circ b)^{-1})^{-1} = (b \circ a)^{-1} = a \circ b = x.$$

Hence

(3.1.8) $\quad (x^{-1})^{-1} = x.$

It follows that $x \to x^{-1}$ is a bijection. By (3.1.2′) and (3.1.7),

(3.1.6′) $\quad x^{-1} = (x * x) \circ x.$

Now we define a groupoid $S(\cdot)$ on the same set $S = G$ by

(3.1.9) $\quad xy = x \circ y^{-1}.$

The product of x and y in $S(\cdot)$ is denoted by xy instead of $x \cdot y$. Then, by (3.1.7), $(xy)^{-1} = (x \circ y^{-1})^{-1} = y^{-1} \circ x = y^{-1}x^{-1}$. Hence

(3.1.10) $\quad (xy)^{-1} = y^{-1}x^{-1}.$

Hence $x \to x^{-1}$ is an involution on $S(\cdot)$. By (3.1.3), (3.1.6) and (3.1.6′), we have

(3.1.11) $\quad x \circ y^{-1} = x^{-1} * y.$

Furthermore,

$$(x * y)^{-1} = (x^{-1} \circ y^{-1})^{-1} = y^{-1} \circ x^{-1} = y * x$$

by (3.1.7) and (3.1.11). Hence

(3.1.12) $\quad (x * y)^{-1} = y * x.$

We show the associativity of $S(\cdot)$:

$$(xy)z = (x \circ y^{-1})^{-1} * z = (y^{-1} \circ x) * z = x \circ (z * y^{-1})$$

by (3.1.1) and (3.1.7),

$$x(yz) = x \circ (y^{-1} * z)^{-1} = x \circ (z * y^{-1})$$

by (3.1.9), (3.1.11) and (3.1.12) whence $(xy)z = x(yz)$.

By (3.1.2), (3.1.7), (3.1.9) and (3.1.11),

$$xx^{-1}x = (xx^{-1})x = (x \circ x)^{-1} * x = (x \circ x) * x = x.$$

(3.1.13) $xx^{-1}x = x$.

Thus $S(\cdot, -1)$ is an involuted regular semigroup. From associativity, $(x^{-1}y)z^{-1} = x^{-1}(yz^{-1})$, which implies

(3.1.14) $(x * y) \circ z = x * (y \circ z)$.

Because of (3.1.1) and (3.1.14) we can replace (3.1.3) by

(3.1.3′) $((y \circ y) \circ x) * y = ((x * x) \circ x) * y$.

Let e be any idempotent in $S(\cdot)$: $e = ee = e \circ e^{-1} = e^{-1} * e$; then $e^{-1} = e^{-1}e^{-1}$ by (3.1.10). Also, by (3.1.13) $e = ee^{-1}e$, $e^{-1} = e^{-1}ee^{-1}$. Now assume (3.1.5) in addition to (3.1.1) through (3.1.4). Then

$$\begin{aligned} e &= ee^{-1}e = e(e^{-1}e^{-1})e = (ee^{-1})(e^{-1}e) = \\ &= (ee^{-1}) \circ (e^{-1}e)^{-1} = (e \circ e) \circ (e^{-1} \circ e^{-1})^{-1} = \\ &= (e \circ e) \circ (e^{-1} \circ e^{-1}) = (e^{-1} \circ e^{-1}) \circ (e \circ e) = \\ &= (e^{-1} \circ e^{-1}) \circ (e \circ e)^{-1} = (e^{-1}e) \circ (ee^{-1})^{-1} = \\ &= (e^{-1}e)(ee^{-1}) = e^{-1}ee^{-1} = e^{-1}. \end{aligned}$$

Therefore we have

(3.1.15) If e is an idempotent in $S(\cdot)$, then $e = e \circ e = e * e$.

Since $e = e^{-1}$ for all idempotents e, we can conclude $S(\cdot)$ is an inverse semigroup by Lemma 1.7 [10 English transl.] but we show here the commutativity of idempotents from $e = e^{-1}$ for the reader's convenience.

Let e and f be idempotents in $S(\cdot)$. By (3.1.15), (3.1.7) and (3.1.5),

$$\begin{aligned} ef &= (e \circ e)(f \circ f) = (e \circ e) \circ (f \circ f)^{-1} = (e \circ e) \circ (f \circ f) = \\ &= (f \circ f) \circ (e \circ e) = (f \circ f) \circ (e \circ e)^{-1} = (f \circ f)(e \circ e) = fe. \end{aligned}$$

Since all idempotents commute, $S(\cdot)$ is an inverse semigroup [3]. The following also holds:

(3.1.16) For all $x \in G$, both $x \circ x$ and $x * x$ are idempotents in $S(\cdot)$ and also in $G(\circ)$ and $G(*)$.

Recall that $\mathscr{B}$ is the class of all bigroupoids $G(*, \circ)$ satisfying (3.1.1) through (3.1.4), and $\mathscr{S}$ is the class of all involuted regular semigroups. Define μ: $\mathscr{B} \to \mathscr{S}$ by

$$\mu(G(*, \circ)) = S(\cdot)$$

where $xy = x \circ y^{-1} = x^{-1} * y$. Assume $\mu(G(*, \circ)) = \mu(G(*', \circ')) = S(\cdot)$. Then $x \circ y^{-1} = xy = x \circ' y^{-1}$ and $x^{-1} * y = xy = x^{-1} *' y$. Since x^{-1} and y^{-1} run through all elements of G, we have

$$x * y = x *' y \text{ and } x \circ y = x \circ' y \quad \text{for all} \quad x, y \in G,$$

hence $G(*, \circ) = G(*', \circ')$. Thus μ is one-to-one. For any $S(\cdot) \in \mathscr{S}$, define $\circ$ and $*$ by $x \circ y = xy^{-1}$ and $x * y = x^{-1}y$. Then $\mu(G(*, \circ)) = S(\cdot)$. Therefore μ is a bijection. Similarly μ': $\mathscr{B}' \to \mathscr{S}'$ is a bijection. It is easy to see that $G(*, \circ)$ is isotopic to $S(\cdot)$.

Theorem 3.2. *Let* $G(*, \circ)$ *be a bigroupoid satisfying* (3.1.1) *through* (3.1.4) *and let* $S(\cdot) = \mu(G(*, \circ))$. *Then the following are equivalent:*

(3.2.1) $G(\circ)$ *is associative.*

(3.2.2) $G(*)$ *is associative.*

(3.2.3) $G(\circ)$ *is commutative.*

(3.2.4) $G(*)$ *is commutative.*

(3.2.5) $S(\cdot)$ *is a semilattice of groups satisfying the identity* $x^2 = 1$.

(3.2.6) $x = x^{-1}$ *for all* x.

(3.2.7) $G(\circ) = G(*)$.

(3.2.8) $G(\circ) = S(\cdot)$.

(3.2.9) $G(*) = S(\cdot)$.

Proof will be done along the following courses:

(3.2.2) $\Leftrightarrow$ (3.2.1) $\Rightarrow$ (3.2.5) $\Rightarrow$ (3.2.6) $\Rightarrow$ (3.2.7) $\Rightarrow$ (3.2.1)

$$(3.2.6) \Leftrightarrow (3.2.3), \quad (3.2.6) \Leftrightarrow (3.2.4)$$

$$(3.2.6) \Rightarrow (3.2.8) \Rightarrow (3.2.1), \quad (3.2.6) \Rightarrow (3.2.9) \Rightarrow (3.2.2)$$

(3.2.1) ⇔ (3.2.2). We have the equivalence of the identities

$$(x \circ y) \circ z = x \circ (y \circ z) \Leftrightarrow xy^{-1}z^{-1} = xzy^{-1} \Leftrightarrow xyz^{-1} = xzy \Leftrightarrow$$

$$\Leftrightarrow zy^{-1}x^{-1} = y^{-1}z^{-1}x^{-1} \Leftrightarrow$$

$$\Leftrightarrow z^{-1} * (y * x^{-1}) = (z^{-1} * y) * x^{-1} \Leftrightarrow$$

$$\Leftrightarrow z * (y * x) = (z * y) * x.$$

Also we have

(3.2.10) $xyz^{-1} = xzy$ for all $x, y, z \in S$.

(3.2.1) ⇒ (3.2.5). Assume $G(*)$ is associative, equivalently $G(\circ)$ is associative. $(x * y) * z = x * (y * z)$ implies $y^{-1}xz = x^{-1}y^{-1}z$, hence we have

(3.2.11) $yxz = x^{-1}yz$ for all $x, y, z \in S$.

Then (3.2.10) and (3.2.11) imply $yxz = yx(z^{-1})^{-1} = yz^{-1}x = zyx$, namely

$$(yx)z = z(yx).$$

Since $S \cdot S = S$, this shows $S(\cdot)$ is commutative. A commutative semigroup S is a semilattice of commutative archimedean semigroups S_α [3]. It can be shown that S_α is a regular semigroup since S is a regular semigroup. But a commutative archimedean semigroup is regular if and only if it is an abelian group. Thus S is a semilattice of abelian groups. Let $x \in S$, say $x \in S_\alpha$. The inverse element x^{-1} is the usual inverse of x in the group. From (3.2.10), $xxx^{-1} = x^3$ implies $x^2 = 1$. Note that commutativity of S follows from (3.2.5). See more [13].

(3.2.5) ⇒ (3.2.6). Obvious.

(3.2.6) ⇒ (3.2.7). Since $x = x^{-1}$, $xy = x \circ y^{-1} = x^{-1} * y$ implies $x \circ y = x * y$ for all $x, y \in G$.

(3.2.7) ⇒ (3.2.1). By (3.2.14), $G(\circ)$ is associative.

(3.2.6) ⇒ (3.2.3). By (3.1.7), we have

$$x \circ y = (x \circ y)^{-1} = y \circ x.$$

(3.2.3) ⇒ (3.2.6). By (3.1.2′) and (3.1.6′),

$$x = x \circ (x * x) = (x * x) \circ x = x^{-1}.$$

(3.2.6) ⇒ (3.2.4). By (3.1.12),

$$x * y = (x * y)^{-1} = y * x.$$

(3.2.4) ⇒ (3.2.6). By (3.1.2) and (3.1.6),

$$x = (x \circ x) * x = x * (x \circ x) = x^{-1}.$$

(3.2.6) ⇒ (3.2.8). $x \circ y = xy^{-1} = xy.$

(3.2.6) ⇒ (3.2.9). $x * y = x^{-1}y = xy.$

(3.2.8) ⇒ (3.2.1) and (3.2.9) ⇒ (3.2.2) are immediate.

Corollary 3.2′. *If* $G(*) = G(\circ)$, *then conditions* (3.1.1) *through* (3.1.4) *imply* (3.1.5) *and they are equivalent to the system of identities:*

$$(xy)z = x(yz), \quad xy = yx, \quad x^3 = x.$$

Let $G(*, \circ) \in \mathscr{B}$ and H a subset of G. If H is closed under both $*$ and $\circ$ then H is closed under the unary operation $x \to x^{-1}$ where x^{-1} is defined by (3.1.6). $H(*, \circ)$ is called a sub-bigroupoid of $G(*, \circ)$, and $H(*, \circ) \in \mathscr{B}$. $H(*, \circ)$ corresponds to an involuted regular subsemigroup $\mu(H(*, \circ))$ of $\mu(G(*, \circ))$. $\mathscr{B}$ is closed under sub-bigroupoids, homomorphic images and direct products while $\mathscr{S}$ is closed under involuted regular subsemigroups, homomorphic images [4] and direct products. Moreover, μ preserves these concepts, that is, $G_1(*, \circ) \subset G_2(*, \circ)$ implies $\mu(G_1(*, \circ)) \subset \mu(G_2(*, \circ))$; if $G_1(*, \circ)$ is homomorphic to $G_2(*, \circ)$ then $\mu(G_1(*, \circ))$ is homomorphic to $\mu(G_2(*, \circ))$;

$$\mu\Big(\prod_{\xi} G_{\xi}(*, \circ)\Big) = \prod_{\xi} \mu(G_{\xi}(*, \circ)).$$

We can say the same thing for μ': $\mathscr{B}' \to \mathscr{S}'$. From Theorem 3.1 we have

Theorem 3.3. *Every bigroupoid has a smallest involuted-regular-semigroup-equivalence relative to* $\mathscr{B}$, *and a smallest inverse-semigroup-equivalence relative to* $\mathscr{B}'$.

Let $\mathscr{C}$ be the class of commutative semigroups satisfying the identity $x^3 = x$. If $H(\cdot)$ is a groupoid, the greatest $\mathscr{B}$-homomorphic image of $H(\cdot)$ is in $\mathscr{C}$ by Theorem 3.2. Then Theorem 3.3. for $H(\cdot)$ becomes an obvious fact:

Corollary 3.4. *Every groupoid has a smallest* $\mathscr{C}$*-congruence.*

Remark 3.5. In Theorem 3.1, (3.1.4) can be replaced by

(3.1.4′) $y * x = ((x * y) * (x * y)) \circ (x * y)$

and (3.1.5) is equivalent to one of the following fourteen identities

(3.1.5.1) $(x \circ x)\, \theta_1\, (y \circ y) = (y \circ y)\, \theta_2\, (x \circ x)$,

(3.1.5.2) $(x * x)\, \theta_1\, (y * y) = (y * y)\, \theta_2\, (x * x)$,

(3.1.5.3) $(x \circ x)\, \theta_3\, (y * y) = (y * y)\, \theta_4\, (x \circ x)$.

(3.1.5.4) $(x \circ x)\, \theta_3\, (x * x) = (x * x)\, \theta_4\, (x \circ x)$

where (θ_1, θ_2) is one of $(\circ, \circ)$, $(*, \circ)$ and $(*, *)$; (θ_3, θ_4) is one of $(\circ, \circ)$, $(\circ, *)$, $(*, \circ)$ and $(*, *)$.

Proof is left to the reader's exercise. For (3.1.4′) and (3.1.5.1) through (3.1.5.3), try the similar proof as Theorem 3.1. For (3.1.5.4), use (1.6).

4. FURTHER OBSERVATION

Consider Kimura and Sen's conditions (1.5) again. Let $\mathscr{K}$ be the class of bigroupoids $K(*, \circ)$ satisfying (1.5), and $\mathscr{G}$ be the class of groups. It is easy to see that $\mu'(\mathscr{K}) = \mathscr{G}$ under μ': $\mathscr{B}' \to \mathscr{S}'$; therefore conditions (1.5) should imply conditions (3.1.1) through (3.1.5). However, we have directly the following:

Proposition 4.1. *$\mathcal{K}$ is a subvariety of $\mathcal{B}'$.*

Proof. (3.1.1) and (3.1.2) are obvious. In [7] it is proved that $x \circ x = y \circ y = x * x = y * y$ for all x, y, and (3.1.14) holds. These imply that (3.1.3′) and (3.1.5) hold. To prove (3.1.4)

$$(x \circ y) * ((x \circ y) \circ (x \circ y)) = (x \circ y) * (x \circ x) =$$

$$= y \circ ((x \circ x) * x) = y \circ x$$

by (3.1.1) and (1.5). Let $\mathcal{A}$ be the class of groups satisfying the identity $x^2 = 1$.

Theorem 4.2. *Every bigroupoid has a smallest group-equivalence relative to $\mathcal{K}$.*

Corollary 4.3. *Every groupoid has a smallest $\mathcal{A}$-congruence.*

Let $H(*, \circ)$ be a bigroupoid. Since $\mathcal{K} \subset \mathcal{B}'$, the smallest $\mathcal{B}'$-congruence on $H(*, \circ)$ is contained in the smallest $\mathcal{K}$-congruence on $H(*, \circ)$. In other words, the greatest $\mathcal{B}'$-homomorphic image of $H(*, \circ)$ is homomorphic to the greatest $\mathcal{K}$-homomorphic image of $H(*, \circ)$. This fact corresponds to the well known result [4] that an inverse semigroup has a smallest group-congruence. Similarly if $H(\cdot)$ is a groupoid, the greatest $\mathcal{C}$-homomorphic image of $H(\cdot)$ is homomorphic to the greatest $\mathcal{A}$-homomorphic image of $H(\cdot)$.

Let $\mathcal{L}$ be the class of groupoids $A(*)$ satisfying (1.2); $\mathcal{R}$ the class of groupoids $A(\circ)$ satisfying (1.2′). $A(*) \in \mathcal{L}$ is isotopic to $G(\cdot) \in \mathcal{G}$ under (f, g, h) where $f(x) = x^{-1}$, $g(y) = y$, $h(x * y) = x^{-1}y$. $A(\circ) \in \mathcal{R}$ is isotopic to $G(\cdot) \in \mathcal{G}$ under (f, g, h) where $f(x) = x$, $g(y) = y^{-1}$, $h(x \circ y) = xy^{-1}$.

Theorem 4.4. *Every groupoid has a smallest group-equivalence relative to $\mathcal{L}$ [$\mathcal{R}$].*

By (1.2), (1.2′), Theorem 3.2 and Corollary 3.2′, we have

$$\mathcal{A} \subset \mathcal{L} \cap \mathcal{R}.$$

Hence the smallest group-equivalence on $H(\cdot)$ relative to $\mathcal{L}$ [$\mathcal{R}$] is contained in the smallest $\mathcal{A}$-congruence on $H(\cdot)$.

Consider a relation between $\mathscr{K}$ and $\mathscr{L}$ [$\mathscr{R}$].

Proposition 4.5. *Let* $A(*) \in \mathscr{L}$ *and* $B(\circ) \in \mathscr{R}$. *Then* $A(*) = G(*)$ *and* $B(\circ) = G(\circ)$ *for some* $G(*, \circ) \in \mathscr{K}$ *if and only if* $A = B = G$ *and*

(4.5.1) $$(x * (x * x)) * y = x \circ ((y \circ y) \circ y)$$

for all $x, y \in A$.

Proof is left to the reader.

Let $H(\cdot, ')$ be a groupoid with a binary operation $\cdot$ and a unary operation $'$. A homomorphism h of $H(\cdot, ')$ to $G(\circ, '')$ is defined to be a homomorphism of $H(\cdot)$ to $G(\circ)$ satisfying

$$h(x') = (h(x))'' \quad \text{for all} \quad x \in H.$$

An equivalence ρ on $H(\cdot, ')$ is induced by a homomorphism in the above sense if and only if ρ is a congruence on $H(\cdot)$ and satisfies

$$x \,\rho\, y \quad \text{implies} \quad x' \,\rho\, y'.$$

Let $\mathscr{I}$ be the class of semigroup $S(\cdot, ')$ with a unary operation satisfying (1.6).

Proposition 4.6. *Any groupoid with a unary operation has a smallest* $\mathscr{I}$*-congruence.*

Independence of conditions

Are conditions (3.1.1) *through* (3.1.5) *independent?*

(4.7) Example of $G(*, \circ)$ which satisfies (3.1.2), (3.1.3), (3.1.4) and (3.1.5) but does not satisfy (3.1.1).

Let $G = \{1, 2, 3\}$ and define $*$ and $\circ$ by the following Cayley table:

$*$	1	2	3
1	1	2	3
2	2	1	2
3	3	2	1

$\circ$	1	2	3
1	1	2	3
2	2	1	2
3	3	2	1

Since $G(*) = G(\circ)$ and it is commutative, we see that $G(*, \circ)$ satisfies the conditions except (3.1.1). In fact

$$(2 \circ 2) * 3 = 1 * 3 = 3, \quad 2 \circ (3 * 2) = 2 \circ 2 = 1$$

hence (3.1.1) is not satisfied. Also it follows from Theorem 3.2 that (3.1.1) fails in this example.

(4.8) Example of $G(*, \circ)$ which satisfies (3.1.1), (3.1.3), (3.1.4) and (3.1.5), but not (3.1.2).

Let $G(*) = G(\circ)$ and assume it is a null semigroup with $|G| > 1$. That is,

$$x * y = x \circ y = a$$

for all $x, y \in G$. Let $b \neq a$. Then $(b \circ b) * b = a \neq b$. Hence $G(*, \circ)$ does not satisfy (3.1.2). It is easy to see that all the other conditions are satisfied.

(4.9) Example of $G(*, \circ)$ which satisfies (3.1.1) through (3.1.4) but not (3.1.5).

By Theorem 3.1 it is sufficient to give an example of an involuted regular semigroup which is not an inverse semigroup.

Let S be the direct product of a group G and a rectangular band $X \times Y$ where X is a left zero semigroup and Y is a right zero semigroup such that $|X| = |Y| > 1$, so X and Y are regarded as the same sets. Let

$$S = \{(x, i, j) : x \in G,\ i \in X,\ j \in Y\}$$

where $(x, i, j)(y, k, l) = (xy, i, l)$. Clearly S is regular. Define

$$(x, i, j)' = (x^{-1}, j, i).$$

Then S is involuted, but not an inverse semigroup.

Problem 1. Is there an example of $G(*, \circ)$ which satisfies (3.1.1), (3.1.2), (3.1.4), (3.1.5) but does not satisfy (3.1.3)? Is there an example of $G(*, \circ)$ which satisfies (3.1.1), (3.1.2), (3.1.3), (3.1.5) but does not satisfy (3.1.4)?

We raise the following question related to Theorem 3.1:

Problem 2. Is it possible to decrease the number of conditions for $G(*, \circ)$? What is the fewest number?

Note that even if the conditions are independent, the number of conditions is not necessarily fewest. For example the three conditions in (1.2) [(1.2′)] are independent, but 3 is not fewest.

A non-trivial right [left] zero semigroup satisfies the first and second conditions in (1.2) [(1.2′)] but not the third. An example which satisfies the first and the third in (1.2′) but not the second is given by

(4.10)

	1	2	3
1	1	2	2
2	2	1	1
3	2	1	1

$$3 \circ (3 \circ 3) \neq 3.$$

An example which satisfies the second and third in (1.2′) but not the first is

(4.11)

	1	2	3
1	1	2	3
2	2	1	3
3	3	3	1

$$(1 \circ 3) \circ (2 \circ 3) \neq 1 \circ 2.$$

The above two examples also show independence of Higman and Neumann's conditions (1.3).

The example (4.10) satisfies the first in (1.3) but not the second: $(x \circ x) \circ ((3 \circ 3) \circ 3) \neq 3$. The example (4.11) satisfies the second in (1.3) but not the first.

Incidentally, Kimura and Sen's conditions (1.5) are also independent. (4.8) is an example which satisfies the first in (1.5) but not the second. Define $G(*)$ to be a left zero semigroup and $G(\circ)$ to be a right zero semigroup. Then $G(*, \circ)$ satisfies the second but not the first. Clearly these examples show the independence of (3.1.1) and (3.1.2).

Problem 3. Study bigroupoids satisfying (3.1.1) through (3.1.5). More generally study bigoupoids satisfying (3.1.1) and (3.1.2).

Acknowledgement. The author is grateful to Professor B.M. Schein for giving him information on references related to this paper. Also the author expresses thanks to Professor G. Pollák and Professor L. Márki for invitation to Proceedings of the conference, 1981. Finally the author thanks Professor Reikichi Yoshida and Professor Miyuki Yamada for their useful remarks about this paper.

REFERENCES

[1] G. Birkhoff, *Lattice Theory,* Amer. Math. Soc., Providence, R.I., 1967.

[2] R.H. Bruck, *A survey of binary systems,* Springer Verlag, Berlin, 1958.

[3] A.H. Clifford – G.B. Preston, *The algebraic theory of semigroups,* Vol. 1, Amer. Math. Soc., Providence, R.I., 1961.

[4] A.H. Clifford – G.B. Preston, *The algebraic theory of semigroups,* Vol. 2, Amer. Math. Soc., Providence, R.I., 1967.

[5] M. Hall, *The theory of groups,* MacMillan, 1959.

[6] G. Higman – B.H. Neumann, Groups as groupoids with one law, *Publ. Math. Debrecen,* 2 (1952), 215–221.

[7] N. Kimura – M.K. Sen, On bigroupoids, *Collected Abstracts,* Special Session of Semigroups, University of California, Davis, 1980, 17–22.

[8] P. Lorenzen, Ein vereinfachtes Axiomensystem für Gruppen, *J. Reine Angew. Math.,* 182 (1940), 50.

[9] B.M. Schein, On the theory of generalized groups, *Dokl. Akad. Nauk. SSSR,* 153 (1963), 296–299 (in Russian); English translation: Soviet Math. Dokl., 4 (1963).

[10] B.M. Schein, On the theory of generalized groups and generalized heaps, *Theory of semigroups and its applications* I, Izdat. Saratov. Univ., (1965), 286–324 (in Russian); English translation: Amer. Math. Soc. Transl., (2) 113 (1979), 89–122.

[11] K. Shoda, *General Theory of Algebra,* Kyoritsu-Shuppan, 1947 (in Japanese).

[12] M. Ward, Postulates for the inverse operations in a group, *Trans. Amer. Math. Soc.*, 32 (1930), 520–526.

[13] R. Yoshida – M. Yamada, On commutativity of a semigroup which is a semilattice of commutative semigroups, *J. Algebra,* 11 (1969), 278–297.

T. Tamura

Department of Mathematics, University of California, Davis, California 95616, USA.

COLLOQUIA MATHEMATICA SOCIETATIS JÁNOS BOLYAI
39. SEMIGROUPS, SZEGED (HUNGARY), 1981.

CORRECTION TO MY PAPER "DENSE EXTENSIONS OF COMMUTATIVE SEMIGROUPS"

L.M. GLUSKIN

In my paper "Dense extensions of commutative semigroups" [*Algebraic Theory of Semigroups* (Proc. Conf. Szeged, 1976), Coll. Math. Soc. J. Bolyai, vol. 20., North-Holland, Amsterdam, 1979, 155–171.] Schein's result on injective semigroups in the category $\mathfrak{K}_{cs}$ of commutative separative semigroups was used (cf. 1.7 in the cited paper). This result turned out to be false: the semigroups, mentioned in 1.7 are injective in the category $\mathfrak{K}_{cr}$ of regular commutative semigroups but there are no injective semigroups in the category $\mathfrak{K}_{cs}$.

Therefore my paper should be corrected as follows:

(1) "$\mathfrak{K}_{cs}$" and "$\mathfrak{K}_c$" should be replaced by "$\mathfrak{K}_{cr}$" on page 159 line 20, page 164 lines 4, 24–25, page 165 line 9; in a similar way, "$\mathfrak{K}_{cs}$ (and, according to 3.1 in the category $\mathfrak{K}_c$)" and "$\mathfrak{K}_c$ (or, equivalently, in the category $\mathfrak{K}_{cs}$ – cf. 3.1)" on page 163 line 29 and page 165 lines 12–13, respectively, should be replaced by "$\mathfrak{K}_{cr}$".

(2) The word "universal" should be replaced by "maximal" on page 165 line 25 and page 167 lines 1–2.

(3) The correct form of Theorem 3.1 is: There are no injective semigroups in the category $\mathfrak{K}_c$.

COLLOQUIA MATHEMATICA SOCIETATIS JÁNOS BOLYAI
39. SEMIGROUPS, SZEGED (HUNGARY), 1981.

CORRECTION AND REMARKS TO OUR PAPER "'EMBEDDING SEMIRINGS INTO SEMIRINGS WITH IDENTITY"

R.D. GRIEPENTROG – H.J. WEINERT

A semiring $S = (S, +, \cdot)$ as considered in [1] is defined to be an algebra such that $(S, +)$ and $(S, \cdot)$ are semigroups, connected by ring-like distributivity. We have to correct one statement of Thm. 1, [1] (and the corresponding statement of Thm. 2 [1]) concerning conditions such that a semiring S is embeddable into a semiring T with a right identity (with identity). For this reason we recall that the set $\mathrm{P}(S)$ of all right translations ρ of $(S, \cdot)$ forms a left distributive seminearring $(\mathrm{P}(S), +, \cdot)$ with respect to the operations

$$x(\rho_1 + \rho_2) = x\rho_1 + x\rho_2 \quad \text{and} \quad x(\rho_1 \cdot \rho_2) = (x\rho_1)\rho_2$$

for all $x \in S$. Let $\widetilde{\mathrm{P}}(S)$ be the additive subsemigroup of $(\mathrm{P}(S), +)$ generated by the set $\mathrm{P}_i(S)$ of the inner right translations of $(S, \cdot)$ together with the identity mapping $\iota = \iota_S \in \mathrm{P}(S)$. Then the corrected version of Thm. 1, [1] reads as follows:

Theorem 1. *Let S be a semiring. Then*

(c) *S is embeddable into a semiring T with a right identity if and only if*

(a) $\widetilde{\mathrm{P}}(S)$ *is contained in the set* $E(S)$ *of endomorphisms of* $(S, +)$. *If this is the case, then*

(b) $(\widetilde{\mathrm{P}}(S), +, \cdot)$ *is a subsemiring of* $(\mathrm{P}(S), +, \cdot)$.

The wrong statement in [1] claims that also (b) ⇒ (c) holds. We note that the proof given in [1] for (b) ⇒ (c) also uses (a) as an assumption; hence it proves in fact (a) ⇒ (c), since (c) ⇒ (a) ⇒ (b) is true and easy to check. Moreover, the same correction applies to Thm. 2, [1], the two-sided version of Thm. 1 which deals with $\widetilde{\mathrm{P}}(S)$ and $\widetilde{\Lambda}(S)$, where the latter is defined analogously for left translations of $(S, \cdot)$. For this theorem, the proof given in [1] remains correct, since the wrong implication (b) ⇒ (c) in Thm. 2 was carried over from the wrong implication (b) ⇒ (c) and its left-right dual of Thm. 1.

Thus it remains to disprove (b) ⇒ (c) or, equivalently, (b) ⇒ (a) in the meaning of Thm. 1 above. Note that the following counterexample deals with a semiring S such that $(S, \cdot)$ is commutative. Further, to make this example as strong as possible, S has a zero 0 (i.e. a neutral of $(S, +)$) satisfying $0x = x0 = 0$ for all $x \in S$ *.

Example. The semiring $S = (S, +, \cdot)$ defined by the tables

$+$	0	a	b	c
0	0	a	b	c
a	a	a	a	a
b	b	b	b	b
c	c	a	b	c

$\cdot$	0	a	b	c
0	0	0	0	0
a	0	a	a	a
b	0	a	a	a
c	0	a	a	a

satisfies (b), but not (a). In particular, S is not embeddable into a semiring T with a right (or with a left) identity.

Proof. Obviously, $(S, +)$ and $(S, \cdot)$ are semigroups, and the distributivity is trivial. In order to check (b), we denote by $\omega = \rho_0$ and $\rho = \rho_a = \rho_b = \rho_c$ the inner right (left) translations of $(S, \cdot)$. Then $\widetilde{\mathrm{P}}(S)$ consists of the mappings

*Note that an additively commutative semiring S always satisfies (a), (b) and (c).

	ω	ρ	ι	$\iota+\rho=\sigma$
0	0	0	0	0
a	0	a	a	a
b	0	a	b	b
c	0	a	c	a,

which is proved by the left one of the following tables:

$+$	ω	ι	ρ	σ
ω	ω	ι	ρ	σ
ι	ι	ι	σ	σ
ρ	ρ	ρ	ρ	ρ
σ	σ	σ	σ	σ

$\cdot$	ω	ι	ρ	σ
ω	ω	ω	ω	ω
ι	ω	ι	ρ	σ
ρ	ω	ρ	ρ	ρ
σ	ω	σ	ρ	σ .

Both tables together show that $(\widetilde{P}(S), +, \cdot)$ is a subalgebra of $(P(S), +, \cdot)$, hence a left distributive seminearring; since $(\widetilde{P}(S), \cdot)$ is commutative, $(\widetilde{P}(S), +, \cdot)$ is a semiring which proves (b). In order to disprove (a) we state that $\sigma \notin E(S)$, which follows from

$$b = b\sigma = (c + b)\sigma \neq c\sigma + b\sigma = a + b = a.$$

We can also disprove (c) directly. Assume that S is a subsemigroup of a semiring T with a right identity e_r. Then according to the last formula we obtain the contradiction

$$b = b(e_r + a) = (c + b)(e_r + a) = c + ca + b + ba = a + b = a,$$

and similarly with $b = (e_l + a)b = (e_l + a)(c + b) = a$ if one assumes that T has a left identity e_l.

REFERENCE

[1] R.D. Griepentrog – H.J. Weinert, Embedding semirings into semirings with identity, *Algebraic theory of semigroups,* (Proc. Conf. Szeged, 1976), Coll. Math. Soc. J. Bolyai, vol. 20, North-Holland, Amsterdam, 1979, 225–245.

R.D. Griepentrog – H.J. Weinert
3392 Clausthal-Zellerfeld, Techn. Univ. Clausthal, Inst. f. Math., Bundesrepublik Deutschland.

PROBLEMS

S. Lajos

1. For a semigroup S, let us define B_0 to be S and B_i to be $B(B_{i-1})$, the multiplicative semigroup of bi-ideals of B_{i-1} $(i = 1, 2, \ldots)$. Which semigroups S have a finite bi-ideal sequence?

2. Let 2^S be the global semigroup (power semigroup) of S. It is known that the bi-ideal semigroup $B(S)$ is a two-sided ideal of 2^S. Put $D(S) = 2^S / B(S)$. What semigroups are of the form $D(S)$ for some S? What semigroups have the property that $S \cong 2^S / B(S)$?

3. Is the set-theoretical union $F(S) \cup B(S)$ a subsemigroup in 2^S where $F(S)$ is the set of all filters of S? In other words, is the product of two filters again a filter or a bi-ideal? [For the definition of a filter, see M. Petrich, Introduction to Semigroups, Charles E. Merrill, Columbus, 1973, p. 30.]

Conjecture: yes.

E.S. Ljapin

1. Let 2^S be the global semigroup of S. What properties of S are inherited by 2^S?

2. Let $\mathscr{A}$ be a semigroup amalgam which is embeddable in a semigroup. Then the amalgamated free product F of $\mathscr{A}$ has the property that every semigroup in which $\mathscr{A}$ is embeddable such that the image of the underlying set of $\mathscr{A}$ is a generating set, is isomorphic to a factor semigroup of F. The set $\Sigma_{\mathscr{A}}$ of all such factor semigroups of F is partially ordered (in view of the natural partial order of congruences on F). F is the greatest element in $\Sigma_{\mathscr{A}}$. Which partially ordered sets can be realized as $\Sigma_{\mathscr{A}}$ for some $\mathscr{A}$? When do there exist minimal elements in $\Sigma_{\mathscr{A}}$ and how are they related to each other? When does there exist a minimal element in $\Sigma_{\mathscr{A}}$ whose underlying set coincides with the underlying set of $\mathscr{A}$ (inner embedding of $\mathscr{A}$)?

J. Meakin

1. Characterize the partially ordered set of $\mathscr{J}$-classes of $\mathscr{J}$-trivial semigroups. Which downwards directed partially ordered sets arise this way?

2. Study the variety of pseudo-semilattices, where by a pseudo-semilattice we mean the biordered set of a pseudo-inverse semigroup. (A pseudo-inverse semigroup is a regular semigroup S in which eSe is an inverse subsemigroup for each idempotent e in S.) Pseudo-semilattices form a variety with respect to $\wedge$ where $\omega^r(e) \cap \omega^l(f) = \omega(e \wedge f)$ for every e, f in a pseudo-semilattice.

D.D. Miller

Define four classes of regular semigroups containing at least one non-idempotent element as follows:

$\mathscr{N}_1 = \{S\text{: every product of non-idempotents is non-idempotent}\}$,

$\mathscr{N}_2 = \{S\text{: every inverse of a non-idempotent is idempotent}\}$,

$\mathscr{N}_3 = \{S\text{: every element of } S \text{ has an idempotent inverse}\}$,

$\mathscr{N}_4 = \{S\text{: } E^2 = S \text{ where } E \text{ is the set of all idempotents in } S\}$.

It turns out that $\mathscr{N}_1 \subsetneqq \mathscr{N}_2 \subsetneqq \mathscr{N}_3 = \mathscr{N}_4$. Is $\mathscr{N}_1$ empty or not?

F. Pastijn

Let B be an infinite band. We say that B has cofinality ω if there exists a strictly increasing sequence of subbands $B_0 \subset B_1 \subset B_2 \subset \ldots$ $\ldots \subset B_n \subset \ldots \subset B$ such that $B = \bigcup_{n=0}^{\infty} B_n$. Is every infinite band of cofinality ω? [It is known that every infinite semilattice is of cofinality ω. (S. Fajtlowicz, *Colloq. Math.*, 42 (1979), 121–122).]

G. Pollák

Let V be a semigroup variety satisfying a non-regular identity $x_1x_2 \ldots x_n = u$ where $x_1, \ldots, x_n$ are different elements of the alphabet. Denote by $G(V)$ the maximal group variety contained in V. Suppose $G(V)$ is finitely based. Does this imply that V is finitely based?

Conjecture: no.

I.S. Ponizovskiǐ

Does the condition that $\mathrm{GL}(n, K) \gneqq S$ is irreducible imply that S is not simple provided S is not a group?

L.N. Ševrin

1. The varieties of all inverse semigroups, groups and semilattices, respectively, are the finitely attainable varieties of inverse semigroups. What are the attainable ones? Are there any more? [V.V. Rasin has proved that if they exist, they must consist of combinatorial inverse semigroups (*Izv. Vysš. Učebn. Zaved. Matematika,* 8 (1978), 80–87).]

2. What can be said about a semigroup S whose subsemigroup lattice Sub S satisfies some non-trivial lattice identity? Is then S a periodic semigroup? What about Sub C where C is the free cyclic semigroup?

L.A. Skornjakov

1. Does $2^R \cong 2^S$ imply $R \cong S$ for arbitrary monoids R, S, where 2^R denotes the global semigroup of R? [Note that in the case R, S are groups the answer is yes (T. Tamura – J. Shafer, *Math. Japon.,* 12 (1967), 25–32) while in the case R, S are semigroups the answer is no (E.M. Mogiljanskaja, *Semigroup Forum,* 6 (1970), 330–333).]

2. Denote by 2_F^R the semigroup of all finite subsets of the monoid R. Does $2_F^R \cong 2_F^S$ imply $R \cong S$ for arbitrary monoids?

3. A mapping $f: A \to [0, 1]$ is called a distribution on the set A if $f(x) = 0$ for all but finitely many x in A and $\sum_{x \in A} f(x) = 1$. Denote by $D(A)$ the set of all distributions on A. If R is a monoid then $D(R)$ is also a monoid for the multiplication defined by $fg(x) = \sum_{yz=x} f(y)g(z)$ for $f, g \in D(R)$. Does $D(R) \cong D(S)$ imply $R \cong S$ for each pair of monoids R, S? [Clearly, an affirmative answer to this problem would imply the same for the previous one, too. For, defining $\text{Supp}\, f = \{x: f(x) \neq 0\}$ we have $\text{Supp}\, fg = \text{Supp}\, f \cdot \text{Supp}\, g$.]

J.-C. Spehner

Let F be the free inverse semigroup on a set X and H, K finitely generated subsemigroups in F. Is then $H \cap K$ finitely generated? [Note that for free groups the answer is yes. (A.G. Howson, On the intersection of finitely generated free groups, *J. London Math. Soc.*, 29 (1954), 428–434).]

K. Todorov

1. Since it seems too difficult to solve Problem 46 of the Sverdlovsk Tetrad (*Semigroup Forum,* 4 (1972), 278, due to B.M. Schein) the following questions are of interest:

a) What is the least natural number in the interval $[1, n^n]$ which does not occur as order of a subsemigroup in $\mathcal{T}_n$?

b) What is the maximal order of the 2-generated subsemigroups in $\mathcal{T}_n$?

2. (B.M. Schein) Describe the maximal commutative subsemigroups of $\mathcal{T}_X$.

A.N. Trahtman

1. Let W be a cover of a locally finite variety. Is W also locally finite?

2. (M.V. Sapir – A.N. Trahtman) Has every finite semigroup either a finite basis of identities or an irreducible one?

SOLUTIONS TO THE PROBLEMS

published in

Algebraic Theory of Semigroups (Proc. Conf. Szeged, 1976), pp. 749–753. Colloq. Math. Soc. J. Bolyai, 20, North-Holland, Amsterdam, 1979.

We report here on the results related to the problems that had been raised at our previous Colloquium on Semigroups. Most of the information presented here were given by the posers of the problems. In the sequel we refer to the problems by the names of their authors and by the numbers (if any).

K. Byleen

The problem was solved in K. Byleen and P. Komjáth, The admissible trace problem for E-unitary inverse semigroups, *Semigroup Forum,* 15 (1978), 235–246. The solution of this problem led to the cited paper of Byleen and Komjáth and to the paper by J. Meakin, The partially ordered set of $\mathcal{J}$-classes of a semigroup, *J. London Math. Soc.,* 21 (1980), 244–256.

J.B. Kim

The author described the $\mathcal{D}$-classes of B_n for $n = 1, 2, 3, 4$.

L. Márki

3. The answer is yes, see W. Rosenow, Analogon eines Satzes von Baer und Levi in der Klasse aller Halbgruppen, *Studia Sci. Math. Hungar.,* 13 (1978), 167–171.

A complete treatment of the problem and related problems appeared in M. Petrich, *Lectures in Semigroups,* J. Wiley and Sons, London–New York–Sydney and Akademie Verlag, Berlin, 1977, Chapter III.